5th Annual International Conference on Material Science and Engineering (ICMSE 2017)

AA006451

IOP Conference Series: Materials Science and Engineering
Volume 275

Xiamen, China
20 – 22 October 2017

Editors:

Roman Szewczyk
Mariya Aleksandrova

ISBN: 978-1-5108-5531-1
ISSN: 1757-8981

Printed from e-media with permission by:

Curran Associates, Inc.
57 Morehouse Lane
Red Hook, NY 12571

Some format issues inherent in the e-media version may also appear in this print version.

Copyright© (2017) by the Institute of Physics
All rights reserved. The material featured in this book is subject to IOP copyright protection, unless otherwise indicated.

Printed by Curran Associates, Inc. (2018)

For permission requests, please contact the Institute of Physics at the address below.

Institute of Physics
Dirac House, Temple Back
Bristol BS1 6BE UK

Phone: 44 1 17 929 7481
Fax: 44 1 17 920 0979

techtracking@iop.org

Additional copies of this publication are available from:

Curran Associates, Inc.
57 Morehouse Lane
Red Hook, NY 12571 USA
Phone: 845-758-0400
Fax: 845-758-2633
Email: curran@proceedings.com
Web: www.proceedings.com

Table of contents

Volume 275

5th Annual International Conference on Material Science and Engineering (ICMSE2017)

20–22 October 2017, Xiamen, Fujian, China

Accepted papers received: 16 November 2017
Published online: 5 January 2018

Preface

Papers

Chapter 1: Nano-scale and Amorphous Materials

IOP Conf. Series: Materials Science and Engineering **275** (2017) 011001 doi:10.1088/1757-899X/275/1/011001

PREFACE OF ICMSE 2017

The 5th Annual International Conference on Material Science and Engineering (ICMSE2017) was held in Xiamen, Fujian, China at October 20-22, 2017. The aim of the conference is to bring together the leading scientists, working in energy, environment, structural, functional, optoelectronic and magnetic materials, biomaterials and nano-scale materials in order to share the latest advances in their research in these fields. We believe that through keynote speeches by outstanding, internationally recognized scholars, in oral sessions and posters presentations, the Conference has reflected an overview of the latest developments in materials science. The topics of ICMSE 2017 ensured an exchange of scientific information and opinions in a field with impressive dynamics, considering the technological progress, theoretical and practical aspects of the research. This forum provides opportunities for the senior and young researchers, as well as representatives to the academia and industry, to exchange new ideas, to establish research or business relations and to find partners for future collaboration. We hope that the conference has constituted significant contribution to the knowledge in this up to date scientific field.

The proceedings provide an opportunity for the readers to engage with a selection of refereed papers that were presented during the ICMSE2017. It intends to collect in four chapters the most recent achievements in Electronic, Optic and Magnetic Materials; Nano-scale and Amorphous Materials; Biomaterials and Functional and Structural Materials, which will be forwarded for publication to IOP Conference Series: Materials Science and Engineering (MSE), and the conference proceedings will be submitted by IOP Conference Series: Materials Science and Engineering (MSE) for EI and ISTP indexing. The papers have been selected for this proceedings based on originality, significance, and clarity for the purpose of the conference.

We would like to thank all participants for their valuable contribution to the Conference and these Proceedings, thus making the program of very high quality. This will re-inforce the position of the future conference editions as a leading forums for sharing the latest developments in materials for broad range of applications. We assessed more than 52 papers, from which 51 were accepted after strict peer-reviewing by few expert referees. This is the right moment to express gratitude to the reviewers for their time, valuable recommendations and constructive evaluation, in order to guarantee the high quality of the Proceedings content. Special thanks to all members of the Technical Program Committee, staff and volunteers for their devoted and professional assistance in the overall organization of the conference. Last, but not least, we would like to acknowledge the sponsors for their support.

We are looking forward to the next 6th Annual International Conference on Material Science and Engineering (ICMSE2018)!

Assoc. Prof. Dr. Mariya Aleksandrova
Technical University of Sofia
Editor of ICMSE2017

Content from this work may be used under the terms of the Creative Commons Attribution 3.0 licence. Any further distribution of this work must maintain attribution to the author(s) and the title of the work, journal citation and DOI.
Published under licence by IOP Publishing Ltd

Committees of ICMSE2017

Editor

Prof. Roman Szewczyk, Industrial Research Institute for Automation and Measurements PIAP, Poland
Dr. Mariya Aleksandrova, Technical University of Sofia, Bulgaria

Honor Chair

Prof. Zhenlu Cui, Fayetteville State University, USA

Co-Chair

Dr. Togay Ozbakkaloglu, the University of Adelaide, Australia
Dr. Malka N. Halgamuge, the University of Melbourne, Australia
Dr. Dariusz Jacek Jakobczak, Koszalin University of Technology, Poland

Technical Program Committee

Dr. Imran memon, Zhejiang University, Pakistan
Dr. Hui Yang, Beijing University of Posts and Telecommunications, China
Dr. Fengxiang Xu, Wuhan University of Technology, China
Dr. Jingyu Yang, Shenyang Aerospace University, China
Dr. Bo Kong, PLA University of Science and Technology, China
Prof. Ming-Chun Tang, Chongqing University (CQU), China
Dr. Muhammad Naufal Bin Mansor, Universiti Malaysia Perlis (UniMAP), Malaysia
Dr. Zhiyu Jiang, University of Chinese Academy of Sciences, China
Dr. Quanbao Ma, University of Oslo, Norway
Dr. Mariya Aleksandrova, Technical University of Sofia, Bulgaria
Prof. Ke Wang, Huaqiao University, China
Dr. Liyuan Sheng, Peking University, China
Prof. Chien-Hung Yeh, Feng Chia University, Taiwan
Dr. Muhammed Enes BAYRAKDAR, Duzce University Computer, Turkey
Dr. Malka N. Halgamuge, the University of Melbourne, Australia
Dr. Hui-Mi Hsu, National Dong Hwa University, Taiwan
Prof. Xue-Qian Fang, Shijiazhuang Tiedao University, China
Dr. Mingchun Tang, Chongqing University, China
Prof. Roman Szewczyk, Industrial Research Institute for Automation and Measurements PIAP, Poland
Prof. Cheng-Yuan Tang, Huafan University, Taiwan
Prof. Antoni Paja, AGH University of Science and Technology, Poland
Dr. Guang Yang, Donghua University, China
Dr. Qiquan Qiao, South Dakota State University, USA
Dr. Yachuang Feng, Chinese Academy of Sciences, China

Content from this work may be used under the terms of the Creative Commons Attribution 3.0 licence. Any further distribution of this work must maintain attribution to the author(s) and the title of the work, journal citation and DOI.
Published under licence by IOP Publishing Ltd

IOP Conf. Series: Materials Science and Engineering **275** (2017) 011002 doi:10.1088/1757-899X/275/1/011002

Prof. Dalin Tang, Worcester Polytechnic Institute, USA
Dr. Shahryar Jafarinejad, College of Environment, Iran
Prof. Yo-Sheng Lin, National Chi Nan University, Taiwan
Dr. Sheikh Ahmad Izaddin Sheikhk Mohd Ghazali, Universiti Teknologi MARA Negeri Sembilan, Malaysia
Dr. Ahmad Fakharian, Qazvin Branch,Islamic Azad University, Iran
Prof. Chi-Wai Chow, National Chiao Tung University, Taiwan
Dr. Mohamad Bin Awang, Universiti Malaysia Terengganu (UMT), Malaysia
Dr. Zhenlu Cui, Fayetteville State University, USA
Dr. Elammaran Jayamani, Swinburne University of Technology Sarawak Campus, Malaysia
Prof. Yong-Cheng Lin, Central South University, China
Dr., Ke Chang, Applied material Pte Ltd., USA
Dr. Muhammad Kamran, COMSATS Institute of Information Technology, Pakistan
Dr. Krzysztof Kuś, University of Warmia and Mazury in Olsztyn, Poland
Dr. Togay Ozbakkaloglu, the University of Adelaide, Australia
Dr. Azhan Hashim, Universiti Teknologi Mara Pahang, Malaysia
Dr. Ahmed Alsheikhy, Northern Border University, Saudi Arabia
Dr. Wael Abdel Samad, Rochester Institute of Technology, USA
Prof. Kung-Chung Hsu, National Taiwan Normal University, Taiwan
Dr. WaiLeong Pang, Multimedia University, Malaysia
Dr. Yousoo Han, University of Maine, USA
Prof. Mu-Chun Wang, Minghsin University of Science and Technology, Taiwan
Dr. Mohd Yuhazri Bin Yaakob, Universiti Teknikal Malaysia Melaka, Malaysian
Prof. Ning Zhang, Beijing Union University, China
Dr. Karol Kyzioł, AGH University of Science and Technology, Poland
Prof. Yuhua Wang, Wuhan University of Science and Technology, China
Prof. Maochieh Chi, WuFeng University, Taiwan
Dr. Corneliu Doroftei, Alexandru Ioan Cuza University of Iasi, Romanian
Dr. Marc Porti, Autonomous University of Barcelona, Spain
Dr. Ting Xie, University of Maryland, USA
Dr. Yian Song, Arizona State University, USA
Dr. Jiliang Li, Purdue University Northwest - North Central Campus, USA
Dr. Tai Thai, La Trobe University, Australia
Dr. Bappa Acherjee, Birla Institute of Technology, India
Dr. Yu-Lin (Eugene) Song, Asia University, Taiwan
Dr. Ng Eng Poh, Universiti Sains Malaysia, Malaysia
Dr. Stefan Talu, the Technical University of Cluj-Napoca, Romania
Dr. Chien-Jen Tang, Feng Chia University, Taiwan
Dr. Timur Sh. Atabaev, Seoul National University, Seoul, South Korea
Dr. Chen Hsu, Semiconductor and Ceramic Material Laboratory, Taiwan
Dr. Somen Bhattacharjee, NSHM Knowledge Campus Durgapur, India
Dr. Byungcheul Kim, Gyeongnam national university of science and technology, Korea
Dr. Niyazi Ugur Kockal, Akdeniz University, Turkey

IOP Conf. Series: Materials Science and Engineering **275** (2017) 011002 doi:10.1088/1757-899X/275/1/011002

Prof. Meilin Liu, Georgia Institute of Technology, USA
Dr. Li Li, University of Queensland, Australia
Dr. Udo Erich Schwingenschl¨ogl, King Abdullah University Of Science and Technology (KAUST), Germany
Dr. Krzysztof Schabowicz, Wroclaw University of Technology, Poland
Prof. Shien-Kuei Liaw, National Taiwan University of Science and Technology, Taiwan
Dr. Rafidah Binti Ali, Universiti Kuala Lumpur, Malaysia
Dr. Syeda Sitwat Batool, COMSATS Institute of Information Technology, Pakistan
Dr. hab. inż. Elżbieta Macioszek, Silesian University of Technology, Poland
Prof. Mohini M. Sain, University of Toronto, Canada
Dr. Dandan Ma, University of Chinese Academy of Sciences, China
Dr. Yonggang Tong, Academy of Armored Forces Engineering, China
Dr. Xiaobo Zhang, Yanjing Institute of Technology, China
Prof. Yandong Qu, Liaoning University of Technology, China
Dr. Chunguo Zhang, Chang'an University, China
Dr. CH. V. V. Ramana, Chirala Engineering College, India
Dr. Yoji Yamato, Software Innovation Center, NTT Corporation, Japan
Dr. Marcela Elisabeta Barbinta - Patrascu, University of Bucharest, Romania
Prof. Partha Pratim Das, Haldia Institute of Technology, India
Dr. Valentino Paradiso, University of Salerno, Italy
Dr. Pello Jimbert Lacha, University of the Basque Country, Spain
Dr. Klimis Ntalianis, Athens University of Applied Sciences (TEI of Athens), Greece
Dr. Antoneta Tomljenovic, University of Zagreb, Croatia
Prof. Gagik Shmavonyan, National Polytechnic University of Armenia, Armenia
Prof. Nitul Dutta, MEF Group of Institutions, India
Dr. Manish Bhardwaj, SRM University, India
Dr. Shaoping Xiao, the University of Iowa, China
Dr. Jiangwei Liu, National Institute for Materials Science, China
Dr. Dariusz Jacek Jakobczak, Koszalin University of Technology, Poland
Dr. P. M. Meshram, Yeshwantrao Chavan College of Engineering (Y.C.C.E), India
Prof. Syed Abdul Rehman Khan, Brasi School of Supply Chain Management, U.S.
Dr. Mahmoud Ebrahimi, University of Maragheh, Iran
Dr. Yoji Yamato, NTT Corporation, Japan
Dr. Hossein Bakhshi Khaniki, Iran University of Science and Technology (IUST), Iran
Dr. Arun Kumar, Amity University, India
Dr. ARBI adnen, ISSAT Kairouan, Tunisia
Dr. Akram Rashid, Air University Islamabad, Pakistan
Dr. Felix Ling Ngee Leh, Universiti Tun Hussein Onn Malaysia (UTHM), Malaysia
Dr. Firman Mangasa Simanjuntak, National Dong Hwa University, Taiwan
Dr. Muhammad Yusof Ismail, Universiti Tun Hussein Onn Malaysia, Malaysia
Dr. Ervina Efzan Mhd Noor, Multimedia University, Malaysia
Dr. Behzad Nematollahi, Swinburne University of Technology, Australia
Dr. Merih Palandoken, Izmir Katip Celebi University, Turkey
Dr. Parthiban Kathirvel, SASTRA University, India

IOP Conf. Series: Materials Science and Engineering **275** (2017) 011002 doi:10.1088/1757-899X/275/1/011002

Dr. Grzegorz Woroniak, Bialystok University of Technology, Poland
Dr. Fankang Li, Oak Ridge National Laboratory, U.S.
Prof. Ying Wang, Hangzhou Dianzi University, China
Prof. Abdelkrim Abourriche, University Cadi Ayaad, Morocco
Dr. Lai Chin Wei, University of Malaya, Malaysia
Dr. Irzaman, Bogor Agricultural University, Indonesia
Prof. Yen-Pei Fu, National Dong Hwa University, Taiwan
Dr. Mohamed M.S. Wahsh, National Research Center, Egypt
Dr., Guohua Xie, Wuhan University, China

IOP Conf. Series: Materials Science and Engineering **275** (2017) 011003 doi:10.1088/1757-899X/275/1/011003

Keynote Speakers of ICMSE2017

Keynote Speaker Ⅰ

Flexible piezoresistive material composed of conductive fillers and polymer

Prof. Luheng Wang

Professor, doctoral supervisor, Shenghua Scholar Distinguished Professor, Central South University, China

Biography:

Luheng Wang received the Ph.D. degree from the Department of Precision Instruments and Mechanology, Tsinghua University, Beijing, China. He is a professor and doctoral supervisor with the School of Information Science and Engineering, Central South University, Changsha, China. He is currently the Shenghua scholar distinguished professor, and the leader of the CSU flexible sensor lab. His current research interest is to study the key mechanical/electrical properties and mechanisms for the novel flexible sensitive materials which possess multi-sensing-functions (e.g. pressure sensitive function, temperature sensitive function, magnetic sensitive function, and gas sensitive function, etc.), and to develop the multi-functional flexible sensor system based on the aforementioned sensitive materials.

Research interests:

Novel flexible sensitive materials, pressure sensitive function, temperature sensitive function, magnetic sensitive function, and gas sensitive function.

Abstract:

The composite composed of the conductive fillers and the polymer, which possesses not only the flexibility but also the piezoresistivity, can be used to develop the flexible pressure sensor. The progress in the research on the piezoresistivity of the composite is reported. The existing theories to explain the piezoresistive phenomena of the composite are summarized, including the “conductive filler gap theory”, the “conducting path theory”, the “conductive phase content theory”, and the “effective conductive path theory”. Furthermore, the limitations and the advantages of the aforementioned theories are compared and analyzed. Finally, the applications of the composite in the development of the flexible pressure sensor are presented.

Content from this work may be used under the terms of the Creative Commons Attribution 3.0 licence. Any further distribution of this work must maintain attribution to the author(s) and the title of the work, journal citation and DOI.
Published under licence by IOP Publishing Ltd

IOP Conf. Series: Materials Science and Engineering **275** (2017) 011003 doi:10.1088/1757-899X/275/1/011003

Keynote Speaker Ⅱ

Emergence of topological and topological crystalline phases in TlBiS2 and TlSbS2

Dr. Udo Erich Schwingenschl¨ogl

Doctor, King Abdullah University Of Science and Technology (KAUST), Germany

Biography:

Dr. Udo Erich Schwingenschl¨ogl was a scientific employee in university of Augsburg, Germany from 07/2001–09/2004. From 09/2004–02/2008, he was a Secondary School Teacher in State of Bavaria, Germany. And from 03/2008–08/2008 he worked as a visiting Professor in University of Bras´ılia (ICCMP), Brazil. And during 09/2008–10/2011, he was an assistant Professor in KAUST, Saudi Arabia. While during 09/2009–11/2012, he was the Program Chair "Materials Science & Engineering" of KAUST, Saudi Arabia. During 11/2011–06/2014, he was the Associate Professor in KAUST, Saudi Arabia, again. During 12/2012–06/2015, he became the Associate Dean "Physical Science & Engineering" in KAUST, Saudi Arabia. And until now, he was the professor of KAUST, Saudi Arabia.

Research interests:

Computational Physics and Materials Science, New Topological Materials

Abstract:

Using first-principles calculations, we investigate the band structure evolution and topological phase transitions in TlBiS2 and TlSbS2 under hydrostatic pressure as well as uniaxial and biaxial strain. The phase transitions are identified by parity analysis and by calculating the surface states. Zero, one, and four Dirac cones are found for the (111) surfaces of both TlBiS2 and TlSbS2 when the pressure grows, which confirm trivial-nontrivial-trivial phase transitions. The Dirac cones at the M points are anisotropic with large out-of-plane component. TlBiS2 shows normal, topological, and topological crystalline insulator phases under hydrostatic pressure, thus being the first compound to exhibit a phase transition from a topological to a topological crystalline insulator.

IOP Conf. Series: Materials Science and Engineering **275** (2017) 011004 doi:10.1088/1757-899X/275/1/011004

Peer review statement

All papers published in this volume of *IOP Conference Series: Materials Science and Engineering* have been peer reviewed through processes administered by the proceedings Editors. Reviews were conducted by expert referees to the professional and scientific standards expected of a proceedings journal published by IOP Publishing.

Content from this work may be used under the terms of the Creative Commons Attribution 3.0 licence. Any further distribution of this work must maintain attribution to the author(s) and the title of the work, journal citation and DOI.
Published under licence by IOP Publishing Ltd

Preparation and Adsorption Property of Imido-acetic Acid Type Chelating Nano-fibers by Electro-spinning Technique

Jiali Yang[1], Lansi Lu[1], Zhu Zhang[1], Minhui Liao[1], Huirong He[1], Lingxing Li[1], Jida Chen[1, *] and Shijin Chen[2]

[1]Chongqing University, Chemistry and Chemical Engineering, Chongqing, China. E-mail:chencqu@cqu.edu.cn,1255409692@qq.com

[2]Bomin Electronics Co., Ltd, Guangdong, China. E-mail:m1598990378@163.com

Abstract: A novel nano-fibrous adsorbent from imino-acetic acid (IDA) and polyvinyl alcohol (PVA) mixture solution was prepared by electro-spinning technique. The nano-fibrous adsorbents with imino-acetic acid functional groups were characterized and demonstrated by fourier transform infrared spectrometry (FT-IR) and the scanning electron microscopy (SEM). The effect of the adsorbents to remove heavy metals such as lead (Pb) and copper (Cu) ions from the aqueous solution was studied. The maximum adsorption percentage (SP) of the metal ions can reach 93.08% for Cu (II) and 96.69% for Pb(II), respectively. Furthermore, it shows that the adsorption procedure of the adsorbents is spontaneous and endothermic, and adsorption rate fits well with pseudo-second-order kinetic model. Most importantly, the reusability of the nanofibers for removal of metal ions was also demonstrated to be used at least five times.

1. Introduction

Heavy metal pollution growing rapidly with industrial growth and development has seriously threatened the public health and safety in recent years [1-4]. Therefore, many techniques (such as precipitation, ion exchange, membrane separation and adsorption) have been applied to deal with heavy metal ions in the waste water. Among them, the adsorption technique has been considered as the effective and promising method due to its low-cost and application convenience [5-8]. The properties of adsorbents (such as surface area, functional group type and density on the adsorbent surface) have significant effect on the adsorption efficiency. Hence, the mophology and the functional group type are the two key factors of adsorbents. Therefore, adsorbents of chelating nano-fibers have been widely studied, such as the adsorption of heavy metal ions from the polluted water. Great surface area, high porosity, and small pore size would have the influence on this efficiency [9,10]. Electrospinning is an efficient method of preparing chelating nano-fibers, which could even yield desirable sub-100-nm spinning fibers [11].Regarding to types of functional groups, imino-acetic acid (IDA) has been widely used for the removal of metal ions as it could simultaneously complex with many kinds of metal ions [12-15]. Unfortunately, most reported IDA type adsorbents were in the form of granule or fiber in micro-meter [16,17], so the applications were limited owing to some disadvantages from adsorbent's dimension. So far, there is no report on IDA type adsorbents in the form of nano-fibers.

The functionalized polymer with imino-acetic acid (IDA) functional group was synthesized in this study. It was applied to prepare IDA type chelating nano-fibers through electro-spinning [18-20]. In order to get nano-fibers with better morphology and more IDA functional groups available, the effects of spinning solution composition, the spinning parameters and the cross-linking conditions on nano-fibers have been investigated and optimized. In addition, the influence of nano-fibers on adsorption of

Content from this work may be used under the terms of the Creative Commons Attribution 3.0 licence. Any further distribution of this work must maintain attribution to the author(s) and the title of the work, journal citation and DOI.
Published under licence by IOP Publishing Ltd

copper and lead ions from the aqueous solutions were investigated and analyzed by the kinetic and thermodynamic models.

2. Experimental

2.1. Materials

PVA polymers (99% hydrolyzed; 87-89% hydrolyzed) were purchased from Shanghai Aladdin Reagent Co. Ltd. (China). Polymer with imino-acetic acid as functional group (IDA) was synthesized in our lab. The other chemicals such as glutaraldehyde (GA), epichlorohydrin (ECH), copper nitrate (Cu(NO3)2·3H2O) and lead nitrate (Pb(NO3)2) were all analytical grade.

2.2. Synthesis of IDA Type Polymer

IDA type polymer was synthesized from polyethyleneimine (PEI) according to the method described previously [15]. Mixing PEI (average Mw 70000), sodium chloroacetate and sodium hydroxide (NaOH) solutions in the molar ratios of 1:1.1:1.1 in 100 mL three-necked flask, and then, it was stirred at 60°C for 12 h. After it was cooled to the room temperature, the solution pH was adjusted to 2, and then it was washed by methanol and dried in the vacuum oven.

2.3. Preparation of IDA Type Chelating Nano-fibers

10 wt% PVA99 solution was prepared at 90°C (solution a). 15 wt% PVA88 solution was provided by dissolving 4.5 g of PVA88 in 30 mL deionized water at 45°C for 4 h (solution b). IDA type polymer was dissolved in sodium hydroxide solution (2mol/L) and it was stirred at 30°C for 2 h to obtain 30wt% homogeneous solution (solution c). The spinning solution was consisted of three solutions mentioned above, and the volume ratio of a:b:c= 3:4:2. The mixture solution was stirred at the room temperature for more than 1h, and then it was spun under the optimized conditions of voltage 24 kV, and its working distance remained 8 cm, and its feeding rate was 0.5 mL/h. The nano-fibers (pristine nano-fibers) were collected finally.

The pristine nano-fiber was pre-cross-linked in GA vapor at 35°C for 18 h (pre-cross-linked nano-fibers), and then it was cross-linked with 20% (mass fraction) ECH solution (pH=3) for 4 h [20-22], then it was washed with the distilled water, and then it was dried in the vacuum oven at the room temperature. The harvest cross-linked nano-fibers (IDA type chelating nano-fibers) was got.

2.4. Analysis and Characterization

The functional groups of nano-fibers (pristine, pre-cross-linked and cross-linked nano-fibers) were determined in the range of 400-4000 cm-1 by FR-IR (Fourier Transform Infrared Spectrometry, IR Affinity-1, SHIMADZU, Japan). The morphology of the nano-fibers was observed by SEM (scanning electron microscope, JSM-7800F, Japan).

To investigate the stability of IDA type chelating nano-fibers in the aqueous solution, the dissolve-loss studies were carried out by the immersion of nano-fibers in the distilled water for 6 h. All experiments were repeated for three times. The dissolve-loss percentage can be calculated by the following equation [23].

$$\text{Dissolve - loss percentage } (\%) = \frac{(m_1\text{-}m_2)}{m_1} \times 100\% \quad (1)$$

Where m_1 and m_2 are the masses of the dry nano-fibers before and after being soaked, respectively.

2.5. Adsorption Experiments

2.5.1. Batch Adsorption Experiments. The effects of adsorption experimental parameters, pH (2-6), time (0-150 mins.), the initial metal ion concentration (50-250 mg/L), and temperature (298-318 K) were studied in the batch mode. In which 0.1 g cross-linked nano-fibers was immersed into 50 mL of

IOP Conf. Series: Materials Science and Engineering **275** (2017) 012001 doi:10.1088/1757-899X/275/1/012001

copper or lead ion solution with stirring for 2.5 h (Fibrous adsorbents loaded Cu(II) or Pb(II) ions). The concentration of metal ions in the solution was determined by atomic absorption spectrometry (AAS, Hitachi, Japan).

The adsorption percentage (SP) and adsorption capacity of metal ions (qe (mg/g)) can be calculated by the following equations [8].

$$SP = \frac{(C_0 - C_e)}{C_0} \times 100\% \qquad (2)$$

$$q_e = \frac{(C_0 - C_e)V}{m} \qquad (3)$$

Where C0 and Ce are the concentration of metal ions in the solution before and after the adsorption (mg/L), respectively. And V (L) is the volume of the solution, m(g) is the mass of nanofibers.

2.5.2. *Adsorption/desorption Experiments.* Fibrous adsorbents loaded Cu(II) or Pb(II) ions were immersed into HCl solution (0.1 mol/L) at the room temperature with stirring. Then, the fibers were filtrated out, and they were washed with the distilled water for three times, and then they were dried in the vacuum oven at the room temperature. The regenerated nanofibers were reused in the next cycle utilization of the adsorption experiments in the same coditions (50 mg/L, pH value of 3 and pH value of 5.5 for Cu(II) and Pb(II) ions solutions, respectively.). Residual ions in solution were determined by AAS. The relative adsorption percentage (RSP) of metal ions on nano-fibers can be calculated by the following equation.

$$RSP = \frac{SP_{(n)}}{SP_{(1)}} \times 100\% \qquad (4)$$

Where SP (1) and SP(n) denote the sorption percentage (%) of adsorption/desorption cycle one and cycle n, respectively.

3. Result and Discussion

3.1. Characterization of Nano-fibers

The functional groups of nano-fibers were characterized by FT-IR (Fig. 1). A broad band at 3300-3500 $cm^{-1}$was assigned to N-H and O-H stretching vibrations of IDA type polymer and PVA, the carboxyl group peaked at 1720 cm^{-1} was attributed to the type of amino acetic acid functional groups. The results showed that the IDA type polymer was successfully synthesized and introduced into the composite nano-fibers. Compared with spectrum (a) and (b), the intensity of peaks at 1096 cm^{-1} and 1250 cm^{-1} corresponding to the stretching motions of C-O-C and C-O in spectrum (b) was increased as the part of hydroxyl groups of pristine nano-fibers cross-linking with GA. In spectrum (c), the sharp peak at 1399 cm-1 was typical stretching vibration of epichlorohydrin. It is observed that the intensity of peak at 1630 cm-1 corresponding to N-H bond was decreased with epichlorohydrin as the cross-linking of imino group, which indicated that N-H bond has not been converted into the type of IDA functional group completely, and the remaining imino group has been successfully crosslinked with ECH [8,21].

IOP Conf. Series: Materials Science and Engineering **275** (2017) 012001 doi:10.1088/1757-899X/275/1/012001

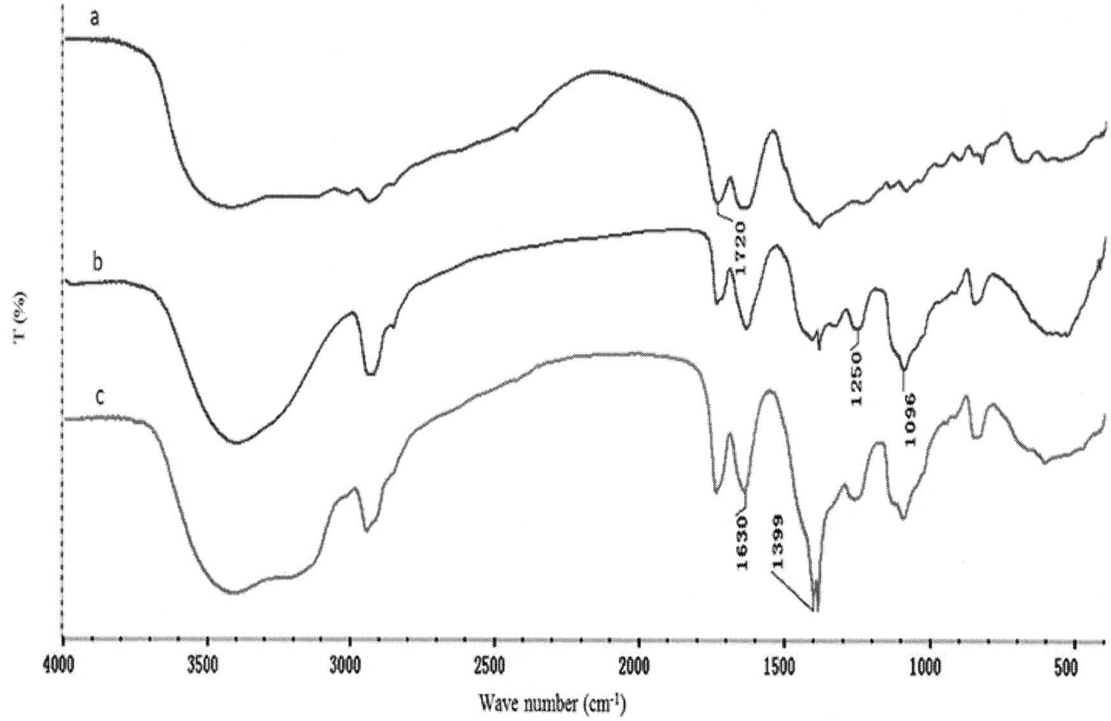

Figure 1. FTIR spectra of IDA type composite chelating nano-fibers: (a) pristine nano-fibers; (b) pre-cross-linked nano-fibers; (c) cross-linked nano-fibers.

Generally, there are two methods to prepare the chelating nano-fibers, one is the modification, namely, the functional group is immobilized onto the nano-fibers surfaces through chemical reaction. And the other method is the employment of the mixture solution which is composed of the functionalized polymers and materials with good spinnability. However, the morphology of the nano-fibers could be destroyed as the functional group is immobilized to nano-fibers surface through chemical reaction [24]. Therefore, electro-spinning mixture solution was used to prepare nano-fibers. The SEM images of pristine, pre-crosslinked and cross-linked nano-fibers were presented in Fig. 2.

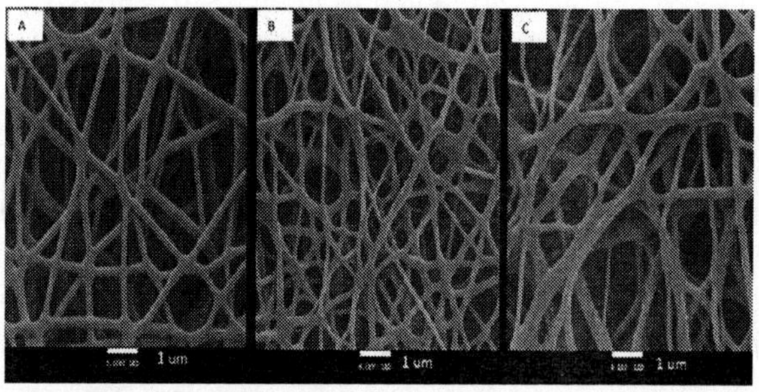

Figure 2. The SEM image of nano-fibers: (A) pristine nano-fibers; (B) pre-cross-linked nano-fibers; (C) cross-linked nano-fibers.

Fig.2 (A) showed that the surface of pristine nano-fibers was very smooth. The diameter of the pristine fibers was quite uniform with average value of (150±77) nm. Fig.2 (B) showed that the morphlogy of nano-fibers pre-crosslinked

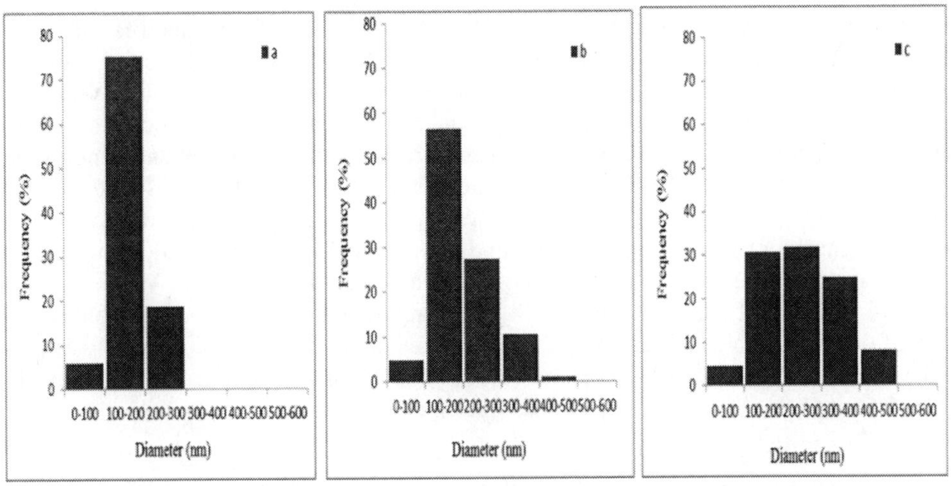

Figure 3. The fiber diameter distribution of nano-fibers: (a) pristine nano-fibers; (b) pre-cross-linked nano-fibers; (c) cross-linked nano-fibers.

With GA vapor, and the diameter of the pre-cross-linked nano-fibers was (190±72) nm. As shown in Fig.2(C), the nano-fibrous morphology almost was the stable, and it still showed the fluffy porous network structure, although the diameter of crosslinked IDA type chelating nano-fibers increased to (250±98 nm) nm. It meant that the chelating nano-fibers were successfully formed without remarkable variation in morphology no matter after pre-cross-linking or cross-linking.

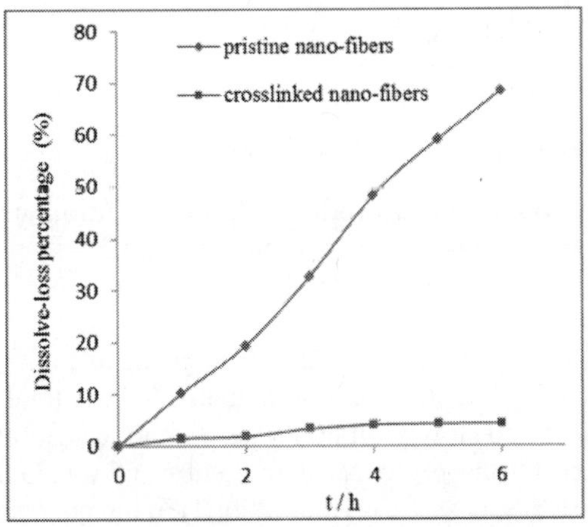

Figure 4. Degree of dissolve-loss percentage of nano-fibers (pristine and cross-linked IDA type composite chelating nano-fibers).

Fig. 4 showed that the dissolve-loss percentage (%) of nano-fibers (pristine and cross-linked with ECH) in the distilled water was more than 6 h. As it can be seen, the dissolve-loss percentage of the cross-linked nano-fibers was considerably decreased. The results suggests that the IDA type composite chelating nano-fiber has been cross-linked effectively, which can increase the water diffusivity resistance and improve the stability of nano-fiber membranes in solution.

3.2. Effect of pH on Adsorption Ability

PH was the important factor which could affect the interaction between adsorbents and metal ions. Since hydroxide would be formed and precipitated as the pH value of copper and lead ion solution was

IOP Conf. Series: Materials Science and Engineering **275** (2017) 012001 doi:10.1088/1757-899X/275/1/012001

above 6, the experiments were conducted while pH value was 2-5 for Cu(II) and pH value was 2-6 for Pb(II), respectively. Fig.5 showed that the adsorption rate (AR) (%) of metal ions on nano-fibers was as the function of the pH. For Cu(II), as pH was 3.0, AR value was the maximum. While for Pb(II), only as pH was 5.5, AR value was the maximum. The adsorption percentage increased gradually, for Cu(II) as pH from 2.0 to 3.0 and for Pb(II) only as pH from 2.0 to 5.5. This phenomenon was resulted from the competitive interaction between H+ and the metal ion for the active sites on the adsorbent. As pH value increased, the concentration of H+ in the solution was reduced, and which resulted in that there was more chance of interaction with the binding sites and metal ions [25]. However, as the pH value further increased, the adsorption percentage decreased rapidly for Cu(II). While as the pH value increased to 6.0 for Pb(II), the SR also showed the reduction, which would probably due to the interaction of OH- ion with Cu(II) and Pb(II), and the free target metal ions available was reduced. To exclude the influence of the hydrolysis of metal ions, the optimal pH value was selected as pH=3.0 for Cu(II), pH=5.5 for Pb(II) for the subsequent adsorption experiments.

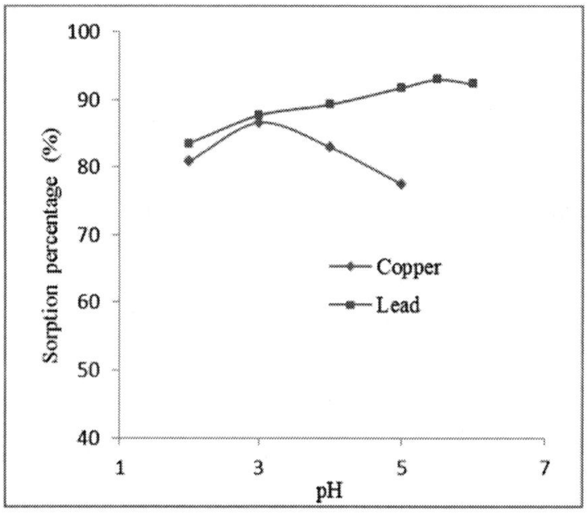

Figure 5. The effect of solution pH on adsorption percentage. (Initial metal ion concentration of 50 mg/L; Temperature of 293 K)

3.3. Effect of Contact Time on Adsorption

The effect of contact time on adsorption percentage was given in Fig.6. The results showed that the AR(%) increased sharply within 10 mins., as more than 70% of total sorption of metal ions was adsorbed, then the AR(%) increased slowly till 80 mins., when the adsorption equilibrium was reached. This trend might be explained by the presence of free sites and available metal ions at first. As the adsorption increased, the available sites of Cu(II) and Pb(II) in the aqueous solution were depleted.

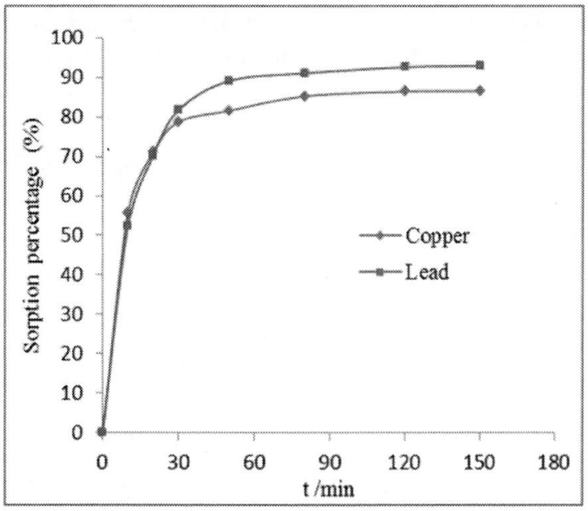

Figure 6. The effect of contact time on adsorption percentage (pH 3.0 for Cu(II) and pH 5.5 for Pb(II); Initial concentration of 50 mg/L; Temperature of 293 K).

To describe the kinetic mechanism of the adsorption process more accurately, the experimental data fitted well with two common models, namely pseudo-first-order (Eq. (5)) and pseudo-second order (Eq. (6)) kinetic models respectively [26]. The fitted parameters were shown in Table 1.

$$\ln(q_e - q_t) = \ln q_e - \frac{k_1}{2.303}t \tag{5}$$

$$\frac{t}{q_t} = \frac{1}{k_2 q_e^2} + \frac{t}{q_e} \tag{6}$$

Where q_e (mg/g) and q_t (mg/g) are the adsorption capacity of metal ions adsorbed at equilibrium and time t (min), respectively. k_1 (min^{-1}) and k_2 (g/mgmin) are the rate constants of pseudo-first-order and pseudo-second-order models, respectively.

Table 1. Kinetic Parameters Of Fibrous Adsorbents For Cu(Ii) And Pb(Ii) Adsorption (Ph=3 For Cu(Ii), Ph=5.5 For Pb(Ii); Initial Metal Ion Concentration Of 50 Mg/L; Temperature Of 293 K)

Metal ion	q_{exp} (mg/g)	Pseudo-first-order model			Pseudo-second-order model		
		k_1 (min^{-1})	$q_{e,cal}$(mg/g)	R^2	k_2(g/mg min)	$q_{e,cal}$ (mg/g)	R^2
Cu(II)	21.64	0.1303	15.12	0.919	0.0124	22.17	0.999
Pb(II)	23.71	0.0925	15.72	0.916	00084	24.51	0.998

By comparing the values of correlation coefficient for pseudo-first-order and pseudo-second-order as shown in Table 1. It was found that the kinetic data of both metal ions fitted with pseudo-second-order model more gretaly. The results suggestted that both concentration of metal ions and unoccupied adsorptive sites were the key factors which influenced on the adsorption performance of the chelated nano-fibers.

3.4. Adsorption Isotherms

The effect of the initial concentration (C0) of metal ions and temperature on the adsorption was presented in Fig.7. It showed that the adsorption capacity of Cu(II) and Pb(II) was increased with the increase of the initial concentration C_0. Without exception, the removal efficiency of metal ions with IDA type chelating nano-fibers also improved with the temperature was increased. The adsorption percentage reached 93.08% for Cu(II), 96.69% for Pb(II) adsorption while the initial concentration was 50 mg/L and the temperature was 40°C. The maximal adsorption capacity was 101.28 mg/g and 111.15 mg/g for Cu(II) and Pb(II), respectively. Which was much larger than that of IDACF[17]. These results indicated that the studied IDA type chelating nanofiber was probably useful for the advanced wastewater treatment.

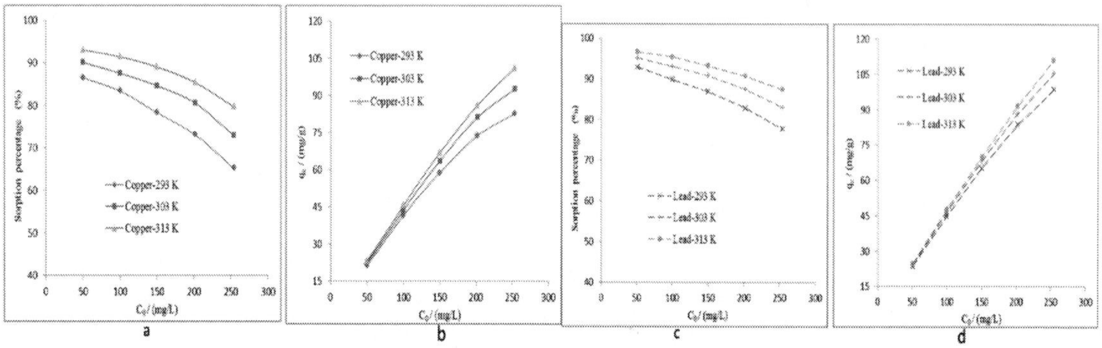

Figure 7. The effect of temperature on adsorption isotherm: (a) adsorption percentage for Cu(II); (b) adsorption capacity for Cu(II); (pH=3.0 for Cu(II); 50 mL) (c) adsorption percentage for Pb(II); (d) adsorption capacity for Pb(II) on IDA type composite chelating nano-fibers (pH=5.5 for Pb(II); 50 mL).

To further understand the adsorption mechanism of IDA type chelating nano-fibers, two isotherm models[27,28], namely Langmuir and Freundlich models were employed to analyze the experimental data as following equations.

$$\frac{C_e}{q_e} = \frac{1}{q_m k_L} + \frac{C_e}{q_m} \tag{7}$$

$$\log q_e = \log k_F + \frac{1}{n}\log C_e \tag{8}$$

where q_e (mg/g) is the adsorption capacity at equilibrium time, C_e (mg/L) is the equilibrium concentration of metal ions in solution, q_m (mg/g) is the theoretical saturation adsorption capacity, and K_L (L/mg) is the Langmuir constant. K_F (mg/g) is the binding energy constant reflecting the affinity of the adsorbents to metal ions and n is the Freundlich parameters related to the adsorption intensity.

Table 2. Isotherm parameters for metal adsorption onto IDA type composite chelating nano-fibers at different temperatures.

Metal ion	T (K)	Langmuir isotherm			Freundlich		
		Qmax (mg/g)	*KL (L/mg)*	*R2*	*KF (mg/g)*	*n*	*R2*
Cu(II)	293	107.5	0.0388	0.999	8.812	1.901	0.961
	303	117.6	0.0521	0.998	11.28	1.919	0.965
	313	128.2	0.0722	0.997	14.62	1.921	0.974
Pb(II)	293	128.2	0.0563	0.996	12.81	1.906	0.988
	303	135.1	0.0792	0.996	16.22	1.926	0.986
	313	138.9	0.113	0.994	20.39	1.960	0.986

As shown in Table 2, the high regression correlation coefficients (R^2>0.99) were presented. It suggested that the adsorption of both Cu(II) and Pb(II) onto IDA type chelating nano-fibers fitted particularly well with Langmuir model compared with Freundlich model. This indicated that the adsorption occurred only on the surface of IDA type nano-fibers, and there existed the monolayer adsorption phenomenon.

3.5. Thermodynamic Parameters

Thermodynamic parameters (such as Gibbs free energy, enthalpy change and entropy change) play vital roles in the adsorption process. It illustrates the degree of spontaneity and feasibility of process. Thermodynamic parameters might be calculated by the following equations [29].

$$\ln K = -\frac{\Delta H^0}{RT} + \frac{\Delta S^0}{R} \tag{9}$$

$$\Delta G = -RT \ln K \tag{10}$$

where R is the gas constant, T is the absolute temperature (K), ΔH^0 is the enthalpy change (kJ /mol), ΔG^0 is the free energy change (kJ /mol), the ΔH^0 can be obtained from the slope of ln$K$ versus plot $1/T$. K (L/mg) is the adsorption equilibrium constant, which equals to Langmuir constant (K_L) owing to the adsorption process fitted well with Langmuir model [30].

Table 3. The parameters of adsorption thermodynamics

Metal ion	ΔG^0 **(KJ/mol)**			ΔH^0 **(KJ/mol)**	ΔS^0 **(KJ/mol K)**
	293 K	***303 K***	***313 K***		
Cu(II)	-25.74	-27.35	-29.11	23.65	0.168
Pb(II)	-26.65	-28.42	-30.28	26.60	0.181

As shown in Table 3, the values of ΔG^0 were negative for both Cu (II) and Pb (II) at the studied conditions, and they were inclined to be more negative with increasing temperature, it indicated that the adsorption process was feasible and the reaction was spontaneous and more favorable at the higher temperatures. The positive value of ΔH^0 also implies that the adsorption process is endothermic. Furthermore, both the ΔH^0 for Cu (II) and Pb(II) are greater than 20.9 KJ, which reveals that the adsorption is a chem.-sorption process due to chelation mechanism.

3.6. Regeneration Characteristics

The reuse of the adsorbent is the key factor in industrial application. To demonstrate the reusability of the IDA type chelating nano-fibers, it was used for five times in the adsorption-desorption of copper and lead ions onto IDA type nano-fibers, it performed almost the same in the same conditions. The relative adsorption percentage of IDA type nano-fibers versus adsorption/desorption cycle was presented in Fig. 8. It showed that RSP (%) could still reach 90% for Cu (II) and 88% for Pb (II) of initial sorption percentage value after five cycles. This indicated that IDA type chelating nano-fibers could be reused without the significant loss in the adsorption property and they could be a very potential adsorbent utilized extensively in aqueous solution.

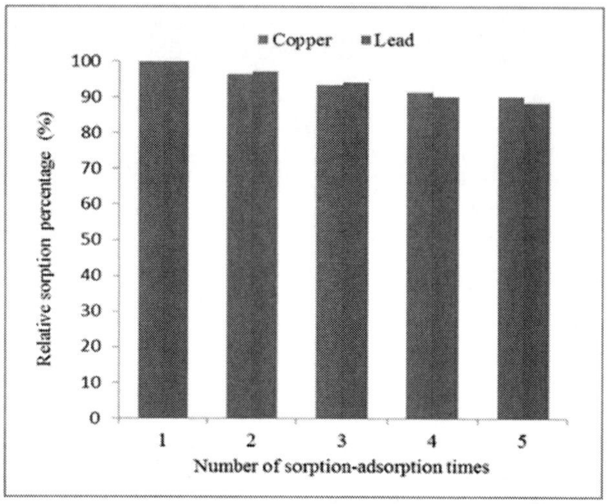

Figure 8. The relationship between the changes of adsorption percentage and adsorption/desorption cycles (pH=3.0 for Cu(II) and pH=5.5 for Pb(II); Initial concentration of 50 mg/L; Temperature of 293 K)

4. Conclusion

In the present work, IDA type chelating nano-fibers was prepared by electro-spinning technique, and the potential of IDA type nano-fibers as the adsorbent for the adsorption of coppper and lead ions from the aqueous solutions was investigated. The maximal sorption percentage (SP) of the metal ions could reach to 93.08% for Cu (II) and 96.69% for Pb (II), respectively. The kinetic data of metal sorption onto the nanofiber membrane fitted well with the pseudo-first-order kinetic model. The comparison of isotherm models showed that Langmuir model could describe the equilibrium data of copper and lead sorption more appropriately than Freundlich model. The thermodynamic parameters show that both of metal ions sorption by the IDA type composite chelating nanofibers is endothermic and spontaneous nature. Furthermore, the adsorption capacity of IDA type nano-fibers can remain almost unchangeable even after being used as adsorption-desorption for five times. This result suggests that the IDA type composite chelating nano-fibers can be a promising adsorbing material for removing heavy metal ions in the aqueous solution.

Acknowledgments

This work was supported by Guangdong Province of Ministry of Education, Ministry of Science and Technology Grants 2012A090300007, as well as supported by the Funds of Large Apparatus and Equipment of Chongqing University 201406150020.

References

[1] L. Bai, H. Hu, W. Fu, J. Wan, X. L. Cheng, L. Zhuge, L. Xiong, Q. Y. Chen, "Synthesis of a novel silica-supported dithiocarbamate adsorbent and its properties for the removal of heavy metal ions, " J. Hazard. Mater, vol. 195, pp.261-275, , august 2011.

[2] M. Alabadi, M. Irani, J. Ismaeili, H. Piri, M. J. Parnian, "Electrospun nanofiber membrane of PEO/Chitosan for the adsorption of nickel, cadmium, lead and copper ions from aqueous solution, "Chem. Eng. J, vol. 220, pp. 237-243, January 2013.

[3] Q. Zhang, K. Xia, Y. Liu, Y. C. Liu, C. Zhang, X. Liu, H. Xu, J. D. Chen, "Genome-wide identification and characterization of cadmium-responsive microRNAs and their target genes in radish (Raphanus sativus L.)roots, "Chem. J. Chinese. Univ, vol. 34, pp. 2667-2673, Seotember 2013.

[4] I. Memon, Q. Ali, A. Zubedi, F. A. Mangi.,"DPMM: dynamic pseudonym-based multiple mix-zones generation for mobile traveler." Multimedia Tools & Applications, pp. 1-30, 2016.
[5] L. Y. Tong, J. D. Chen, Y. Y. Yuan, Z. H. Cui, M. F. Ran, Q.Zhang, J. Bu, F. Q. Yang, X. H. Su, H. Xu, "A Novel Lead Ion-Imprinted Chelating Nanofiber: Preparation,Characterization, and Performance Evaluation, "J. Appl. Polvm. Sci, vol. 132, pp. 41507, September 2014.
[6] .F. L. Fu, Q. J. Wang, "Removal of heavy metal ions from wastewaters: a review. "Environ. Manage, vol. 92, pp. 407-418, March 2011.
[7] S. Abbbasizadeh, A. R. Keshtkar, M. A. Mousavian, "Preparation of a novel electrospun polyvinyl alcohol/titanium oxide nanofiber adsorbent modified with mercapto groups for uranium(VI) and thorium(IV) removal from aqueous solution. "Chem. Eng. J, vol. 220, pp. 161-171, January 2013.
[8] L. R. Rad, A. Momeni, B. F. Ghazani, M. Irani, M. Mahmoudi, B. Noghred, "Removal of Ni2+ and Cd2+ ions from aqueous solutions using electrospun PVA/zeolite nanofibrous adsorbent. "Chem. Eng. J, vol. 256, pp. 119-127, November 2014.
[9] S. Wu, F. Li, H. Wang, L. Fu, B. Zhang, G. Li, "Effects of poly (vinyl alcohol) (PVA) content on preparation of novel thiol-functionalized mesoporous PVA/SiO2 composite nanofiber membranes and their application for adsorption of heavy metal ions from aqueous solution."Polym, vol. 51, pp. 6203-6211, October 2010.
[10] F. Ruggieri, A. A. D'Archivio, D. Camillo, L.Lozzi. M. A. Mercorio, S. J. Santucci, "Optimization in control: methods and applications. "Sep. Sci, vol. 38, pp. 1402-1410, August 2002.
[11] Hai. T. Zhang ; Q. Wang ; Z. Y. Chen ; F. L. Wei ,"Dynamics and Feedback Control of Electrospinning Processes." IEEE Transaction on Control Systems Technology, vol. 25, no.2, pp. 2017.
[12] T. B.Qian, W. Q.,"Effect of lignocellulose-derived inhibitors on growth and hydrogen production by Thermoanaerobacterium thermosaccharolyticum W16. "Techno. of Water. Treat, vol. 12, pp. 300-304, November 2009.
[13] L.Yang, Y. Li, L. Wang, Y. Zhang, X. Ma, Z. Ye, "Preparation and adsorption performance of a novel bipolar PS-EDTA resin in aqueous phase. "J. Hazard. Mater, vol. 180, pp. 98-105, March 2010.
[14] I. Memon, "Authentication User's Privacy: An Integrating Location Privacy Protection Algorithm for Secure Moving Objects in Location Based Services." Wireless Personal Communications, vol. 82, no.3, pp. 1-16, 2015.
[15] Y. B. Li, B. J. Gao, R. X. Wang, R. K. Du, "Cardiomyocyte-derived adiponectin is biologically active in protecting against myocardial ischemia-reperfusion injury. "J. Funct. Polvm, vol. 23, pp. 236-242, December 2009.
[16] S. Kagava, H. Miyazaki, Y. Inoue, T. Kato, H. Yanai, W. Kamichatani, T. Wajiwara, M. Saito, K. Tohda, "Chelating fibers prepared with a wet spinning technique using a mixture of a viscose solution and a polymer ligand for the separation of metal ions in an aqueous solution. "J. Hazard. Mater, vol. 203-204, pp. 370-373, December 2011.
[17] J. X. Qian, "A statistical analysis of coal mine accidents caused by coal dust explosions in China. " Study on preparation and application of iminodiacetic acid chelate fiber. Thesis, Beijing. Institute of. Technology, vol. 22, pp. 528-532, 2009.
[18] I. Memon, L. Chen, A. Majid, M. Q. Lv, I. Hussain, G. C. Chen, "Travel Recommendation Using Geo-tagged Photos in Social Media for Tourist." Wireless Personal Communications, vol. 80, no.4 ,pp. 1347-1362, 2014
[19] E. Baştürk, S. Demir, Ö. Danış, M. V. Kahraman, "Covalent immobilization of α-amylase onto thermally crosslinked electrospun PVA/PAA nanofibrous hybrid membranes. "J. Appl. Polvm. Sci, vol. 127, pp. 349, May 2012,.
[20] L. L.Wu, X. Yuan, S. Sheng, "Immobilization of cellulase in nanofibrous PVA membranes by electrospinning. "J. Membr. Sci, vol. 250, pp. 167-173, March 2005.

[21] C. Tang, C. D. Saquing, J. R. Harding, S. A. Khan, "In Situ Cross-Linking of Electrospun Poly(vinyl alcohol) Nanofibers. "Macromolecules, vol. 43, pp. 630-637, December 2009.
[22] Y. H. Yuan, "Study on the adsorption of copper ions by epichlorohydrin-crosslinked chitosan/sodium alginate adsorbent. " Preparation, Structure and Propertiesof ModifiedPolymers Based on Chitosan and Sodium Alginate Ph.D. Thesis, South China University of Technology, vol. 40, pp. 148-154, July 2012.
[23] M. A. Ghalia, Y. Dahman, "Radiation crosslinking polymerization of poly (vinyl alcohol) and poly (ethylene glycol) with controlled drug release. "J. Polvm. Res, vol. 22, pp. 1-9, November 2015.
[24] X. Li, C. Zhang, R. Zhao, X. Xu, X. T. Jia, C. Wang, L. Li, "Efficient adsorption of gold ions from aqueous systems with thioamide-group chelating nanofiber membranes. "Chem. Eng. J, Vol. 229, pp. 420-428, August. 2013.
[25] Y. Zhang, Y. Chen, C. Wang, Y. Wei, "Immobilization of 5-aminopyridine-2-tetrazole on cross-linked polystyrene for the preparation of a new adsorbent to remove heavy metal ions from aqueous solution. "J. Hazard. Mater, vol. 276, pp. 129-137, July 2014.
[26] Y. S. Ho, G. McKay, "Comments on using of "pseudo-second-order rate equation" in Journal of Radioanalytical and Nuclear Chemistry, Volume 283. " Process. Saf. Environ. Prot, vol. 76, pp. 332, August 2010.
[27] S. J. Allen, G. Mckay, J. F. Porter, "Adsorption isotherm models for basic dye adsorption by peat in single and binary component systems. "J. Collid. Interface. Sci, vol. 280, pp. 322, December 2004.
[28] A. R. Keshtkar, M. Irani, M. A. Moosavian, "Removal of uranium (VI) from aqueous solutions by adsorption using a novel electrospun PVA/TEOS/APTES hybrid nanofiber membrane: comparison with casting PVA/TEOS/APTES hybrid membrane. "J. Radioanal. Nucl. Chem, vol. 295, pp. 563-571, January 2013.
[29] M. Irani, A. R. Keshtkar, M. A. Moosavian, "Removal of cadmium from aqueous solution using mesoporous PVA/TEOS/APTES composite nanofiber prepared by sol–gel/electrospinning. "Chem. Eng. J, vol. 200, pp. 192-201, August 2012.
[30] E. S. Kunarti, "Production of metal ion imprinted polymer from mercapto–silica through sol–gel process as selective adsorbent of cadmium. "Desalination, vol. 251, pp. 83-89, February 2010.

IOP Conf. Series: Materials Science and Engineering **275** (2017) 012002 doi:10.1088/1757-899X/275/1/012002

The Research Progress of Zr Based Amorphous Composite Materials

Xiaohong Su*, **Huie Hu and Xiaodong Kong**

Naval university of engineering,China.Email:surrey0908@sina.com
Naval university of engineering,China.Email:huhuie@163.com
Naval university of engineering,China.Email:kxd63813@hotmail.com

Abstract: Based on the introduction and forms of the second phase the Zr based amorphous composites was classified.This paper introduced the research work of the endogenous phase-Zr based amorphous composite materials, the particles-Zr based amorphous composites, fiber-Zr based amorphous composite materials, the skeleton-Zr based amorphous composite materials, and prospected the development prospect of Zr based amorphous composites.

1.Introduction

Zr base amorphous alloys have excellent mechanical properties such as high strength, high elasticity, high toughness, high hardness, wear resistance, low thermal expansion coefficient and excellent corrosion resistance, etc [1-9], at the same time, Zr base amorphous alloy has the high glass forming ability, relatively simple preparation technology and equipment, is currently one of the research of amorphous alloy. But the alloy with low plasticity at room temperature show the typical brittle fracture, hardly happen macroscopic plastic strain, which seriously limits the application of the alloy as the engineering material, and makes its superior performance cannot get to play. The introduction of the second phase as strengthening phase to Zr base amorphous alloy to form Zr base amorphous composite material, which strengthened its room temperature plasticity and further improved other properties, also has opened up a broad application prospect for Zr base amorphous alloy applications in engineering materials.

2.The Research Progress of Zr Base Amorphous Composite Material

Based on the introduction of the second phase in Zr base amorphous alloy the Zr base amorphous composite materials can be divided into endogenous phase-Zr base amorphous composites and plus phase-Zr base amorphous composite materials, And the plus phase-Zr base amorphous composite materials can be divided into the particle-Zr base amorphous composites, the fiber-Zr base amorphous composites, skeleton - Zr base amorphous composite materials, and other.

2.1 Endogenous Phase-Zr Base Amorphous Composite Materials

By adjusting the alloy composition and process conditions, the crystallization of amorphous alloy or primary crystallization phase as the second phase was introduced into Zr base amorphous alloy, which was uniformly distributed in the alloy matrix to prevent shear zone extending and inspire the toughening effect of multiple shear bands formation. Fan et al [10-11] found $Zr_{53}Ti_5Ni_{10}Cu_{20}Al_{12}$ and $Zr_{60}Cu_{20}Pd_{10}Al_{10}$ have good plasticity at room temperature, and the nanocrystalline's exist in amorphous matrix by high resolution electron microscopy (sem), thus forming nanocrystalline-Zr base amorphous composite materials. The composite material has macroplasticity which due to stress con-

Content from this work may be used under the terms of the Creative Commons Attribution 3.0 licence. Any further distribution of this work must maintain attribution to the author(s) and the title of the work, journal citation and DOI.
Published under licence by IOP Publishing Ltd

centration near the nanocrystalline, which leads to the generation of multiple shear zone and the single shear zone's difficulty to expand rapidly [12].

The amorphous with metastable structure have the trend of crystalline transition spontaneously in the process of heating. Therefore, by reasonable control of annealing process, some amorphous transform to nanocrystalline phase with size in a few nanometers to hundreds of microns, which evenly dispersed in the amorphous matrix to enhance the effect of matrix [13]. Q Wang, D K Wang et al [14] produced Zr base amorphous matrix composites enhanced by 10.5% vol nanocrystalline of the size of 2-10 nm by controlling the process conditions annealing treatment Zr41.2Ti13.8Cu12.5Ni10Be22.5 amorphous alloy. By study on the crystallization behavior of Zr56.6Cu17.3Ni12.5Al9.6Ti4 amorphous alloy with temperature, Xing Dawei [15] found that with the increase of annealing temperature and annealing time increases, the crystallization phase increased obviously, and the amorphous phase decreased significantly.

Meanwhile the primary crystallization phase could be separated out in the cooling process of the alloy melt by controlling the cooling rate. With the cooling process, the rest of the freezing melt into amorphous structure, forming a crystallization phase distributed evenly in the amorphous matrix of composite structure. Liu Jinmin et al [16] prepared in-situ martensite CuZr phase enhanced Zr50.5Al9Ni13.05Cu27.45 amorphous alloy composite materials by using copper spray casting process. Because the large number of coherent twin boundary of martensite CuZr phase can effectively prevent localized shear band extension, prompt and initiation of multiple shear zone when deformation, in order to delay the initiation and propagation of crack. The mechanical properties of composite materials greatly improve, the compression fracture strength and strain respectively reached 2190 MPa and 8%.

By adding high melting point elements into Zr base amorphous alloy ,the plastic branch crystal separate out, which can inhibit the effect of shear zone rapidly expanding in the amorphous matrix, so as to make the mechanics of composite materials is improved. By adding Nb, Johnson et al developed a new alloy for Zr56.2Ti13.8Nb5.0Cu6.9Ni5.6 Be12.5, the Nb act witn Zr and Ti to form the branch crystal of β-Zr(Ti),which Set on the fully amorphous matrix with volume fraction of 25%,and the size of the branch crystal is 100 microns. By adding element of Hf and Ag, Huo Juntao et al [17] prepared the bulk amorphous composite materials Zr-Cu-Al-Hf(Ag). Compressive test shows that thestrength and plastic of (Zr50Cu40Al10)98Hf2 is better than Zr50Cu40Al10, the σmax is 2002MPa, the maximum plastic strainεp is 8%.The experiment confirmed that the element Hf promote the formation of branch crystal in the Zr base amorphous alloy

2.2 *Particle Phase-Zr Base Amorphous Composite Materials*

By adding particle phase into Zr base amorphous matrix can prevent the expansion of the shear zone, and promote the generation of fracture surface and the additional shear band, keep Zr base amorphous composite materials in high strength and high plasticity. The plus particle phase mainly includes two kinds of refractory metal particles and ceramic particles. Through the analysis of the mechanical properties of Amorphous composites reinforced by particles it's found that: if the enhancement phase particles is toughness, compared with the system of pure amorphous composites plasticity significantly increased; if the reinforced phase is brittle particles, the plasticity increase of amorphous composites is not big, but the strength will be higher than toughness reinforced particles of amorphous composite materials. It's because the plastic improvement mechanism of the two kind of composite materials is different [18-19]. Currently the die casting and liquid infiltration casting process are mainly adopted for preparing particle reinforced composites.

More research for plus toughness particles reinforced Zr base amorphous composites focused on such as adding metal W, Ta, Mo, Nb and TiNb [20-26]. The study found that the Zr base amorphous composites with W particle size less than 10 microns, volume fraction of 60% changed its yield stress less, and raised the plastic strain greatly, the total variable should be increased by 350%.By scanning electron microscopy (sem) in-situ tensile test combined with finite element simulation, Z. Zhu et al [24] reveals that the plastic deformation of Ta particles caused the shear zone is evenly distributed on the substrate around the particles, which triggered an initiation and development of the secondary shear zone, and helps to improve the plasticity of Zr54Ta8Cu15.6Ni12.4Al10 amorphous composites. By microstructure observation H. Choi-Yim et al found around a certain thickness of amorphous phase was generated arround the enhanced phase Ta, Nb, Mo particles.

The main research for plus brittle particles reinforced Zr base amorphous composites focused on ceramic particles ZrC,SiC,WC,ZrO2,etc[27-32]. Kato et al in 1997 prepared for the first piece of Zr base amorphous composite material, will by adding join ZrC particles into the amorphous alloy matrix Zr55Al10Ni5Cu30,and found that with the adding of ZrC particles, the young's modulus, the tensile strength, the microhardness and the plastic strain of the composites increased significantly. Then Conner et al reported that Zr57Nb5Al10Cu15.4Ni12.6 amorphous composites reinforced by SiC, WC ceramic particles has relatively lower tensile strength than W, Ta toughness particles as strengthening phase; the WC particles reinforced amorphous composite has plastic strain ratio up to 3-5%,but the addition of SiC particles did not improve the plastic of the amorphous composite, In view of Choi-Yim and Wu on research results of reinforcement of SiC particles to amorphous matrix composites, Liu Jinmin [32] concluded that the interface microstructure changes have a significant impact on mechanical properties of the composites by studying the different content of SiC particles reinforced Zr50.5Al9Ni4.05Cu36.45 amorphous alloy composite material. The limited Interfacial reaction products ZrC can enhance the bonding strength of interface, passing stress effectively, which hinder the expanding of localized shear band, prompted multiple shear zone of initiation, help to improve the strength and plastic of the alloy. While too much interface reaction can weaken the bonding strength of interface, make the alloy crack quickly along the interface during compression deformation, characterized by brittle fracture.

2.3 Fiber-Zr Base Amorphous Composite Materials

By adding ductile metal fiber into amorphous matrix can prevent the expansion of the shear zone and promote fracture surface area, and the additional shear band to keep amorphous composite materials in high strength and high plasticity. Infiltration casting process is a used preparation methods of fiber reinforced Zr base amorphous composite materials, according to join the fiber morphology the composite material can be divided into continuous fiber reinforced and short fiber reinforced two kinds.

The main research for continuous fiber reinforced phase focused on W, Mo and Ta wire [33-48], which mainly more research on W silk as enhanced phase, its from 40-1000 microns in diameter, volume can be as high as 80%.The composite has more excellent compression performance because the W wires have good combination of matrix and the help under compression loading the formation of shear band. Conner, Cheng Huanwu et al systematically studied the mechanical behavior and fracture characteristics of the W/Zr base amorphous alloy composite materials, the results showed that with the increase of W wire volume content of the elastic modulus of the composite increased, and the compression plasticity first increases and then decreases with the increase of fiber volume content, the fracture mode of the composite with the increase of fiber content from the shear fracture to longitudinal splitting.

Because Zr base amorphous alloy under the action of the enhancement of the W wire can produce self-sharpening effect caused by the localized shear fracture behavior in the process of penetrating the destruction of the object, at the same time has high density, can be used to withstand the impact of structural materials such as armor-piercing core. G Rong, Lei Bo et al carried out a series of high speed penetration test on W wire/Zr base amorphous composites of, studied the self-sharpening behavior and penetration of the material. The good mechanical properties and potential defense applications of W wire reinforced Zr base amorphous composite makes it one of the most widely studied of amorphous alloy composite material.

In view of the limitations of high strength and high modulus of one-way continuous fiber reinforced amorphous composites only along the fiber direction, S T Deng, Ma Guangcai et al [49-51]. added the short fiber as reinforced phase such as stainless steel capillary, carbon nanotubes to the Zr base amorphous alloy, in which the strengthen phase randomly distributed to pile arrangement, and the composites exhibit isotropic; At the same time the introduction of the short fiber can achieve lower volume percentage to introduce more interface, so than the same volume of continuous fiber reinforced composite materials have greater compressive plastic strain. In CNTs-Zr base amorphous composite materials, along with the increase in CNTs adding quantity, the young's modulus, the shear modulus,the bulk modulus and the hardness of amorphous composites increased.

2.4 Skeleton-Zr Base Amorphous Composite Materials

The amorphous composites with endogenous phase enhancement, particles enhancement and fibers enhancement all can't limit the expansion of the amorphous shear zone from the direction of 3d.But in the amorphous composites reinforced by skeleton, matrix and the skeleton reinforced phase interact with each other within the 3d space, which effectively reduce the stress concentration, can enhance the deformation resistance and the plastic of amorphous composites, at the same time can realize the composites' isotropic [20]. Now the main research for skeleton-Zr base amorphous composite materials focused on W skeleton and SiC skeleton reinforced Zr base amorphous composites [52-57].

Zhang et al studied the quasi static compression performance of Zr base amorphous composites reinforced by W skeleton of 80% volume content, it is concluded that the compressive strength of the composites can be as high as 3450Mpa, the fracture strain reached 50%, the strength and plasticity are far higher than that of W particles and W wire enhanced amorphous composite materials. Li et al studied the quasi-static and dynamic compression mechanical properties of the W skeleton amorphous composites in contrast, and analyzed the influence of W skeleton volume fraction on mechanical properties. they found that composite materials' dynamic compression strength higher than quasi static, consistent with W silk reinforced amorphous composites, which shows the positive strain rate sensitive effect of composite materials; As W skeleton volume fraction increases, the pore is dense, plastic of composite material, the better.

2.5 Others

By introducing a component with large mixing enthalpy, the alloe melt with difficult miscibility gap was prepared. Under the condition of rapid solidification, melt structure is rapid freezing, thereby the two phase or multiphase amorphous composite structure was obtained. Differences in performance between different phase, can be used to extend the effective way to change the internal cracks of amorphous, so as to improve the mechanical properties of amorphous.

Sun Linlin et al [58] found that the uniform amorphous matrix separated into 2 kinds of amorphous alloy. when the typical amorphous structure material $Zr48Cu36Ag8Al8$ alloy kept warm at 703K for 20 min. Wang Zhongyuan et al [59] found that alloy melt liquid phase separation occurred during the solidification process when they added the Co-Cu binary metastable immiscible alloy into the Zr-Ce binary stable immiscible alloy, the Co and Cu elements respectively distributed in rich Zr and Ce in the liquid phase, and eventually rich Zr-Co and rich Ce-Cu two liquid phase were formed. Under the condition of rapid solidification, rich Zr-Co and rich Ce-Cu two liquid phase formed by the liquid-liquid separation of quaternary alloy respectively Glass changed to form the Zr and Ce based amorphous phase, eventually formed two-phase amorphous alloy composite materials.

3.Prospect

Zr base amorphous composite materials with high specific strength, specific stiffness, high hardness, high wear resistance an, corrosion resistance and other excellent physical properties would have broad in the aerospace, precision instruments and equipment manufacturing, chemical industry and other fields application prospects. Zr base amorphous alloy has similar mechanical performance parameters to human bone, and at the same time has many characteristics, such as good biocompatibility with the human body environment, So Zr base amorphous composite material could be used for the preparation of medical scalpels, artificial teeth, artificial bone and biological sensing material. Moreover, with the enhancement function of W wire, Zr base amorphous composite material has a self-sharpening effect in the in the process of penetrating the destruction of the object because of the local shear fracture behavior in the process of deformation,that makes W fiber-Zr base amorphous composites penetrator have a great application prospect in the field of national defense.

4. Acknowledgment

This work was financially supported by the Natural Science Foundation of China under Grant No.51575522 and the Basic Research Foundation of the College of Science in Naval University of Engineering.

5.References

[1] Wang Weihua, Wang Wenkui. The discowery and development of new multicomponent bulk amorphous alloy materials[J].Physics,1998,27(7):398-403;
[2] Inoue A.High strength bulk amorphous alloy with low critical cooling rates[J],Mater Trans,1995,36(7):866-875;
[3] Bruek H A, Chitmanl, Rosokis A,et al. Quasi-static constitutive behavior of Zr41.25Ti13.75Cu12.5Ni10Be22.5 bulk amorphous alloys [J]. Seripta Metall Mater, 1994, 30(4):429-434;
[4] Conner R D, Rosakis A J, Johnson W L, et al.Fraeture toughness determination for a beryllium-bearing bulk metallic glass [J]. Seripta Mater.1997,37(9):1373-1378;
[5] Inoue A. Stabilization of metallic supercooled liquid and bulk amorphous alloys[J].Aeta mater,2000,48:279-306;
[6] Peter J Blau. Friction and wear of a Zr-based amorphous metal alloy under dry and lubricated conditions.Wear,2001,250:431-434;
[7] M Z Ma, R P Liu, Y. Xiao, et al. Wear resistance of Zr-based bulk metallic glass applied in bearing rollers [J]. Materials Science and Engineering A. 2004,386:326-330;
[8] Qin Xiujuan. Corrosion resistance of Zr41.5Ti14Cu13Ni10Be22.5 bulk amorphous alloy[J].J Mater Sci Technol, 2003, 15(1):93-94;
[9] Yang Zhigang, Wang Tiejun, Wang Gaochao, et al. Elecetrochemical property of a Zr-based amorphous alloy [J]. Chinese Journal of Rare Metals, 2004,28(1):25-28;
[10] Fan C, Li C, Inoue A. Deformation behavior of Zr-based bulk nanocrystalline amorphous alloys[J].Physical Review B,2000,61(6):3761-3763;
[11] Fan C, Inoue A. Ductility of bulk nanocrystalline composites and metallic glasses at room temperature[J].Applied Physics Letters,2000,77:46;
[12] Sun Xianhu. Deformational Behavior And Mechanisms of Ti-based In-situ Metallic Glass Composites at Room Temperature [D], Tiyuan, Taiyuan University of Technology,2016;
[13] Lu Ke. Phase transformation from an amorphous alloy into nanocrystalline materials [J]. Acta Metallurgica Sinica,1994,30(1):1-21;
[14] Q. Wang, D.K. Wang. High temperature homogeneous plastic flow behavior of a Zr based bulk metallic glass matrix composite [J]. Journal of Alloys and Compounds,2010,495:50-54;
[15] Xing Dawei, Huang Yongjiang, Shen Jun. Crystallization Behaviors of ZrCuNiAlTi4 Bulk Amorphous Alloy during Continuous Heating [J]. Rare Metal Materials and Engineering, 2007,36(7):1181-1184;
[16] Liu Jinmin, Yuan Xiaoguang. Effects of Cooling Rate on Microstructure and Mechanical Prop erties of ZrAlNiCu Alloy [J]. Special Casting & Nonferrous Alloys, 2010,30(10):883-886;
[17] Huo Juntao. The Formation and Mechanical Behavior of Zr-Cu -Al-Hf(Ag) BMG Composites [D]. Harbin, Harbin Institute of Technology, 2009;
[18] Hu Zhuangqi, Zhang Haifeng. Recent progress in the area of bulk amorphous alloys and composites [J]. Acta Metallurgica Sinica, 2010, 46(11):1391-1421;
[19] Zhang Xinqiang. In-situ Micro Mechanical Behavior Study of Porous W/Zr-based Metallic Composite [D].Beijing, Beijing Institute of Technology, 2015;

[20] Choi-Yim H, Schroers J, Johnson W L. Microstructures and mechanical properties of tungsten wire/particle reinforced ZrNbAlCuNi metallic glass matrix composites [J]. Applied Physics Letters, 2002, 80:1906-1908;

[21] Qiu Keqiang, Ren Yinglei. Tungsten Particulate Rreinforced Zr Based Bulk Metallic Glass Composites [J]. Rare Metal Materials and Engineering, 2006,35(1):66-69;

[22] Conner R D, Choi-Yim H, Johnson W L, Mechanical properties of Zr57Nb5Cu15.4Ni12.6 metallic glass matrix particulate composites [J]. Journal of Materials Research,1999,14(8):3292-3297;

[23] Z. Zhu, H. Zhang, Z. Hu, et al. Ta-particulate reinforced Zr-based bulk metallic glass matrix composite with tensile plasticity [J]. Scripta Materialia, 2010(62):278–281;

[24] Chu Zhenhua, Yuan Guangyin, Kato Hidemi, et al. Toughening Mechanism of Zr-Based and Cu-Based Bulk Metallic Glasses Composites by TiNb Particle[J].Rare Metal Materials and Engineering, 2013,42(6):1154-1158;

[25] Z.H. Chu, H. Kato, G.Q. Xie, et al. Enhanced Mechanical Properties by Cyclic compression in TiNb/Zr-Based Metallic Glassy Composite[J].Metallurgical and Materials Transactions A, 2013,44A:4595-4600;

[26] Choi-Yim H, Conner R D， et al. Processing, microstructure and properties of ductile metal particulate reinforced Zr57Nb5Cu15.4Ni12.6 bulk metal lie glass composites [J]. Acta Materialia, 2002, 50 (10):2737-2745;

[27] Hidemi Kato, Akihisa Inoue. Synthesis and Mechanical Properties of Bulk Amorphous Zr-Al-Ni-Cu Alloys Containing ZrC Particles[J].Mater Trans JIM,1997,38(9):793-800;

[28] Choi-Yim H, Busch R, et al. Synthesis and characterization of particulate reinforced Zr57Nb5Al10Cu15.4Ni12. 6 bulk metallic glass composites [J]. Acta Materialia, 1999,47(8):455-462;

[29] Wu Xiao-feng, Qiu Ke-qiang,Zhang Hai-feng, et al. Effect of SiC on the glass forming ability and thermal stability of Zr55Al10Ni5Cu30 bulk metallic glass[J]. Acta Metallurgica Sinca,2003,9(4):414-418;

[30] Liu Jinmin, Yuan Xiaoguang, Zhang Haifeng. Microstructure and deformation behavior of Zr-based bulk metallic glass composite reinforced with SiC particle[J].Journal of Shenyang University of Technology,2010,32(2):157-161;

[31] Zhang Jian. The Synthesis of ZrO2/Zr-Based Bulk Metallic Glass Composites and Study of Their Mechanical Properties [D]. Shanghai, Shanghai Jiao Tong University, 2009;

[32] Chu Zhenhua, Huang Bin, Yuan Guangyin, et al. Microstructure and Fracture Behavior of Zr55Cu30Al10Ni5 Bulk Metallic Glass and Its Composites Containing ZrO2[J].Rare Matal Materials and Engineering,2011,40(5):0765-0768;

[33] Conner R D, Dandliker R B, Johnson W L. Mechanical properties of tungsten and steel fiber reinforced Zr41.25Ti13.75Cu12.5Ni10Be22.5 metallic glass matrix composites [J]. Acta Materialia,1998,46(17):6089-6102;

[34] Conner R D, Dandliker R B, Scruggs V, et al. Dynamic deformation behavior of tungsten-fiber/metallic-glass matrix composites[J].International Journal of Impact Engineering,2000,24(5):435-444;

[35] H Zhang, Z F Zhang, Z G Wang,e t al. Effects of Tungsten Fiber on Failure Mode of Zr-Based Bulk Metallic Glassy Composite [J]. Metallurgical and Materials Transactions A, 2006,37A:146-154;

[36] H Zhang, Z F Zhang, Z G Wang,et al. Fatigue damage and fracture behavior of tungsten fiber reinforced Zr-based metallic glassy composite [J]. Materials Science and Engineering A, 2006, 418:146–154;

[37] J H Chen, Y Chen, M Q. Jiang, Direct Observation on the Evolution of Shear Banding and Buckling in Tungsten Fiber Reinforced Zr-Based Bulk Metallic Glass Composite [J]. Metallurgical and Materials Transactions A, 2014,45A:5397-5408;

[38] Ma Weifeng. Study on the Interfacial Behavior and Property of W Fiber Reinforced Zr-Based Amorphous Matrix Composites [D]. Xi′an, Northwestern Polytechnical University, 2006;
[39] Cheng Huanwu, Cai Hongnian, Wang Lu. Research on Fracture Mode Under Dynamic Tension of W Fiber/Zr-based Amorphous Alloy Matrix Composites[J].Journal of Materials Engineering, 2007,7:67-69;
[40] Cheng Huanwu, Cai Hongnian, Wang Lu. Dynamic deformation of tungsten fiber reinforced Zr-based amorphous matrix composites[J].Ordnance Material Science and Engineering,2008,31(2):23-26;
[41] Chen Demin, Wang Gang, Sun Jianfei. Deformation behavior of tungsten wires enhanced zr-based bulk metallic glass composite at high strain rate [J]. Acta Metallurgica Sinica, 2006, 42(9):1003-1008;
[42] Chen Chunsheng, Li Jinshan, Ma Weifeng. Interfacial Characteristics and Mechanical Properties of W Fiber Reinforced ZrTiCuNiBe Amorphous Matrix Composite [J]. Rare Metal Materials and Engineering, 2007, 36(8):1390-1393;
[43] R B Dandliker, R D Conner, W L Johnson. Melt infiltration casting of bulk metallic-glass matrix composites [J]. J Mater Res,1998,13:2896-2901;
[44] G Rong, D W Huang, M C Yang. Penetrating behaviors of Zr-based metallic glass composite rods reinforced by tungsten fibers [J]. Theoretical and Applied Fracture Mechanics, 58(2012):21–27;
[45] X.W. Chen*, L.M. Wei, J.C. Li. Experimen tal research on the long rod penetration of tungsten-fi er/Zr-based metallic glass matrix compo site into Q235 steel target[J].International Journal of Impact Engineering, 79(2015): 102-116;
[46] Wang Zhihua. Wf/Zr-based Bulk Metallic Glass Composites and Its High-velocity Impact Behavior [D]. Nanjing, Nanjing University of Science and Technology, 2006;
[47] Lei Bo, Huang Dewu, Yang Mingchuan. An Experimental Study on the Sell-Sharpening Behavior of Armour Piercer Using Tungsten Fiber Composite Material[J].Transactions of Shenyang Ligong University, 2007,27:72-76;
[48] Choi-Yim H, Seung-Yub Lee, Conner R D. Mechanical behavior of Mo and Ta wire-reinforced bulk metallic glass composites [J]. Scripta Materialia,2008,58(9):763-766;
[49] S T Deng, H Diao, Y L Chen, et al. Metallic glass fiber-reinforced Zr-based bulk metallic glass [J]. Scripta Materialia,2011,64:85–88;
[50] Ma Guangcai, Fu Huameng, Wang Zheng. Study On Fabrication and Properties of 304 Stainless Steel Capillary Tubes/Zr53.5Cu26.5Ni5Al12Ag3 Bulk Metallic Glass Compisites [J]. Acta Metallurgica Sinica, 2014,50(9):1087-1094;
[51] Zan Bian, Ru Ju Wang, Wei Hua Wang, et al. Carbon-Nanotube-Reinforced Zr-Based Bulk Metallic Glass Composites and Their Properties [J]. Adv Funct Mater, 2004,14(1):55-63;
[52] Zhang H F, Wang A M, Li H, et al. Quasi-static compressive property of metallic glass/porous tungsten bi-continuous phase composite[J].Journal of Material Research, 2006, 21(6):1351-1354;
[53] Wang Hai-feng, Li Wen. Deformation Behavior of W/Zr-based Amorphous Composite under Compression[J].Transactions of Shenyang Ligong University, 2010, 29(1):80-82;
[54] Xue Y F, Cai H N, Wang L, et al. Strength-improved Zr-based metallic glass/porous tungsten phase composite by hydrostatic extrusion[J].Applied Physics Letters,2007,90: 881-901;
[55] Xue Y F, Cai H N, Wang L, et al. Dynamic compressive deformation and failure behavior of Zr-based metallic glass reinforced porous tungsten composite[J].Material Science and Engineering,2007, A445-446: 275-280;
[56] Liu Weihua, Wang Lu, Cheng Huanwu. Dynamic mechanical characteristics of SiC skeleton/Zr based amorphous matrix composites[J].Ordnance Material Science and Engineering, 2005,28(3):16-18;
[57] Liu Jun, Wang Lu, Cheng Huanwu. Fracture Behavior of Zr-Based Bulk Amorphous Alloy Under High-Speed Impact[J].Journal of Beijing Institute of Technology, 2010,30(7):873-877;

[58] Sun Linlin, Wang Jun. Phase Separation and Microstructure Evolution of Zr48Cu36Ag8Al8 Bulk Metallic glass in the Supercooled Liquid Region[J].Rare Metal Materials and Engineering,2016,45(3):567-570;
[59] Wang Zhongyuan, He Jie. Liquid-Liquid Phase Separation and Formation of two Glassy Phases in Zr-Ce-Co-Cu Immiscible Alloys [J], Acta Metallurgica Sinica,2016;

IOP Conf. Series: Materials Science and Engineering **275** (2017) 012003 doi:10.1088/1757-899X/275/1/012003

Effects of Growth Temperature on Electrocatalytic Properties of Three-dimensional Sulfur-doped Graphene Foam

Jinhao Zhou[a] and Yuanfu Chen

State Key Laboratory of Electronic Thin Films and Integrated Devices, University of Electronic Science and Technology of China, Chengdu 610054, P. R. China
Email: [a]zhousuiyuan@163.com

Abstract: Three-dimensional sulfur-doped graphene (3DSG) has been grown by chemical vapor deposition (CVD) using nickel foam as substrate and thianthrene as the solid carbon and sulfur source. The results show that the growth temperature of 3DSG has significant influence on the electrocatalytic activity for hydrogen evolution reaction (HER). The 3DSG sample under optimal growth temperature of 800 °C has the best electrocatalytic activity. The excellent electrocatalytic activity can be attributed to its well-designed porous, conductive and 3D sulfur-doped graphene structure.

1. Introduction

Design and engineering of high-efficiency electrocatalysts for hydrogen evolution reaction (HER) is of importance for hydrogen energy research [1]. Pt-based electrocatalysts as the most active HER catalysts are expensive which hindered their applications [2]. Therefore, transition-metal-based catalysts such as Fe, Co, Ni, Mo, W and transition-metal sulfides, selenides, phosphide have been intensively studied as the promising HER catalyst [3-7]. Recently, metal-free carbon-based catalysts have been studied as promising HER catalysts [8]. Among them, graphene have been drawn much attentions due to its high electronic conductivity, high mechanical strength and large surface area [9]. Furthermore, it has been demonstrated that doping graphene with heteroatoms such as B, N, S, P is an effective strategy to improvement its electrocatalytic activity [10-13]. On the other hand, most of graphene-based catalysts are based on 2D graphene sheets and the ineffective packing of 2D graphene sheets inevitably decreases the number of catalytically active sites. Therefore, design and fabrication graphene with 3D porous structures as well as heteroatoms doping still remains challenge.

Herein, we demonstrated a facile method to prepare three-dimensional sulfur-doped graphene (3DSG). The 3DSG was synthesized by chemical vapor deposition (CVD) using nickel foam as substrate and thianthrene ($C_{12}H_8S_2$) as the solid carbon and sulfur source. The free-standing 3DSG can be directly used as binder-free electrocatalyst and it exhibited significant improvement HER performance than pristine graphene foam. The excellent electrocatalytic activity can be attributed to its well-designed porous, conductive and flexible 3D sulfur-doped graphene structure. Furthermore, the effects of growth temperature on the electrocatalytic performance of 3DSG was also studied.

Content from this work may be used under the terms of the Creative Commons Attribution 3.0 licence. Any further distribution of this work must maintain attribution to the author(s) and the title of the work, journal citation and DOI.
Published under licence by IOP Publishing Ltd

2. Experimental

2.1. Synthesis of 3DSG and 3DGF

The 3DSG was synthesized by CVD which is similar with our previous reports [14, 15]. Briefly, the commercial nickel foams were firstly washed in dilute HCl and deionized water, then dried and loaded onto a tube furnace. The thianthrene ($C_{12}H_8S_2$) with 200 mg was placed in the upstream of the tube furnace. The nickel foams were annealed at 1000 °C for 30 min under H_2 (30 sccm) and Ar (270 sccm) and then cooled down to the growth temperature. Then the thianthrene was heated up to 140 °C and introduced to the substrate for 30 min growth. The 3DGF was synthesized using 50 sccm CH_4 as carbon source. The Ni foam template was etched away using etching solution contains 1 M $FeCl_3$ and 0.5 M HCl. The 3DSG synthesized at growth temperature 700, 800, 900 °C denote as 3DSG700, 3DSG800 and 3DSG900, respectively.

2.2. Characterizations

The morphologies, structures and elemental compositions of the samples were characterized with field emission scanning electron microscope (SEM, JSM-7000F, JEOL), transmission electron microscope (TEM, Tecnai F20 at 200 kV), Raman spectrometer (Renishaw, 532 nm) and X-ray photoelectron spectroscopy (XPS, Kratos XSAM800, Al Ka radiation (144 W, 12 mA, 12 kV)).

2.3. Electrochemical Measurements

The electrochemical measurements were carried on by using CHI 660D electrochemical workstation (CH Instruments, Inc.). A saturated calomel electrode (SCE) and graphite rod were used as reference and counter electrodes, respectively. The as prepared graphene foams were directly used as binder-free working electrodes. The effective electrochemical geometric area was 0.78 cm^2 and the mass loading was 2.68 mg/cm^2. The electrolyte was 0.5 M H_2SO_4 and degassed by bubbling with Ar for 2 h. The polarization curves were collected with linear sweep voltammetry at a sweep rate of 0.5 mV s^{-1}. The electrochemical impedance spectroscopy (EIS) test was performed at a potential of -0.35V versus RHE with the frequency ranging from 0.1 Hz to 0.1 MHz.

3. Results and Discussion

The morphologies and microstructures of 3DSG800 was characterized by SEM and TEM. As shown in Figure 1a, the 3DSG possess porous structure and the pore diameter is between 200 and 600 μm. The corresponding HRTEM image (Figure 1b) shows one to few layer of the graphene film folded over itself, indicating that the graphene foam comprises of less than 10 atomic layer of carbon.

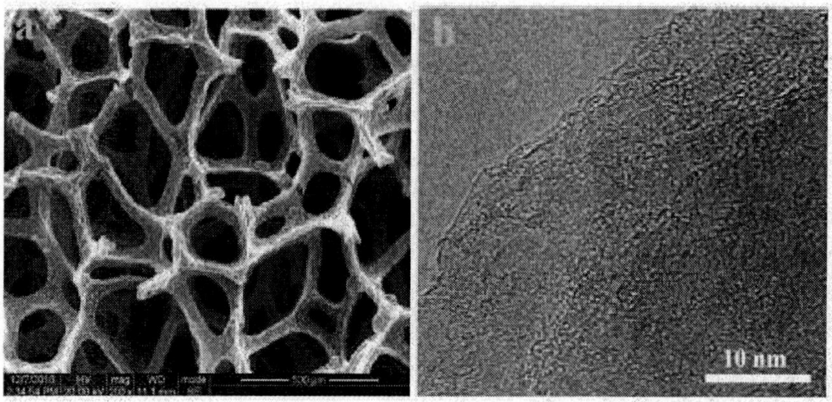

Figure 1. SEM image (a) and TEM image (b) of 3DSG

IOP Conf. Series: Materials Science and Engineering **275** (2017) 012003 doi:10.1088/1757-899X/275/1/012003

Raman was utilized to characterize the structure and defect level of the prepared graphene foams. As shown in Figure 2, the Raman spectrum of all the prepared samples shows a typical G-band (ca. 1580 cm^{-1}) and 2D-band (ca. 2710 cm^{-1}) which represents the carbon sp^2 bonding in graphene [16]. The D-band is related to the disorder and defect and it was not observed in 3DGF which indicates its perfect graphitic structure with a high crystallinity [26]. In contrast, the D-band at ca. 1343 cm^{-1} appeared in all the 3DSG samples, which can be mainly attribute to the introduction of sulfur atoms. Furthermore, the intensity of D-band becomes larger with lower growth temperature, indicates that the 3DSG contains more disorder or defects at lower growth temperature.

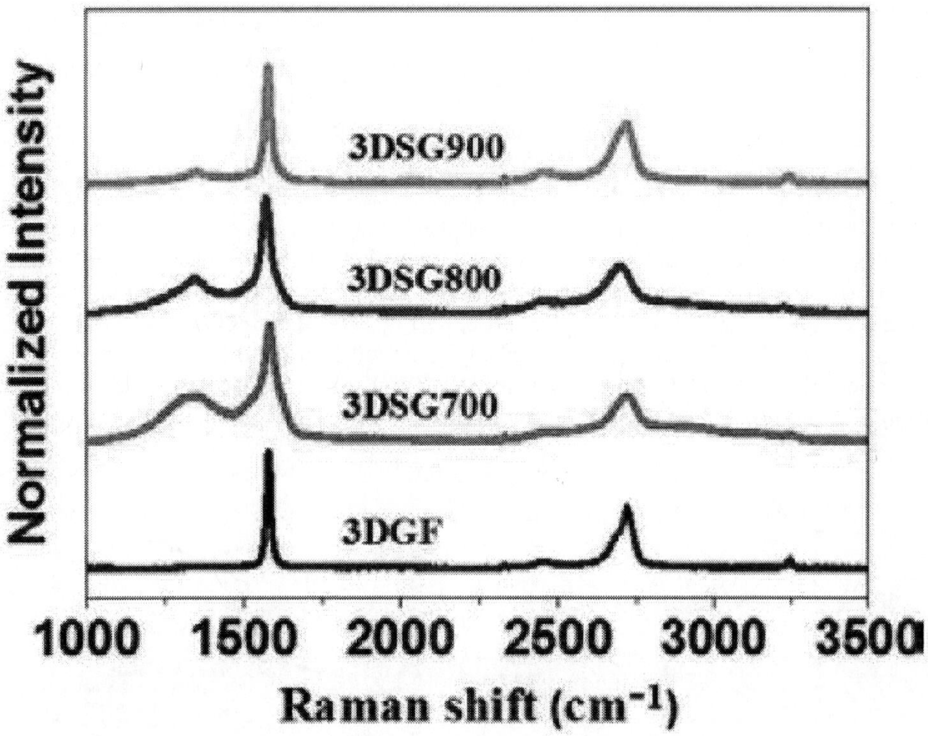

Figure 2. Raman spectra of 3DGF and 3DSG at different synthesized temperature.

The XPS measurements were performed to investigate the elemental composition in 3DSG. Figure 3a shows the high-resolution C 1s spectra of 3DSG800 which has a strong band of C-C species (284.6 eV) with a weak shoulder of C-S/C-O groups (285.2 eV) [17-19]. The high-resolution S 2p spectra of 3DSG800 (Figure 3b) can be deconvoluted into S-H (161.8 eV), C-S-C (163.5 eV), C=S (164.7 eV), C-SO_2-C (168.2 eV) and C-SO_3-C (169.4 eV) [17-21]. The S contents of 3DSG800 were measured to be 2.8 at%. These results confirm that S atoms have been successfully doped into the graphitic structures of graphene foam.

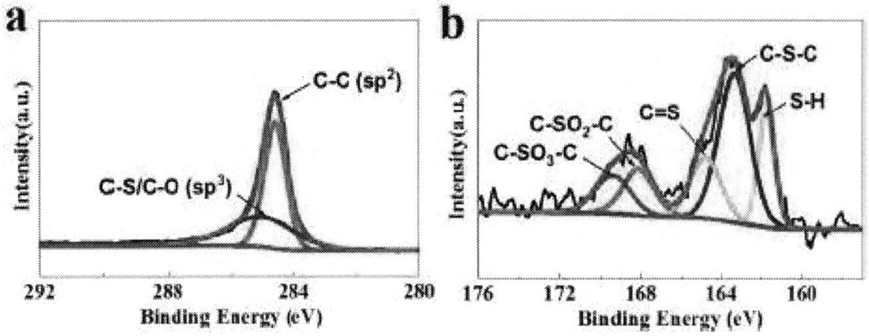

Figure 3. High-resolution XPS C 1s (a) and S 2p (b) spectra of 3DSG800.

To evaluate the effects of growth temperature on the HER performances of the 3DSG, the free-standing graphene foams were directly used as binder-free working electrodes and were tested in typical three-electrode setup. For reference, HER performance of 3DGF and commercial Pt catalyst (20 wt% Pt/C load on glassy carbon electrode) was also investigated. As shown in Figure 4a, all of the 3DSG samples show more positive onset potential than that of 3DGF, indicating that the catalytic activity of 3DGF is largely enhanced after S-doping. The operating overpotentials at 10 mA/cm^2 current were measured to be 419 mV for 3DSG700, 376 mV for 3DSG800 and 435 mV for 3DSG900. The 3DSG grown at 800 °C shows the best catalytic activity. It is further confirmed by Tafel plots (Figure 4b) and the 3DSG800 shows the lowest Tafel slope of 96 mV/dec. It indicates that optimization of growth temperature is important to prepare S-doped graphene foam towered catalytic activity.

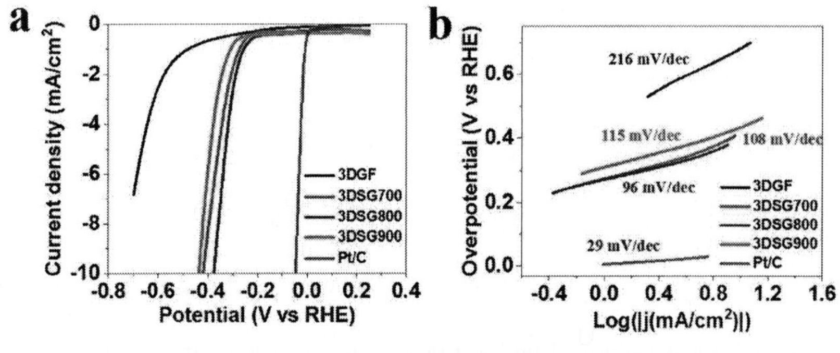

Figure 4. Polarization curves (a) and corresponding Tafel plots (b) of 3DGF, 3DSG and commercial 20% Pt/C catalysts in 0.5 M H_2SO_4.

The HER kinetics of 3DGF and 3DSG were further investigated though electrochemical impedance spectroscopy (EIS). As shown in Figure 5, the charge transfer resistance (R_{ct}) of 3DSG800 is about 117 Ω cm^2, which is much lower than that of 3DSG700 (205 Ω cm^2), 3DSG800 (471 Ω cm^2) and 3DGF (1904 Ω cm^2), suggesting the easier electron transfer at the 3DSG800/electrolyte interface [22].

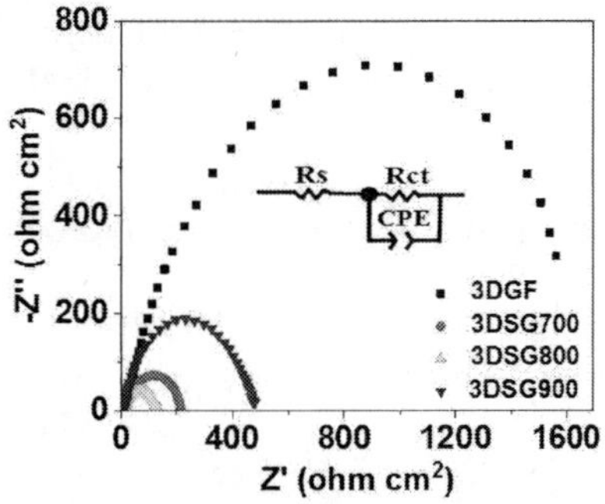

Figure 5. Nyquist plots of HER on 3DGF and 3DSG.

4. Conclusions

In summary, we have successfully synthesized 3DSG by CVD and the effects of growth temperature on the electrocatalytic performance of 3DSG were also studied. The as prepared free-standing 3DSG possess high crystallinity and high porosity and it can be directly used as an electrocatalytic electrode for HER. The 3DSG exhibited significant improvement HER performance than 3DGF. Furthermore, we demonstrate that optimization of growth temperature is important to prepare 3DSG towered high electrocatalytic activity. These results suggest that rational design and prepare graphene combing with 3D porous structure and heteroatom doping is an effective way to develop highly active metal-free HER catalysts.

5. Acknowledgments

The research was supported by the National Natural Science Foundation of China (Grant No. 51372033), the National High Technology Research and Development Program of China (Grant No. 2015AA034202), and the 111 Project (Grant No. B13042).

6. References

[1]Fei H, Dong J, Arellano-Jimenez MJ, Ye G, Kim ND, Samuel ELG, et al., Atomic cobalt on nitrogen-doped graphene for hydrogen generation, Nat Commun, 6 (2015)
[2]Liu X, Dai L, Carbon-based metal-free catalysts, Nat Commun, 1 (2016) 16064
[3]Gong M, Wang D, Chen C, Hwang B, Dai H, A mini review on nickel-based electrocatalysts for alkaline hydrogen evolution reaction, Nano Res, 9 (2016) 28-46
[4]Shi Y, Zhang B, Recent advances in transition metal phosphide nanomaterials: synthesis and applications in hydrogen evolution reaction, Chem Soc Rev, 45 (2016) 1529-41
[5]Jamesh MI, Recent progress on earth abundant hydrogen evolution reaction and oxygen evolution reaction bifunctional electrocatalyst for overall water splitting in alkaline media, J Power Sources, 333 (2016) 213-36
[6]Safizadeh F, Ghali E, Houlachi G, Electrocatalysis developments for hydrogen evolution reaction in alkaline solutions – A Review, Int J Hydrogen Energ, 40 (2015) 256-74
[7]Zeng M, Li Y, Recent advances in heterogeneous electrocatalysts for the hydrogen evolution reaction, J Mater Chem a, 3 (2015) 14942-62
[8]Hu C, Dai L, Carbon-Based Metal-Free Catalysts for Electrocatalysis beyond the ORR, Angew Chem Int Edit, 55 (2016) 11736-58
[9]Zhou W, Jia J, Lu J, Yang L, Hou D, Li G, et al., Recent developments of carbon-based electrocatalysts for hydrogen evolution reaction, Nano Energy, 28 (2016) 29-43
[10]Sathe BR, Zou X, Asefa T, Metal-free B-doped graphene with efficient electrocatalytic activity for hydrogen evolution reaction, Catalysis Science & Technology, (2014)
[11]Huang X, Zhao Y, Ao Z, Wang G, Micelle-Template Synthesis of Nitrogen-Doped Mesoporous Graphene as an Efficient Metal-Free Electrocatalyst for Hydrogen Production, Sci Rep-Uk, 4 (2014)

IOP Conf. Series: Materials Science and Engineering **275** (2017) 012003 doi:10.1088/1757-899X/275/1/012003

[12]Jiang H, Zhu Y, Su Y, Yao Y, Liu Y, Yang X, et al., Highly dual-doped multilayer nanoporous graphene: efficient metal-free electrocatalysts for the hydrogen evolution reaction, J Mater Chem a, 3 (2015) 12642-5

[13]Zheng Y, Jiao Y, Li LH, Xing T, Chen Y, Jaroniec M, et al., Toward Design of Synergistically Active Carbon-Based Catalysts for Electrocatalytic Hydrogen Evolution, Acs Nano, 8 (2014) 5290-6

[14]He J, Chen Y, Lv W, Wen K, Li P, Qi F, et al., Highly-flexible 3D Li2S/graphene cathode for high-performance lithium sulfur batteries, J Power Sources, 327 (2016) 474-80

[15]Qi F, Li P, Chen Y, Zheng B, Liu J, Zhou J, et al., Three-dimensional structure of WS 2 /graphene/Ni as a binder-free electrocatalytic electrode for highly effective and stable hydrogen evolution reaction, Int J Hydrogen Energ, (2017)

[16]Ferrari AC, Basko DM, Raman spectroscopy as a versatile tool for studying the properties of graphene, Nat Nanotechnol, 8 (2013) 235-46

[17]Liu X, Zhou W, Yang L, Li L, Zhang Z, Ke Y, et al., Nitrogen and sulfur co-doped porous carbon derived from human hair as highly efficient metal-free electrocatalysts for hydrogen evolution reactions (vol 3 pg 8840, 2015), J Mater Chem a, 3 (2015) 10135

[18]Yu X, Zhang M, Chen J, Li Y, Shi G, Nitrogen and Sulfur Codoped Graphite Foam as a Self-Supported Metal-Free Electrocatalytic Electrode for Water Oxidation, Adv Energy Mater, 6 (2016)

[19]Jeon I, Zhang S, Zhang L, Choi H, Seo J, Xia Z, et al., Edge-Selectively Sulfurized Graphene Nanoplatelets as Efficient Metal-Free Electrocatalysts for Oxygen Reduction Reaction: The Electron Spin Effect, Adv Mater, 25 (2013) 6138-45

[20]Islam MM, Subramaniyam CM, Akhter T, Faisal SN, Minett AI, Liu HK, et al., Three dimensional cellular architecture of sulfur doped graphene: self-standing electrode for flexible supercapacitors, lithium ion and sodium ion batteries, J Mater Chem a, (2017)

[21]Lu Y, Chen J, Wang A, Bao N, Feng J, Wang W, et al., Facile synthesis of oxygen and sulfur co-doped graphitic carbon nitride fluorescent quantum dots and their application for mercury(II) detection and bioimaging, J Mater Chem C, 3 (2015) 73-8

[22]Ito Y, Cong W, Fujita T, Tang Z, Chen M, High Catalytic Activity of Nitrogen and Sulfur Co-Doped Nanoporous Graphene in the Hydrogen Evolution Reaction, Angew Chem Int Edit, 54 (2015) 2131-6

Divalent Metal Ions Induced Osteogenic Differentiation of MC3T3E1

Guoshou Wang[a], Wenta Su[b*], Pohung Chen[b] and Teyang Huang[a]

[a]Department of Orthopedic Surgery Mackay Memorial Hospital, Taipei, Taiwan
[b]Department of Chemical Engineering and Biotechnology, National Taipei University of Technology, Taipei, Taiwan
1 Sec. 3, Chung-Hsiao E. Rd, Taipei 10608, Taiwan
[*]Corresponding author. Tel.: +886-2-27712171 ext. 2554; Fax: +886-2-27317117
E-mail: f10549@mail.ntut.edu.tw

Abstract: Biomaterial scaffolds blended with biochemical signal molecules with adequate osteoinductive and osteoconductive properties have attracted significant interest in bone tissue engineering regeneration. The divalent metal ions can gradually release from the scaffold into the culture medium and then induced osteoblastic differentiation of MC3T3E1. These MC3T3E1 cells expressed high activity of alkaline phosphatase, bone-related gene expression of collagen type I, Runx2, osteopontin, osteocalcin, and significantly enhanced deposited minerals on scaffold after 21 days of culture. This experiment provided a useful inducer for osteogenic differentiation in bone repair.

1. Introduction

Biomedical material containing inducing agent is an implant substitutes placed into body for damaged tissue to repairing. Many natural and synthetic polymers are employed as tissue scaffolds, which are biodegradable and bioresorption by the body after being implanted.

Bone mineral containing 36.7% calcium and 16% phosphorus, and seventy percent of bone consists of hydroxyapatite and the tooth enamel composed of almost ninety percent hydroxyapatite. However, various divalent ions such as magnesium, strontium, zinc can also promote bone cell maturation, increase the amount of new bone, and can inhibit the activity of osteoclasts [1-3].

In generally, ceramic or polymer composite scaffolds believed to mimic the natural bone tissue. The complex requirements for clinical bone regeneration including biocompatibility, mechanical support, osteoconduction and osteroinduction have been widely investigated [4]. Chitosan has structural characteristics similar to glycosamines, and derived from the alkali-deacetylation of chitin. Chitosan is biocompatible, non-antigenic, nontoxic and biodegradation in vivo, and had studied for many biomedical applications [5].

In this study, we investigate the possible influence of different divalent metal ions for the osteogenic differentiation of MC3T3E1 cells. Various divalent metal phosphates synthesized by wet precipitation and then blended with chitosan solution to form 3D scaffolds. Cellular proliferation was measured by MTT assay, alkaline phosphatase (ALP) activity assay, real-time polymerase chain reaction and Alizarin red S staining were performed to determine cellular matrix maturation and mineralization.

Content from this work may be used under the terms of the Creative Commons Attribution 3.0 licence. Any further distribution of this work must maintain attribution to the author(s) and the title of the work, journal citation and DOI.
Published under licence by IOP Publishing Ltd

2. Materials and Methods

2.1. Synthesis of Divalent Metal Phosphate

Every *divalent metal* phosphate was synthesized via wet precipitation using 0.3M MgCl2, 0.8 M SrCl2, 0.33 M Ba(NO3)2, and 0.1 M Zn(NO3)2 with 0.2 M Na2HPO4 as agitation solution for one day, rinsed with distilled water, and dried in the oven. Resulting synthesized products were magnesium phosphate (MgP), strontium phosphate (SrP), barium phosphate (BaP), and zinc phosphate (ZnP). Calcium phosphate (CaP) was purchased from Sigma.

2.2. Fabrication of porous chitosan scaffolds containing divalent metal phosphate

Chitosan (MW 400 000; 98.5% deacetylation) was obtained from C&B Industrial Co., Ltd. (Taiwan). Porous chitosan scaffolds were prepared using the freeze-drying method. A 3% (w/v) chitosan solution was prepared by dissolving chitosan in 0.167 N acetic acid. Divalent metal phosphate scaffolds were prepared by adding 1.0% divalent metal phosphate into the chitosan solution. All solutions were poured into a cylinder mold and kept in a freezer at −20 °C for one day. The frozen samples were then lyophilized in a freeze dryer (FDU-1200 EYELA, Japan) at −50 °C and 11.2 Pa for 8 h. Subsequently, all scaffolds were neutralized with 1.25 N sodium hydroxide solution in a shaker for 30 min and then thoroughly rinsed with deionized water.

2.3. Cellular proliferation by MTT assay

To evaluate the cellular viability of SHEDs in chitosan scaffolds with or without divalent metal phosphate, MC3T3E1 cells were seeded into the scaffolds at 2×10^4 cells/scaffold with DMEM medium supplemented with 10% FBS and 1% penicillin and streptomycin. The medium was replaced every 3 days for 21 days. For measured proliferated cells, the whole scaffold was treated with 5 mg/mL MTT at 37 °C for 24 h. The cultivated medium was removed fully, and formazan was solubilized in dimethyl sulfoxide. The metabolized MTT was evaluated in terms of its optical density in a spectrophotometer at 570 nm (Thermo Scientific Multiskan FC).

2.4. Alkaline phosphatase assay

Approximately 500 μL PBS containing 0.1 M glycine, 1 mM $MgCl_2$, and 0.5% Triton X-100 (pH 10.5) was added to each sample to rupture the cell membranes and release the ALP molecules. The samples incubated for 1 h; after which, 100 μL of the supernatant extracted and transferred into microcentrifuge tube wrapped in aluminum foil, and added 200 μL of p-nitrophenyl phosphate solution (Sigma) to each tube. The tubes were placed in a water bath at 37 °C for 30 min, followed by an ice bath at 0 °C for 10 min to reduce the reaction rate. Approximately 50 μL of 3N sodium hydroxide solution was added to each microcentrifuge tube to stop the reaction. The collected solutions were placed into a 96-well microplate for ALP assay using an ELISA reader (Thermo Scientific Multiskan FC) at a wavelength of 405 nm.

2.5. Reverse transcription polymerase chain reaction (RT-PCR)

Total RNA from differentiated cells was extracted with Trizol reagent (Ambion®, Life Technologies™, USA). First-strand cDNA was synthesized from 500 ng of the RNA by using a Super Script®III reverse transcriptase kit (Invitrogen). RT-PCR was performed according to the protocols of Invitrogen™ Corporation (USA). RT-PCR was performed using a Smart Quant Green Master Mix with dUTP and ROX according to the protocols of Protech Technology Enterprise Co., Ltd. (Taiwan). Sixty cycles of quantitative RT-PCR (Lightcycler 480, Roche) were performed for the target genes. The control experiments were conducted with the housekeeper gene GAPDH. The PCR primers of collagen type I, Runx2, osteopontin (OPN), osteocalcin (OCN), and GAPDH listed in Table 1.

Table1. Primer sequence.

Gene name	F: 5′- TGCTTGAATGTGCTGATGACAGGG -3′ R: 5′-TCCCCTCACCCTCCCAGTAT-3′
Collagen type I	F: 5- CTGAGGTAACTTGCTAACG -3′ R: 5’-ATCAATACACTAAGAAATGTTTCAAGG -3′
Osteocalcin	F: 5′- CTCTGCCTTAAACACACATTG -3′ R: 5′- TTCCCTTTGCCCACCTC -3′
Osteopontin	F: 5- AGCCAATGATGAGAGCAA -3′ R: 5’- TTCAGGTGTTTATCTTCTTCCTTAC-3′
GAPDH	F: 5’-ATGAGAAGTATGACAACAGCC- 3’ R: 5’-AGTCCTTCCACGATACCAA -3’

2.6. Calcium quantification
The calcium (Ca) deposited for bone nodule formation was stained using Alizarin red S dye. All samples washed twice with PBS and immersed into 95% ethanol solution for 30 min, and then stained with 1% Alizarin red S for 10 min, and then solubilized with 10% cetylpyridinium chloride. Images of stained cells were captured using a confocal microscope (Leica, TCS-SP5-AOBSK), the amount of Ca was calculated using standard solutions, and absorbance at 570 nm was measured (Thermo Scientific Multiskan FC).

2.7. Statistical analysis
All results were performed at least in triplicate experiments for different samples. Data were analyzed by ANOVA and are presented as means ± standard deviation (SD). Statistical comparisons performed in which $p < 0.05$ was considered significant.

3. Results
All fabricated chitosan scaffolds containing divalent metal phosphate have an 8 mm diameter and 6 mm height and exhibited a 3D porous fibrous structure. The SEM image of the scaffold cross-section exhibits a homogenous structure with regular porosity, a pore size of approximately 150 μm to 200 μm, and excellent pore interconnections, and all tested metal ions are continuously and steadily released from the scaffold (data not shown).

3.1. Cell proliferation and ALP activity
MC3T3E1 cells proliferating in the chitosan scaffolds containing divalent mental phosphate were measured by MTT assay, and pure chitosan scaffold was used as control. Figure 1 shows the graph of cellular viability on the scaffolds after 1, 7 and 14 days of culture with and without divalent mental phosphate supplement. The number of living cells in the scaffolds increased continuously with time, indicating that the scaffolds are biocompatible for cellular growth. Based on the result, Mg, Ca, Sr, and Zn can enhance cellular proliferation, but Ba supplement slightly inhibited cellular proliferation compared with control.

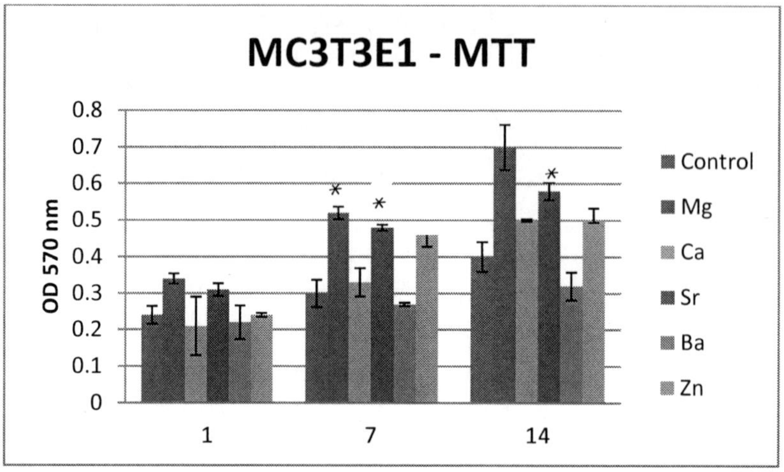

Figure 1. Proliferation of MC3T3E1 cells seeded on the scaffolds with and without divalent mental phosphate supplement after 1, 7, and 14 days of culture. (*P<0.05)

ALP is among the most widely recognized functional markers of early osteoblastic differentiation. ALP activity was measured in MC3T3E1 differentiation among scaffolds containing different divalent mental phosphates cultured for 14 days. Among all tested scaffolds, Mg, Sr, and Zn significantly upregulated ALP activity, showing high expression for 14 days of culture; however, Ba had slower ALP expression than that of control during cultivation period (Fig. 2).

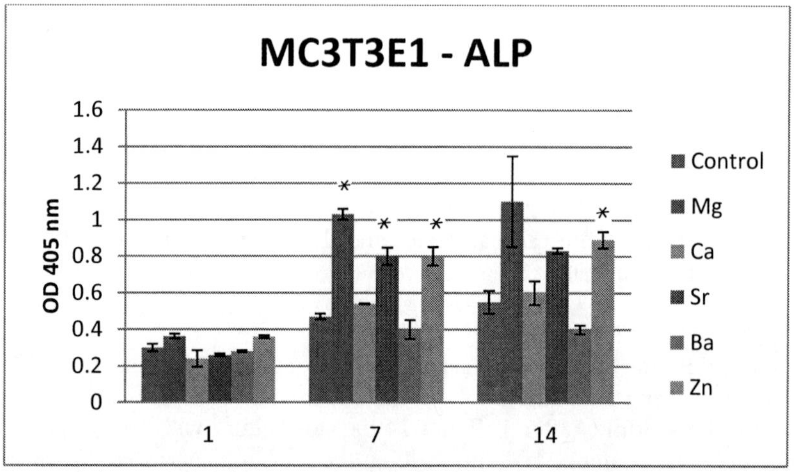

Figure 2. ALP activity expression from MC3T3E1 on various chitosan scaffold with and without divalent mental phosphate supplement during osteogenic differentiation. Values are expressed as mean ± SD (n=3), *P<0.05

3.2. Osteoblastic gene expression

The RT-PCR results show that bone-related genes can be detected in culturing system. Figure 3 shows the expression levels quantified in the expression folds by the housekeeping gene GAPDH as control for bone specific proteins in the scaffolds after 1, 7 and 14 days culture. The MC3T3E1 differentiated cells produce maximal levels of type I collagen on 7 days cultivation, after which they appear to decrease gradually (Fig. 3a). The highest expression of Runx2 (21.9 folds) was noted after 7 days of culture (Fig. 3b). The highest OPN expression (30.6 folds) and terminal marker gene OCN (26.3 folds) were observed in Mg samples after 14 days of culture (Fig. 3c and Fig. 3d).

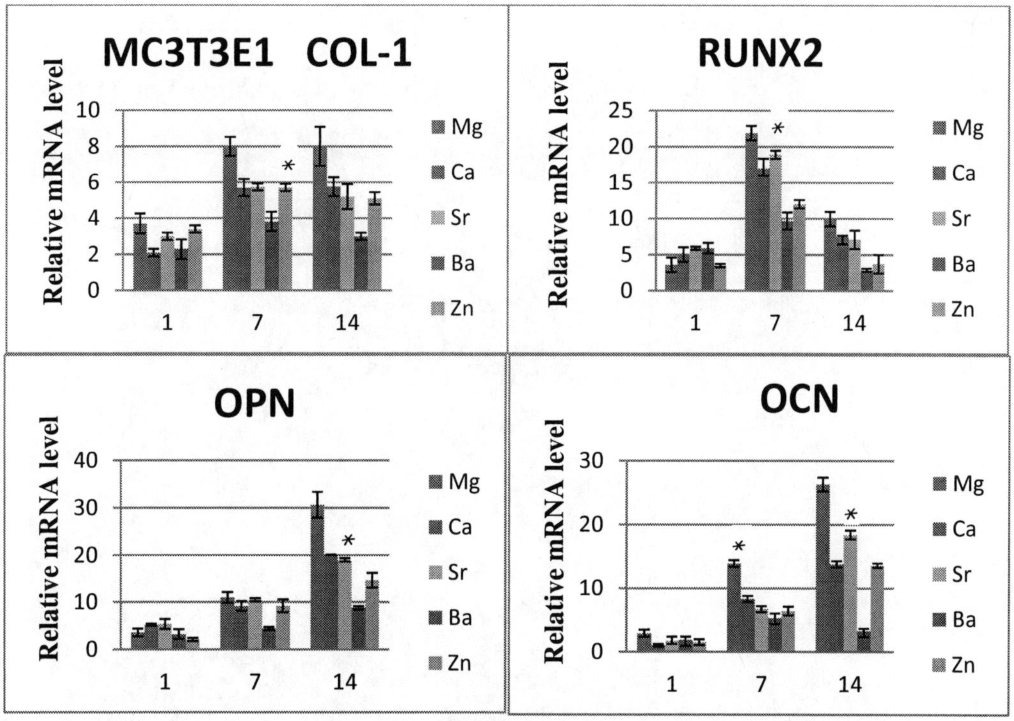

Figure 3. Relative mRNA expression level of COL-I (a), RUNX2 (b), OPN (c) and OCN (d) on various chitosan scaffolds during osteogenic differentiation of SHEDs. Values are expressed as mean ± SD (n=3), * $p < 0.05$.

3.3. Calcium quantification in the mineralized matrix

Alizarin red S was used to stain the intracellular Ca as well as the Ca-binding proteins and proteoglycans, and this method is useful in evaluating bone differentiation. Images of Alizarin red S staining on various chitosan scaffolds with and without divalent metal phosphate after 1, 7, and 14 days of culture are shown in Fig. 4a. Quantitative result shown in Fig. 4b indicated that Ca content increased continuously during 14 days of differentiation. Both Mg and Zn exhibited the largest increment among all the scaffolds. Ca depositions in Mg, Ca, Sr, and Zn-chitosan scaffolds containing divalent mental phosphate were higher than the depositions in pure chitosan scaffold, but Ca depositions in Ba scaffold is lower level and a negative effect. The formation of CaP salts or mineral deposition is a primary function of osteoblast cells, and the scaffold containing divalent metal phosphate can produce highly mineralized deposition during osteogenic differentiation, except for Ba.

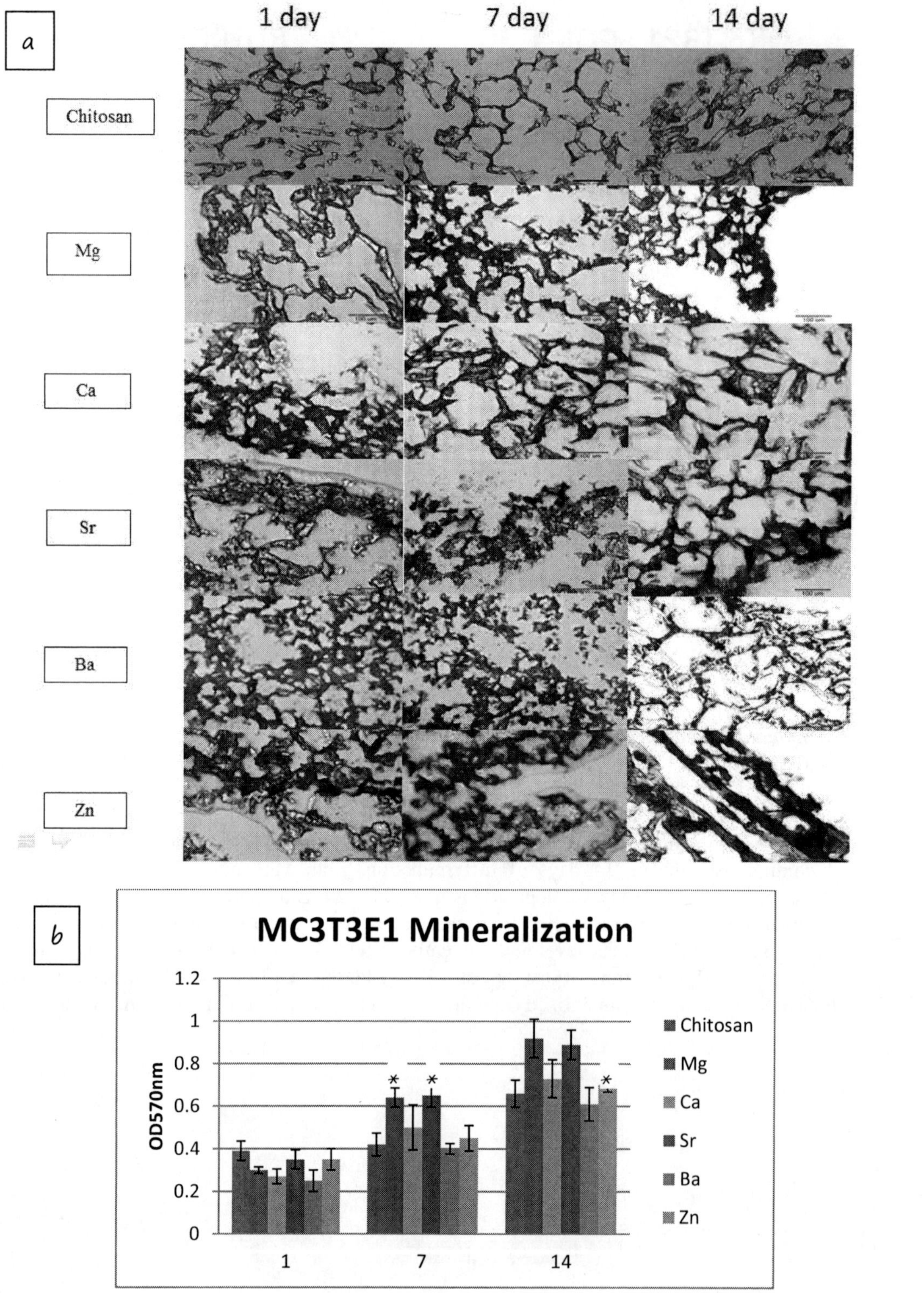

Figure 4. Alizarin red S stained on various chitosan scaffolds (a) and quantification of calcium deposition during osteogenic differentiation (b). Values are expressed as mean ± SD (n=3), *P<0.05. Scale bar is 20 μm.

4. Discussion

The implants possess osteoconduction capabilities, allowing bone cell migration and bone deposition; however, osteoconductive materials cannot completely regenerate larger scale of bone defects. Therefore, osteoinductive material stimulation for osteogenesis of mesenchymal stem cells is needed for bone tissue engineering [6, 7]. Chitosan scaffold containing divalent mental phosphate mimicked bone mineral chemistry, and the trace ions released from scaffold can regulate bone metabolism [8].

ALP is a major osteogenic marker with a crucial function in mineralization; it is considered as an early-stage marker of osteogenic differentiation [9]. Biomineralization, which occurs in the late stages of osteogenesis, is a marker of fully differentiated stem cells [10]. All the tested samples had appeared that divalent metal phosphates can induce MC3T3E1 to osteogenic differentiation especially Mg. In addition, qualitative assay of mineralization was performed through Alizarin red S staining, and these Ca depositions on scaffold were higher than that on pure chitosan scaffold, but Ba inhibited the Ca deposition, as Fig. 4b.

In this study, Mg, Sr, and Zn showed higher osteoconductivity over culture time, high gene expression of bone-related, and higher degree of mineral deposition than conventional Ca supplement. However, the Ba appeared a negative effect in osteogenic differentiation.

5. Conclusions

This study demonstrated that divalent mental phosphate had been confirmated can induce effectively MC3T3E1 to osteogenic differentiate continuously into mature bone. Especially, Mg is greatest inducer for osteogenic differentiation and Ba is a negative agent. Therefore, understanding these different differentiated behaviors will be useful for the design and development of strategies for tissue engineering.

6. Acknowledgment

The authors would like to thank the National Science Council of Taiwan MOST 105-2622-E-027-031-CC3 and NTUT-MMH Joint Research Program NTUT-MMH-106-08 for their financial support in this study.

7. Reference

[1] Luo T, Wu C, Zhang Y. The in vivo osteogenesis of Mg or Zr-modified silicate-based bioceramic spheres. J Biomed Mater Res A 2012; 100:2269-2277.

[2] Wu C, Zhou Y, Lin C, Chang J, Xiao Y. Strontium-containing mesoporous bioactive glass scaffolds with improved osteogenic/cementogenic differentiation of periodontal ligament cells for periodontal tissue engineering. Acta Biomater 2012; 8:3805-3815.

[3] Zhu Y, Li X, Yang J, Wang S, Gao H, Hanagata N. Composition–structure–property relationships of the CaO–MxOy–SiO_2–P2O_5 (M = Zr, Mg, Sr) mesoporous bioactive glass (MBG) scaffolds. J Mater Chem 2011; 21:9208-9218.

[4] Su WT, Pan YJ. Stem cells from human exfoliated deciduous teeth differentiate toward neural cells in a medium dynamically cultured with Schwann cells in a series of polydimethylsiloxanes scaffolds. J Neural Eng 2016; 13:46005.

[5] Su WT, Wu PS, Ko CS, Huang TY. Osteogenic differentiation and mineralization of human exfoliated deciduous teeth stem cells on modified chitosan scaffold. Mater Sci Eng C 2014; 41:152-160.

[6] Kim TG, Shin H, Lim DW. Biomimetic scaffolds for tissue engineering. Adv Funct Mater 2012; 22:2446-2468.

[7] Hoppe A, Güldal NS, Boccaccini AR. A review of the biological response to ionic dissolution products from bioactive glasses and glass-ceramics. Biomaterials 2011;

32:2757–2774.

[8] Su WT, Shih YA, Ko CS. Effect of chitosan conduit under a dynamic culture on the proliferation and neural differentiation of human exfoliated deciduous teeth stem cells. J Tissue Eng Reg Med 2016; 10:507-517.

[9] Hall SL, Dimai HP, Farley JR. Effects of zinc on human skeletal alkaline phosphatase activity *in vitro*. Calcif Tissue Int 1999; 64:163-172.

[10] Su, W.T.; Chou, W.L.; Chou, C.M. Osteoblastic differentiation of stem cells from human exfoliated deciduous teeth induced by thermosensitive hydrogels with strontium phosphate. *Mater. Sci. Eng. C* **2015**, *52*, 46-53.

The Improvement of Utilization Ratio of Metal Organic Sources for the Low Cost Preparation of MOCVD-synthesized YBCO Films based on a Self-heating Technology

Ruipeng Zhao[1], Qing Liu[1], Yudong Xia[2], Bowan Tao[1] and Yanrong Li[1]

[1]State Key Lab of Electronic Thin Films and Integrated Devices, University of Electronic Science and Technology of China, Chengdu 610054, PR China
[2] School of Physical Science and Technology, Southwest Jiaotong University, Chengdu 610031, PR China
Email: zhaoruipeng1990@sina.com

Abstract: We have successfully applied metal organic chemical vapor deposition (MOCVD) to synthesize biaxially textured $YBa_2Cu_3O_{7-\delta}$ (YBCO) superconducting films on the templates of $LaMnO_3$/epitaxial MgO/IBAD-MgO/solution deposition planarization (SDP) Y_2O_3/Hastelloy tape. The YBCO films have obtained dense and smooth surface with good structure and performance. A new self-heating method, which replaced the conventional heating-wire radiation heating method, has been used to heat the Hastelloy metal tapes by us. Compared with the heating-wire radiation heating method, the self-heating method shows higher energy efficiency and lower power consumption, which has good advantage to simplify the structure of the MOCVD system. Meanwhile, the utilization ratio of metal organic sources can be increased from 6% to 20% through adopting the new self-heating method. Then the preparation cost of the YBCO films can be also greatly reduced.

1. Introduction

In the past few years, $REBa_2Cu_3O_{7-\delta}$ ((REBCO), RE= rare earth elements) high temperature superconducting tapes have been amply researched by many research teams for its application in electric power field [1-3]. Because of its high current carrying capacity and large irreversible field, REBCO high temperature superconducting tapes have great potential for application in the aspects of superconducting cable, strong magnet, superconducting motor, generator, superconducting current limiter and so on [4-6]. Metal organic chemical vapor deposition (MOCVD) has been proved as one of the most promising techniques to massively prepare biaxially textured REBCO superconducting thin films on the surface of biaxially textured substrates. However, the low utilization ratio of metal organic sources makes it difficult to reduce the fabrication cost of REBCO films. Therefore, a new self-heating method was used to prepare the MOCVD-synthesized YBCO superconducting films to improve the utilization ratio of metal organic sources and reduce the cost. For the new self-heating method, the current is introduced to the metal tapes through the brush [7], as is shown in figure 1.

Nowadays, the research and preparation of the buffer template, which is used to prepare the YBCO superconducting films, is mainly based on the ion beam assisted deposition (IBAD) [8] and rolling assisted biaxially textured substrates (RABiTS) [9] technology. Compared to RABiTS, the IBAD process has the better advantage in the cost and product performance. Therefore, the IBAD technology is widely used in the world. In our experiments, the templates of $LaMnO_3$/epitaxial MgO/IBAD-MgO/solution deposition planarization (SDP) Y_2O_3/Hastelloy tape are adopted to prepare YBCO films,

Content from this work may be used under the terms of the Creative Commons Attribution 3.0 licence. Any further distribution of this work must maintain attribution to the author(s) and the title of the work, journal citation and DOI.
Published under licence by IOP Publishing Ltd

as is shown in figure 2. In this paper, we focus on the development of the heating technology to improve the utilization ratio of metal organic sources. What's more, the structure, morphology and performance of the prepared YBCO films based on the novel heating method have been also investigated.

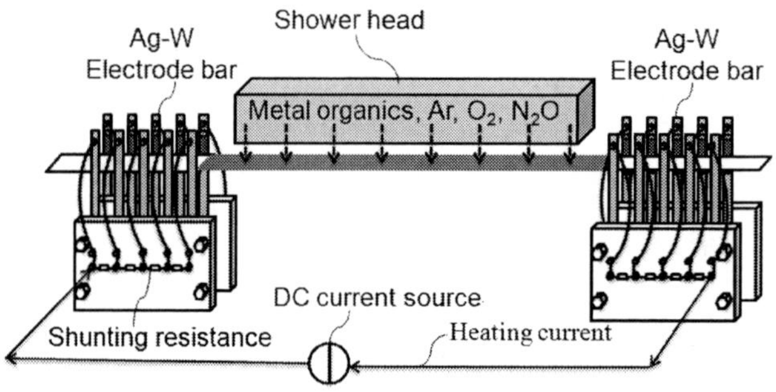

Figure 1. The schematic diagram of the self-heating method for the preparation of YBCO films

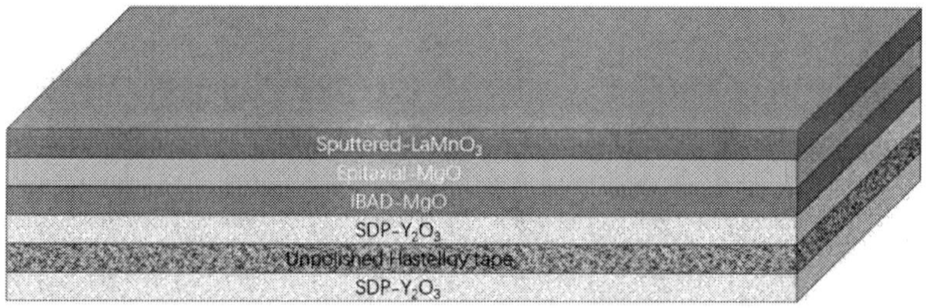

Figure 2. The schematic diagram of the buffer layers on the Hastelloy tape

2. Experimental

In the experiments, the precursors of metal organic sources were prepared by dissolving $Y(tmhd)_3$, $Ba(tmhd)_2\cdot(1,10\text{-phenanthroline})_2$ and $Cu(tmhd)_2$ (tmhd: 2, 2, 6, 6-tetramethyl-3, 5-heptanedionate) in tetrahydrofuran (THF). The tapes of $LaMnO_3$/epitaxial MgO/IBAD-MgO/solution deposition planarization (SDP) Y_2O_3/Hastelloy tape, with flat surface and good structure, were used as the templates (LMO template). The YBCO films were quickly prepared on the LMO template by metal organic chemical vapor deposition (MOCVD). Firstly, the prepared precursors were quickly atomized by a home-made nozzle and evaporated rapidly by the heating evaporator. Then, the vapors of metal organic sources, which were mixed with the argon, oxygen and nitrous oxide, were transported into the reaction chamber and reacted to grow YBCO films on the surface of the heating tapes. At last, the prepared YBCO superconducting tapes were annealed in flowing oxygen at 500 °C and furnace cooled to room temperature.

X-ray diffraction (XRD, Bede D1 system) patterns were measured using Cu K_α radiation (λ=1.5406 Å) in the mode of θ–2θ step-scan, ω-scan, φ-scan and *Chi*-scan for the prepared samples. The RMS roughness of LMO template layer was characterized by atomic force microscope (AFM, Seiko SPA300HV). The surface morphology was characterized by a scanning electron microscope (SEM, JEOL7500F). And the composition of thin films was characterized by an energy dispersive spectrometer (EDS, Oxford INCA). The thickness was measured by a step profiler (Veeco Dektak 150). Moreover, the critical current density (J_c) and critical current (I_c) at 77 K and 0 T were obtained by the Leipzig J_c–scan system and four-probe-method measurement system.

3. Results and Discussion

The used LMO template is characterized by the AFM and SEM system. Figure 3a shows the AFM image of $LaMnO_3$ template for the preparation of YBCO superconducting film. The root-mean-square (RMS) roughness of $5\times5\mu m^2$ is 2nm, which meets the demand of the preparation of biaxially textured YBCO films. Figure 3b is the SEM image of $LaMnO_3$ film morphology, which shows that its surface is very dense and smooth. Therefore, the used LMO templates are very suitable for the growth of the YBCO superconducting thin film.

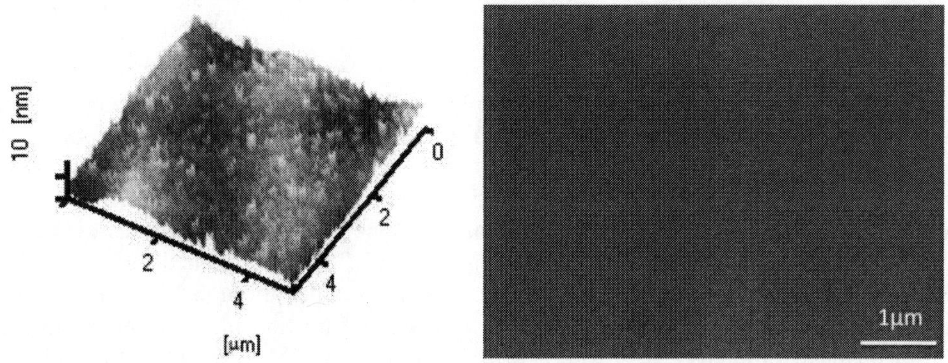

(a) (b)

Figure 3. (a) The AFM image of the $LaMnO_3$ template, R_{rms}=2nm; (b) The SEM image of the $LaMnO_3$ template

As we adopt the new self-heating method, it avoids the problem that the temperature of the shower is too high due to using the radiation heating methods. So the shower head can be very close to the surface of template tapes. Thus, the YBCO films prepared by the heating-wire radiation heating method and the self-heating method respectively are characterized by the XRD θ-2θ step-scan, as is shown in figure 4. The diffraction peaks of YBCO *(00l)* can be clearly observed in the curves of θ–2θ patterns, which indicate that the grains of YBCO films are purely *c*-axis-oriented growth. However, the intensity of YBCO *(00l)* diffraction peaks of YBCO film based on the self-heating method is obviously stronger than that of the radiation heating method. It indicates that the YBCO films prepared by the self-heating method crystallize better. Moreover, the XRD χ-scan of YBCO (102) has also been performed, which is used to characterize the *c*-axis-oriented grains and the *a*-axis-oriented grains. For the two χ-scan curves, there is only one peak around 56.6°, representing the diffraction of (102) plane of *c*-axis oriented YBCO grains, which indicate that all the prepared YBCO grains aligned their *c*-axis perpendicular to the surface of the LMO templates.

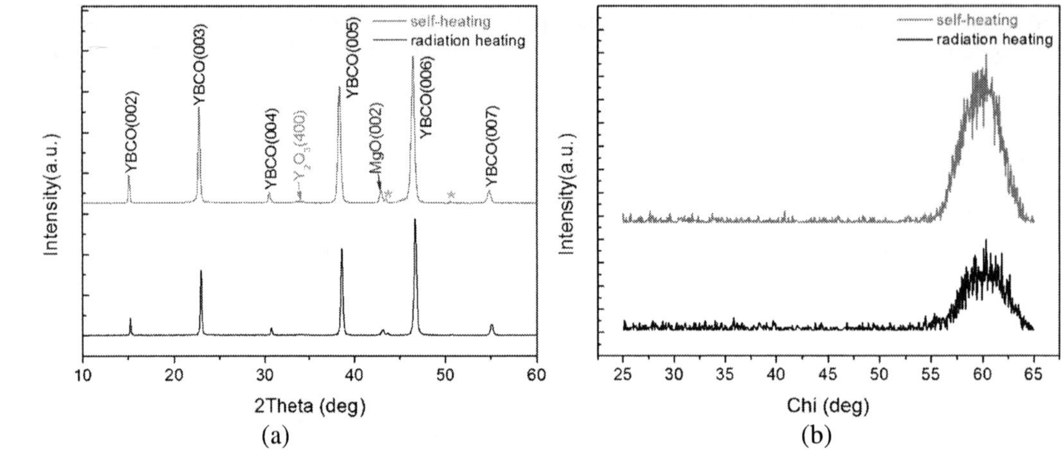

Figure 4. (a) The XRD θ-2θ scanning patterns of YBCO films based on the different heating methods; (b) The XRD χ-scan curves of YBCO (102) based on the different heating methods

What's more, the XRD ω-scan of YBCO (005) and the XRD φ-scan of YBCO (103) have also been performed, which are used to characterize out-of-plane and in-plane textures of YBCO films. Based the radiation heating method, the rocking curve of YBCO film obtained through YBCO (005) ($\Delta\omega$) yields the FWHM value of 1.35° and the φ-scan curve obtained through YBCO (103) ($\Delta\varphi$) yields the FWHM value of 3.18° in figure 5a. Similarly, for the self-heating method, the rocking curve of YBCO film obtained through YBCO (005) ($\Delta\omega$) yields the FWHM value of 1.27° and the φ-scan curve obtained through YBCO (103) ($\Delta\varphi$) yields the FWHM value of 2.8° in figure 5b. It can be seen that the FWHM values of YBCO films prepared based on the novel self-heating method is even smaller than that of YBCO films prepared based on the radiation heating method, which also indicates that the biaxial texture can be better improved through adopting the self-heating technology.

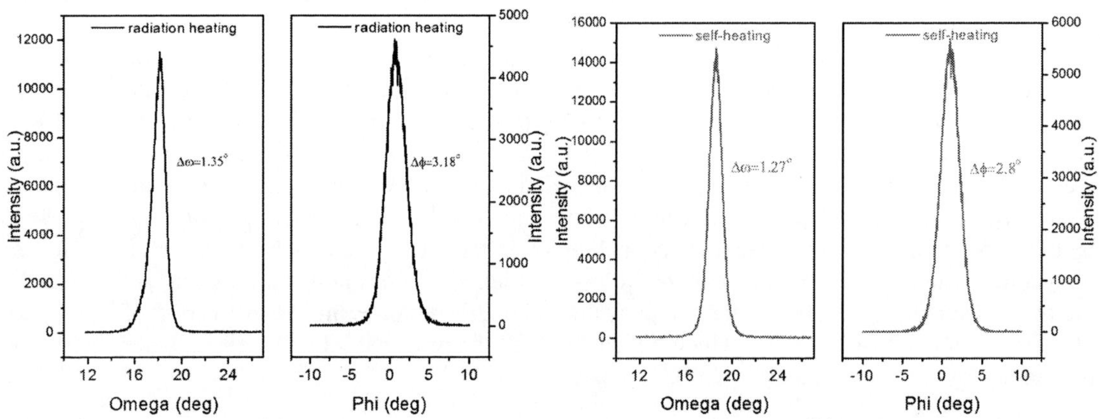

Figure 5. (a) The XRD ω-scan andφ-scan curves YBCO films based on the radiation heating method. (b) The XRD ω-scan andφ-scan curves YBCO films based on the self-heating method

The surface morphologies of the superconducting films are observed by the SEM system. Figure 6 shows the morphology images of these prepared samples based on the different heating method. There are some lager outgrowths formed on the surface of YBCO film prepared by the radiation heating method. In comparison, the surface of YBCO film prepared by the self-heating method is flat and there are not basically large outgrowths. For these samples prepared by the different heating method, the critical current density (J_c) at 77K and 0T are measured by the induction and four-probe methods and the corresponding measured values are shown in Table 1. Owing to the narrow distribution of grain alignments, the J_c of the sample prepared based on the self-heating method is 5 MA/cm^2 (0 T, 77 K), which is higher than the J_c of 3.6 MA/cm^2 based on the radiation heating method. Meanwhile, depending on the amount of metal organic sources and the thickness of the corresponding YBCO films, the utilization ratio of metal organic sources can be increased from 6% to 20% through adopting the new self-heating method. And the deposition rate of YBCO films can be greatly increased.

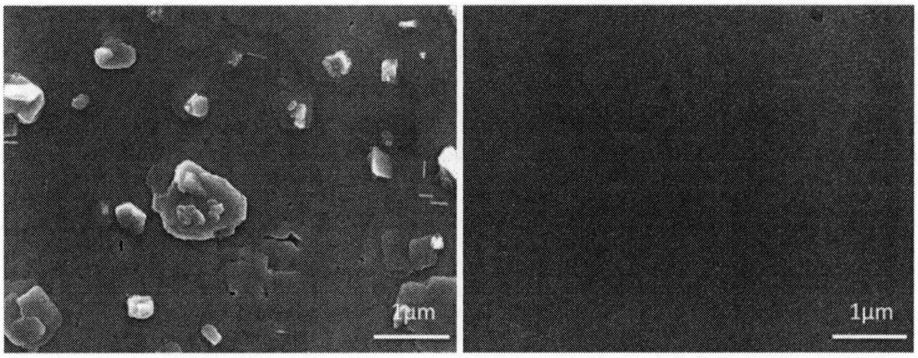

(a) (b)

Figure 6. (a) The SEM image of YBCO film based on the radiation heating method. (b) The SEM image of YBCO film based on the self-heating method

Table 1. The critical current density and utilization ratio of metal organic sources based on the different heating method

Heating method	Deposition rate (nm/min)	Utilization ratio	Critical current density (MA.cm^{-2})
Radiation heating	300	~6%	3.6
Self-heating	>600	~20%	5.0

4. Conclusion

In this study, a novel self-heating method is used to prepare YBCO films by us. Different from the conventional radiation heating method, the shower head can be very close to the metal tapes without excessive temperature. Therefore, the utilization ratio of metal organic sources can be greatly improved. It is concluded that the YBCO film with high J_c at a high metal organic sources utilization ratio of 20% could been successfully prepared.

5. References

[1] Rupich M W, Li X P, Sathyamurthy S, Thieme C L H, DeMoranville K, Gannon J and Fleshler S 2013 *IEEE Trans. Appl. Supercond.* Second Generation Wire Development at AMSC 23 6601205

[2] Zhang Y F, Lehner T F, Fukushima T, Sakamoto H and Hazelton D W 2014 *IEEE Trans. Appl. Supercond.* Progress in Production and Performance of Second Generation (2G) HTS Wire for Practical Applications 24 7500405

[3] Selvamanickam V, Chen Y, Xiong X, Xie Y, Zhang X, Rar A, Martchevskii M, Schmidt R, Lenseth K and Herrin J 2008 *Physica C.* Progress in second-generation HTS wire development and manufacturing 468 1504-1509

[4] Sung H J, Park M, Go BS and Yu I K 2016 *Supercond. Sci. Technol.* A study on the required performance of a 2G HTS wire for HTS wind power generators 29 054001

[5] Chen X Y, Jin J X, Xin Y, Shu B, Tang C L, Zhu Y P and Sun R M 2014 *IEEE Trans. Appl. Supercond.* Integrated SMES Technology for Modern Power System and Future Smart Grid 24 3801605
[6] Zheng Y B, Wang Y S, Pi W, Ju P and Wang Y S 2014 *Physica C.* Current distribution among layers of single phase HTS cable conductor 507 59-64
[7] Zhang F, Zhao R P, Xue Y, Wang H, He Y Y, Zhang P, Tao B W, Xiong J and Li Y R 2016 *Appl. Phys. A.* Self-heating technique of metallic substrate for reel-to-reel and double-sided deposition of $YBa_2Cu_3O_{7-\delta}$ films 122 81
[8] Jin H J, Kim G, Jo W, Shin HY, Yoon S, Kim K and Kim M 2015 *IEEE Trans. Appl. Supercond.* Spatial Investigation on Structural Properties of $GdBa_2Cu_3O_{7-x}$ Coated Conductors Grown on IBAD-MgO Based Stainless-Steel Substrates 25 6604704
[9] Sieger M, Hanisch J, Pahlke P, Sparing M, Gaitzsch U, Iida K, Nast R, Reich E, Schultz L, Holzapfel B and Huhne R 2015 *IEEE Trans. Appl. Supercond.* $BaHfO_3$-Doped Thick $YBa_2Cu_3O_{7-\delta}$ Films on Highly Alloyed Textured Ni-W Tapes 25 6602604

Synthesis and Characterization Carbon Nanotubes Doped Carbon Aerogels

Yuelong Xu Meifang Yan and Zhenfa Liu*

Institute of Energy Resources, Hebei Academy of Sciences, Shijiazhuang 050081, China

Hebei Engineering Research Center for Water Saving in Industry, Shijiazhuang 050081, China

*Institute of Energy Resources, Hebei Academy of Sciences, Shijiazhuang 050081, China.

Email: xudalong.cool@163.com

Abstract: Polycondensation of phloroglucinol, resorcinol and formaldehyde with carbon nanotube (CNT) as the additives, using sodium carbonate as the catalyst, leads to the formation of CNT - doped carbon aerogels. The structure of carbon aerogels (CAs) with carbon nanotubes (CNTs) were characterized by X-ray diffraction and scanning electron microscopy. The specific surface area, pore size distribution and pore volume were measured by surface area analyzer. The results show that when the optimum doping dosage is 5%, the specific surface area of CNT - doped carbon aerogel is up to 665 m^2 g^{-1} and exhibit plentiful mesoporous.

1. Introduction

Carbon aerogels were first reported by Pekalaand coworkers in the 1990s [1-2], which exhibits plentiful three-dimensional cross-linked pores. carbon aerogels have become a new kind of aerogel materials, which has gotten more and more attentions because of the advantages of low thermal conductivity, low density, high electrical conductivity and plentiful porosity [3-5].These specific physicochemical properties lead to a wide application of carbon aerogels in the catalyst supports field, supercapacitors and batteries fields, gas and metal adsorption fields [6-7].

The traditional synthesis method of carbon aerogels is the condensation-polymerization of resorcinol and formaldehyde with sodium carbonate as the catalyst[8].Those as - prepared carbon aerogels exhibit more centered pore size distribution, however which has lower specific surface area and poor electrochemical properties. Nowdays, researchers paid more and more attentions to the modified carbon aerogels in order to improve the physicochemical properties. Metal doped carbon aerogels have been reported, which exhibits superior electrochemical properties; organic polymer doped carbon aerogels have been prepared, which exhibits porous structure and excellent adsorption properties; strong base activated carbon aerogels are used in the supercapacitors and batteries fields because of excellent the electric double-layer capacitance behavior[9-11].

Carbon fiber doping has been reported for preparation of CAs in recent decades, which could improve the mechanical strength and electrochemical properties of carbon aerogels. However the textural properties exhibit poor improvements. Carbon nanotube doped carbon aerogels have been few reported. In this paper, CNT - doped CAs were synthesized via a sol–gel process by condensation-polymerization of phloroglucinol, resorcinol and formaldehyde using sodium carbonate

Content from this work may be used under the terms of the Creative Commons Attribution 3.0 licence. Any further distribution of this work must maintain attribution to the author(s) and the title of the work, journal citation and DOI.
Published under licence by IOP Publishing Ltd

as catalyst with carbon nanotube (CNT) as the additives.

2. Experimental

2.1. Materials

The chemical reagents used are analytical grade phloroglucinol, resorcinol, formaldehyde (30%), carbon nanotube (CNT) and sodium carbonate (C). Ultima IV X-ray diffraction instrument (XRD, Rigaku, Japan), Scanning electron microscopy (SEM, S-4800-I, Hitachi, Japan), Specific surface analyzer ASAP2420 (micromeritics, USA).

2.2. Synthesis and Characterization of Carbon Aerogels

Carbon aerogels were synthesized by polycondensation of phloroglucinol (P), resorcinol (R) and formaldehyde (F), using sodium carbonate as catalysts (C), with carbon nanotube (CNT) as additives, and deionized water as the solvent. The molar ratio of P / R was 0.15, the molar ratio of (P + R)/F was 0.5, and the molar ratio of (P + R) / C was 500. The concentration of raw material (P + R + F) in solution was 45 wt%. The concentrations of carbon nanotube (CNT) in solution were 2 wt%, 5 wt%, 10wt%, respectively. Phloroglucinol, resorcinol, formaldehyde, carbon nanotube (CNT) and sodium carbonate (C) catalyst were added into deionized water. Ultrasonic oscillator was used to dissolve the reactants to form a homogeneous solution in test tubes. Putting test tubes in water bath and keeping 50°C for some time until the hydrogels were obtained. The synthetic route of carbon aerogels is shown in figure 1. The hydrogels were soaked in acetone , and the acetone was exchanged once a day for 4 times. Drying was carried out at room temperature. The dried doped aerogels were carbonized at 900°C for 3 h with a heating rate of 2°C /min under a flowing nitrogen atmosphere (100 mL/min).

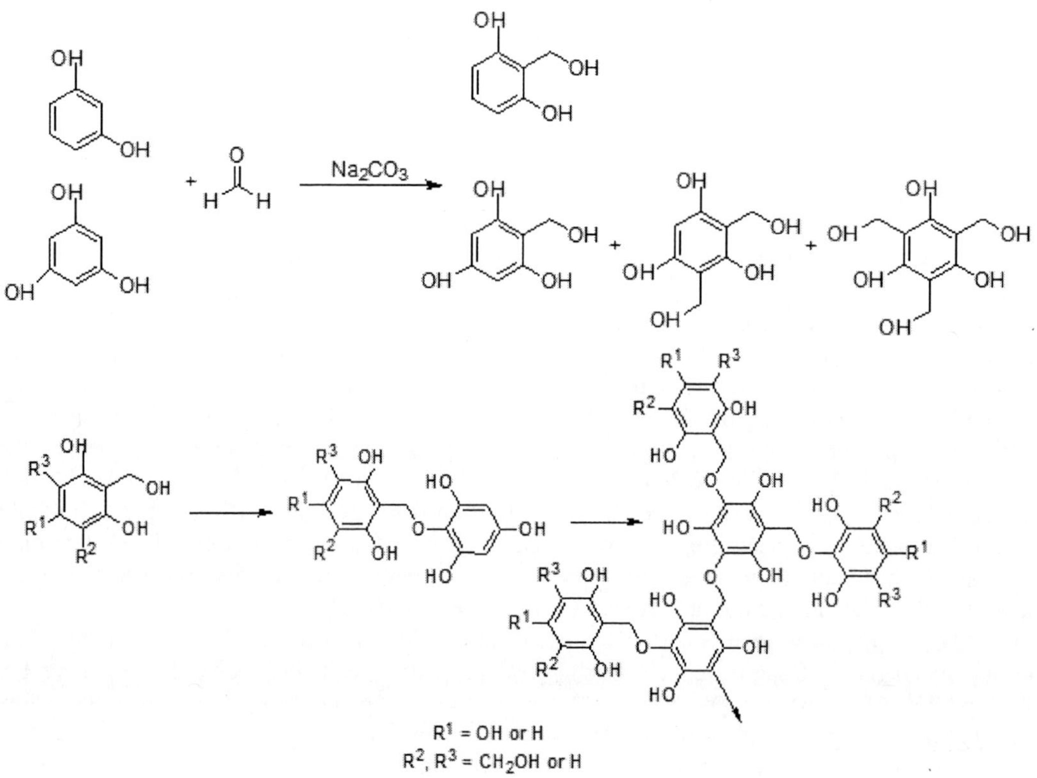

Figure1. The synthesis route of CNT - doped CAs

2.3. Characterization

The doped carbon aerogels were characterized by X-ray diffraction (XRD) and scanning electron microscopy. The Brunauer–Emmett–Teller (BET) method was used to calculate the specific surface areas in a relative pressure range from 0.0 to 1.0. The pore volumes and pore size were caculated from the adsorption-desorption isotherms using the Barrett–Joyner–Halenda (BJH) model. The total pore volumes (Vt) were estimated from the adsorbed amount at a relative pressure (P/P0) of 0.998.

3. Results and Discussion

3.1. XRD Patterns of Modified Carbon Aerogels

The XRD patterns of CNTs - doped CAs with different additives are shown in figure 2. As shown in figure 2, the two obvious peaks at the 2θ of 23.5 and 43.8 are assigned to the (002) peak of graphite carbon and (101) peak of the crystal plane diffraction, respectively, which is the characteristic peak of carbon aerogels. And there are no difference among the different quantities of CNTs in figure 2.

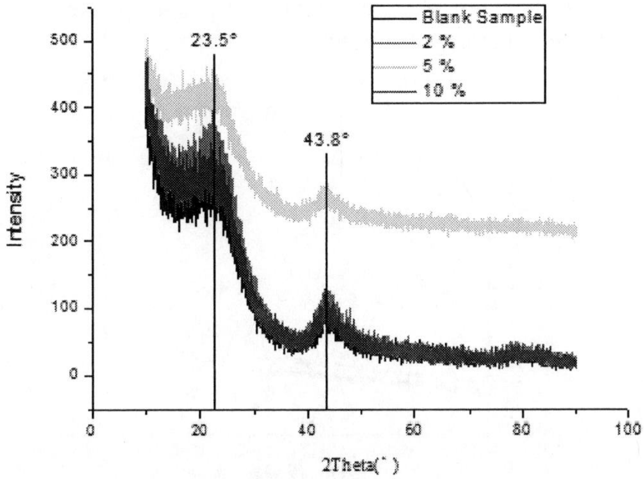

Figure 2. XRD patterns of CNTs - doped CAs

Figures 3A–D show SEM-images of CNTs - doped CAs. As shown in Figures 3A–D, Figures 3B and 3C have the obviousthree-dimensional cross-linked pores which is indicative of the presence of mesopores in CNTs - doped CAs. As shown in Figures 3D, the dense pores were caused by the reunion of CNTs. Figures 3B and 3C demonstrate that the CNTs doped method could be conducive to the generation of micropore and improve the specific surface area.

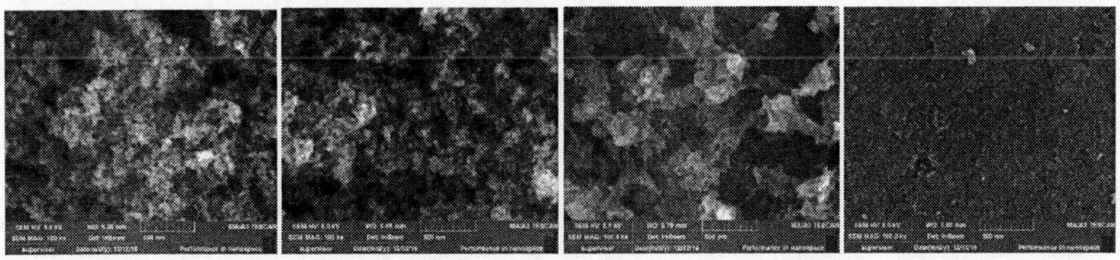

Figure 3. SEM—images of CNTs - doped CAs

3.2. Pore structure of Modified Carbon Aerogels

The specific surface area (S_{BET}), total pore volume (V_{total}) and BJH desorption average pore diameter

IOP Conf. Series: Materials Science and Engineering **275** (2017) 012006 doi:10.1088/1757-899X/275/1/012006

($D_{average}$) of modified carbon aerogels was listed in Table 1. As shown in table 1, when CNTs is 5 wt%, the doped carbon aerogels have the highest specific surface area, up to 665 m^2/g, and the largest total pore volume, up to 1.253 cm^3/g. When the doped dosage is higher than 5 wt%, CNTs reunion together which effects the porous texture.

Table 1. Textual Propertiesof CNTs - doped CAs

Entry	$S_{BET}(m^2/g)$	$S_{micro}(m^2/g)$	$D_{average}(nm)$	$V_{total}(cm^3/g)$
Blank Sample	501	279	5.61	0.464
2 %	579	314	5.65	0.547
5 %	665	451	23.21	1.253
10 %	482	268	5.43	0.436

The N_2 adsorption–desorption isotherms of CNTs - doped CAs are shown in figure 3. As shown in figure 4, all the CNTs - doped CAs exhibit type-IV isotherms indicative of the existence of well-developed porous structure respond, which are corresponding to the classification of IUPAC. The isotherms of all the modified carbon aerogels exhibit the same hysteresis loop shape, which is corresponding to the evidence of mesopores existence.

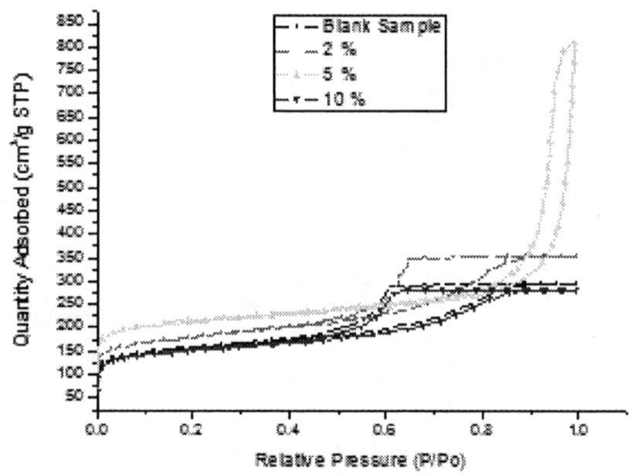

Figure 4. Nitrogen adsorption – desorption isotherms.

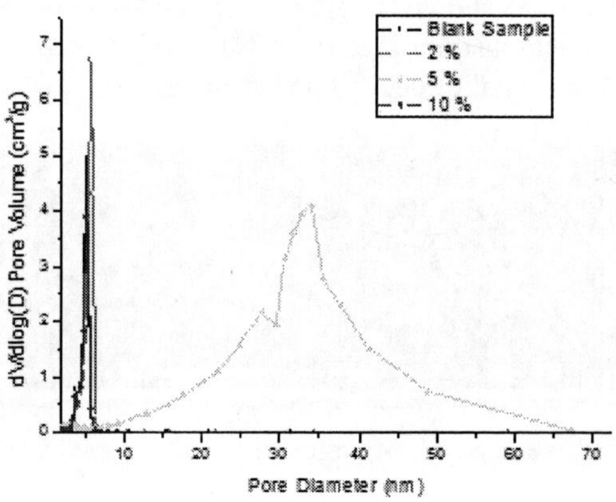

Figure 5. Pore size distribution

The drawings of the pore size distribution of CNTs - doped CAs were shown in figure 4. As shown in figure 5, the pore size distribution of CNTs - doped CAs are belong to normal distribution, and the pore size range is from 4 nm to 70 nm, which is the evidence of mesopore distribution. The mesopore distribution is benefit for the application in the adsorption of heavy metal ions and chromatographic separation field.

4. Conclusions

Carbon nanotube doped carbon aerogels were prepared through polycondensation of phloroglucinol, resorcinol and formaldehyde with carbon nanotube as the additives, using sodium carbonate as the catalyst; The doped carbon aerogels of 5wt% CNTs have the highest specific surface area, up to 665 m^2/g; The pore size of doped carbon aerogels is controlled, and the pore size range is from 4 nm to 70 nm. This work opens a new avenue for obtaining modified carbon aerogels by an eco-friendly method .

5. Acknowledgment

This work was financially supported by the Natural Science Foundation of Hebei Province (15291407D) and (16214402D).

6. References

[1] Pekala RW, Organic aerogels from the polycondensation of resorcinol with formaldehyde *J. Mater. Sci.*, **vol.24**, 1989, pp. 3221 - 3227.

[2] Pekala RW, Alviso CT and Lemay JD, In: Hench LL, West JK (eds), Chemical processing of advanced materials. *Wiley, New York*, 1992, pp. 671 - 683.

[3] Lijuan Zhao, Jinghong Zhao, Xijing Sun, Quan Li, Jinrong Wu and Aiping Zhang, Enhanced thermoelectric properties of hybridized conducting aerogels based on carbon nanotubes and pyrolyzed resorcinol – formaldehyde resin, *Synthetic Metals*, **vol.205**, 2015, pp. 64 – 69.

[4] Sun Z, Yang X, Zhang G, Zheng S, Frost RL A novel method for purification of low grade diatomite powders in centrifugal fields. *Int J Miner Process* **vol.125**, 2013, pp. 18–26.

[5] Zhang J, Kong L-B, Cai J-J, Li H, Luo Y-C, Kang L Hierarchically porous nickel hydroxide/mesoporous carbon composite materials for electrochemical capacitors. *MicroporousMesoporous Mater.* **vol.132**, 2010, pp. 154–162.

[6] X. Yuan, Y.-J. Chao, Z.-F. Ma, X. Deng, Preparation and characteriza-tion of carbon xerogel (CX) and CX – SiO composite as anode materialfor lithium-ion battery, *Electrochem. Commun.* **vol.10**, 2007, pp. 2591–2595.

[7] F. Seker, K. Meeker, T.F. Kuech, A.B. Ellis, Surface chemistry of prototypical bulkII–VI and III–V semiconductors and implications for chemical sensing, *Chem.Rev.* 100 (2000) 2505–2536.

[8] P.V. Kamat, Photoinduced transformations in semiconductor metal nanocomposite assemblies, *Pure Appl. Chem.* **vol.74,** 2002, pp. 1693–1706.

[9] Liu, R., Wu, D., Feng, X. &Muellen, K. Nitrogen-doped ordered mesoporous graphitic arrays with high electrocatalytic activity for oxygen reduction. *Angew. Chem. Int. Edn.* **vol.49**, 2010,pp.2565–2569.

[10] Qu, L., Liu, Y., Baek, J-B. & Dai, L. Nitrogen-doped graphene as eficient metal-free electrocatalyst for oxygen reduction in fuel cells. *ACS Nano.* **vol.4**, 2010, pp.1321–1326.

[11] Lukowski, M. A.; Daniel, A. S.; Meng, F.; Forticaux, A.; Li, L. S.; Jin, S. Enhanced Hydrogen Evolution Catalysis from Chemically Exfoliated Metallic MoS 2 Nanosheets. *J. Am. Chem. Soc.* **vol.135**,2013, pp.10274–10277.

Hydrothermal Synthesis, Characterization and Raman Vibrations of Chalcogenide SnS Nanorods

Shujaat Ali[1,2], Fengping Wang[1,*], Saba Zafar[3] and Tahir Iqbal[2]

[1]Department of Physics, School of Mathematics and Physics, University of Science and Technology Beijing, Beijing 100083, P. R. China.
[2]Department of Physics, Hafiz Hayat Campus, University of Gujrat, Gujrat 50700, Pakistan.
[3]School of Science Changchun University of Science and Technology, Changchun, 130022, P. R. China.
*Corresponding author Email: fpwang@ustb.edu.cn

Abstract: One dimensional nanostructure have been considered promising candidates for optoelectronics and energy storage devices. However, cost-effective approach for large-scale production using wet chemical method is still a big challenge. Here, single crystalline monodispersed stannous sulfide (SnS) nanorods in the range of 400nm to micrometer have been synthesized using solution-based method. The morphology of the prepared nanorods was controlled by tuning the parameters, including the concentrations of reagents, reaction time and temperature. The resultant nanorods were characterized using XRD, FESEM, and Raman spectroscopy techniques. Four vibrational modes 141 cm^{-1}, 165.3 cm^{-1}, 218 cm^{-1} and 289.6 cm^{-1} were observed by Raman spectroscopy, which is consistent with nano orthorhombic SnS herzenbergite phase.

1. Introduction

Last decade, the advancement in nanotechnology has been expanding rapidly and applying successfully in practical applications. Researcher have been paying attention to controlled shaped morphology of nanomaterials after growth, particularly solution-phase fabrication of one-dimensional nanostructures of binary semiconductors materials. Nanostructures of metal chalcogenides such as CdS, CdSe, ZnS, SnS, etc., have attracted considerable interest having their application in optical, electronic and superconductor devices [1, 2]. Narrow band gap IV–VI (IV = Sn, Ge, Pb; VI = S, Te, Se) semiconductors exhibiting infrared (IR) and near-infrared (NIR) optical activity have been extensively studied. Among them, tin chalcogenide based materials such as tin sulfide semiconducting nanomaterial has attained immense importance owing to its interesting potential applications such as photodetectors, optoelectronics, and data storage[3, 4]. SnS is a narrow bandgap semiconductor with an optical band gap of direct transitions 1.32–1.5 eV and indirect transitions 1–1.3 eV [5]. Additionally, it has the advantage that its constituent elements are abundant in nature and not possessing any health and environmental hazards. Also the orthorhombic herzenbergite modification of SnS possesses a layered structure consisting where the Sn and S atoms are tightly bonded in a layer and layers are bonded by weak Van der Waals forces [6] Importantly, SnS exhibits both the p- and n-type conduction depending on the concentration of tin [7]. The narrowband gap and the interesting structural property of SnS make it a potential candidate for solar absorber in thin film solar cell and near-infrared (NIR) detector [8, 9], as photovoltaic materials [10] and in holographic recording

Content from this work may be used under the terms of the Creative Commons Attribution 3.0 licence. Any further distribution of this work must maintain attribution to the author(s) and the title of the work, journal citation and DOI.
Published under licence by IOP Publishing Ltd

medium [11]. Moreover, it has the advantage of its constituent elements being abundant in nature and not posing any health and environmental hazards. These phenomena demonstrate that SnS has appropriate electrical and optical features for use in solar cells [8].

Tin sulfides show a variety of phases, such as SnS, SnS_2, Sn_3S_4 and SnS, due to the versatile coordinating characteristics of tin and sulfur [12]. Nowadays, there is a great deal of current interest in the synthesis of nanostructures with specific nano-morphologies and structures [13–15]. Inorganic materials with different morphologies, which have the same chemical composition, may exhibit different properties [16]. In recent years, SnS nanostructures have been prepared by various methods, such as hydrothermal method [5], solvothermal method [17, 18], aqueous solution method [19], and polyol route [20].

In this paper, we developed a very simple solution based method for synthesis of large scale of SnS nanorods. A series of experiments are carried out to observe the growth mechanism of 1D SnS NRs. Phonon modes are calculated by Raman spectroscopy.

2. Experimental

2.1. Sample Preparation

All chemicals were of analytical grade (Beijing Chemicals Co., Ltd.) and were used without further refinement. Tin (II) chloride ($SnCl_2.2H_2O$) and sodium sulfide ($Na_2S.9H_2O$) were taken as tin and sulfur sources respectively, and ethylene glycol was used as a solvent. In a typical synthesis, 0.226 g of $SnCl_2.2H_2O$ and 0.678 g of Na_2S were dissolved ultrasonically in 20 ml of ethylene glycol respectively. The stoichiometric ratio of 3:1 $Na_2S.9H_2O$ and $SnCl_2{\cdot}2H_2O$ solutions were used for precipitation. The $SnCl_2.2H_2O$ solution was added dropwise in Na_2S solution under magnetic stirring. The solution turns to dark brown color with the addition of tin (II) chloride solution. The final participation transferred into 40 ml Teflon-lined stainless autoclave. The autoclave was placed in the oven at a temperature of 250°C for 20 h and then cooled down naturally. Gray powder was collected and washed for several times with distilled water and ethanol to eliminate the impurities before drying at 70°C for 9 h.

2.2. Characterization

The identification of crystal structure was performed by X-ray diffractometer (XRD) using Cu Kα radiation (λ = 1.5406 Å), with 2θ ranging from 10° to 650°. The elemental composition and structural characterization of the as-synthesized powder were analyzed by energy dispersive analysis (EDS) (INCA X-MAX 50, England), Scanning Electron Microscopy (SEM) (Zeiss-SUPRA55). Confocal Raman micro-spectroscopy (Horiba Jobin Yvon HR 800 Raman spectrometer France) was used to analyze the phonon modes of as-synthesized SnS nanorods.

3. Results and Discussion

The morphology of as-prepared nanomaterial was firstly investigated by FE-SEM. Fig. 1(a, b) shows the low and high magnification FE-SEM images which demonstrate the high yield of 1D rod-like structure of SnS. These synthesized SnS nanorods are well shaped and randomly oriented over the copper tape. The crystallinity of SnS nanorods has the width of about 400 nm and the length is in several micrometers range. The crystal structures of the as-synthesized SnS nanorods were analyzed using X-ray diffraction as shown in Fig. 2(a). The crystal structures of the as-synthesized SnS nanorods were analyzed using X-ray diffraction as shown in Fig. 2(a). All the diffraction peaks are in good agreement with orthorhombic tin chalcogenide herzenbergite parameters a = 0.4329 nm, b = 1.1192 nm, c = 0.3984 nm and $\alpha = \beta = \gamma = 90^{\circ}$ (JCPDF Cards 39-0354). Moreover, no signals from other materials are observed in the spectra. The strong and sharp diffraction peaks indicate that the product is well crystallized. Debye–Scherrer formula is used to calculate the crystallite size of the particles which is in the range of 300–400 nm. Elemental composition of the SnS nanorods is

IOP Conf. Series: Materials Science and Engineering **275** (2017) 012007 doi:10.1088/1757-899X/275/1/012007

measured through EDS analysis as illustrated in Fig. 2 (b). The EDS spectrum has contained only Sn and S elements, which is in good agreement with the XRD result, which ratifies the pure configuration of SnS without any impurities such as other binary sulfides.

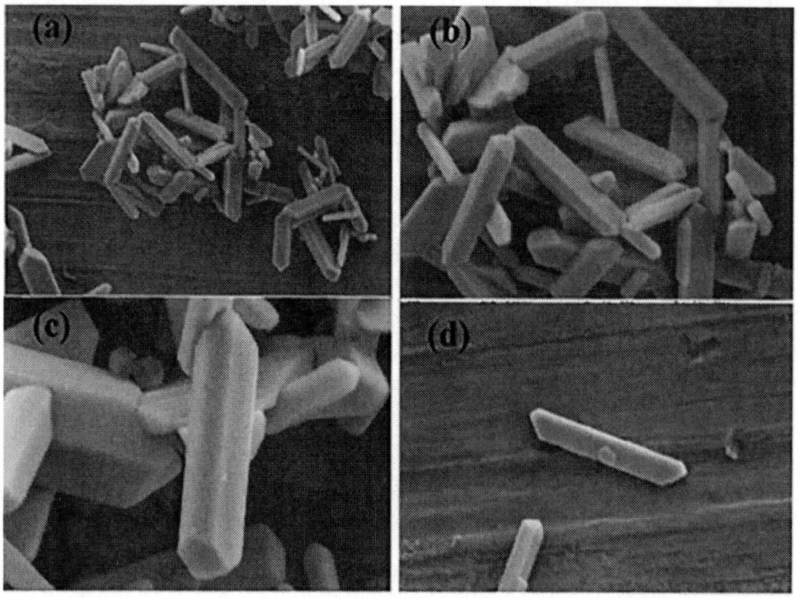

Figure 1. (a, b) Low magnification FESEM images of nanorods, **(c, d)** high magnification FESEM images of single SnS nanorod.

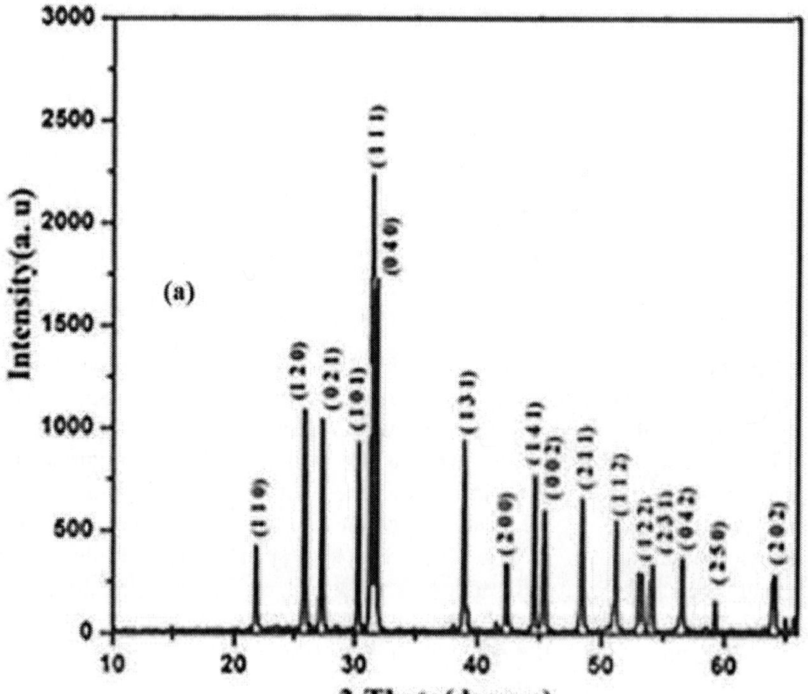

IOP Conf. Series: Materials Science and Engineering **275** (2017) 012007 doi:10.1088/1757-899X/275/1/012007

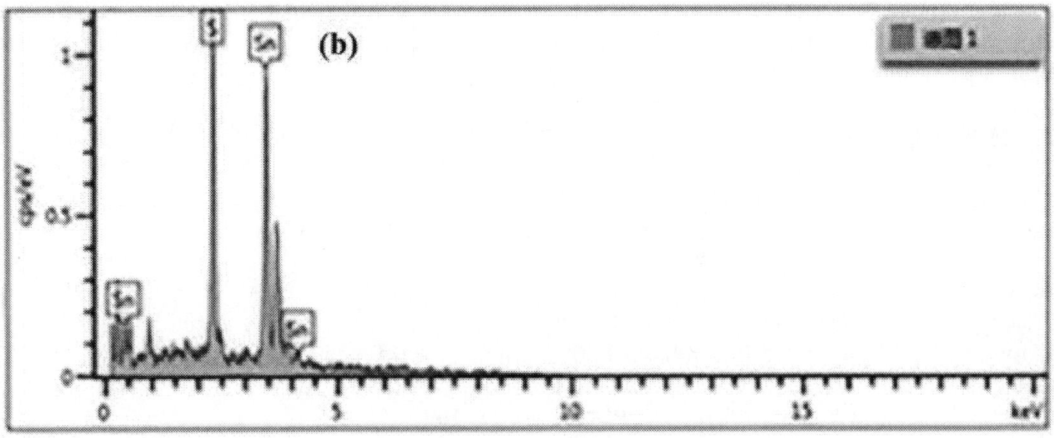

Figure 2. **(a)** X-ray diffraction pattern and **(b)** EDS spectrum of as-synthesized SnS nanorods.

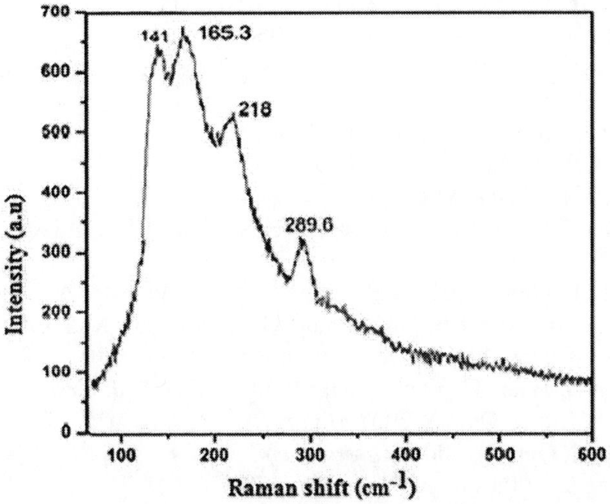

Figure.3. Raman spectrum of SnS nanorods at 373K.

Furthermore, structural and vibrational modes of the prepared tin monosulfide nanoparticles were measured by Raman spectroscopy. Raman spectroscopy is one of the advanced techniques to examine the compositions and chemical states. Fig. 3 indicates the Raman spectrum of the as-prepared SnS nanorods. The dominant Raman peaks are at 141 cm^{-1}, 165.3 cm^{-1}, 218 cm^{-1} and 289.6 cm^{-1}. The lattice vibrations at 141 cm^{-1} correspond to the B_{2u} mode, the peak at 165.3 cm^{-1} to the B_{3g} mode, the peak at 218 cm^{-1} to the A_g mode and the peak at 289.6 cm^{-1} to the B_{2g} mode of SnS nanoparticles at room temperature. Tin sulfide has 21 optical active modes of which 12 are Raman active modes, 7 are infrared active modes and 2 are inactive. According to the reported literature A_g, B_{2g} and B_{3g} are Raman active modes while B_{2u} is the infrared active mode of SnS nanostructure [21].these phenomena shows that the as-synthesized nanostructures are promising materials for energy storage applications and solar cells.

4. Conclusions

In summary, this paper presents synthesize of the single crystalline SnS nanorods using hydrothermal process. The FESEM results had exhibited that the length of prepared SnS nanorods was in the range of several micrometers, the width was about 300-400 nm and the average crystallite size was estimated to be approximately 25 nm. Raman spectroscopy analyses indicated that the as-prepared nanorods mainly exhibit SnS phase. Such materials with controlled morphologies may exhibit a special performance in solar cell and energy storage applications.

5. Acknowledgments
This research work has been performed with the financial support of University of Science and Technology Beijing (fundamental development fund and chancellor scholarship program) and the Fundamental Research Funds for the Central Universities: FRF-BR-09-007A.

6. References
[1] Lakowicz J R, Gryczynski I, Piszczek G, Murphy C J 2002 *J. Phys. Chem.* B **106** 5365.
[2] Bowers M J, McBride J R, Rosenthal S J 2005 *J. Am. Chem. Soc.* **127** 15378.
[3] Dhas N A, Zaban A, Gadenken A 1999 *Chem. Mater.* **11** 806.
[4] Biswas S, Kar S, Chaudhuri S 2005 *J. Phys. Chem.* B **109** 17526.
[5] Jiang T, Ozin G A 1998 *J. Mater. Chem.* **8** 1099.
[6] Zhu H, Yang D, Zhang H 2006 *Mater. Lett.* **60** 2686.
[7] Cheng S, Chen Y, He Y, Chen G 2007 *Mater. Lett.* **61** \ 1408.
[8] Thangaraju B, Kaliannan P 2000 *Appl. Phys.* **33** 1054.
[9] Nnebesny K W, Col lins G E, Lee P A, Chau L K, Danziger J, Osburn E, Armstrong N R 1991 *Chem. Mater.* **3** 829.
[10] Tanusevski A 2003 *Semicond. Sci. Technol.* **18** 501.
[11] Jiang T, Ozin G A 1998 *J. Mater. Chem.* **8** 1099.
[12] Lijima S 1991 *Nature (London)* **354** 56.
[13] Chopra N G, Luyken R J, Cherrey K, Crespi V H, Cohen M L, Louie S G, Zettl A 1995 *Science* **269** 966.
[14] Rao A M, Richter E, Bandow S, Chase B, Eklund P C, Williams K A, Fang S, Subbas wamy K R, Menon M, Thess A, Smalley R E, Dresselhaus G, Dresselhaus M S 1997 *Science* **275** 187.
[15] Yu S, Wu Y, Yang J, Han Z, Xie Y, Qian Y 1998 *Chem. Mater.* **10** 2309.
[16] An C, Tang K, Shen G, Wang C, Yang Q, Hai B, Qian Y 2002 *J. Cryst. Growth* **244** 333.
[17] Li Q, Ding Y, Wu H, Liu X, Qian Y 2002 *Mater. Res. Bull.* **37** 925.
[18] Hu H, Yang B, Zeng J, Qian Y, 2004 *Mater. Chem. Phys.* **86** 233.
[19] Liu Y, Hou D, Wang G, 2003 *Chem. Phys. Lett.* **379** 67.
[20] Shen G, Chen D, Tang K, Huang L, Qian Y, Zhou G, 2003 *Inorg. Chem. Commun.* **6** 178.
[21] Nikolic P M, Mijkovic P L, Mehajlovic B, Lavrencic 1977 *J. Phys. C. Solid Status Phys.* 10 1289.

Rapid and Convenient Separation of Chitooligosaccharides by Ion-Exchange Chromatography

Yuxiao Wu[1], Wei-Peng Lu[2], Jianing Wang[3], Yunhua Gao[4] and Yanchuan Guo[5,*]

[1]Technical Institute of Physics and Chemistry, Chinese Academy of Sciences, Beijing 100190, PR China; University of Chinese Academy of Sciences, Beijing100049, PR China. Email: wuyuxiao@mail.ipc.ac.cn
[2]Technical Institute of Physics and Chemistry, Chinese Academy of Sciences, Beijing 100190, PR China. Email: luweipeng@mail.ipc.ac.cn
[3]Technical Institute of Physics and Chemistry, Chinese Academy of Sciences, Beijing 100190, PR China. Email: wangjianing@mail.ipc.ac.cn
[4]Technical Institute of Physics and Chemistry, Chinese Academy of Sciences, Beijing 100190, PR China. Email: yhgao@mail.ipc.ac.cn
[5,*]Technical Institute of Physics and Chemistry, Chinese Academy of Sciences, Beijing 100190, PR China. Email: yanchuanguo@mail.ipc.ac.cn

Abstract: Pervious methods for separation of highly purified chitooligosaccharides was time-consuming and labor-intensive, which limited the large-scale production. This study developed a convenient ion-exchange chromatography using the ÄKTA™ avant 150 chromatographic system. Five fractions were automatically collected under detecting the absorption at 210 nm. The fractions were analyzed by high-performance liquid chromatography. It proved that they primarily comprised chitobiose, chitotriose, chitotetraose, chitopentaose, and chitohexaose, respectively, with chromatographic purities over 90%. The separation process was rapid, convenient and could be monitored on-line, which would be benefit for the mass production of chitooligosaccharides.

1. Introduction

Chitosan, a linear polysaccharide consisting of β (1–4) linked *N*-acetyl-*D*-glucosamine (GlcNAc) or D-glucosamine (GlcN) residues, is prepared from chitin by partial deacetylation. Chitosan is biodegradable, non-toxic, biocompatible, and is widely used in food, medicine, agriculture, wastewater treatment, biological materials, and other fields [1–3]. Chitooligosaccharides (COS) are the degradation oligomers of chitosan. COS have a low degree of polymerization (DP), usually DP < 20, leading to a high solubility in water and be easily absorbed by organisms. COS therefore show superior bioactivity over chitosan, including antioxidant [4], antimicrobial [5, 6], antitumor [7, 8], anti-inflammatory [9], and immunomodulatory [10] effects.

Using highly purified COS in bioassays is of significance to knowing which molecules are causing the effects, and even the mechanisms, of observed bioactivities. To date, however, reports on the separation of COS are scarce. Several studies have separated COS by size exclusion chromatography [11], ion-exchange chromatography [12–14], or affinity chromatography [15]. In these works, the separated COS were detected based on chromogenic reactions, such as the Imoto method, the phenol–sulfuric acid method, thin-layer chromatography (TLC), or a bicinchoninic acid (BCA) reagent kit, owing to the absence of chromophores of COS [7,11–14]. For high resolution, the fractions were pooled

Content from this work may be used under the terms of the Creative Commons Attribution 3.0 licence. Any further distribution of this work must maintain attribution to the author(s) and the title of the work, journal citation and DOI.
Published under licence by IOP Publishing Ltd

with a volume as small as possible and then the fractions were monitored using the above methods to draw an elution curve. Usually, hundreds of tubes of fractions were obtained, meant that hundreds of thousands of chromogenic reaction should be repeatedly done. This was time- and labor-intensive, and has limited the large-scale production of COS. This study developed a convenient ion-exchange chromatography for the separation of COS. The ultraviolet detector was used in the separation of COS. Although COS have no characteristic absorption in the ultraviolet (UV)–visible wavelength region, there are several hydroxyl and amine groups in the sugar ring. The n→σ* transitions of the lone pair electrons in the heteroatoms of these groups (–OH, $-NH_2$) induce a strong absorption band near 200 nm. This can be used as the detection signal in the separation of COS mixtures — a mix of a homologous series. The entire separation process was rapid, convenient and could be monitored on-line, which would be benefit to industrial production.

2. Experiment

2.1 Materials

Chitooligosaccharide(COS) mixtures with a degree of deacetylation of >95% and a molecular weight of 1,500 Da were purchased from Zhejiang Golden-Shell Pharmaceutical Co., Ltd. (Zhejiang, China). Mixed chitosan oligomer standards containing chitobiose, chitotriose, chitotetraose, chitopentaose, and chitohexaose were obtained from Shanghai Hui-cheng Biological Technology Co., Ltd. (Shanghai, China). All other chemicals were of analytical grade and purchased from Beijing Chemical Co., Ltd. (Beijing, China).

An ÄKTA™ avant 150 chromatographic system equipped with Unicorn™ 6.1 software, SP Histrap HP column (5ml), Sephadex G-10 gel packing were purchased from GE Healthcare Co., Ltd. (Uppsala, Sweden).

2.2 Preparation of Chitooligosaccharide Mixture

Low degree-of-polymerization (DP) chitooligosaccharides (LCOS) were prepared from the above COS mixtures by ethanol precipitation. First, 10 % (w/v) COS aqueous solution was added to three times the volume of ethanol. The mixture was stirred at room temperature for 4 h and kept at 4 °C overnight. Subsequently, the mixture was filtered through a Buchner to remove the insoluble high-DP oligomers; the supernatant was rotary evaporated under vacuum to remove the ethanol. The residual aqueous solution was lyophilized.

2.3 Separation of Low Degree-of-polymerization Chitooligosaccharides

The LCOS (80 mg) were dissolved in 1 mL of an acetate buffer (50 mmol/L, pH 4.6). The solution was filtered through a microporous membrane (0.45 μm) and injected into the ÄKTA™ avant 150 chromatographic system through a SP Histrap HP cation-exchange column (5ml) for separation. After washing the column with 3 column volume (cV) of HAc–NaAc buffer (50 mmol/L, pH 4.6), it was eluted with a mixture of NaCl (2 mol/L) and HAc–NaAc buffer (50 mmol/L, pH 4.6) using a linear gradient (0–60% NaCl) at a flow rate of 10 mL/min. Eluate was detected in real time by a ultraviolet (UV) detector at 210 nm and fractions were automatically collected using fraction collector according to the peak values. The start level and the end level of the peak fractionation were both 200 mAU.

The fractions were desalted by gel filtration. The fractions were loaded on a Sephadex G-10 (26 mm × 50 cm) column and eluted with deionized water at 8 mL/min. Fractions were monitored by the UV detector at 210 nm and collected. The remaining solution was concentrated and lyophilized.

2.4 Identification of Chitooligosaccharides

HPLC analysis was performed using a Shimadzu LC-10AT_{VP}Plus instrument (Shimadzu, Kyoto, Japan) with a differential refractive-index detector. COS were separated on an Asahipak NH_2P-50E column (4.6 mm × 25 cm) (Shodex, Shanghai, China). The optimized HPLC separation was carried out with a mobile phase consisting of acetonitrile/water (70/30, v/v) at a flow rate of 1.0 mL/min. The column temperature was set at 25 °C.

3. Results and Discussion

3.1 Separation of Chitooligosaccharides

COS could be separated by the cation-exchange column on the basis of their differing numbers of amino groups. Under acidic conditions, the positively charged amino groups of the COS interact with the sulfopropyl (–$CH_2CH_2CH_2SO_3$–) group of the chromatographic packing. The COS were then desorbed by the high NaCl concentration of the eluant. Longer oligosaccharide chains resulted in more amino groups on the COS and were more difficult to desorb from the column. UV detection is commonly used in separation processes; however, COS have no characteristic absorption in the UV-visible region, so the strong absorption band near 200 nm, owing to n→σ* transitions of the lone pair electrons in the hydroxyl and amine groups, was used as the detection signal in the separation of the COS mixture.

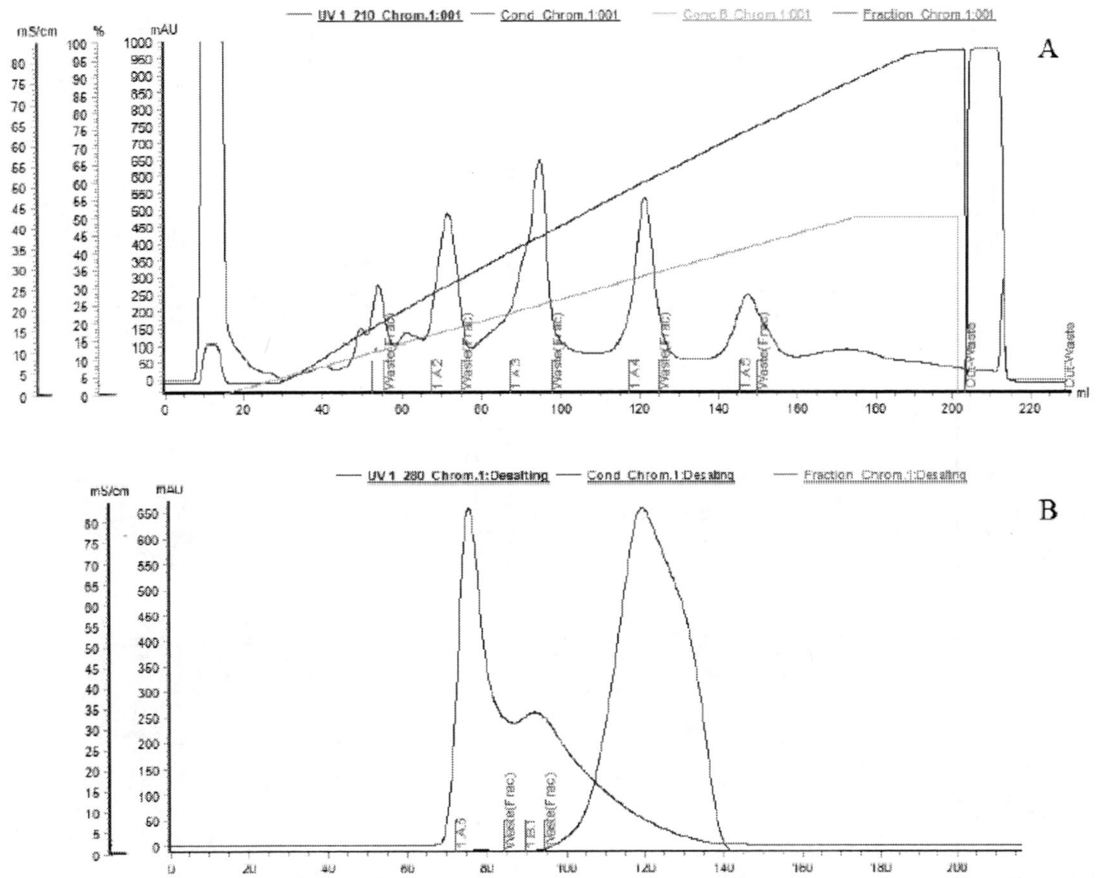

Figure 1. Chromatograms for cation-exchange via SP Histrap HP (A) and desalting via Sephadex G-10 (B). The chromatograms show the absorption signal at 210 nm in AU, the conductivity signal in mS/cm, the concentration of 2M NaCl solution in % and the fractionation collector.

The prepared LCOS were separated on the SP Sepharose HP column. Chromatograms A in figure 1 is the elution curve for cation-exchange via SP Histrap HP. The lines of different color (blue, brown and green) illustrated the absorption signal at 210 nm in AU (UV 1 210 chrom.1), the conductivity signal in

IOP Conf. Series: Materials Science and Engineering **275** (2017) 012008 doi:10.1088/1757-899X/275/1/012008

mS/cm (Cond chrom.1), the concentration of 2M NaCl solution in % (Conc B chrom.1), respectively. After washing the column, the oligomers were eluted with a mixture of NaCl and HAc–NaAc buffer. With the increase of the concentration of 2M NaCl solution, the conductivity was increase and the oligomers were desorbed. Five fractions were automatically collected when the level of the peak were higher than 200 mAU. Chromatograms B in figure 1 is the elution curve for desalting via Sephadex G-10. The results show that the conductance of each fraction was reduced to 5 mS/cm after the desalting: the NaCl in the fraction was completely removed.

3.2 Identification of the Fractions

Within a homologous series, the DP and composition of the COS can be analyzed by hydrophilic-interaction chromatography. The retention of a COS is based on the hydrophilic interactions between the stationary phase and its hydroxyl and amino groups. In general, the more hydroxyl and amino groups a COS possesses, the greater the number of hydrogen-bonding interactions with the stationary phase, i.e., the longer the oligosaccharide chain length, the longer the retention time.

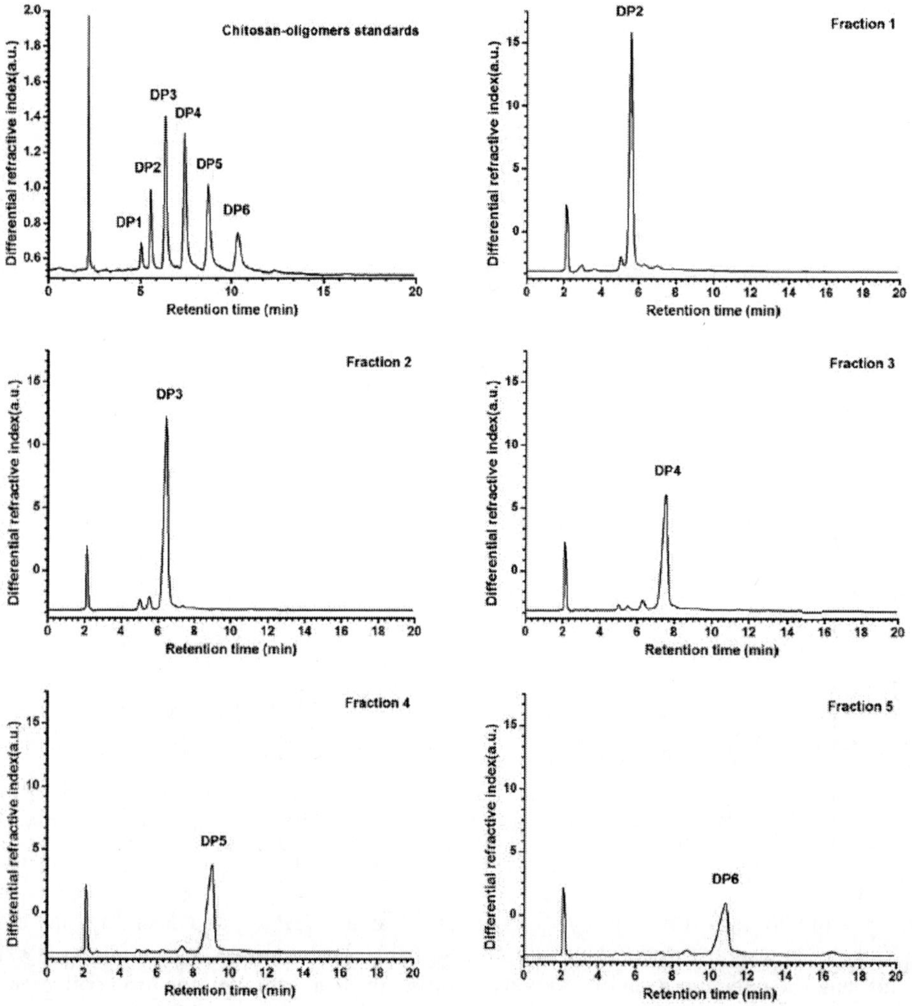

Figure 2. HPLC spectra of the chitosan oligomer standards and each separated fraction. The retention times of glucosamine di- to hexa-mers were 5.56, 6.43, 7.54, 9.01, and 10.83 min, respectively.

Table 1. The Retention Time of Chitosan-Oligomers and Fractions

Standards	glucosamine	chitobiose	chitotriose	chitotetraose	chitopentaose	chitohexaose
Retention time (min)	4.95	5.49	6.34	7.38	8.80	10.39
Fractions 1		5.56				
Fractions 2			6.43			
Fractions 3				7.54		
Fractions 4					9.01	
Fractions 5						10.83

Chitosan oligomer standards that contained chitobiose, chitotriose, chitotetraose, chitopentaose, and chitohexaose were used to calibrate the retention times. Figure 2 illustrates the DP distribution and composition of each separated fraction and its comparison with the standards. Each of the five separated fractions had an individual peak in the HPLC spectra, the retention times of which were 5.56, 6.43, 7.54, 9.01, and 10.83 min, respectively. As depicted in Table 1, these peaks matched well with those of the standards, corresponding to dimers (chitobiose), trimers (chitotriose), tetramers (chitotetraose), pentamers (chitopentaose), and hexamers (chitohexaose), respectively. Assuming that the oligomers give a similar response per mass unit, according to the relative peak areas of the HPLC spectra, the contents of the basis oligomer in fractions 1–5 were 94.87%, 91.96%, 89.11%, 91.81%, and 87.11%, respectively.

4. Conclusions

We have developed a convenient ion-exchange chromatographic method for the separation of pure COS. UV detection is used in the separation of COS. Using an ÄKTA™ avant 150 chromatographic system, five glucosamine oligomers with single DPs were isolated by ion-exchange and desalted by gel filtration. The isolated fractions contained 94.87% dimer, 91.96% trimer, 89.11% tetramer, 91.81% pentamer, and 87.11% hexamer, respectively. The entire separation process is rapid, convenient and could be monitored on-line, which would be benefit to industrial production.

5. Acknowledgment

We gratefully acknowledge the financial support for this work by the Technical Institute of Physics and Chemistry (Grant No.Y4AH011M15) and the State Natural Sciences Fund, China (Project No. 21506236).

6. References

[1] I. F. Uchegbu, M. Carlos, C. McKay, X. L. Hou and A. G. Schatzlein, Polym. Int. 63, 7, 1145-1153, 2014.
[2] A. Domard, Carbohydr. Polym. 84, 2, 696-703, 2011.
[3] A. Jimtaisong and N. Saewan, Int. J. Cosmetic Sci. 36, 1, 12-21, 2014.
[4] M. D. L. Trinh, D. H. Ngo, D. K. Tran, Q. T. Tran, T. S. Vo, M. H. Dinh and D. N. Ngo, Carbohydr. Polym. 103, 502-509, 2014.
[5] B. Liu, X. Y. Wang, C. S. Pang, J. W. Luo, Y. Q. Luo and R. C. Sun, Carbohydr. Polym. 92, 2, 1078-1085, 2013.
[6] X. L. Liu, W. S. Xia, Q. X. Jiang, Y. S. Xu and P. P. Yu, J. Agric. Food Chem. 62, 1, 297-303, 2014.
[7] C. N. Xiong, H. G. Wu, P. Wei, M. Pan, Y. Q. Tuo, I. Kusakabe and Y. G. Du, Carbohydr. Res. 344, 15, 1975-1983, 2009.
[8] Z. G. Luo, X. X. Dong, Q. Ke, Q. W. Duan and L. Shen, Oncol. Lett. 8, 1, 361-366, 2014.
[9] J. C. Fernandes, H. Spindola, V. de Sousa, A. Santos-Silva, M. E. Pintado, F. X. Malcata and J. E. Carvalho, Mar. Drugs 8, 6, 1763-1768, 2010.

[10] Y. S. Lin, M. F. Wu, Y. Takamori, Y. Okamoto and S. Minami, J. Exp. Nanosci. 9, 8, 860-870, 2014.
[11] Z. Ji, Y.X. Jiang, X.Q. Li, Y.L. Li, S.S. Chen, J. Shanghai Ocean Uni., 22, 634-640, 2013.
[12] X. L. Wei, Y. F. Wang, J. B. Xiao and W. S. Xia, Int. J. Biol. Macromol. 45, 4, 432-436, 2009.
[13] K. C. Li, R. G. Xing, S. Liu, R. F. Li, Y. K. Qin, X. T. Meng and P. C. Li, Carbohydr. Polym. 88, 3, 896-903, 2012.
[14] K. C. Li, S. Liu, R. G. Xing, H. H. Yu, Y. K. Qin, R. F. Li and P. C. Li, J. Sep. Sci. 36, 7, 1275-1282, 2013.
[15] F. Le Devedec, L. Bazinet, A. Furtos, K. Venne, S. Brunet and M. A. Mateescu, J. Chromatogr. A 1194, 2, 165-171, 2008.

IOP Conf. Series: Materials Science and Engineering **275** (2017) 012009 doi:10.1088/1757-899X/275/1/012009

Synthesis and Characterization of Hyaluronic Acid Modified Colloidal Mesoporous Silica Nanoparticles

Wenbiao Zhang[1,2,3], Yu Wang[1,2,3], Zhen Li[1,2,3], Wanxia Wang[1,2,3], Honghao Sun[1,2,3] and Mingxing Liu[1,2,3,*]

[1]Key Laboratory of Fermentation Engineering (Ministry of Education), [2]Hubei Provincial Cooperative Innovation Center of Industrial Fermentation, [3]Hubei Key Laboratory of Industrial Microbiology, Hubei University of Technology, Wuhan 430068 China

[*]Corresponding author: lmxing@mail.hbut.edu.cn

Abstract: The colloidal mesoporous silica nanoparticles functionalized with hyaluronic acid (CMS-HA) were successfully synthesized by grafting hyaluronic acid onto the external surface of the amino-functionalized mesoporous silica nanoparticles ($CMS-NH_2$). Moreover, the paticle properties of CMS-HA were characterized by fourier transform infrared spectroscopy (FT-IR), dynamic light scattering (DLS) and transmission electron microscopy (TEM). The nanomaterials were negatively charged and had a relatively uniform spherical morphology with about 100 nm in diameter, which could make it more compatible with blood. So the results suggested that the CMS-HA might be a critical nanomaterial for applying in target drug delivery system.

1. Introduction

Spherical mesoporous silica particles with nanoscale as a promising drug delivery system have attracted much attention owing to their excellent properties including high loading capacity, high colloidal stability, facile surface modification ability and excellent biocompatibility [1]. However, although considerable progress has been made, it is still necessary and urgent to overcome the lack of target selectivity of drugs [2]. Here, the intention is to develop a effective material to achieve high targeting efficiency and ensure the high dispersity of CMS in physiological condition [3].

Hyaluronic acid (HA) attached to the surface of nanoparticles can promote specific target and improve drug delivery efficiency via the overexpression of the transmembrane glycoprotein CD44 on cancer cells [4]. HA has been widely reported as a targeting ligand in biomedical field due to its excellent biocompatibility, biodegradability and targeting specificity [5]. Furthermore, HA acting as gatekeepers can be grafted onto various nanoparticles for controlled drugs release [6] since it can respond to hyaluronidase (HAase) which is found to be upregulated in many tumor matrices and endolysosomes [7].

Herein, a novel drug delivery systems based on CMS modified with HA (CMS-HA) has been synthesized via grafting HA onto amino functionalized colloidal mesoporous silica nanoparticles ($CMS-NH_2$) into one system by an amidation reaction (figure 1). Furthermore, doxorubicin (Dox) will be loaded into the CMS-HA and the drug loading efficiency and controlled release properties will be further researched. The CMS-HA which combined the advantages of CMS and HA as a drug carrier might make the delivery process more controllable [8].

Content from this work may be used under the terms of the Creative Commons Attribution 3.0 licence. Any further distribution of this work must maintain attribution to the author(s) and the title of the work, journal citation and DOI.
Published under licence by IOP Publishing Ltd

2. Experimental Section

2.1. Materials

Tetraethyl orthosilicate (TEOS, 98%), 3-aminopropyltriethoxysilane (APTES, 98%), hexadecyl trimethyl ammonium Chloride (CTAC, 97%) and hyaluronic acid (HA, Mw=10-20 KDa) were obtained from Aladdin Chemistry Co. (Shanghai). N-hydroxysuccinimide (NHS, 98%) and 1-ethyl-3-(3-(dimethylamino) propyl) carbodiimidehydrochloride (EDC•HCl, 99%) were supplied by J&K Chemical Co. (Beijing). Triethanolamine (TEA) was obtained from Sinopharm Chemical Reagent Co., Ltd. (China).

2.2. Synthesis of Amino-functionalized Colloidal Mesoporous Silica Nanoparticles (CMS-NH_2)

The CMS-NH2 was synthesized according to the published method with minor alterations [9]. Water (64 mL, 3.55 mol), ethanol (10.5 mL, 0.179 mol) and 25 wt % CTAC solution (10.4 mL, 7.86 mmol) were premixed, followed by adding TEA (12.4 mL, 0.093 mol). The mixture prepared as stock solution was stirred at room temperature until all TEA was dissolved, and the pH value of the solution was about 10.5. Typically, the stock solution (20mL) was heated to 60°C, to which TEOS (1.454 mL, 6.5mmol) was slowly added dropwise within 2-3 min under stirring, and then APTES (0.152 ml, 0.65 mmol) was added dropwise into the solution 15 min after adding TEOS. The reaction mixture was cooled to room temperature after reacting for 2 h. Finally, the resulting products were centrifuged (10000 rpm, 15 min), washed with ethanol for two times. Extraction of the template from the products was performed by the means of refluxing for two times in a solution containing 5 mL of concentrated hydrochloric acid in 45 mL of ethanol at 60°C for 2 h, and then the CMS-NH2 was centrifuged (10000 rpm, 15 min), washed with ethanol and distilled water, and redispersed in distilled water (5 mg/mL) for following use.

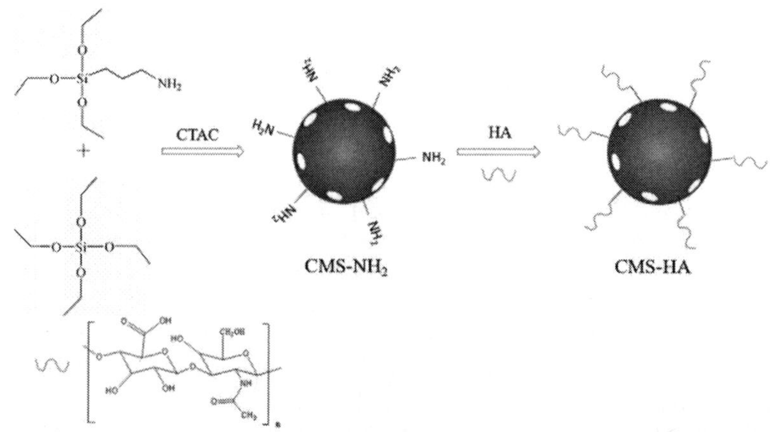

Figure 1. The synthesis route of mesoporous silica nanoparticles modified with hyaluronic acid (CMS-HA).

2.3. Synthesis of Hyaluronic Acid Functionalized Mesoporous Silica Nanoparticles (CMS-HA)

The CMS-HA was synthesized by the amidation reaction between CMS-NH2 and HA. Briefly, 15 mg HA was hydrated in 3mL distilled water at room temperature and activated by EDC (9 mg) and NHS (6 mg) for 1 h. Then, 5mL of CMS-NH2 suspension (5 mg/mL) was added to the mixed solution. The solution was stirred for another 24h at rome temperature. Finally, the obtained samples were centrifuged, washed with water and ethanol, and redispersed in distilled water to preserve.

2.4. Characterizations

The zeta potential and particle size of the nanoparticles were determined with a Malvern Zetasizer Nano ZS90. The zeta potential and hydrodynamic diameter were measured at 25°C and pH 7. Transmission electron microscopy (TEM) was performed on a JEOL JEM-2100F transmission electron microscope. Fourier transform infrared (FT-IR) spectrum was obtained on Nicolet Nexus 470 using the KBr pellet technique.

3. Results and Discussion

The functionalization of HA onto the surface of CMS was validated by different methods. The FT-IR spectra of HA, CMS, CMS-NH_2, CMS-HA were shown in figure 2. Absorption peaks observed at 1080 cm^{-1}, 800 cm^{-1} and 960 cm^{-1} could be attributed to typical vibration bands of siliceous materials (curve B, C, D). Compared with HA and CMS (curve A, B), a new absorption peak at 1516 cm^{-1} was observed in the sample CMS-NH_2, which can be ascribed to N-H bending vibrations of amino groups (curve C). The appearance absorption band at around 1408 cm^{-1} can be attributed to the C=O stretching of the carboxyl groups within the HA (curve D), which indicated that HA has been successfully grafted on CMS nanoparticles.

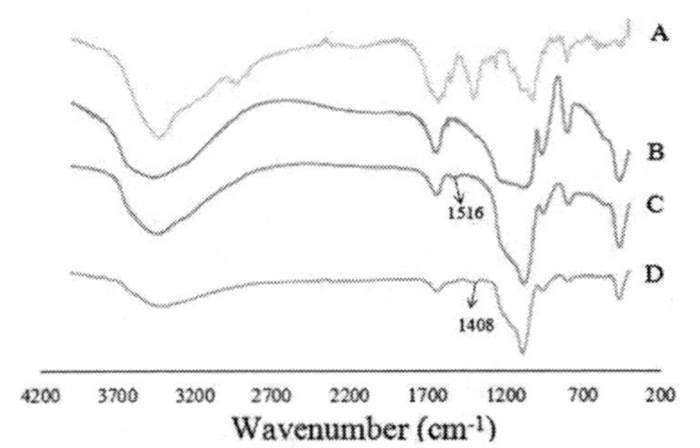

Figure 2. The FT-IR spectra of HA (A), CMS (B), CMS-NH_2 (C), CMS-HA (D).

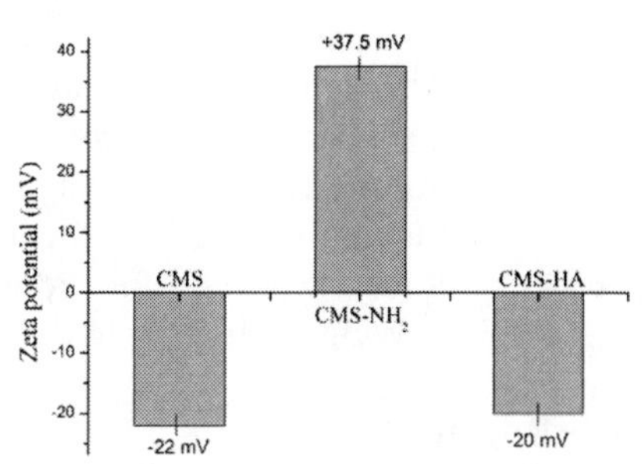

Figure 3. Zeta potential analysis of CMS, CMS-NH_2 and CMS-HA.

Meanwhile, the particle size and the zeta potential of the nanoparticles were measured by DLS. As indicated in figure 3 and table 1, the zeta potential of CMS, CMS-NH_2 and CMS-HA were -22 mV, +37.5 mV, and -20 mV, respectively. CMS-HA are negatively charged due to the ionizing of the carboxyl groups within HA at pH=7.0 [10]. Moreover, the average size of the CMS-NH_2 and CMS-HA were 151 nm and 508.4 nm, respectively. These results confirmed that HA has been successfully conjugated to CMS-NH_2 by the charge inversion and the growing size.

Table 1. DLS measurements of CMS-NH_2 and CMS-HA.

Materials	Size (nm)	PDI	Zeta (mV)
CMS-NH_2	151 ± 1.3	0.101 ± 0.053	37.5 ± 2.6
CMS-HA	508.4 ± 2.2	0.543 ± 0.046	-20 ± 1.7

The morphology of CMS-NH_2 and CMS-HA were analyzed by TEM (figure 4). The spherical morphology and the wormhole arrangement mesoporous of CMS-NH_2 were obviously observed with the diameter of 90 nm in figure 4(a). After grafting HA onto the surface of the CMS, the mean diameter growed to 100 nm. Though there were no obvious differences in shape and diameter presented between CMS-NH_2 and CMS-HA, it was hard to detect the mesoporous structure of CMS-HA and the clear border could be obviously observed in figure 4(b), which confirmed that HA were efficiently functionalized on the CMS-NH_2. In this study, the size measured by DLS was slightly larger than that obtained from TEM on account of the hydration layer in aqueous condition [11].

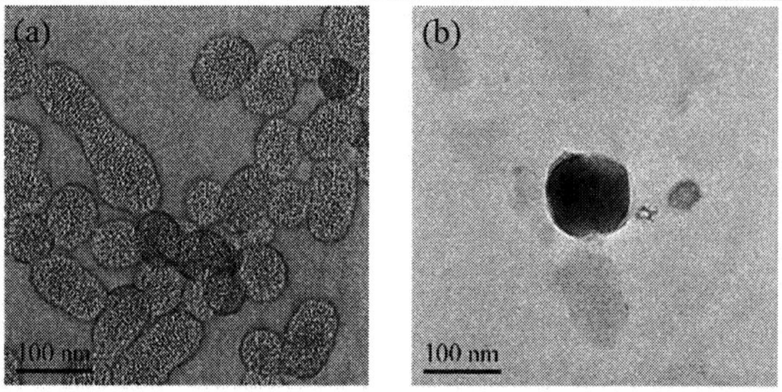

Figure 4. TEM of CMS-NH_2 (a) and CMS-HA (b).

4. Conclusions

A novel colloidal mesoporous silica nanoparticles modified with hyaluronic acid were successfully synthesized by grafting hyaluronic acid onto the external surface of amino functionalized silica nanoparticles using amide bond as a cross linker. The CMS-HA presented negatively charged and had a relatively uniform spherical morphology with about 100 nm in diameter, which might be more compatible with blood. In addition, the carboxyl groups on the surface of CMS-HA could be combined with various types of functional groups. Therefore, CMS-HA would be a promising nanomaterial for applying in target drug delivery system.

5. Acknowledgments

This research was financially supported by National Natural Science Foundation of China (No. 81201197).

6. References

[1] Yamamoto E and Kuroda K 2016 *B. Chem. Jpn.* **89** 501-9
[2] Yang D, Wang T, Su Z, Xue L, Ran M and Zhang C 2016 *Acs. Appl. Mater. Inter.* **8** 22431-41
[3] Ma M, Chen H, Chen Y, Zhang K, Wang X, Cui X and Shi J 2012 *J. Mater. Chem.* **22** 5615-21
[4] Gotte M and Yip G 2006 *Cancer. Res.* **66** 10233-7
[5] Prestwich G D 2011 J. *Control. Release* **155** 193-9
[6] Song S, Chen F, Qi H, Li F, Xin T, Xu J, Ye T, Sheng N, Yang X and Pan W 2014 Pharm. Res. **31** 1032-45
[7] Yu M, Jambhrunkar S, Thorn P, Chen J, Gu W and Yu C 2013 *Nanoscale* **5** 178-83
[8] Zhao Q, Geng H, Wang Y, Gao Y, Huang J, Wang Y, Zhang J and Wang S 2014 Acs. Appl. Mater. *Interfaces* **6** 20290-9
[9] Kobler J, Möller K and Bein T 2008 *Acs. Nano.* **2** 791-9
[10] Choi K Y, Min K H, Na J H, Choi K, Kim K, Park J H, Kwon I C and Jeong S Y 2009 *J. Mater. Chem.* **19** 4101-7
[11] Cauda V, Argyo C and Bein T 2010 *J. Mater. Chem.* **20** 8693-9

Enzyme Functionalized AuNPs and Glucometer-based Protein Detection

Tao Dai, Jie Fang, Wen Yu and Guoming Xie*

Key Laboratory of Medical Diagnostics of Education, Department of Laboratory Medicine, Chongqing Medical University, 400016, China
*E-mail: guomingxie@cqmu.edu.cn

Abstract: We here developed a novel method for protein detection by using protein aptamer-functionalized magnetic beads for protein recognition and invertase -functionalized AuNPs catalyze sucrose generate glucose that can be detected by a glucometer.First,the invertase and DNA probe P2 are immobilized onto the gold nanoparticles (I.P2@AuNPs). Next protein aptamer P1 are immobilized onto the streptavidin-coated Magnetic beads (P1@MB).P1 and P2 can complementary to form double-stranded DNA. When target protein presence, P1 combine with target and release I/P2@AuNPs. Then magnetic separation, take supernatant fluid and add sucrose after a period of reaction, detection of glucose concentration by glucometer, thus achieve the sensitive and selective detection of the target protein.

1. Introduction

To our knowledge proteins play critical roles in a variety of life activities [1]. So highly sensitivity and selectivity protein detection methods are particularly important in clinical medicine, disease diagnosis [2], biology and food safety [3]. Over the past decades, the detection of different types of protein has attracted many researchers' interests. In order to improve the sensitivity and selectivity of the methods, a mass of biomaterials, such as gold nanoparticles [4], silver nanoparticles [5] have been used to the detection of protein. They have favourable biocompatibility [6] and can be easily connected to the biological macromolecules. In addition, the functionalized magnetic beads (MBs) were also applied for the protein detection, because of the DNA aptamers or antibodies can be easily immobilization onto the surface of MBs with the effect of chemical coupling or physical adsorption [7]. Meanwhile the separation of magnetic beads are very convenient and effective, This is good for the following study.

Recently,various signal amplification strategies based on different enzymes[8], sucrase is a kind of common enzyme.Not only sucrase has a strong catalytic ability but sucrase catalytic sucrose generate glucose can be combined with our common testing instrument - glucometer. With the help of these reagents and instrument, we constructed a sensitive, selective and convenient method for detection of protein.

2. Materials and Methods

2.1. Materials and Reagents

Streptavidin-coated magnetic beads (1 um in average diameter) were purchased from NEW ENGLAND BioLabs Inc. All oligonucleotide sequences and 1×PBS were purchased from Sangon Biotechnology. Co, Ltd (Shanghai, China). Grade VII invertase from baker's yeast (S. cerevisiae), and

Content from this work may be used under the terms of the Creative Commons Attribution 3.0 licence. Any further distribution of this work must maintain attribution to the author(s) and the title of the work, journal citation and DOI.
Published under licence by IOP Publishing Ltd

gold acid chloride trihydrate, were the products of Sigma-Aldrich. Ultrapure water (Milli-Q 18.2 M, Millipore System Inc.) was employed in all runs. All other chemicals were of reagent grade and used as received.

2.2. Methods

This study aimed at the development of sensitive and selective method for the quantitative detection of proteins. A schematic of this proposed concept is presented in **Scheme 1**. First the streptavidin-coated magnetic beads were conjugated with biotin-labeled DNA probe P1 (P1@MBs), P2 and invertase were immobilization on the AuNPs (I.P2@AuNPs).The P1@Mbs and I.P2@AuNPs form I.P2 P1@MBs. When the target protein presence the aptamer (P1) will combine with the targets and release I.P2@AuNPs. Then magnetic separation, take the supernatant and add sucrose solution after reaction a period of time,use glucometer (ACCU-CHEK Active) mesure the consentration of glucose.

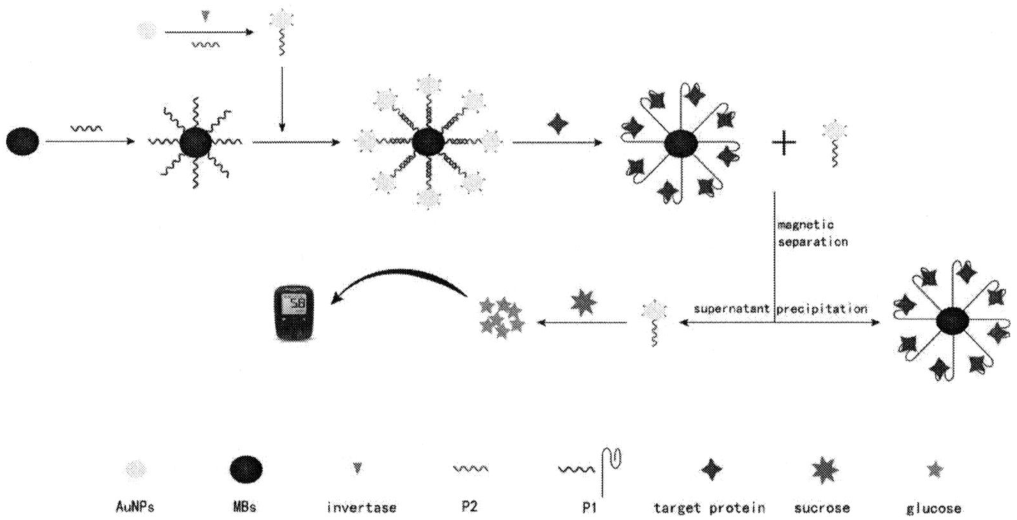

Scheme 1. Illustrate the process of experiment

2.2.1 Preparation of I.P2@AuNPs. Gold nanoparticles (26 nm) were prepared as previous work [9,10]. Briefly, 10 μL of $HAuCl_4$ solution was added into 96 mL of ultrapure water, and the mixture was boiled with vigorous stirring. Then, 4 mL of trisodium citrate solution (1%, m/v) was quickly injected into the boiling solution. The solution was kept stirring and heating to its color changed from pale yellow to wine red. After stopt stirring and cooled to room temperature, the resulting solution was stored in brown glass bottles at 4 °C for future use.

As shown in **scheme 1** The I.P2@AuNPs were synthesized with AuNPs, P2 and invertase according to the literature [11,12,13].First, invertase (80μL, 30 mg/mL) was added into 400 μL of the AuNPs solution. The resulting mixture was incubated for 20 min at room temperature with gentle shaking. During this process, invertase were covalently bound to AuNPs via the dative binding between AuNPs and free -SH groups of the invertase. Then, DNA probe P2(20 μL, 1.0 μM) dissolved into 10 mM Tris-HCl buffer including 10 mM TCEP and 0.1 M NaCl (pH 7.4) was added to the mixture, and incubated overnight. Finally, the resulting mixture was centrifuged (13,400g) for 20 min at 4°C and stored at 4 °C (400 μL1× PBS, pH 7.4) until use.

2.2.2Preparation of I.P2 P1@MBs. A portion of 0.96 mL 1 mg/mL solution of streptavidin-coated magnetic beads (MBs) was placed close to a magnetic rack for 1 minute. The clear solution was discarded and replaced by 0.96 mL of PBS. This buffer exchange procedure was repeated twice. Then, 32 μL of 100 μM biotin-DNA (P1) was added to the MB solution and well mixed for 30 minutes at room temperature. After that, the MBs were washed twice using buffer to remove excess P1.

Dissolved the precipitation get P1@MBs solution. Later, 32 μL of 100 μM I.P2@AuNPs were added to the P1@MBs solution and well mixed for 30 minutes at room temperature. Using magnetic separation and 1× PBS buffer washing three times to remove excess I.P2@AuNPs. then the as-prepared I.P2 P1@MBs were dispersed in 1×PBS.

First the streptavidin-coated magnetic beads were conjugated with biotin labeled DNA probe P1 according to the protocol suggested by the manufacturer. Next, magnetic separation and removed the supernatant, resuspension the precipitate in 1×PBS and get P1@MBs solution.Then, excess I.P2@AuNPs are added to the P1@MBs solution, after an hour of reaction,the excess I.P2@AuNPs were washed three times by 1×PBS buffer.The I.P2 P1@MBs were dispersed in 1×PBS untill use. All the processes shown in **scheme 1**.

3. Results

First we characterization of gold nanoparticles by transmission electron microscopy (Hitachi-7500. Japan) and the result shown in **figure 1**. It reveals that the synthesis of AuNPs have a more uniform diameter size.Then we optimized the invertase concentration of synthesis of invertase@AuNPs.The concentration of invertase from 30 mg/mL to 70 mg/mL was used in the expriment.Next characterization of the invertase@AuNPs by UV-Vis spectra(Thermo Nanodrop-1000, USA) shown in **figure 2.** In the subsequent experiments We choose the concentration of invertase as 30 mg/mL.

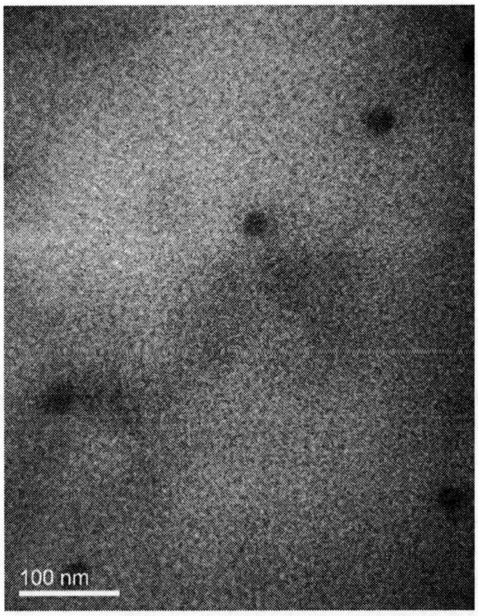

Figure 1. The transmission electron microscopy map of gold nanoparticles.

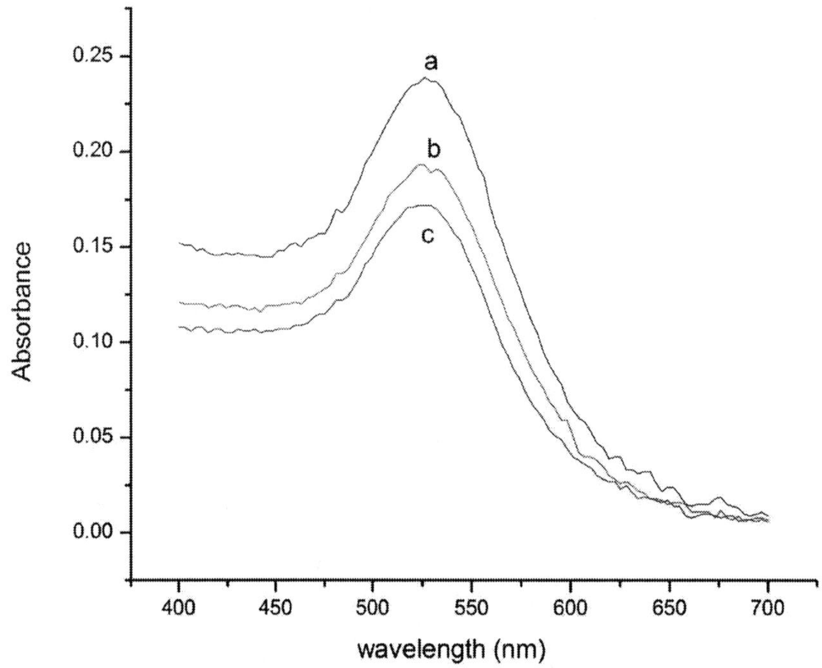

Figure 2. UV-Vis spectra of invertase@AuNPs .The invertase concentration of synthesis of invertase@AuNPs: (a) 30 mg/mL, (b) 50 mg/mL, (c) 70 mg/mL.

4. Conclusions

In this study I.P2 P1@MBs were prepared via streptavidin-coated magnetic beads and invertase labelled AuNPs through P1 and P2 hybridization.Furthermore,we optimized the invertase concentration of synthesis of invertase@AuNPs and utilized UV-Vis spectra characterization to get the best concentration 30 mg/mL. What we composited are great potential to detect proteins.

5. References

[1]. Vaart A V D, Bursulaya B D, Brooks C L, et al. Are Many-Body Effects Important in Protein Folding?. Journal of Physical Chemistry B, 2000, 104(40):9554-9563.
[2]. Chan H N, Xu D, Ho S L, et al. Ultra-sensitive detection of protein biomarkers for diagnosis of Alzheimer's disease. Chemical Science, 2017, 8.
[3]. Garber E A E, Venkateswaran K V, O'Brien T W. Simultaneous multiplex detection and confirmation of the proteinaceous toxins abrin, ricin, botulinum toxins, and Staphylococcus enterotoxins A, B, and C in food. Journal of Agricultural & Food Chemistry, 2010, 58(11):6600-6607.
[4]. [4]Mao S, Lu G, Yu K, et al. Specific protein detection using thermally reduced graphene oxide sheet decorated with gold nanoparticle-antibody conjugates. Advanced Materials, 2010, 22(32):3521-6.
[5]. Kahraman M, Sur I, Culha M. Label-free detection of proteins from self-assembled protein-silver nanoparticle structures using surface-enhanced Raman scattering. Analytical Chemistry, 2010, 82(82):7596-7602.
[6]. Shukla R, Bansal V, Chaudhary M, et al. Biocompatibility of gold nanoparticles and their endocytotic fate inside the cellular compartment: a microscopic overview. Langmuir the Acs Journal of Surfaces & Colloids, 2005, 21(23):10644.

IOP Conf. Series: Materials Science and Engineering **275** (2017) 012010 doi:10.1088/1757-899X/275/1/012010

[7]. Erdem A, Congur G, Eksin E. Voltammetric Aptasensor Based on Magnetic Beads Assay for Detection of Human Activated Protein C. Methods in Molecular Biology, 2016, 1380:163.

[8]. Chen S U, Hu J X, Ke H Z, et al. Enzyme Assisted Signal Amplification and Its Applications in Biomolecule Detection. Journal of Analytical Science, 2016.

[9]. Jin R, Wu G, Li Z, et al. What controls the melting properties of DNA-linked gold nanoparticle assemblies?. Journal of the American Chemical Society, 2003, 125(6): 1643-1654.

[10]. Liu J, Lu Y. Preparation of aptamer-linked gold nanoparticle purple aggregates for colorimetric sensing of analytes. NATURE PROTOCOLS-ELECTRONIC EDITION-, 2006, 1(1): 246.

[11]. Wang J, Liu G, Munge B, et al. DNA-Based Amplified Bioelectronic Detection and Coding of Proteins. Angewandte Chemie International Edition, 2004, 43(16): 2158-2161.

[12]. Hill H D, Mirkin C A. The bio-barcode assay for the detection of protein and nucleic acid targets using DTT-induced ligand exchange. NATURE PROTOCOLS-ELECTRONIC EDITION-, 2006, 1(1): 324.

[13]. Zhang B, Liu B, Tang D, et al. DNA-based hybridization chain reaction for amplified bioelectronic signal and ultrasensitive detection of proteins. Analytical chemistry, 2012, 84(12): 5392-5399.

IOP Conf. Series: Materials Science and Engineering **275** (2017) 012011 doi:10.1088/1757-899X/275/1/012011

High Concentration of Benzyladenine Solution Stimulates Anthers for Inducing Callus in *Ricinus Communis* L.

Ying Liu[1,a], Shuying Yan, Fuguang Yang, Dongliang Li, Jianian Tang, Guoxuan Liu, Shiwan Lin, Sufang Niu[2,b,*] and Yali Yang[3,c,*]

[1]Faculty of Agricultural Science, Guang Dong Ocean University, Haida Road #1, Mazhang District, 524088, Zhanjiang, Guangdong, P. R. China.
[2]Fisheries College, Guang Dong Ocean University, Zhanjiang 524088, P.R. China.
[3]Department of Basic Medicine, School of Medicine, Jiaying University, Meizhou, Guangdong, P.R. China
E-mail: [a]liuying85168@126.com; [c,*]wolf0487@126.com; [b,*]187041607@qq.com;

Abstract: An high-frequency protocol for induction of callus from anther explants of *Ricinus communis* was described. When anther explants of *R. communis* was cultured directly onto medium containing 6-benzylaminopurine (BA) induced formation of only poor quality callus that had a low induction frequency of anther callus (10.67%). However, treating the anther explants with high concentrations (7.5-120 mg/L) of BA solution for short time periods (5-80 min) helped to improve the induction frequency and enhance the quality of the callus formation significantly. The best callus induction (41.25%) was observed when anther explants were treated with 15 mg/L BA solution for 10 min before being inoculated onto hormone-free Murashige and Skoog (MS) medium for 30 days. In order to further optimize the culture system, after treated with 15 mg/L BA for 10 min, anther explants were inoculated on the hormone-free MS medium contained concentrations of sodium nitroprusside (SNP). The results showed that SNP significantly promoted the response of callus induction, especially when 8 mg/L SNP was applied, the the highest percentage of callus induction (60.37%) were gained.

1. Introduction

Since anthers of *Datura innoxia* have been used for tissue culture to obtain haploid plants successfully, study of anther culture has developed rapidly, and many countries, including China, have conducted research work in this area [1]. Up to now, about 300 haploid plants have obtained from all kinds of crops by in vitro culture of anthers [2].

Anther culture, combined with conventional hybrid breeding, distant crossbreeding, mutation breeding and transgenic technology, has become one of the more extensive and effective methods of biotechnology in crop breeding. So far, more than 40 kinds of intact plants have been gained via anther culture method in China [2].

Castor (*Ricinus communis* L.) is an annual or perennial herb of *Euphorbiaceae* and accounted as one of the top ten oil crops in the world [3]. Castor oil from seed is an important industrial oil with extensive applications [4]. However, the primary limitation for its large-scale cultivation as an oil crop is the low and inconstant seed yields due to nature heterozygous of *R. communis* plants. In order to breed advanced *R. communis* varieties for large scale cultivation and produce more seeds, haploid breeding of the species might be a possible solution. Haploid breeding has been considered as an advanced strategy for breeding new varieties with improved traits. The haploid plants can be obtained

Content from this work may be used under the terms of the Creative Commons Attribution 3.0 licence. Any further distribution of this work must maintain attribution to the author(s) and the title of the work, journal citation and DOI.
Published under licence by IOP Publishing Ltd

by using anther culture, and then the chromosomes can be doubled to obtain the purely diploid cultivars [2]. The characters of offspring of the excellent homozygous lines will not be separated. What's more, the breeding period can be significantly shorten [2]. Therefore, establishment of tissue culture system for efficient callus induction of castor anther explants will accelerate the breeding progress in *R. communis*.

In this study, we have developed an efficient callus induction procedure for *R. communis* by utilizing high concentration of BA solution to deal with the anther explants. Details of the experiments are described in this paper.

2. Materials and Methods

2.1. Preparation of Anther Explants

Male flowers (The blossoms had not opened) of castor were surface-sterilized for 60 s with 75% (v/v) ethanol, immersed in 2% (v/v) sodium hypochlorite (NaClO) for 20 min and finally rinsed 5 times in sterile distilled water. The perianths were removed from the male flowers and anther explants could be gained by carefully separated from stamens with sterile dissecting needle.

2.2. Preparation of 6-benzylaminopurine (BA) Solution and Treating the Anther Explants with BA Solution

6-benzylaminopurine (BA) (Sigma-Aldrich Co., St. Louis, MO, USA) was dissolved in 1 mol/L NaOH water solution, adjusted to final pH value of 5.8-6.0 and diluted to various concentrations of 0, 7.5, 15, 30, 60 and 120 mg/L. The prepared BA solution was sterilized by filtration using 0.22 μm water filters (Millipore, USA) prior to use.

Anther explants were soaked in glass bottles containing different concentrations (0, 7.5, 15, 30, 60 and 120 mg/L) of BA solution for various time periods (0, 5, 10, 20, 40 and 80 min). After treatment, the explants were briefly placed on sterile dry filter paper in petri dish to absorb excess moisture.

2.3. Direct Induction of Callus Using Anther Explants of R. communis L.

For inducing callus, anther explants were inoculated on hormone-free MS medium after treated with BA solution for various time periods. For comparison, anther explants were also treated using conventional methods and inoculated directly on MS medium containing different concentrations of BA (0, 0.5, 1, 2 and 4 mg/L) as reported previously [3,4]. The percentage of induction of callus was recorded after 30 days of culture.

2.4. Preparation of Culture Medium and culture Maintenance

Uniform culture conditions were applied in all experiments. Basal MS formula was used for all tissue culture experiments. Mediums used in our experiment contained 2.5% sucrose and 0.7% agar that were adjusted to pH 5.8-6.0 with 1 mol/L NaOH, and then being autoclaved at 1.4 kg cm-2 for 20 minutes. All culture treatments were kept at 25 ± 1°C under a 12 h photoperiod of 60-80 μmoL m-2s-1 intensity (cool white fluorescent tubes).

2.5. Evaluation of the Results and Data Analysis

All experiments were based on a completely randomized factorial design and repeated three times with 25-30 replicates per treatment. Statistical analysis of the data was carried out using SPSS 17.0 software, and data in the same column followed by different letters were significantly different at $p \leq 5\%$ level as determined by Duncan's multiple range test. The results were expressed as means ± SD (standard deviation) of three independent experiments.

3. Results

3.1. Induction of Callus from Anther Explants with Conventional Culture Methods

Anther explants without BA solution treatment were inoculated onto MS medium containing different concentrations of BA as reported previously [3,4]. The concentration of BA in the medium influenced the response of callus induction observably (Table 1). Of the different concentrations of BA (0, 0.5, 1, 2 and 4 mg/L) tested, when there was no BA added into callus-inducing medium, callus could not be induced from anther explants of R. communis (Table 1 and Fig. 1A); and the the highest percentage of callus induction (10.67%) were observed when 1 mg/L BA was applied (Table 1 and Fig. 1B). However, when BA was used at concentration higher than 1 mg/L, the induction rate of callus dropped sharply. What's more, most of the induced callus were very tiny and underdeveloped (Table 1 and Fig. 1B).

Table 1. Effects of BA concentration on callus induction of castor anthers*

Number	BA concentration (mg/L)	Induction rate of callus (%)
1	0	0c
2	0.5	8.45±0.39b
3	1	10.67±0.45a
4	2	7.34±0.42b
5	4	4.21±0.35

*Data in the same column followed by different letters are significantly different at $p \leq 5\%$ level as determined by Duncan's multiple range test.

3.2. Induction of Callus from Anther Explants by Treated with BA Solution before Culture

To study the effect of time duration of BA treatment on callus induction, anther explants treated with 15 mg/L BA solution for various time periods before inoculation of explants on the hormone-free MS medium. The results showed that time duration of the treatment significantly influenced the response of callus induction (Table 2). Treatment with 15 mg/L BA solution for 10 min was the most suitable and it achieved the highest regeneration percentage (41.25%) (Table 2 and Fig. 1C). The percentage of induction of callus was directly proportional to time duration of BA treatment when the time periods of BA treatment was not longer than 10 min. However, when the time duration of BA treatment was longer than 10 min, i.e. 20 min, the callus induction frequencies decreased markedly, and then the percentage of callus induction was 35.42% (Table 2). Therefore, treatment with solution for 15 min was the optimum time periods for the induction of callus, and the callus seemed very healthy and vigorous.

Table 2. Effects of treating anther explants with 15 mg/L BA solution for various time durations on callus induction*

Number	Treating duration (min)	Induction rate of callus (%)
1	0	0e
2	5	19.31±3.45d
3	10	41.25±2.32a
4	20	35.42±1.96b
5	40	28.63±2.68c
6	80	17.72±2.51d

* Data in the same column followed by different letters are significantly different at $p \leq 5\%$ level as determined by Duncan's multiple range test.

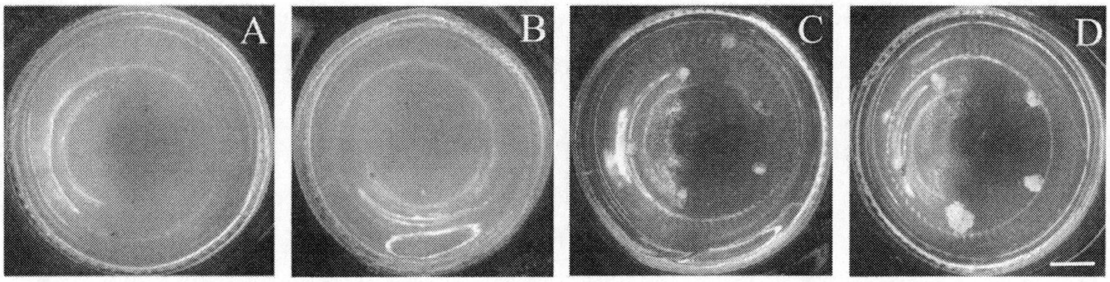

Figure 1. Direct induction of callus from anther explants of *R. communis.*
(A) Anther explants were inoculated on hormone-free MS medium for 30 days; (B) anther explants were inoculated on MS medium supplemented with 1 mg/L 6-BA for 30 days; after being treated with 15 mg/L TDZ solution for 10 min, anthers explants were inoculated on (C) hormone-free MS medium; (D) MS medium contained 8 mg/L SNP for 30 days (bar =1 cm).

Furthermore, anther explants were treated with various concentrations of BA solution for 10 min before being inoculated onto hormone-free MS medium. The concentrations of BA solution significantly influenced the response of callus induction (Table 3). The application of 15 mg/L BA resulted in the highest percentage of callus induction (41.25%) (Table 3 and Fig. 1C). The percentage of induction of callus was directly proportional to the concentration of BA solution when the concentration of TDZ was not higher than 15 mg/L (Table 3). However, when TDZ was used at concentrations higher than 15 mg/L, the percentage of callus induction was dramaticlly decreased (Table 3). Therefore, 15 mg/L BA solution was the optimum choice for the induction of callus, and the callus seemed healthy and lively (Table 3 and Fig. 1C). Finally, it's clear that treatment with BA solution was more effective than conventional methods for induction of callus from anther explants (Tablel and Table 3).

Table 3. Effects of concentrations of BA solution on induction callus of castor anthers*

Number	Concentrations of BA solution (mg/L)	Induction rate of callus (%)
1	0	0e
2	7.5	22.46±1.69c
3	15	41.25±2.32a
4	30	34.22±2.08b
5	60	23.54±3.17c
6	120	16.31±2.24d

* Data in the same column followed by different letters are significantly different at $p \leq 5\%$ level as determined by Duncan's multiple range test.

3.3. Influence of SNP in the MS Medium on the Culture Response of Anther Explants

After being treated with 15 mg/L BA for 10 min, anther explants were inoculated onto MS medium containing different concentrations of SNP for tested of their culture response. Date summarized in Table 4 indicated that SNP supplemented to the MS medium had positive effects, facilitating the induction rate of callus. When 8 mg/L SNP was applied, the the highest percentage of callus induction (60.37%) were gained (Table 4 and Fig. 1D). However, when SNP was used at concentration higher than 8 mg/L, the callus induction was inhibited and the induction efficiency was gradually decreased.

Table 4. Effects of concentrations of SNP on callus induction of castor anthers*

Number	Concentrations of SNP (mg/L)	Induction rate of callus (%)
1	0	41.25±2.32d
2	2	47.42±1.43c
3	4	54.26±2.64b
4	8	60.37±2.78a
5	16	50.61±3.49bc
6	32	37.19±2.57d

* Data in the same column followed by different letters are significantly different at $p \leq 5\%$ level as determined by Duncan's multiple range test.

4. Results

Haploid breeding helps to modify traits of the species and cultivate new varieties. Whilst tissue culture methods for inducing callus from anther explants are very important and a prerequisite for haploid breeding. Conventional culture methods for inducing callus from anther explants of R. communis included direct inoculation of the explants on a medium containing cytokinin such as BA at low concentrations (usually 0.5-4 mg/L) [3,4]. These methods showed low efficiency of inducing callus and the callus were hardly to further enlargement and growth. However, treatment of anther explants with BA solution at high concentrations (7.5-120 mg/L) in this study for a short time periods (5-80 min) before inoculation on hormone-free MS medium increased the frequency of callus induction and caused the formation of bigger callus as compared with the conventional methods.

BA pertains to the cytokinin family of hormones. cytokinin is an essential factor for the induction of callus formation in most cases of plant tissue culture [5,6]. The success of experimental results given in this paper implied that the induction of callus formation might not requires cytokinin during the whole culture period; when the process of cell division for the formation of callus was activated, the presence of cytokinin was no longer necessary. On the contrary, long presence of cytokinin in the medium might have only negative effects. What's more, it has been well documented in many academic journals and text books that cytokinin has the effects of inhibition growth and development of plant organs and tissues [7,8].

In our previous study, high concentrations (30 mg/L) of BA treatment for 20 min before inoculating the explants on hormone-free MS medium promoted the buds regeneration frequency from hypocotyl explants in soybean [7]. These results suggested that the new culture method might be applicable to other kinds of plant species and different types of explant.

SNP is usually used as NO donor in plant tissue culture studies [9]. NO might interact with auxins for the regulation of plant cell division, dedifferentiation and redifferentiation [10]. Furthermore, NO was involved in the auxin response during adventitious root formation and the development of buds and shoots might also be promoted [11]. Results in this study confirmed the effect of SNP on the promotion of callus induction, but meanwhile the results demonstrated that the concentration of SNP was vitally important. Such as 16-32 mg/L SNP causes negative effects in the present study.

In conclusion, an efficient callus formation protocol was established for *R. communis* using anther explants. These methods have the potential to facilitate the haploid breeding.

5. Summary

An efficient callus formation protocol was developed for R. communis by using anther explants. The best effect of callus induction was gained from anther explants treated with 15 mg/L BA solution for 10 min, followed by 30 days of culture on MS medium contained 8 mg/L SNP.

6. Acknowledgements

This work were supported by the Program for Scientific Research Start-up Funds of Guangdong Ocean University (R17023), the Project of Science and Technology of Zhangjiang City (2016B01004), the Project for Innovation and Strong School of Department of Education of Guangdong Province (2016KQNCX067), and the National Innovation and Entrepreneurship Training Program for College Students (201610566019).

7. References

[1] Hu D F. Advances in plant flower breeding. Beijing: China Agricultural Science and Technology Press, 1996, 22-42.

[2] Sa R N, Chen Y S, Huang F L, *et al.* Research progress of plant anther culture technology. Journal of Inner Mongolia University for Nationalities, 2008, 23(6): 650-653.

[3] Shao Z M, Chen Y S, Huang F L, *et al.* Effects of low temperature pretreatment and light conditions on callus induction of castor anthers. Journal of Inner Mongolia University for Nationalities, 2012, 27 (2): 189-193.

[4] Li G R, Huang F L, Wang W Y, *et al.* Optimization of callus induct conditions of *Ricinus communis* anthers. Journal of Inner Mongolia University for Nationalities. 2012, 27(6): 670-673.

[5] Khemkladngoen N, Cartagena J, Shibagaki N, *et al.* Adventitious shoot regeneration from juvenile cotyledons of a biodiesel producing plant, *Jatropha curcas* L. Journal of bioscience and bioengineering, 2011, 111(1): 67-70.

[6] Kumar N, Reddy M P. Thidiazuron (TDZ) induced plant regeneration from cotyledonary petiole explants of elite genotypes of *Jatropha curcas*: a candidate biodiesel plant. Industrial Crops and Products, 2012, 39: 62-68.

[7] Liu Y, Yu L, Zhang Q, *et al.* High concentration short duration treatment of benzyladenine stimulates adventitious bud regeneration from hypocotyl explants in soybean. Advanced Materials Research, 2013, 647: 331-337.

[8] Liu Y, Tong X, Hui W, *et al.* Efficient culture protocol for plant regeneration from petiole explants of physiologically mature trees of *Jatropha curcas* L. Biotechnology & Biotechnological Equipment, 2015, 29(3): 479-488.

[9] Petřivalský M, Vaníčková P, Ryzí M, *et al.* The effects of reactive nitrogen and oxygen species on the regeneration and growth of cucumber cells from isolated protoplasts. Plant Cell, Tissue and Organ Culture, 2012, 108(2): 237-249.

[10] Ötvös K, Pasternak T P, Miskolczi P, *et al.* Nitric oxide is required for, and promotes auxin-mediated activation of, cell division and embryogenic cell formation but does not influence cell cycle progression in alfalfa cell cultures. The Plant Journal, 2005, 43(6): 849-860.

[11] Pagnussat G C, Simontacchi M, Puntarulo S, et al. Nitric oxide is required for root organogenesis. Plant Physiology, 2002, 129(3): 954-956.

Formation of Polyelectrolyte Complex Colloid Particles between Chitosan and Pectin with Different Degree of Esterification

Hui Wang*, **Hongyuan Sun and Jieyu He**

Department of Food Science and Technology, Hainan Tropic Ocean University, Sanya, China.
Email: hwan072@163.com

Abstract: The effects of degree of esterification, pectin/chitosan ratio and pH on the formation of polyelectrolyte complex colloid particles between chitosan (CS) and pectin (PE) were investigated. Low methoxyl pectin (LPE) was achieved by de-esterifying high methoxyl pectin (HPE) with pectin methyl esterase. Turbidity titration and colorimetric method was used to determine the stability of complex colloid particles. The structure and morphology of complex particles were characterized by FTIR and TEM. When pectin solution was dropped into chitosan solution, complex colloidal dispersion was stable as PE/CS mass ratio was no more than 3:2. Colloidal particles of HPE-CS complex coagulated at larger ratio of PE/CS than LPE-CS. The maximum complex occurred at pH 6.1 for HPE-CS and pH 5.7 for LPE-CS, and decreasing pH leaded to the dissociation of complex particles. Electrostatic interactions between carboxyl groups on pectin and amino groups on chitosan were confirmed by FTIR. Colloidal particle sizes ranged from about 100 nm to 400 nm with spherical shape.

1. Introduction

Electrostatic interactions occurring between polyanions and polycations result in polyelectrolyte complex (PEC) or forming an insoluble solid. A series of special surface properties, excellent emulsifying properties or different scales of PEC can be obtained based on the different reaction degree of materials [1]. The encapsulation characters of PECs directly promote research and development of controlled release system. PECs have been applied in the field of agriculture, biotechnology, food flavoring, biomedicine, cosmetics and textiles [2]. PEC reaction degree can be controlled by different reaction conditions, such as pH, ionic strength, the ratio of polyanion and polycation, temperature, etc [3]. If core materials exist in the system, different sizes of encapsulating particles, such as microcapsules or nanocapsules are formed. Nanocapsules with small particle size are easy to disperse and suspend in water to form a stable colloidal solution with unique properties, and thus have more wide applications [4].

PECs of natural polysaccharides are excellent controlled release carrier due to rich material source, security and stability, mild reaction conditions, being able to maintain stability and activity of embedding materials, being degradable or soluble under normal physiological conditions, environmentally responsibility [5]. Pectin is composed of partly methylated galacturonic acid with good film-forming property. According to degree of esterification (DE), pectin can be divided into low methoxyl pectin (LPE, DE < 50%) and high methoxyl pectin (HPE, DE > 50%). Pectin extracted from fruit peel is mostly HPE, and LPE is mainly prepared from de-esterification from HPE. Enzymatic preparation of LMP is by removing methyl ester groups from galacturonic acid of HPE backbone

Content from this work may be used under the terms of the Creative Commons Attribution 3.0 licence. Any further distribution of this work must maintain attribution to the author(s) and the title of the work, journal citation and DOI.
Published under licence by IOP Publishing Ltd

through pectin methyl esterase (PME). Chitosan and pectin can form complex by electrostatic attraction, which is often applied in the controlled release of active materials and in the preparation of membranes and microspheres with pH sensitive properties. Charge density in macromolecules significantly affects the final complexes, therefore pectin molecular chain with different DE affect complexation with chitosan due to different number of negative ions [6].

This paper studied the preparation of complex colloidal particles by chitosan and pectin which was extracted from banana peel and de-esterified to form LPE by PME. The effects of DE of pectin, PE/CS ratio and pH value on the colloid particles were examined. The study is intended to provide foundation for encapsulation of active material by complex colloidal particles in the next step.

2. Materials and Methods

2.1. Materials

Banana peel was from Brazil banana in Sanya farm (Hainan province, China). Novoshapes pectin methyl esterase was obtained from Novozymes China. Chitosan (degree of acetylation of more than 95%) was obtained from Qiyun Biotechnology Co., Ltd (Guangdong Province, China) with molecular weight of 4×105 by viscosity method. Other reagents were analytical grade.

2.2. Experimentals

2.2.1. Preparation and Purification of Pectin. Pectin was extracted from banana peel by acid method after pretreatment of peel. The extraction solution was centrifuged and the supernatant was divided into two parts, one of which was recorded pectin solution 1. The other part was diluted with equal volume of water followed by adjusting pH to 4.5 and PME was added into the solution. The solution was mixed at 45 ℃ for 70 min followed by quickly inactivation of enzyme by boiling solution for 3 min, and cooled. The solution was recorded as pectin solution 2. Two pectin solutions were concentrated, decolored by macroporous resin, subjected to protein removal by Sevag method, then dialyzed with deionized water in dialysis bag with molecular weight cut-off of 12 kDa for 24 hours, precipitated with ethanol, dried to achieve pectin 1 and pectin 2 respectively.

2.2.2. Determination of DE, Intrinsic Viscosity [η] and Viscosity Average Molecular Weight Mη of pectin. DE of pectin samples was determined by titration method.

Measurement of M_η: 5 mg/mL pectin solution was prepared with 0.1 M phosphate buffer solution as solvent. 10 mL pectin solution was pipeted to Ostwald U-tube viscometer in water bath at 30°C. The flow times of pectin solution with different concentrations were recorded by a stopwatch. M_η was calculated by the following relationships: Relative viscosity (η_r) = t/ t_0, Specific viscosity (η_{sp}) = η_r-1, Reduced viscosity (η_{red}) = η_{sp}/C, Huggins equation: $\eta_{sp}/C = [\eta] + k'[\eta]C$, Intrinsic viscosity $[\eta] = \eta_{sp}/C$ (when C→0), M_η: Mark-Houwink-Sakurada relationship $[\eta] = kM_\eta^\alpha$ (where k = 2.34×10^{-5}, a = 0.8224).

2.2.3. Effect of PE/CS Ratio on the Formation of Stable Complex Colloidal Dispersion. The pectin solution (1 mg/ml) and 1 mg/mL chitosan solution of acetic acid (1% (w/v)) were prepared and pH was adjusted to 5.5. Pectin solution was added to chitosan solution while stirring at room temperature, and the changes of the system in the process of complex formation were observed. The conditions to form a stable colloidal dispersion were determined.

The complex dispersions with different PE/CS ratio were centrifuged and settled down. The supernatants were used to determine the content of chitosan by colorimetric method [19]. Chitosan in supernatant did not form colloidal particles with pectin.

2.2.4. The Effect of pH on the Formation of Stable Complex Colloidal Dispersion by Turbidity Titration. The pectin solution (1 mg/ml) and chitosan solution of acetic acid (1 mg/ml) were mixed, pH of the mixture was adjusted to 11-12. The absorbance of the solution was measured at 500 nm by

UV-VIS spectrophotometer with the change of pH. No absorbance of pectin and chitosan solutions was recorded at this wavelength.

2.2.5. Particle Size Measurement. Particle size was determined by using a Malvern Mastersizer 2000, equipped with Malvern application software version 4.00 (Mastersizer 2000, Malvern Instruments Ltd., UK). The instrument is based on the principle of laser light-scattering and capable of measuring size down to 0.01 μm.

2.2.6. FT-IR Measurement. Dried samples were ground with KBr in 1:20 ratio, pressed into pellets and measured by FTIR spectrometer (IRAffinity–1, Japan). The samples were scanned over a range of 400 cm-1 - 4000 cm-1 (30 scans).

2.2.7. Transmission Electron Microscopy. A small amount of colloid dispersion was placed in copper grid directly and dyed with phosphotungstic acid, blotted with filter paper. The image was acquired using transmission electron microscope (HitachiH - 600) at 200 kV.

3. Results and Discussion

3.1. DE, intrinsic viscosity [η] and Mη of Pectin

Table 1 shows DE, intrinsic viscosity [η] and M_η of two types of pectin. Pectin 1 was extracted from banana peel with acid method and DE was 64.5%, Pectin 2 was the de-esterification product of pectin 1 and DE was 42.2%. The results demonstrate that banana peel pectin was HPE and LPE was achieved by PME hydrolysis.

Table 1. The characteristics of two types of pectin.

Pectin	DE (%)	[η] (l/g)	M_η
Pectin 1 (HPE)	64.51±4.19	0.3431	116680
Pectin 2 (LPE)	42.27±2.97	0.2806	91159

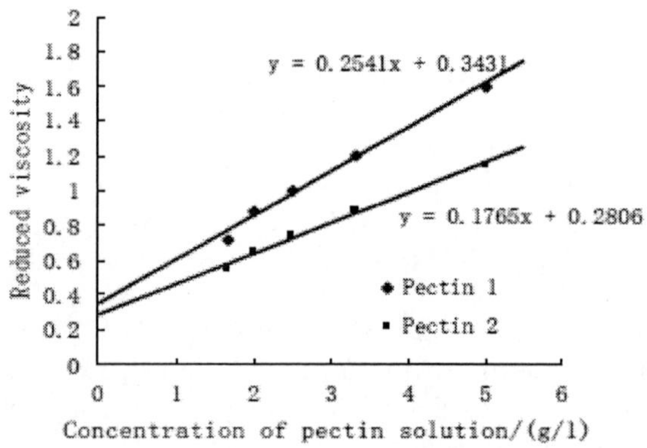

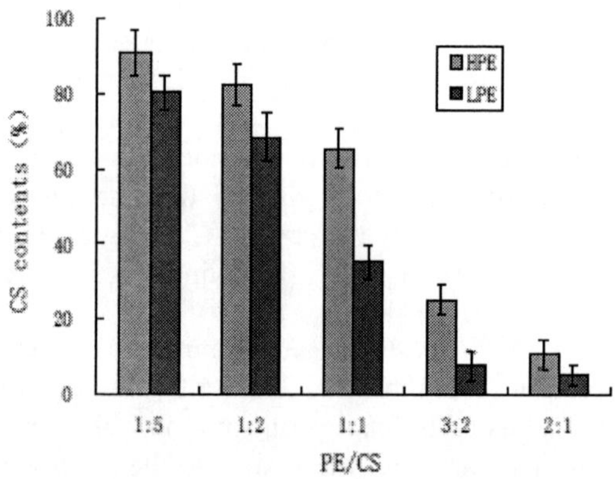

Figure 1. Linear relationship between reduced viscosity and concentration of pectin
Figure 2. The CS contents in supernatant after centrifugation of PE-CS complex solution.

Figure 1 is Huggins fitting line for pectin 1 and pectin 2. Intrinsic viscosity [η] can be obtained by plotting reduced viscosity against polymer concentration on extrapolation to infinite dilution. [η] is the eigenvalues of the polymer and indirectly reflects the shape and size of the polymer. When the testing conditions are fixed, [η] values are only determined by the molecular weight of the sample. M_η of the pectin is obtained by Mark-Houwink-Sakurada equation (see table 1). The results show that M_η of banana peel HPE is 1.17×10^5 and LPE is 9.0×10^4, less than HPE. This may be due to elimination reaction during enzymatic hydrolysis.

3.2. The effect of PE/CS Ratio on the Stability of Complex Colloid Dispersion

Stable colloidal dispersion of pectin and chitosan complex can be formed when their concentrations are less than 5 mg/mL, otherwise complex colloid is easy to coagulate. Table 2 presents the change of complex system when pectin was dropped into chitosan solution. It can be seen that stable colloidal dispersion was formed in certain mass ratio range of PE/CS, beyond this range, colloidal dispersion coagulated. This is because the negatively charged carboxylic groups on pectin molecular chains interacted with positively charged amino groups of chitosan, hydrophobic polyelectrolyte complexes were generated. When less pectin was added, too much unreacted amino groups surrounded outside the core of complex to form hydrophilic shell, since like charges repel each other, coagulation of complex particles was hampered and stable colloidal dispersion was formed. When pectin content was gradually increasing, unreacted amino groups decreased and positive charges outside colloid particles decreased, thus phase separation occurred at certain PE/CS ratio. Colloid particles coagulated at 2:1 of HPE/CS, while coagulated at 3:2 of LPE/CS. The results show that different charge density in pectin molecular chain affects the formation of polyelectrolyte complexes, because the LPE molecules have more carboxylic groups than HPE, less LPE were needed to completely interact with chitosan.

Table 2. The effect of PE/CS ratio on the stability of complex system.

PE/CS	1:10	1:5	1:2	1:1	3:2	2:1
HPE	√	√	√	√	√	×
LPE	√	√	√	√	×	×

√ indicates stable complex colloid dispersion;
× indicates coagulation of complex colloid

The CS contents in supernatant were determined after centrifugation of PE-CS complex solution to settle down colloid particles. The results are shown in Figure 2. The CS contents in the supernatant decreased gradually with the increase of PE, which means that reacted chitosan increased. Complex degree of LPE was larger than that of HPE at the same PE/CS ratio. Almost all of CS was present in the solution at 1:5 of PE/CS since very few complex particles formed. The unreacted CS in the supernatant was much less at 3:2 of LPE/CS, reacted CS percentage reached 92.51%, and reacted CS percentage at 2:1 of HPE/CS was 89.23%.

Figure 3 show the relationship between the specific viscosity of supernatant solutions of PE-CS complex and the weight percentage of pectin in pH 5.5. The specific viscosity of the supernatant solutions continuously decreased to a minimum value with increase in the weight percentage of pectin in the complex and then increased as the weight percentage of pectin increased. The weight percentage of pectin at minimum value of specific viscosity for HPE-CS was larger than LPE-CS, which means there were more HPE complexed with CS than LPH. The results coincided with Figure 2.

3.3. The Effects of pH on the Stability of Complex Colloid Dispersion.

Figure 4 is the absorbance values of solution varying with the change of pH at 3:2 of PE/CS. It shows that the absorbance of system had almost no change in alkaline solution, the absorbance increased sharply when pH decreased to a certain value. The maximum absorbance for HPE was at pH 6.1, and for LPE was at pH 5.7. This is because protonation of CS amino groups increased when pH deceased, and the electrostatic interaction between amino groups and carboxylic acid became more and more strong, finally achieved full complexation. With pH further decreased, dissociation of carboxylic acid decreased, charge density decreased, which leaded to partial dissociation of complex particles and absorbance of solution decreased. The pH value at the maximum absorbance for HPE was larger than LPE.

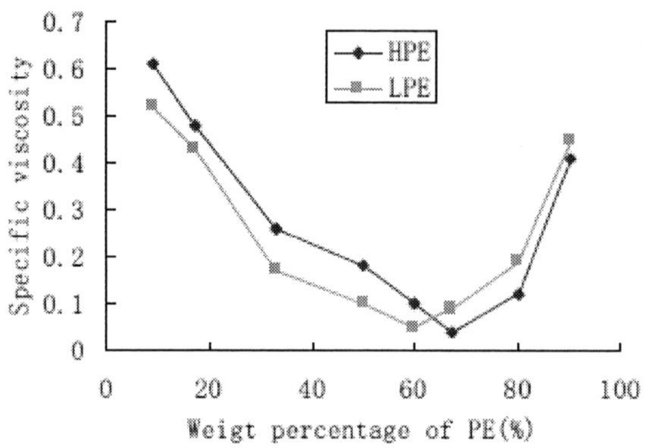

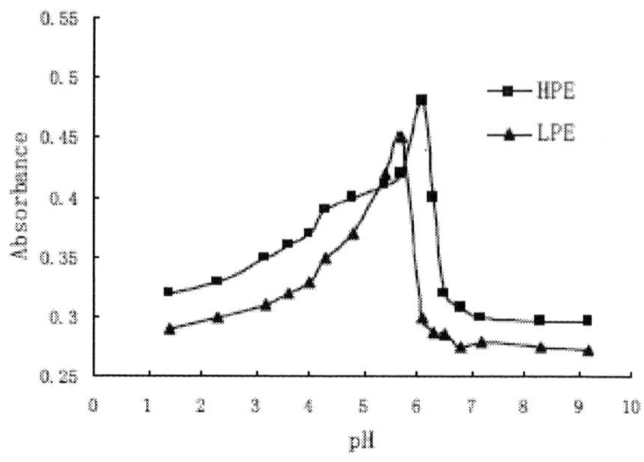

Figure 3. Specific viscosity of supernatant solutions of PE-CS complex in pH 5.5.
Figure 4. The effect of pH on absorbance of complex solution.

3.4. The Effects of NaCl Concentration on Complex Colloid Particle Size.

There are two cases for polyelectrolyte complexes to the response of the ionic strength, coagulation and dissolved, depended on the structure of the polyelectrolyte. 10% NaCl solution was added to the complex colloid dispersion (The weight ratio of PE/CS was 3:2) and the particle size of the colloid were measures after half hour stirring. The results are shown in Table 3. It can be seen that the particle size of LPE-CS was larger than HPE-CS and partile size of both complex decreased with the increase of NaCl concentration. The HPE-CS particles dissolved when NaCl concentration was 60 mg/mL and the system became clear solution and LPE-CS particles dissolved at 70 mg/mL of NaCl concentration. Ionic strength has more effect on HPE-CS colloid dispersion than on LPE-CS.

Table 3. The effect of NaCl concentration on PE-CS complex particle size with NaCl 10%.

NaCl (mg/ml)		0	10	20	30	40	50	60	70
Particle size (nm)	LPE-CS	432	440	426	395	347	284	206	Particles dissolved
	HPE-CS	323	351	324	286	237	198	Particles dissolved	Particles dissolved

3.5. FT-IR Analysis of PE/CS Complex.

A general review of the FTIR spectra of HPE, LPE, CS, HPE-CS and LPE-CS are shown in Figure 5. The obtained spectral data were analyzed by comparing the following characteristic regions: O-H stretching (3100 ~3600 cm^{-1}), C-H stretching (2800 ~3000 cm^{-1}), carboxylic group stretching region (1200 ~1800 cm^{-1}) and C-O stretching (1000 ~1300 cm^{-1}), which reflects the monosaccharide composition of the pectin. Two peaks at 1745 and 1620 cm^{-1} in the carboxylic groups region assigned to the C=O stretching of methylated carboxyl groups, and to anti-symmetric stretching modes of COO^-, respectively. The spectrum of HPE and LPE reveal that a decrease of 1745cm^{-1} band and increase of 1620 cm-1 band were related to the decrease of pectin DE, thus the area ratio of these two absorption peaks can indicate the DE. The infrared analysis of the PE-CS complex samples revealed some changes occurred. N-H stretching of CS was in 1647 cm^{-1}, and it shifted to lower wavenumber to 1640 cm^{-1} in PE-CS. a decrease of 1745cm^{-1} band and disappear of 1620 cm^{-1} band demonstrate that electrostatic interactions occurred in PE-CS complex [20].

3.6. Morphology of Complex Colloid Particles.
The TEM images of LPE-CS particles in stable colloid are depicted Figure 6. It is clear that the colloid particles were monodispersed and the majority of particles were basically spherical. The particles had a wide size distribution with a diameter range of 100 nm ~ 400 nm. There was no agglomeration of particles, which means that the system was stable. There is no much difference between the morphology of LPE-CS and HPE-CS colloid particles.

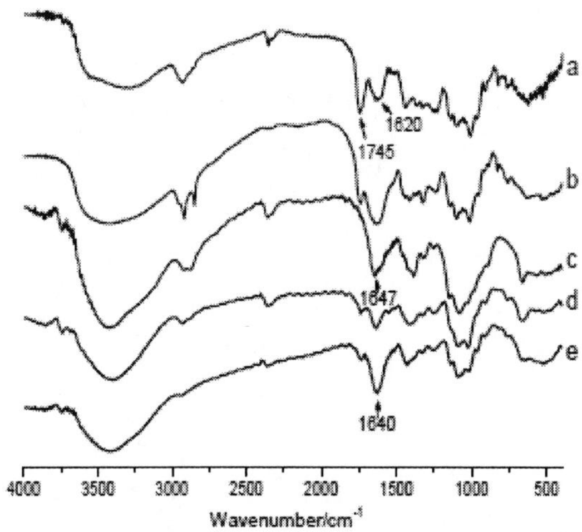

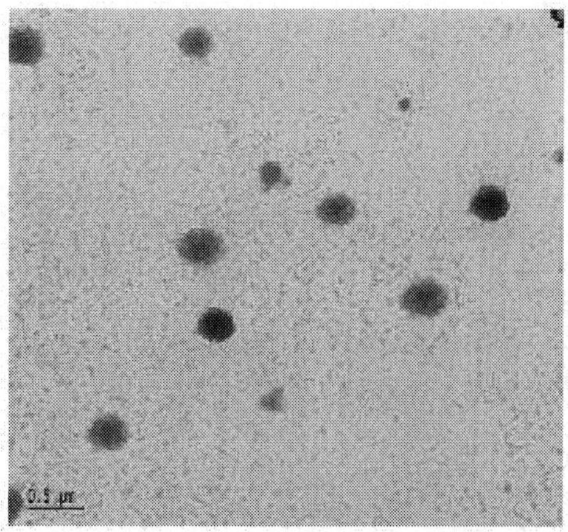

Figure 5. FTIR spectrum of a) HPE, b) LPE,c) CS, d) HPE-CS, e) LPE-CS
Figure 6. TEM image of LPE-CS complex colloid (LPE/CS ratio 3:2, pH5.5).

4. Conclusion
DE of pectin, solution concentration, ratio of PE/CS and pH value have effect on complex reaction of negatively charged pectin and positively charged chitosan. The formation of stable complex colloidal dispersion is due to the charges on the particle surface. Particle surface charges were neutralized when PE and CS thoroughly complexed, and phase separation occurred. DE of pectin reflects the number of carboxylic acid on the molecule. LPE has more carboxylic acid groups than HPE, and has more negative charges in solution after dissociation. HPE reacted with more chitosan than LPE when they are same amount and HPE reached maximum complexation with chitosan in higher pH than LPE. Complexation between pectin and chitosan was mainly via electrostatic interaction, colloid particles were basically spherical and particle sizes were about 100 nm to 400 nm.

5. Acknowledgements
We acknowledge the financial support by the National Natural Science Foundation of China (Grant No. 51363018); Hainan Natural Science Foundation (Grant No. 317153).

6. References
[1] Gucht, J., Spruijt, E., Lemmers, M., et al. Polyelectrolyte complexes: Bulk phases and colloidal systems. *Journal of Colloid and Interface Science* 2011, 361(2): 407-422.
[2] Hedayati, R., Jahanshahi, M., Attar, H. Fabrication and characterization of albumin-acacia nanoparticles based on complex coacervation as potent nanocarrier. *Journal of Chemical Technology and Biotechnology* 2012, 87(10): 1401-1408.
[3] Chollakup, R., Beck, J.B., Dirnberger K. Polyelectrolyte Molecular Weight and Salt Effects on the Phase Behavior and Coacervation of Aqueous Solutions of Poly(acrylic acid) Sodium Salt and Poly(allylamine) Hydrochloride. *Macromolecules* 2013, 46(6): 2376-2390.
[4] Jayakumar, R., Menon, D., Manzoor K, C., et al. Biomedical applications of chitin and chitosan

based nanomaterials—A short review. *Carbohydrate Polymers* 2010, 82(2): 227-232.

[5] da Silva, M.A., Bierhalz, A.C.K. & Kieckbusch, T.G. Modelling natamycin release from alginate/chitosan active films. *International Journal of Food Science & Technology* 2012, 47(4): 740-746.

[6] Davidenko, N., Blanco, M.D., Peniche, C., et al. Effects of different parameters on the characteristics of chitosan-poly(acrylic acid) nanoparticles obtained by the method of coacervation. *Journal of Applied Polymer Science* 2008, 111(5): 2362-2371.

IOP Conf. Series: Materials Science and Engineering **275** (2017) 012013 doi:10.1088/1757-899X/275/1/012013

Anti-bacteria Effect of Active Ingredients of *Cacumen Platycladi* on the Spoilage Bacteria of Sauced Pork Head Meat

Xiao Li[1,2,3,4], Lingyi Xu[1,2,3,4], Yuqian Cui[1,2,3,4], Meixia Pang[1,2,3,4], Fang Wang[1,2,3,4] and Jinghua Qi[1,2,3,4,*]

[1]College of Food Science and Engineering, Beijing Key Laboratory of Detection and Control of Spoilage Organisms and Pesticide Residues in Agricultural Products, Beijing 102206, China

[2]Food Chemistry teaching team, Beijing University of Agriculture, Beijing 102206, China

[3]Pig innovation team, Modern agricultural industry technology system Beijing innovation team, Beijing 100102,China

[4]High Level Talents of Beijing Universities Cross Training "Real Training Plan" Project, Beijing 102206, China

*Email:abc960718@sina.com

Abstract: Extraction and anti-bacteria effect of active ingredients of *Cacumen Platycladi* were studied in this paper. Extraction combined with ultrasonic was adopted. The optimum extraction condition was determined by single factor test; the anti-bacteria effect of active ingredients and minimum inhibitory concentration(MIC) were valued by Oxford-cup method. The results indicated that kaempferol was the active ingredients of *Cacumen Platycladi* whose optimum extraction condition for ethanol concentrations were sixty-five percent and twenty minutes with ultrasonic assisted extraction.; the active ingredients of *Cacumen Platycladi* had anti-bacteria effect on *Staphylococcus, Proteus, Bacillus, Serratia* and MIC was 0.5 g/mL,0.5 g/mL,0.0313 g/mL and 0.0625 g/mL. The active constituent of *Cacumen Platycladi* is kaempferol which has obvious anti-bacteria effect and can be used to prolong the shelf-life of Low-temperature meat products.

1. Introduction

Low-temperature meat products are processed with the sterilization condition of 68~72 °C which can retain mostly nutritive value and flavour with short shelf life [1,2]. Sauced pork head meat is a kind of Low-temperature meat products with special flavour and short shelf life about 10 days which is popular with public. Previous researches of Sauced pork head meat have demonstrated that *Staphylococcus, Proteus, Bacillus and Serratia* are the main microorganisms leading to rot [3-5].Therefore the preservatives used in Low-temperature meat products have got more attention on prolonging shelf life in recent years [6].

The food preservatives are classified as natural and chemical, with the improvement of people's life level, Consumers prefer the natural, safe, efficient preservatives to chemical preservatives which is harmful [7]. At present, species [8,9] and Chinese herbal medicine [10,11] were mostly used as natural bacteriostatic agent. In this paper, the anti-bacteria effect of active ingredients of *Cacumen Platycladi* was investigated which is a medicinal plant in China. The previous researches showed that volatile oils, flavonoids, tannin, organic acid and inorganic elements were the main chemical substances of

Content from this work may be used under the terms of the Creative Commons Attribution 3.0 licence. Any further distribution of this work must maintain attribution to the author(s) and the title of the work, journal citation and DOI.
Published under licence by IOP Publishing Ltd

Cacumen Platycladi [12,13]. And volatile oils and flavonoids have different antibacterial effects [14,15]. Kaempferol is a kind of flavonoid which is the main chemical substance of *Cacumen Platycladi* [16-19].In this paper, anti-bacteria effect of kaempferol was studied aiming to extend the shelf-life of Sauced pork head meat.

2. Material and Methods

2.1. Sample

Cacumen Platycladi purchased from Mengzheng parity pharmacy in Beijing, China was smashed by Pulverizer (Tianjin Tasite Instruments Co., Ltd, China) and 40 meshes.

2.2. Test Strains

All the test strains including *Staphylococcus* (X12), *Proteus* (T4), *Bacillus* (X10) and *Serratia* (N11) were separated from putrid Sauced pork head meat.

2.3. Extraction of Active Substances

Five grams of sample were soaked in ethanol for 15 hours and extracted with ultrasonic (Kunshan Ultrasonic Instruments Co., Ltd, China) of 480 W and 50 °C.Finally the supernatant was taken to evaporate until paste after centrifuged with 3000 rpm for 5 minutes and then preserved at 4°C until tested.

2.3.1. Standard Curve of Kaempferol. Standard solution of kaempferol(97%;Aladdin Industrial Corporation, Shanghai, China) was prepared for 0.0500 mg/mL. Ultraviolet absorption wavelength was valued by spectrum scanning standard solution. Standard curve was drawn by detecting absorbance values of standard solution of different concentration [20].

2.3.2. Concentration of Ethanol. Five grams of sample were extracted with 125 mL ethanol [21] of 60%, 65%, 70%, 75% and 80% respectively with ultrasonic for 30 minutes. Finally, the optimum concentration of ethanol was determined by the yield of kaempferol .

2.3.3. Time of Ultrasonic Extraction. Five grams of sample were extracted with 125 mL ethanol of optimum concentration [21], extraction time with ultrasonic were 20minutes, 30minutes, 40minutes, 50minutes and 60minutes respectively. Finally, the optimum extraction time was determined by the yield of kaempferol.

2.3.4. Determination of the Yield of Kaempferol. The paste of 0.00125 g/mL was suspended with 70% ethanol whose yield of kaempferol was determined by standard curve.

2.4. Bacteriostatic Experiments

All the strains preserved at -80°C were thawed at room temperature and mixed with sterile nutrient broth of 50 mL which was prepared by dissolving of 19 g/L in water ,next the sterile nutrient broth(Beijing land bridge technology CO., LTD, China)inoculated with strains was incubated for 12 h at 37 °C which was repeated two times.

The sterile nutrient agar (Beijing land bridge technology CO., LTD, China)which was prepared by dissolving of 33 g/L in water was inoculated with bacterial suspension of 1×10^9 CFU/mL diluted by plate counts method [22] (1 mL bacterial suspension in 100 mL agar medium) and poured into sterile petri dishes. Two sterile Oxford cups were placed on the surface of agar medium and marked with C and CK (C: antibacterial solution; CK: blank control), then 100 μL sterile water and antibacterial solution which was prepared by dissolving the paste of 1 g/mL in sterile water were added into Oxford cups respectively, Finally all the plates were incubated for 16~18h at 37°C.Thereafter the anti-bacteria effect of active substances was determined by measuring the diameter of inhibition zone in the plates.

2.5. *Determination of Minimum Inhibitory Concentration*

Minimum inhibitory concentration (MIC) was determined by Oxford-cup method [23]. The sterile nutrient agar prepared by dissolving of 33 g/L in water was inoculated with bacterial suspension (1 mL bacterial suspension in 100 mL agar medium) and poured into sterile petri dishes. Eight sterile Oxford cups marked numbers were placed on the surface of agar medium symmetrically and steadily, next 100 μL antibacterial solution of different concentrations were added into Oxford cups respectively and sterile water was blank control. All plates were incubated for 16~18h at 37°C. The lowest concentration that had no inhibition zone compared with blank control was regarded as MIC.

2.6. *Statistic Analysis*

Data analysis and charts were made by Microsoft Office Excel 2007, significant difference analysis was valued by *IBM SPSS Statistic* 21.

3. Results and Discussions

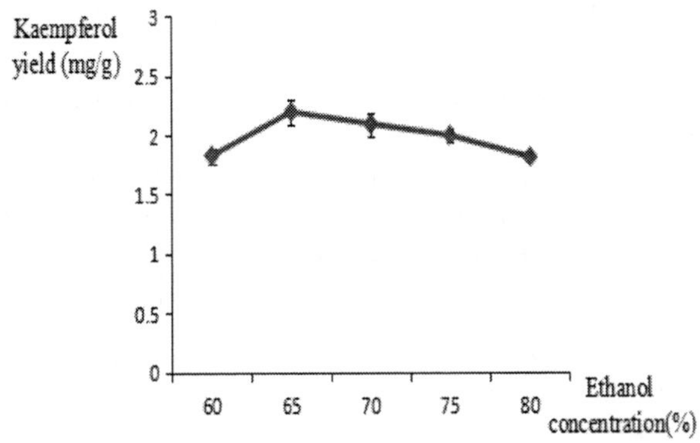

Figure 1. Effect of concentration of ethanol on Kaempferol yield.

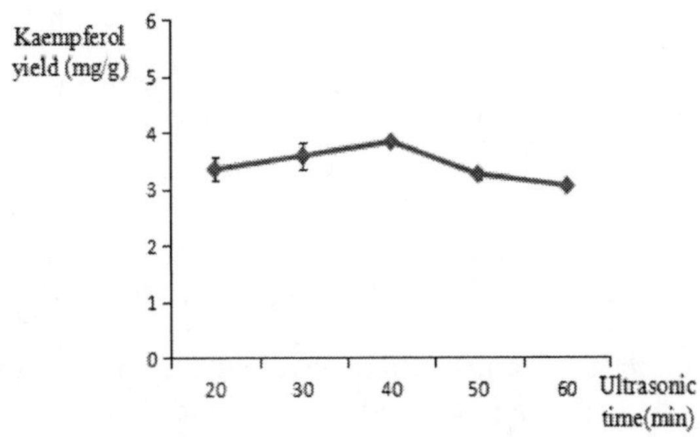

Figure 2. Effect of ultrasonic extraction time on Kaempferol yield.

Figure.1 demonstrated that the concentration of ethanol had an obvious impact on the yield of kaempferol ($P < 0.05$), with the increase of concentration, the kaempferol yield increased first and then decreased. The highest yield of Kaempferol was 2.19 mg/g with 65% ethanol.

Kaempferol is a kind of flavonols and slightly dissolved in water, easily dissolved in hot ethanol, exsisted in *Cacumen Platycladi* with free form mainly and kaempferol-7-O-glucosides[24].As is showed in figure 1, the reason that kaempferol yield decreased when the concentration of ethanol over 65% was that higher ethanol concentration restrained the hydrolysis of kaempferol-7-O-glucosides.Finally,the concentration of ethanol used in extraction was determined with 65%.

Figure 2 indicated that ultrasonic time had no obvious impact on the yield of kaempferol ($P > 0.05$), so ultrasonic extraction time was 20 minutes and the yield of Kaempferol was 2.31mg/g. Compared to

IOP Conf. Series: Materials Science and Engineering **275** (2017) 012013 doi:10.1088/1757-899X/275/1/012013

ethanol extraction, ultrasonic can break cells and obtained higher yield .As is shown in Figure 1 and Figure 2,the extraction condition of active substances of *Cacumen Platycladi* was determined as follows: solid-liquid ratio of 1:25,65% ethanol, temperature of 50°C,ultrasonic time of 20 minutes, power of 480W.

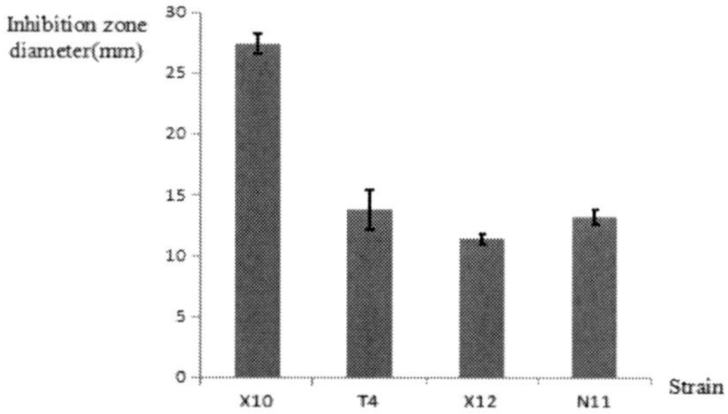

Figure 3. The anti-bacteria effect of active ingredients of Cacumen Platycladi.
X10=*Bacillus*, T4=*Proteus*, X12=*Staphylococcus*, N11=*Serratia*

As is showed in figure 3, the active ingredients of *Cacumen Platycladi* had different anti-bacteria effects on all the strains, especially the highest anti-bacteria effect on X10 with inhibition zone of 27.4 mm, While the anti-bacteria effect on X12 was weakest whose diameter of inhibition zone was only 11.37 mm.

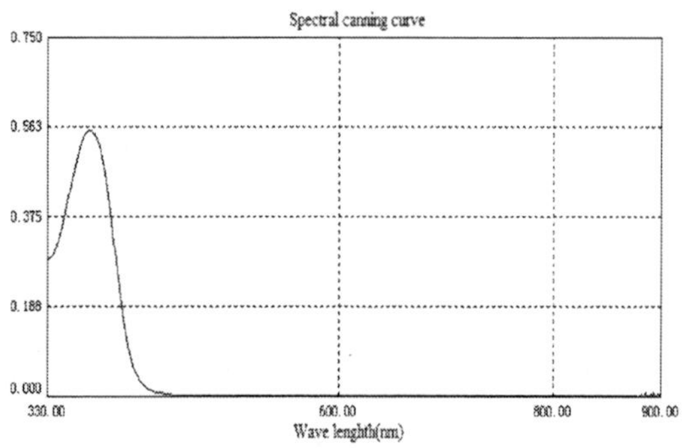

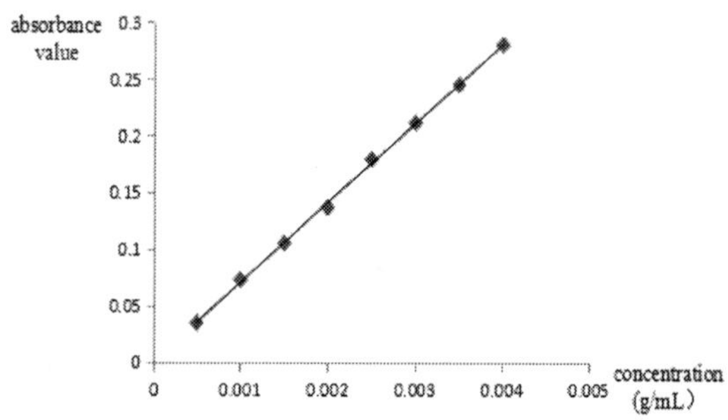

Figure 4. spectral scanning curve of Kaempfero; **Figure 5.** The standard curve of Kaempferol

As was shown in figure 4, ultraviolet absorption wavelength of Kaempferol was 369 nm.
Figure 5 demonstrated linear regression equation: $y = 69.92872\,x + 0.0007$ $R^2=0.9996$

Study demonstrated that kaempferol was the main active substance in the extractive [25,26]whose molecular structure included much hydroxy, which can enter the cells of bacteria and fungus to destroy the plasma membrane and restrain the synthesis of nucleic acid [13].thereby the active ingredients of *Cacumen Platycladi* can prolong the shelf-life of Sauced pork head meat through inhibiting microorganism.

Table 1. Determination of minimum inhibitory concentration of active substances of *Cacumen Platycladi.*

Strains	The concentration of antibacterial solution (g/mL)						
	1.0	0.5	0.25	0.125	0.0625	0.0313	0.0157
X10	+++	++	++	++	++	++	-
T4	+	+	-	-	-	-	-
X12	+	+	-	-	-	-	-
N11	+	+	+	+	+-	-	-

"d" means diameter of inhibition zone, d≥20mm ="+++"; 15mm≤d≤20mm ="++"; 10mm≤d≤15mm ="+"; d≤10mm ="+-";"-"means no inhibition zone.

Table 1 demonstrated that MIC of active ingredients of *Cacumen Platycladi* was 0.0313 g/mL, 0.5 g/mL, 0.5 g/mL and 0.0625 g/mL respectively about X10, T4, X12 and N11.active ingredients showed higher anti-bacteria effect on X10 than others. While with the concentration of effective component decreased, the anti-bacteria effect on T4 and X12 changed significantly.

4. Conclusion

The results showed that the extraction condition of *Cacumen Platycladi* was determined as follows: solid-liquid ratio 1:25, 65% ethanol, ultrasonic temperature of 50 °C, ultrasonic time of 20 minutes and power of 480 W. The active ingredients of *Cacumen Platycladi* showed anti-bacteria effect on *Proteus, Staphylococcus, Serratia, Bacillus* and their MIC was 0.0313 g/mL, 0.5 g/mL, 0.0625 g/mL and 0.5 g/mL respectively.

The study results provide reliable data for the use of natural plant bacteriostatic agent on low-temperature meat products. But effect components and bacteriostatic mechanisms need further research.

5. References

[1] Yuan X Q. Research on quality changes of stewed meat in seasoning, 2012.
[2] SB/T 10481-2008. Requirements of quality and safety on Pasteurized meat products, China.
[3] Li Y L. Isolation and identification of spoilage bacteria in low-temperature meat products, 2009. J. Inner Mongolia Agricultural Science and Technology, (3):62-63.
[4] Xiao X. Bacterial diversity and the inhibitory mechanism of antibacterial plant extract against specific spoilage organism in Yao meat, 2013.
[5] Lei M, Jiang R R, Ju R H and Xiao Y. Microbiology detection of pasteurized meat products and
[6] Zhang D Q and He Z. Research progress of the preservation technology of processed meat products, 2008.J.Meat Research, (5):3-7.
[7] Fan J S. Functional food additive, 1993. J. Tianjin Science and Technology Press, 318.
[8] Tian F, et al. Study on antibacterial activity of 7 spices on Pseudomonas, Staphylococcus and Brochothrix, 2017. Journal of Beijing University of Agriculture, 32(2):10-14.
[9] Liu L, Kong B H, et al. The inhibition effect of spices extraction on Listeria monocytogenes in culture medium and in chilled pork,2008.Science and Technology of Food Industry, 29(9):87-90.
[10] Liu F, Cao X Z, et al. Study on bacteriostatic activity of total flavonol from angelica sinensis,2015.J.China Food Additives,(1):60-63.
[11] Zhou J X. Reviews on research progresses, actual problem and prospects on natural food preservatives from plant materials, 2006. J. Food Science, 27(1):263-268.
[12] Chen X F, et al.. The research progress on the chemical composition, physical activity and hair loss prevention function of Cacumen Biotae, 2010. J. Chinese Wild Plant Resources, 29(3):1-5.

[13] Asif Ahmad, Muhammad Kaleem, et al. Therapeutic potential of flavonoids and their mechanism of action against microbial and viral infections-A review,2015.J.Food Research International,77:221-235.
[14] Liu W J, Ao C J, et al.. Antibacterial activity of plant extracts and its mechanism of action, 2016. Chinese Journal of Animal Nutrition, 28(8):2344-52.
[15] Wang Y X. Optimization of extraction process and evaluation of anti-bacteria effect of total flavonoids in Syringa leaves, 2013.
[16] Shan M Q, Qian W, et al.. Analysis of flavonoids in Platycladi Cacumen by UPLC-MS, 2011. J. China Journal of Chinese Materia Medica, 36(12):1626-29.
[17] Zhang M J. Separation and analysis of flavonoids in chinese traditional and herbal drugs, 2002.
[18] Zhao Y G, Chang L X, et al.. Research on flavonoids in Platycladi Cacumen by ultrasonic extraction, 2005. Journal of Hebei Agricultural University, 28(2):54.
[19] Wang L P. Studies on extraction technology and aglucosidase inhibitory activity of the Platycladi Cacumen flavonoid, 2012.
[20] Chen Z G, et al.. Determination of quercetin content in Raw drug by ultraviolet spectrophotometry, 2016 .J. Progress in Veterinary Medicine, 37(1):122-124.
[21] Cui B. Study on the extraction of purification technology of flavone of Mangosteen, 2012.
[22] GB 4789.2-2010, National food safety standard food microbiological examination: plate counts, China.
[23] Li Y, Liu J, et al.. Antioxidation and antibacterial property of flavonoids compounds extracted from Nitraria leaves, 2015. J. Food Industry, 36(3):94-98.
[24] Ma M, Wang L, et al. Optimization of extraction technology of kaempferol from Sophora japonica by orthogonal test, 2014. J. China Pharmacy, 25 (43):4069-71.
[25] Zhou Y, Guo B S, Zhneg Y, et al. Identification of the antibacterial activity from Lohan Kuo fruit, 2008. J. Lishizhen Medicine and Material Medical Research, 19(6):1544-46.
[26] Zhang G Q, Liu Y H, et al.. Determination of the total content of flavonoids in the Platycladi Cacumen, 2001. Journal of northwest of pharmacy, 16(3):107-108.
[27] Liu W J, Ao C J, et al.. Antibacterial activity of plant extracts and its mechanism of action, 2016. Chinese Journal of Animal Nutrition, 28(8):2344-52.

IOP Conf. Series: Materials Science and Engineering **275** (2017) 012014 doi:10.1088/1757-899X/275/1/012014

Antibacterial Activity of Anthraquinone from Aloe on Spiced Pig Head

Lingyi Xu[1,2,3,4], Xiao Li[1,2,3,4], Yuqian Cui[1,2,3,4], Meixia Pang[1,2,3,4], Fang Wang[1,2,3,4] and Jinghua Qi[1,2,3,4*]

[1]College of Food Science and Engineering, Beijing Key Laboratory of Detection and Control of Spoilage Organisms and Pesticide Residues in Agricultural Products, Beijing 102206, China.
[2]Food Chemistry teaching team, Beijing University of Agriculture, Beijing 102206, China.
[3]Pig innovation team,Modern agricultural industry technology system Beijing innovation team,Beijing 100102,China .4High Level Projects of Beijing Universities Cross Training, Beijing 102206,China
*Email:abc960718@sina.com

Abstract: [Objective] To optimize the extraction of anthraquinone from Aloe by ultrasonic extraction and its antibacterialactivity. [Method]The influences of different extraction time and ethanol concentration, on anthraquinone contentwere evaluated by asingle factor experiment. And anthraquinone content was determined by ultraviolet spectrophotometry. The bacteriostasis of anthraquinone on spiced pig head's common putrefying bacteria: Staphylococcus, Serratieae, Bacillus, Proteus and the minimal inhibitory concentration (MIC) were studied by oxford plate assay system. [Result]The best extraction time was 30 minutes and the best ethanol concentration was 80%. The antibacterial activity of the Aloe anthraquinone on Staphylococcus Aureus, Bacillus Proteus is obviously, the minimum inhibitory concentrations were 0.0625 g/mL, 0.05 g/mL, 0.125 g/mL respectively and no inhibitory effect on Serratieae. [Conclusions] The anthraquinones from Aloe can inhibit a part Of spoilage bacteria inspiced pig heads.

1. Introduction

Low-temperature meat productsare popular with the public, because of its rich nutrition, strong flavor and rich sense of mastication. However, in order to maintain its flavor, the processing of sterilization conditionneeds to be controlled at 68-70 degrees for 30 minutes.This leads to some spores and some heat-resistant bacteria are not all killed, and their higher water activity was also beneficial for putrefying bacteria to grow and reproduce, which causes that the low-temperature meat products are more perishable than other meat products and affects their shelf life [1-2].

To prolong the shelf life of low-temperature meat products, people generally choose to add some synthetic antibacterial agents, but with the improvement of living standards, people's health awareness gradually increased, more and more pollution-free, pollution-free natural products are demanded [3-9].

Aloe is a liliaceous plant, cold in property, bitter in flavor, rich in nutrients, available for daily consumption, and the main chemical are carbohydrate, anthraquinone compounds, enzyme (polypeptide), amino acids, vitamins and steroid compounds and other compounds [9-10]. The

Content from this work may be used under the terms of the Creative Commons Attribution 3.0 licence. Any further distribution of this work must maintain attribution to the author(s) and the title of the work, journal citation and DOI.
Published under licence by IOP Publishing Ltd

anthraquinone from aloe has high medicinal value and can be bactericidal, antibacterial, antiinflammatory, detoxification, and promote wound healing, enhance immunity, beauty and so on.

In the food industry, the traditional method of anthraquinone from Aloe is used ethanol or alkaline aqueous solution to extract, but it has many disadvantages, such as many processes, large losses, long cycle, low yields ofextraction. Ultrasonic extraction has many advantages, such as high extraction rate, easy operation, fast extraction speed and less loss of active components. It has been widely used in the extraction of natural products [11-15].This experiment wasused ethanol as extraction agent, and extracted the anthraquinones form Aloe by ultrasonic extraction method. The extraction of anthraquinones in Aloe was determined by using 1,8-dihydroxyanthraquinone as the reference substance. The extraction process was optimized. It can provide scientific basis for further research and production. The oxford plate assay system was used to investigate the inhibition of anthraquinone compounds form Aloe to the main spoilage bacteria, Staphylococcus, Serratia, Bacillus and Proteus in spiced pig head, for the future, added the Aloe as a bacteriostatic agent into the low-temperature meat products to extend its shelf life.

2. Materials and Methods

2.1 Materials

Aloe block are purchased at Mengzheng Pharmacy. All organic reagents came from Beijing land bridge Company and all Inorganic reagents came from Beijing Chemical Works. The indicator bacteria were isolated and purified from spiced pig head by Beijing University of Agriculture.

2.2 Alcohol Extract

Took out the dried Aloe block and grinded into powder by pulverizer, after 40 mesh sieve, it was placed in a sealed place at normal temperature.

2.2.1 Ethanol Concentration. The 2.5g samples were divided into five groups on average.Each group according to the solid-liquid ratio of 1:25, were added for 50%, 60%, 70%, 80%, 90% ethanol, soaking at room temperature, overnight. The samples were extracted by ultrasonic power 480Wat 30 ℃ for 40 min. The extraction with ethanol was repeatedthrice. The ethanol extracts were filtered and evaporatedby using a rotary evaporator at 50*°C*, kept at 4 *°C*.

2.2.2 Ultrasonic Time. The 2.5g samples were divided into five groups on average.Each group according to the solid-liquid ratio of 1:25, were added for60% ethanol soaking at room temperature, overnight. Each group was extracted by ultrasonic power 480W at 30 °C for 20min, 30min, 40min, 50min, 60 min. The extraction with ethanol was repeatedthrice.The ethanol extracts were filtered and evaporatedby using a rotary evaporator at 50°C, kept at 4 °C.

2.3 *Ultraviolet Spectrophotometry*

1,8-dihydroxyanthraquinone standardwas accurately weighed10mg and dissolved in 100mL methanol, which was the standard solution of 1,8-dihydroxyanthraquinone.Added 0.5 magnesium acetate-methanol solution as chromogenic agent, with 0.5Magnesium Acetate-methanol solution as blank sample, the standard solution was spectrally scanned in 300-700nm [16], to determine the optimum absorption wavelength.

The standard solution was removed 0.5mL, 1.0mL, 1.5mL, 2.0mL, 2.5mL, 3.0mL, 3.5mL, 4.0mL and added 0.5 magnesium acetate-methanol solution as chromogenic agent, then constant-volumed in I0mL volumetric flask. With 0.5 Magnesium Acetate-methanol solution as a blank solution, it was measured the absorbance. Absorbance (A) and concentration (C) was used as coordinates and fitted equation of linear regression.

2.4 *Actication of Culture*
Strains (Staphylococcus, Serratieae, Bacillus, Proteus) were recovered to normal temperature and poured into 50mL sterilized nutrient broth,cultured 12 h at 30 ° C. And then the strain added to new nutrient broth medium with 2% inoculation amount, activated again.

2.5 *Bacteriostat of Anthraquinone by Oxford Cup Method*
The sterilenutrient agar was inoculated with microbial cells (1 mLof microbial cell suspension in 100 mL agar medium) at 46 ° C andpoured into sterile petri dishes. Each dishe was about 15-20mL.

The sterilized Oxford cup was placed on the carrier plategently. 100 μL of 1 g / mL of Aloe active ingredients and distilled water were added to the Oxford cups, respectively. Three sets of parallel observation were made. After incubating12hat 30°C,the antibacterial activity was evaluated by measuring the inhibition zones.

2.6 *Determination of Minimum Inhibitory Concentration*
1g / mL of Aloe active ingredients were diluted 2 times into different mass concentration gradient. 100mLAloeextract of different concentrations were added in different Oxford cups on the same plate, and the purified distilled water was used as blank control. Three sets of parallel observation were made. After culturing 12hat 30°C, the antibacterial activity was evaluated by measuring the inhibition zones. The lowest concentration of the extract that the inhibition zone was the same as the ck was regarded as MIC [17]. Data Processing software: SPSS 14.0.

3. Results and Discussion

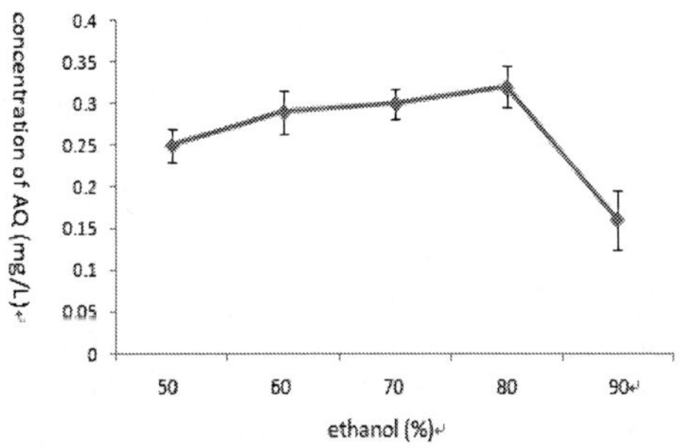

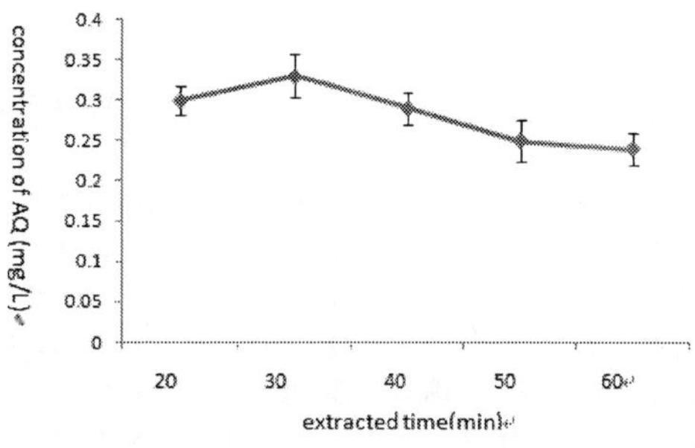

Figure 1. The effect of ethanol concentration extraction of Anthraquinone; **Figure 2.** The effect of extracted time on extraction of Anthraquinone

The as seen in figure.1, with the increase of ethanol concentration, the concentration of anthraquinone in the extract is also slowly increasing, and when the increase to 80%, the concentration of anthraquinone drops rapidly. When the concentration of ethanol was 80%, the concentration of the extract was the highest. The reason is that the form of anthraquinone in Aloe is conjugated anthraquinone glycoside and free anthraquinone, and the content of conjugated anthraquinone glycoside is higher. The conjugated anthraquinone glycosides are dissolved in water easily, while the free anthraquinone is soluble in organic solvent easily [18].So when the ethanol concentration is less than 80%, with the increase of the concentration of ethanol, the free anthraquinone is soluble more.Total concentration of anthraquinone is gradually increased. Ut when the ethanol concentration is higher than 80%, anthraquinone glycosides which dissolved in water easily are dissolved decreased, making the concentration of the anthraquinonedecreased.

The as seen in figure.2, before 30min, the concentration of anthraquinone increases with the increase of extracted time. After 30min, the concentration of anthraquinone decreases slowly with the increase of extracted time.The reason may be that some anthraquinones are unstable in heat. With the increase of extraction time, it will lead to decomposition [19].

Table 1. Inhibitory zone diameter of anthraquinonein Aloe

strain	Inhibitory zone diameter/mm
Proteus	12±0.2
Staphylococcus	15±0.5
Bacillus	12±0.2
Serratieae	0

Note: 0 means no inhibition zone

As can be seen from table 1, 1g/mL of anthraquinonein Aloe has obvious inhibitory effect on Proteus, Staphylococcus and Bacillus, but it has no bacteriostasis effect on Serratieae.

Table 2. MIC of anthraquinonein Aloe

anthraquinonein Aloe (g/mL)		1	0.05	0.25	0.125	0.0625	0.03125
Inhibitoy zone diameter /mm	Proteus	12±0.2	10±0.1	0	0	0	0
	Bacillus	15±0.5	15±0.2	14±0.5	9±0.3	6±0.1	0
	Serratieae	12±0.2	12±0.4	12±0.2	6±0.1	0	0

Note: 0 means no inhibition zone

As can be seen from table 2,theanthraquinones in Aloe has well bacteriostasis effect on Staphylococcus which minimum inhibitory concentration is 0.0625g/mL and the maximum inhibition zone is 15mm;the effect of Bacillus is second which minimum inhibitory concentration is 0.125g/mL and the maximum inhibition zone is 13mm;the bacteriostasis effect on Proteus is ordinary which minimum inhibitory concentration is 0.5g/mL and the maximum inhibition zone is 12mm.The reason may be that Staphylococcus and Bacillus are gram positive bacteria, while Proteus and Serratieae belong to gram negative bacteria and the anthraquinonein Aloe is better for gram positive bacteria.

The components of anthraquinones in Aloe are complex. There are more than 20 kinds of substance s [20] at least,such as aloin, Aloesin, Aloe-emodin and so on. The antibacterial mechanism of anthraquinonein Aloe might be two. One is to change the permeability of cell membrane leading to the cell damage: bacteria are through the selective permeability of cell to absorb nutrition from the outside world for its growth and reproduction. When the permeability of the cell membrane changing, the leakage of nucleic acids causes the cell deformation and nutrient flowing will result irreversible damage to bacteria themselves, and make bacterial die ultimately [21-22].The other one is to inhibit the respiratory metabolism of bacteria: the growth and reproduction of bacteria cannot be separated from respiratory metabolism. Respiratory metabolism is the carbohydrates being oxidized and discomposed to release energy. Anthraquinone in Aloe can inhibit the Krebs cycle of oxidation of sugar and inhibit the synthesis of ATP and NADH in cell, which obstructs the oxidation of sugar. The respiratory metabolism of bacteria is affected, and then the bacteriacan even lead to death [23-25].

4. Conclusion

In the single factor analysis of ultrasonic extraction, it can be concluded that 80% ethanol concentration is the best concentration of extraction and the extraction time of 30min is the best time. The main component of the extraction liquid is anthraquinone.

The anthraquinonein Aloe is better for gram positive bacteria in bacteriostasis effect. And with the decrease of the concentration of active ingredients, the inhibitory effect gradually decreased, which indicated that high concentration extracts have certain bacteriostatic effect.

5. References

[1] Jos H.J and Huisin't Veld Microbial and biochemical spoilage of foods: anoverview. International 1996*Journal of Food Microbiology* **33** 1-18

[2] Gram L and Dalgaard P Fish spoilage bacteria problems and solutions 2002*Current Opinion in Biotechnology* **13** 262-266

[3] Cao Yin Extraction process of anthraquinone from Aloe 2017*TEXTILE AUXILIARIES***34**46-49.

[4] W ANG Wei,WU Nan,FU Yu-Jie and ZU Yuan-Guang Antimicrobial Activities of MenthahaplocalyxBriq. EssentiaOil 2007 *BULLETINOFBOTANICAL RESEARCH* **27** 626-629

[5] Zhang guo quan Study on the extraction of the active substance from aloe barba densisandits antimicrobial effect 2011 *Shan Xi: Northwest A & F University*1-2

[6] ZHAI Mei-zhi,LI Xiao-wing,LIN QI-ying,XIE LIAN-hui Extract of inhibiting composition and inhibitory activity from walnut leaves 2003 *Journal of Northwest Forestry University* **15(4)** 89-91

[7] LIU Zhi-xiong, WANG Li jun, YANG Ya zhen, BAI Xue and LI Ren jie Research on bacteriostasis effect of Aloe 2007*JOURNAL OF SHANGHAI UNIVERSITY OF ENGINEERING SCIENCE* **21(3)** 219-221.

[8] GANZLER K, SZONAL I and SALGO.A. Effetive sample preparation method for extracting biologically active compounds from different matrices by a microwave technique 1990 *Journal of Chromatrgr* **520** 257-262.

[9] ZHANG Lei et al. Study on the Antibacterial Activity of Aloe Microwave Extracts 2012 *Journal of Anhui Ari. Sci* **40 (21)** 11032-11034

[10] ZHOU Mo-Lin, SHANGGUAN Xin-chen, ZHAN Ya-qiong and YANG Wu-ying Optimization of Ultrasonic-assisted Extraction for total Anthraquinones from Aloe Leaves by Response Surface MethodologyF 2013*Acta Agriculturae Universitatis Jiangxiensis* **35 (5)** 1055-1062

[11] Viljoen A M and Wyk B V. A Chemotaxonomic and morphological appraisal of Aloe series Purpurascetes,Aloe section AnguiAloe and their hybrid, Aloe broomii 2001*Biochemical Systematic and Ecology* **2 9(6)** 621-631

[12] Lambert R.J.W., Skandamis P.N., Coote, P.J., et al A study of minimum inhibitory concentration and mode of action of oregano essential oil, thymol and carvacrol 2001*Journal of Applied vticrobiology* **91** 453 462.

[13] Vilkhu K, Mawson R, Simons L, et al Applications and opportunities for ultrasound assisted extraction in the food industry a review 2008 *Innovative Food Science and Emerging Tecchnologies* **9 (2)**161-169.

[14] Wang J,Sun B,Cao Y, et al Optimisation of ultrasound-assisted extraction of phenolic compounds from wheat bran 2008*Food Chemistry* **106 (2)** 804-810.

[15] Valero M.and Salmeron M CAntibacterial activity of 11 essential oils against Bacillus cereus in tyndallized carrot broth 2003 *International Journal of Food Microbiology***85**73-81.

[16] Cha T L, Qiu L, Chen C T, et al. Emodin down -regulates androgen receptor and inhibits pro-statecancer cell growth 2005*Cancer research* **65(6)** 2287-2295.

[17] Steenkamp V, Stewart M J. Medicinal applications and toxicological activities of Aloe products 2007 *Pharmacological Biology* **45(1)** 411-420.

[18] Zhao Xia. Study on Extraction of the Anthraquinone from Aloe by ultrasonic method 2013 *Fiber and its Wide Watch* **13** 20

[19] M Vmatoru. An overview- of the ultrasonically assisted extraction of broactive principles from herbs 2001 *Ultrasonics Sonochemistry* **8** 303-313

[20] L Panwnyk, E Bearfoy, J P Lornmer and T J Nason. The extraction of rutin from flower buds of Sophora japonica 2001 Ultrasomcs Sonochemistry **8**299-301

[21] Kieth Poole. Multidrug resistance in gram-negative bacteria2001 C*rent Opinion in biology* 500-508

[22] Vivek K. Bajpai, Ajay Sharma and Bokyung Moon Chemical Composition analysis and antibacterial mode of action of taxes cuspidata leaf essential oil against foodbrone pathogens 2014 *Journal of food safety* **34** 9-20.
[23] National Committee for Clinical Laboratory Standards, Performance standards forantimicrobial susceptibility testing; ninth informational supplement 2008 *NCCLS document M100-S9* 120-126.
[24] Wei Jin, Ying-Feng Wang, Ri-Li Ge, et al. Simultaneous analysis of multiple bioactive constituents in Rheum tanguticum Maxim. ex Balf. By high-performance liquid chromatography coupled to tandem mass spectrometry 2007*Rapid Communications in mass Spectrometry* **21** 2351-2360.
[25] Ultee A.. Adaptation of the food-borne bacillus cereus to carvacrol 2000 *Archives of Microbiology* **174(4)** 233-238.

IOP Conf. Series: Materials Science and Engineering **275** (2017) 012015 doi:10.1088/1757-899X/275/1/012015

Nitrite Contents in Fresh Vegetables of Different Families and Genus

Yuqian Cui[1-4], Xiao Li [1-4], Lingyi Xu [1-4], Meixia Pang [1-5], Jinghua Qi [1-4] and Fang Wang [1-4]

[1]College of Food Science and Engineering, Beijing Key Laboratory of Detection and Control of Spoilage Organisms and Pesticide Residues in Agricultural Products, Beijing 102206, China
[2]Food Chemistry teaching team, Beijing University of Agriculture, Beijing 102206, China
[3]Pig innovation team, Modern agricultural industry technology system Beijing innovation team, Beijing 100102, China
[4]High Level Talents of Beijing Universities Cross Training" Real Training Plan". Project, Beijing 102206, China
[5]E-mail: abc960718@sina.com

Abstract: The aim of this study is firstly aimed at investigating the contents of nitrite in common consumed vegetables according to families and genus classification. The vegetables were randomly collected and analyzed in quartile sampling according to GB5009.30-2016.The vegetables were analyzed by the software of Spss20.0 and statistically significant Duncan multiple comparisons. The data indicates that the nitrite contents in different families and different genus vegetables in same family were significant ($P<0.01$). A relatively high nitrite concentration was observed in Chenopodiaceae which is 0.5920mg/kg dry weight. A relatively low nitrite concentration was observed in Dioscoreaceae that contentration is 0.0032mg/kg dry weight. The nitrite contents of different genus are large, in which the relatively high concentration samples were red beet root (0.886mg/kg dry weight), peanut (0.7485mg/kg dry weight), corn kernels (0.7119mg/kg dry weight), Lotus root (0.592mg/kg dry weight).

1. Introduction

The different nitrite contents in vegetables are resulted from the type of vegetables, and environmental factors such as cultivate type, composition of soil, light intensity, temperature and moister, growth density, duration of growth period, harvesting time, storage time, edible plant portion and use of nitrogen fertilizer [1]. Vegetables can also produce nitrite by oxidation of endogenous nitrogen oxides, and nitrate can be reduced by nitrite reductase to produce nitrite [2]. Nitrite itself is relatively non-toxic but as precursor material of nitrosamines poses a threat to human health. Nitrite in the acidic environmentcan form a strong carcinogenic nitrosamine [3]. Nitrosamines have a strong carcinogenic effect, which can cause esophageal cancer, stomach cancer, liver cancer and colorectal cancer [4]. This study describes the nitrite contents in vegetables of 15 families and 46 genus.

Content from this work may be used under the terms of the Creative Commons Attribution 3.0 licence. Any further distribution of this work must maintain attribution to the author(s) and the title of the work, journal citation and DOI.
Published under licence by IOP Publishing Ltd

2. Material and Method

2.1. Chemicals and Reagent
Sodium Nitrite (CAS:7632-00-0). Potassium Ferrocyanide, Zinc Acetate Dihydrate, Aceticacid, Sodium Borate, Chlorane, p-aminobenzene sulfonic acid, N-1-Naphthylethylene diamine dihydrochloride, All of the chemical reagents were obtained from China National Accord Medicines.

2.2. Samples
A total of edible vegetable samples was freshly collected from Supermarket. They were cleaned in tap water and deionized water three times. The edible portion were cut into small pieces and then were placed in the homogenizer (JT-C). Fresh vegetables were placed in a refrigerator at -4°C and detected in 12 hours.

2.3. Nitrite Extraction
Homogenized sample (5g) was weighed out and placed into 50mL beaker, 12.5mL 50g/L saturated Borax solution was added in the sample. Later the sample was transferred to the conical flask in 70°C water and mixed. The mixed solution was heated for 15mins in water bath kettle. After 15mins, the flask was transformed to the cold-water bath so as to let the solution cool to room temperature. Potassium ferrocyanide solution (5mL) was added to the flask, Zinc Acetate solution (5mL) was added followed by to precipitate the protein. The solution constant volume to scale line in water. Finally, the solution with shaking stand for 30 another mins. After removing the upper fat, the extracts were filtered through φ18cm filter-papers. The first zone of filtrate was discarded in order to overcome possible nitrate contamination from the filter-papers.

2.4. Determination of Nitrite
Absorbance was measured at 538 nm using a UV/VIS Spectrophotometer (TU-1810) with digital readout screen where absorbance of the sample was displayed. Absorbance reading for each sample was taken three times and the average of the reading recorded.

2.5. Data Processing Method
The original data was entered into Excel and analyzed statistically with SPSS software for windows Version 20.0 and drawed with Excel later.

Table 1. Nitrite content of vegetables in different families (mg/kg wet weight).

family	Genus	Mean±standarddeviation
Chenopodiaceae	3	3.856±0.002
Amaranthaceae	1	3.754±0.162
Convolvulaceae	3	3.856±0.002
Nymphaeaceae	1	3.036±0.000
Umbelliferae	5	2.856±0.002
Zingiberaceae	4	2.456±0.007
Gramineae	3	2.356±0.003
Cruciferae	11	2.271±0.010
Leguminosae	9	2.246±0.003
Compositae	6	2.094±0.0020
Cucurbitaceae	8	2.012±0.010
Liliaceae	9	1.320±0.0090
Solanaceae	8	1.302±0.0010
Araceae	1	0.122±0.003
Dioscoreaceae	1	0.021±0.0030

Table 2. Nitrite content of vegetables in different families (mg/kg dry weight).

family	Genus	Mean±standarddeviation
Nymphaeaceae	1	0.5920±0.0007
Chenopodiaceae	3	0.5205±0.0004
Gramineae	3	0.3602±0.000
Convolvulaceae	2	0.3300±0.000
Leguminosae	7	0.315±0.0000
Zingiberaceae	2	0.263±0.0000
Umbelliferae	3	0.2463±0.000
Araceae	1	0.2401±0.0010
Amaranthaceae	1	0.162±0.0020
Liliaceae	8	0.158±0.0010
Cruciferae	3	0.134±0.0000
Compositae	3	0.108±0.0010
Solanaceae	3	0.108±0.0010
Cucurbitaceae	5	0.101±0.0010
Dioscoreaceae	1	0.0032±0.0000

3. Results and Discussion

The contents of nitrite (Wet Weight) in 15 family vegetables in are shown in table 1. The result is different significantly in 15family ($P<0.05$). The limit standard nitrite content in vegetable is less than or equal to 4.0mg/kg (fresh weight). The data in table 1 indicates that exceeding standard nitrite content vegetables are beans (4.602mg/kg wet weight) in Leguminosae, endive (4.151mg/kg wet weight) in Compositae, spinach (4.691mg/kg wet weight) in Chenopodiaceae, swamp cabbages (4.080mg/kg wet weight) in Convolvulaceae.

In order to eliminate the nitrite content difference caused by the different water content in the different family vegetables, the fresh weight of all the vegetables is changed to dry weight. The contents of nitrite (Dry Weight) in 15 family vegetables are shown in table 2.

3.1. Nitrite Content in Liliaceae

The low levels of nitrite of onion and garlic found in this study could be as a result of chemical components (figure.1). Organic sulfides in onion and garlic can block the formation of Nitrosamines [5] Sulfydryl and nitrite can form nitrous esters to remove nitrite effectively. The nitrite content of onion in early growth period was 0.1279mg/kg dry weight, while late growth period onion was 0.0408mg/kg dry weight in the same variety. which is due to the high absorption and transformation intensity of nitrogen in the plant growth and development period.

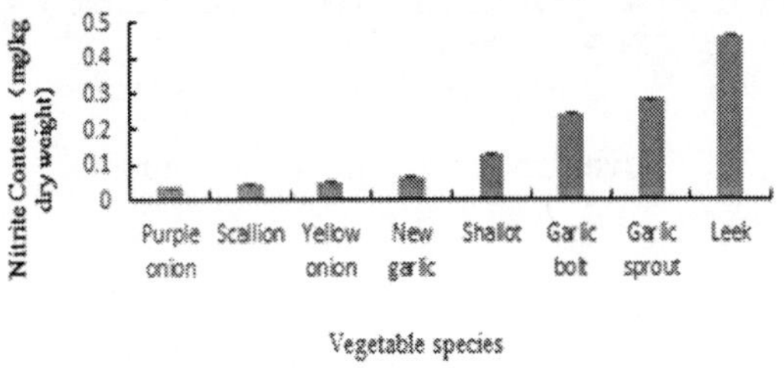

Figure 1. Nitrite content in Liliaceae

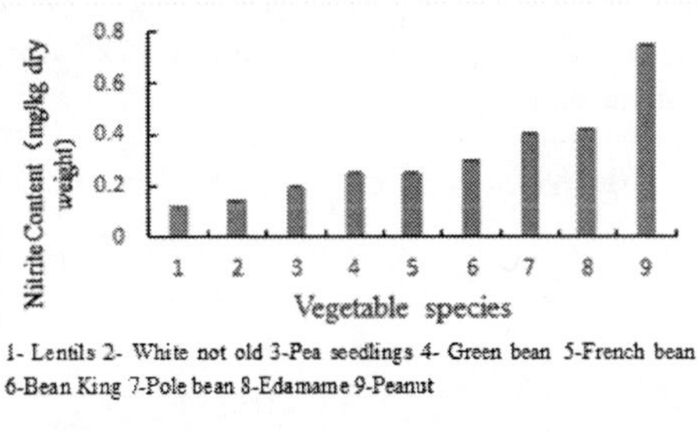

Figure 2. Nitrite content in Leguminosae

3.2. Nitrite Content in Leguminosae

In the oxidation conditions and the participation of soil microorganisms, ammoniums produce nitrification and transform to nitrate ions, in which part of nitrate ions are assimilated and part of nitrate ions in the shape of denitrification in the soil, the formed nitrite ions are also absorbed by the plant.In the same leguminous plants, significant differences(figure 2) in content are caused by different breeds genotype, bean king (0.3012mg/kg dry weight), white not old (0.1442mg/kg dry weight). One of the main reasons may be different genotypes caused by genetic variation in the long-term cultivation and breeding process [6].

3.3. Nitrite Content in Cruciferae
The results (figure.3) showed that the nitrite content of brassica and radish in Cruciferae exist significant difference (P=0.194). The consequence of Cabbage mustard's quality deteriorated arising from high nitrogen treatment and content decrease of Vitamin C and sugar [7]. The fresh radish nitrite content is below 1mg/kg wet weight, because the amount and activity of nitrite reductase is much more than nitrate reductase in vegetable tissues and more stable and inactivate. It's also related to the process that the nitrite nitrogen is reduced to NH+4 in time [8].

3.4. Nitrite Content in Compositae
In the Asteraceae (figure.4), the endive is high because it's suitable for fertilizer and more nitrogen [9]. The reason of nitrite content in Chrysanthemumcoronarium and Lobular Chrysanthemum is different is that the environment and nitrate reductase activity [10].

3.5. Nitrite Content in Cucurbitaceae
The result (figure.5) showed that the nitrite contents of three different cucumber were significantly affected by genotype and environmental factors followed by ($P<0.05$). Nitrite nitrogen compounds formed by free nitrate nitrogen in the action of nitrate reductase conversion cannot completely convert to ammonium salts into the nitrogen metabolism cycle, resulting in more nitrite accumulate in the fruit [11].The nitrite in the Loofah and Wax gourd were not detected almost because the nitrate reductase is little in the fruit and nitrate will be reduced to nitrite under the action of nitrate reductase and NADH most in the leaves [8].

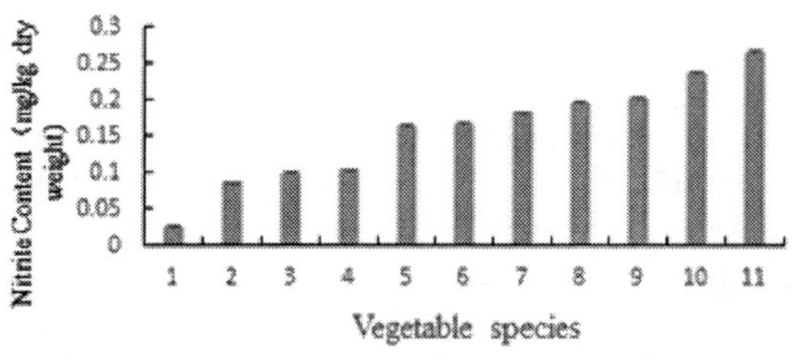

1- Carrot 2- Cole 3-Red radish 4- Cherry radish 5-Turnip 6-Cabbage 7-Chinese cabbage 8-Purple cabbage 9-Pakchoi 10-Milk lettuce 11-Cabbage mustard

Figure 3. Nitrite Content in Cruciferae

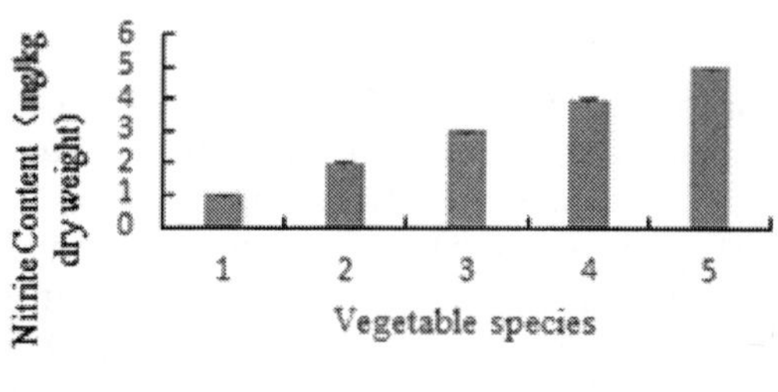

1-Lobular Chrysanthemum 2- Lettuce 3-Leaf Lettuce 4-Endive 5- chrysanthemumcoronarium

Figure 4. Nitrite content in Compositae

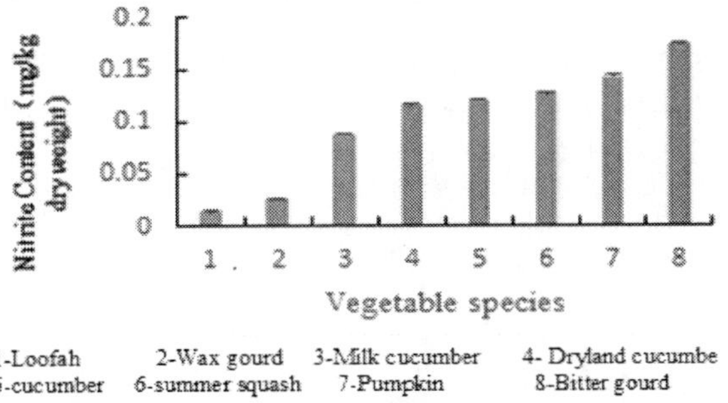

Figure 5. Nitrite content in Cucurbitaceae

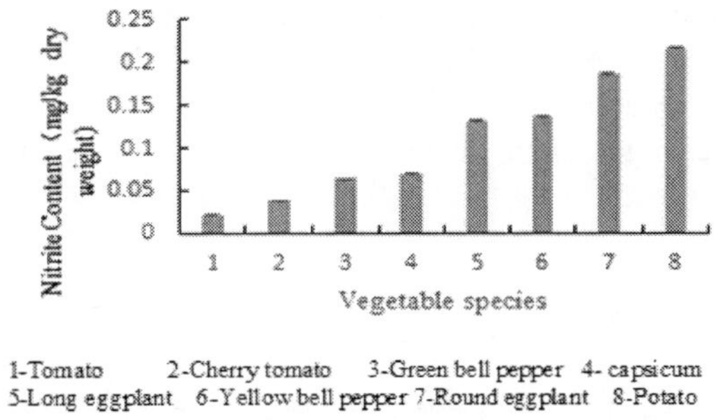

Figure 6. Nitrite content in Solanaceae

3.6. Nitrite Content in Solanaceae

In the Solanaceae, the highest nitrite Content is potato(0.2151mg/kg dry weight, figure.6), Round eggplant (0.185mg/kg dry weight) followed by, the lowest is pepper(0.0638mg/kg dry weight) and tomatoes(0.0216 mg/kg dry weight). The nitrite content in Solanaceae is relatively is little because of antioxidant active ingredients such as polyphenols and flavonoids in pepper[12] and lycopene, vitamin C and various antioxidant enzymes in Tomato. The mechanism of action is mainly to inhibit the radicals [13].

3.7. Nitrite Content in Gramineae

Bamboo shoot(figure.7) is the kind of vegetable of low accumulation in nitrite which are associated to reduction of sugar and vitamin C. The results showed that Maize is 0.7119mg/kg dry weight. It's attributed to contained reductase and the absorbing of nitrogen [14]. The nitrite content of wild rice stem is related to the change of reducing sugar whose content has a trend of rise first then fall in its growing season, so the nitrite content of it change with reducing sugar [15].

3.8. Nitrite Content in Umbelliferae

Celery (figure.8) is a nitrate-accumulating vegetable [16], The higher nitrate content in celery stems may be the propagation of nitrate-reducing bacteria in celery [17]. The dry matter content of nitrite in celery is relatively low because it contains a lot of antioxidant component such as volatile oil and flavonoids [18]. The relatively high nitrite content in Coriander is attributed to the strong ability of absorbing fertilizer.

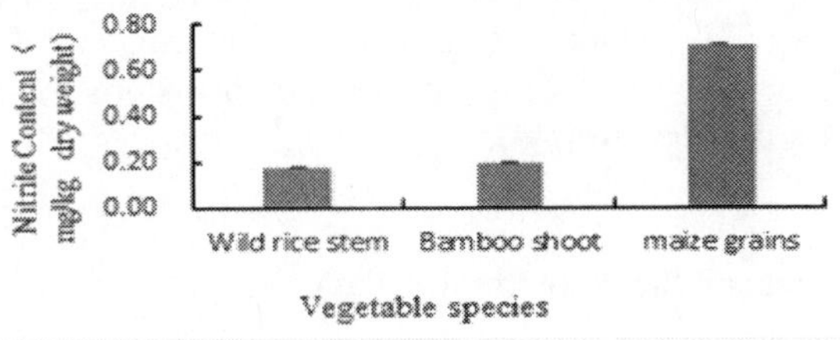

Figure 7. Nitrite content in Gramineae

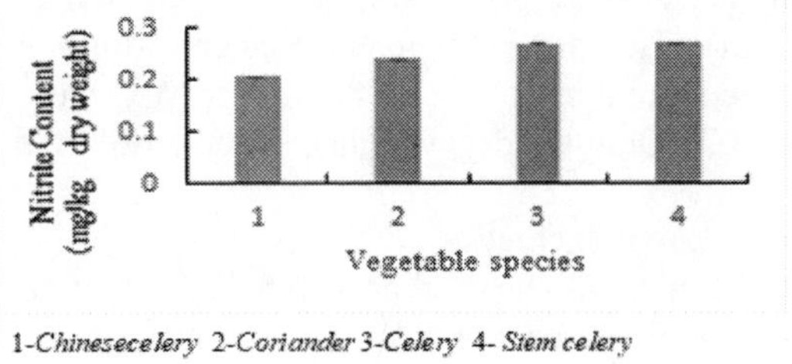

Figure 8. Nitrite content in Umbelliferae

3.9. Nitrite Content in Zingiberaceae

Although ginger (figure.9) has diarylheptanoids as the antioxidant substance [19],Ginger is a kind of plant that require high nitrogen. In this paper, Ginger nitrite is 0.4651mg/kg dry weight, while the Tender ginger is only 0.0732mg/kg dry weight. The fact indicates the Tender ginger contains much more antioxidant activity.

3.10. Nitrite content in Convolvulaceae

Purple potato and sweet potato are suitable for potassium fertilizer condition (figure.10). The potassium as an activator of nitrate reductase, so as to significantly improve the activity of nitrate reductase and accelerate the reduction of nitrate into nitrite though increasing the supply of potassium [20].

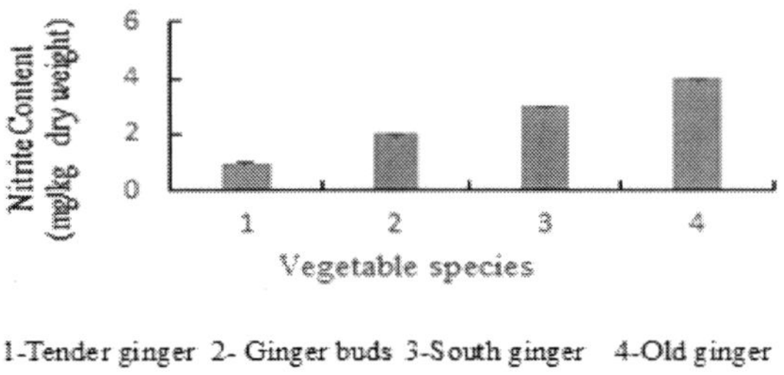

Figure 9. Nitrite content in Zingiberaceae

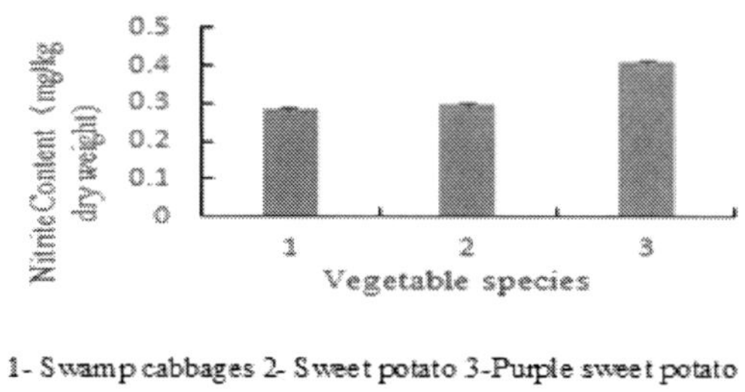

Figure 10. Nitrite content in Convolvulaceae

3.11. Nitrite Content in Chenopodiaceae

Spinach (figure.11) could accumulate higher content of nitrite with the higher nitrate reductase activity [21]. The activity of nitrate reductase was increased with the increase of nitrogen application rate. The content of nitrite in beet leaves is higher than that of beetroot, which is related to the early transport of fertilizer from root to leaf [22].

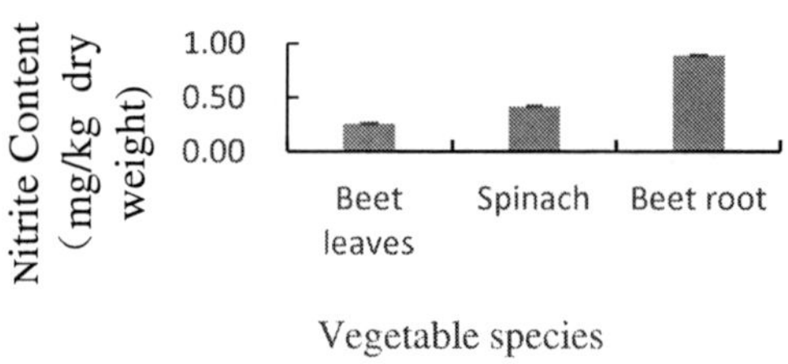

Figure 11. Nitrite content in Chenopodiaceae

4. Conclusion

The average content of nitrite in different families was different. The highest content was Ny mphaeaceae (0.5920mg/kg dry weight), the lowest was the Dioscoreaceae (0.0032mg/kg dry w eight). The different genus concentrations are large, in which the relatively high concentration include red beet root (0.886mg/kg dry weight), peanut (0.7485mg/kg dry weight), corn ke rnel s (0.7119mg/kg dry weight), Lotus root (0.592mg/kg dry weight).

5. Ackowledgements

I wish to thank the teacher who help me in experiment and the laboratory manager who give me a hand in experiment equipment.

6. References

[1] Correia M and Barroso Â Contribution of different vegetable types to exogenous nitrate and nitrite exposure 2010 Food Chemistry.**120** 960–966
[2] Wright M and Davison K Nitrate accumulation in crops and nitrate poisoning in animals 1964 Advances Agronomy **16** 197-247
[3] JR Kelley and JM Duggan Gastric cancer epidemiology and risk factors 2003 Journal of Clinical Epidemiology **56** 1-9.
[4] WU Yingzhen Risk assessment of Nitrite in Foods 2009 Animal Husbandry and Feed Science **30** 62-63

[5] Chang Qing Study on the effect of Chinese onion on removing nitrite 2008 Science and Technology of Food Industry **6** 175-176
[6] Lin Guanjie Pactors affecting the Nitrate content in vegetables 1995 Journal of Shanghai Agricultural College 1
[7] Chen Riyuan Effect of nitrogen nutriment on the growth and quality of Chinese Kale 2005 Transactions of the CSAE 143-146.
[8] Wu Ping Plant Nutrition Molecular Physiology 2001 (Bei Jing: Science Press)
[9] Tong Jun Cultivation techniques of Endive 2014 Jilin agriculture19 60
[10] Jin Yazhong Study on Nitrate Content and Renson of Different Kids of leafy Vegetable's Leaf 2010 Journal of Heilongjiang Bayi Agricultural University 22 1-3
[11] Zhao Chunbo Nitrogen Efficiency Types Analysis on Differences Cucumber Varieties 2015 Changchun. Doctor 's Degree of Jilin Agricultural University
[12] Zhang Jing The research progress of chemical composition of pepper 2009 Chinese Traditional Patent Medicine 31 1906-1912
[13] Shi Lisha Study on antioxidant activity of tomoto in vitro 2012 Guangdong Agricultural Sciences 131-133
[14] Zhang Hongtian The Study of different Cultivation Methods and amount of Nitrogen Aplication on Corn Growth and Nutrient Absorption 2005 Daqing: Master's Degree of Heilongjiang Bayi Agricultural University
[15] Zhang Meiling Effects of Fertilization and Preservative on the Quality and Physiology of the Culm for Few Flower Wild rice on the Mountains 2008 Hefei: Master's Degree of Anhui Agricultural University
[16] Cheng Longjun. 2003. The Biochemical Changes during Stem Swelling in Zizania latifolia. Acta horticulture Sinica30(6):741.
[17] Chen Jing Dynamic relationship between nitrite content and microorganisms in celery during storage 2014 Food science and Technology 39 63-66
[18] Dong Xiao Research Progress of Natural Western-style Spice-Petroselinum crispum 2013 China Condiment 38 4-7
[19] Yang Leixiang. 2009. Antioxidative and cytotoxic properties of diarylheptanoids isolated from Zingiber officinale, China Journal of Chinese Materia34(3):319-323
[20] Zhang Weixia Cultivation technology of parsley ate on storage root quality of edible weet potato 2016 Humic Acid 24-28
[21] Zhang Meng Physiological Differences of Spinach Cultivars with Different Nitrate Levels 2016 Acta Botanica Boreali Occidentalia Sinica 36 2469-2476.
[22] Du Yongcheng Effect of nitrogen fertilization on nitrate reductase and nitrite reductase activities of sugar beet 2012 Plant Nutrition and Fertilizer Science 18 717-72

IOP Conf. Series: Materials Science and Engineering **275** (2017) 012016 doi:10.1088/1757-899X/275/1/012016

Research on the Titaniferous Slag Smelting Process and Furnace

Wei Lin[1] and Piyu Guo[2]

[1]CISDI Engineering Co., Ltd, NO.1, Saidi Road, Jinyu Avenue, Yubei District 401122, Chongqing, China, Wei.A.Lin@cisdi.com.cn.

[2]Chongqing Sino French Tangjiatuo Wastewater Treatment Co., Ltd, Tieshanping Street, Taipingchong Village, Jiangbei District400026, Chongqing, China, gpyfff@126.com

Abstract: The article mainly introduces the raw material of the titaniferous slag smelting and the conventional and new metallurgical process of the titaniferous slag smelting. Through analyzing the different fuctions of the furnaces in the two smelting process, the article summarized the advantage of new process compared with the conventional one.

1. Introduction

The titaniferous slags are divided into high titanium type and low grade titanium type, and the high titanium slags which are produced by using titanium concentrates are industrial materials containing TiO_2 (75% or higher). On the other hand, the low grade titanium slags contain TiO_2 (45%~50%) which are produced by using vanadium titanium magnetite. The main production equipments are the arc furnaces which are normally divided into circular type and rectangular type. According to the different furnace sealed types, the arc furnace can also be divided into three types: airtight, half closed and open types.

The titaniferous slags are the main material to produce titanium dioxide, titanium sponge and electrode which have a very wide range of applications. The titanium dioxide product can be widely used in coatings, plastic, papermaking, printing ink, chemical fiber, rubber, cosmetics, food and other industries [1].

2. The Raw Material of the Titaniferous Slag Smelting

The raw material of smelting titaniferous slag are mainly mineral aggregates and reducing agents.

2.1. The Mineral Aggregates

There are mainly three kinds of mineral aggregates, which are titanium concentrates, rutile and titanium iron ores. Among them the rutile and titanium concentrates contain more TiO_2, and the titanium iron ores contains less TiO_2. There are almost 965 million tons of titanium concentrates reserved in our country which mainly located in Sichuan, Hainan, Hebei, Yunnan, Guangdong and Guangxi province. 97% of titanium concentrates are rock ores which are 837.85 million tons. And 3% of titanium concentrates are placer which are 22.51 million tons. The composition of the typical titanium concentrates and titanium iron ores are shown in table 1. The titanium slags contain different TiO_2 because of using the ore materials with different TiO_2 grades. The common product acid dissolved titanium slags containing TiO_2 73% - 77% demand that the titanium concentrate materials contain 46.5% TiO_2 or higher.

Content from this work may be used under the terms of the Creative Commons Attribution 3.0 licence. Any further distribution of this work must maintain attribution to the author(s) and the title of the work, journal citation and DOI.
Published under licence by IOP Publishing Ltd

Table 1: The main composition of the titanium concentrates and titanium iron ores

Items	TFe	TiO_2	V_2O_5	S	SiO_2	Al_2O_3	MgO	CaO
Titanium concentrates	33.59	47.20	0.25	0.015	1.56	0.43	1.20	0.20
Titanium iron ores	56.74	11.84	0.60	0.08	3.35	3.0	1.40	0.08

2.2. *The Reducing Agents*

The main types of the reducing agents are gas, charcoal, anthracite, coke and etc. The reactivity and resistivity of the charcoal is very good, but its price is high because of the lack of resources. The charcoal is not used widely because of the high volatile flux and ash content. The reactivity and resistivity of the gas and anthracite is lower than the charcoal, also they have large-scale reserves, but they are not widely used due to the high volatile flux content and difficulty control in arc furnace smelting process. The coke has low reducibility but it is used widely in titanium slag smelting because of the low conductivity, large-scale reserves, a high carbon and low volatile flux content. In conclusion, the best reductant of the titanium slag smelting is the coke.

3. The Process and Furnace of the Titaniferous Slag Smelting

Number The conventional titanium slag smelting usually choose the titanium concentrates as raw material. The main furnace of the conventional smelting process shown in figure 1 is circular or rectangle arc furnace equipment.

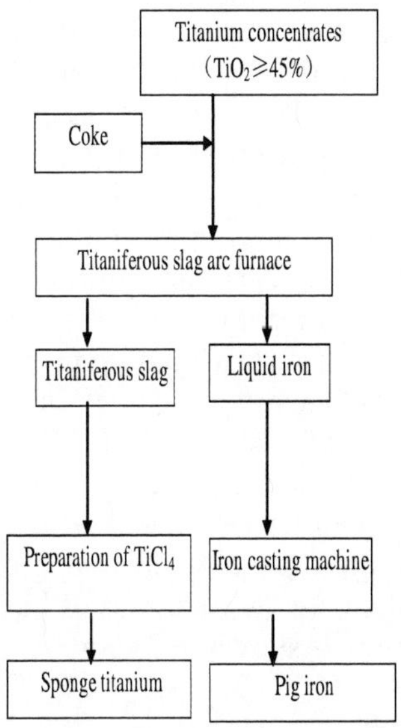

Figure.1: The conventional titaniferous slag smelting process.

The titanium concentrates are beneficiated from the titanium magnetite, so the conventional titanium slag smelting process need beneficiating operation and the request to feedstock is more rigorous. In recent years, many domestic mining companies research and test new technology process which use directly vanadium titanium magnetite as raw material. The famouse titaniferous slag producing company in Panzhihua city has a new technology process proved by the practice which adopt the new arc furnaces. It is shown in figure 2.

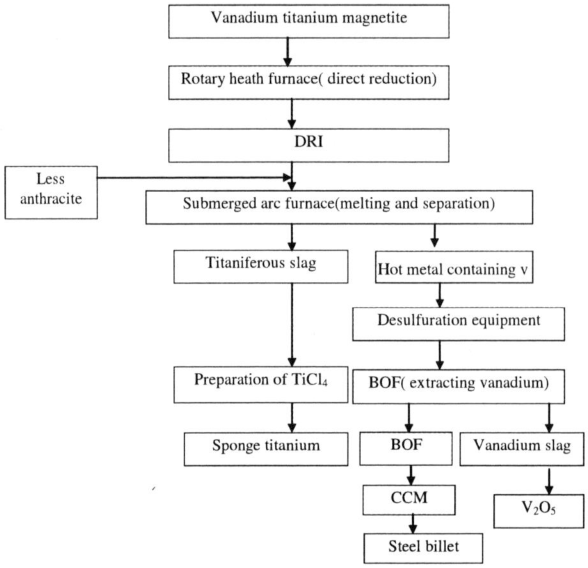

Figure.2: The new titaniferous slag smelting process.

In this new process, the vanadium titanium magnetites are directly reduct by rotary hearth furnace, then the DRI are added into the submerged arc furnace. The fuction of the titaniferous slag arc furnace is mainly melting DRI and separating titaniferous slag. As a result, the furnace output a lot of hot metal containing vanadium and titanium slags. The new process compared to conventional titanium concentrate smelting process has the following features: 1) the power consumption per ton of slags gets 500-600 kwh lower than the power consumption in the conventional process. The new process also decreases the electrode consumption and reduces production cost. 2) The products of the submerged arc furnace are mainly hot metal containing vanadium and titanium slags, but the titanium slags have 20%-50% TiO_2. Compared with the conventional process of which titanium slags have more than 75% TiO_2. So in the new process the titaniferous slags need deep processing and further enrichment of TiO_2.

The new smelting process can also change the types of smelting equipment to adapt to different scales of production. For the large-scale smelting, the shaft furnace or rotary kiln can be used instead of the rotary hearth furnace to product the direct reduced iron[2], and in the follow-up process the large rectangle submerged arc furnace can be uesed instead of conventional circular arc furnace to product the titanium slag and the liquid iron. Such the melting process is adopted by one New Zealand steel company of which the annual output are 700,000 tons of liquid iron and 700,000 tons of titanium slag contained 35% TiO_2. Its process flow is shown in figure 3.

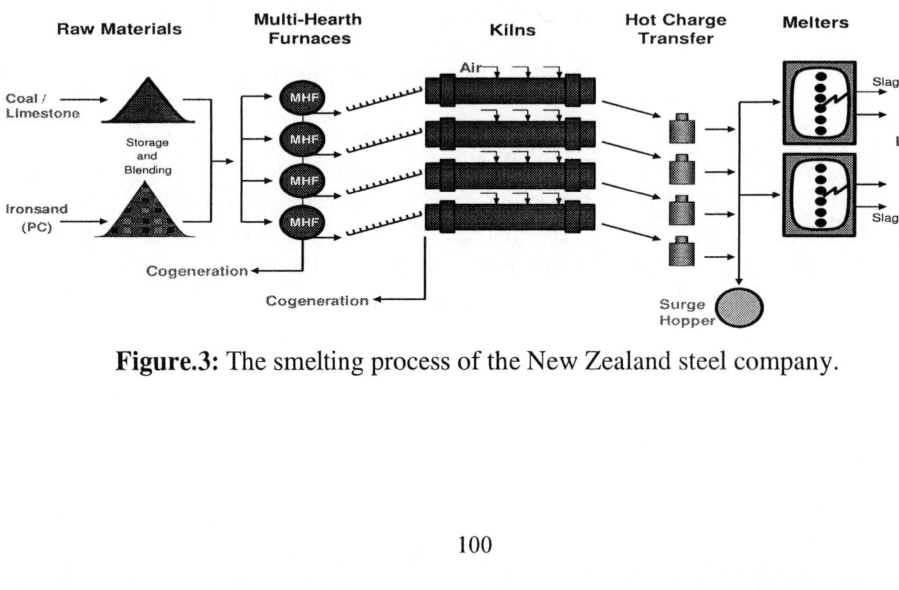

Figure.3: The smelting process of the New Zealand steel company.

The main smelting equipments of the New Zealand steel company are 4 sets of ₵4.6 x 65 m rotary kiln and 2 sets of 42 MW rectangular submerged arc furnace. In the rectangular arc furnace, the action is mainly melting process of DRI and separating process of titaniferous slag. The smelting process of the New Zealand steel company increases the rate of vanadium reduction and decreases the rate of titanium reduction with controlling MFe in the slag which has obtained more vanadium in the hot metal and more titanium in the slag.

4. Conclusions

By comparison beteen the new smelting process and the conventional process, conclusions are drawn as below.

1) The raw material of smelting titaniferous slag are mainly the titanium concentrates and the titanium magnetites, which are used in different technological processes. The coke is commonly choosed as the main reducing agent.

2) In the new technological process of smelting titanium slag which chooses the titanium magnetites as the raw material, the submerged arc furnace mainly smelt DRI and separate the titaniferous slag, of which the process are deep reduction reactions. The key factors are controlling less MFe content in the slag and reducing more vanadium into the liquid iron.

3) The power consumption per ton of slags in the new process gets 500-600 kwh lower than the power consumption in the conventional process.

4) The new smelting process could obtain more vanadium in the hot metal and more titanium in the slag compared with the conventional process.

5. References

[1] Shaoli.Yang, Jifu.Sheng, The titanium slag and pig iron smelting technology using titanium iron ore. Metallurgy Industry t, 47, 2006.

[2] Mansheng.Chu, Zhengen.Liu, YuTang, The special metallurgical resources and noncoking smelting technology. Iron and Steel Technology, 141, 2014.

Effect of MUF/Epoxy Microcapsules on Mechanical Properties andFractography of Epoxy Materials

Zhuo Ni[1], Yuhao Lin[2] and Xuexiao Du[3]

[1]College of Chemistry and Chemical Engineering, Shenzhen University Shenzhen,China e-mail: zhuoni@szu.edu.cn
[2]College of Chemistry and Chemical Engineering, Shenzhen University Shenzhen, China e-mail:823394522@qq.com
[3]College of Chemistry and Chemical Engineering, Shenzhen University Shenzhen,China e-mail:xuexiaodu@szu.edu.cn

Abstract: Melamine-urea-formaldehyde (MUF) microcapsules were synthesized, morphology, shell thickness, average diameter and interface morphology were studied by scanning electron microscope (SEM). The spherical MUF microcapsules are size normal distribution without adhesion and accumulation, being compact, rough and uneven with a thickness of 3.2μm and a core contents is approximate 70%. A latent imidazoleas the curing agent for a cross-linking chemical reaction for cracking repairing .A good dispersion of MUF microcapsules and a good interfacial bonding are obtained. Effects of MUF microcapsule size and content on bending property and dynamic mechanical propertywere investigated. Both bending strength and storage modulus of the composite are considerably reduced with an increasing addition of the microcapsules whereas the glass transition temperatures are almost not influenced. Significant toughening effects of MUF microcapsules on the epoxy composites are observed at the conditions of different content and size of microcapsule especially at low microcapsule contents and small microcapsule sizes.

1. Introduction

In this paper, MUF microcapsules with a core of modified epoxy adhesive were synthesized. A latent imidazole catalyst with heat sensitivity and stability was used as curing agent for a design of self-repairing system, which was added together with MUF microcapsules into epoxy matrix in order to fabricate applicable microcapsule/epoxy self-healing composite. Morphology, shell thickness, average diameter of the MUF microcapsules and fracture morphology of the microcapsule/epoxy composite material were studied by SEM. Effects of microcapsule size and content on the bending property, the dynamic mechanical property and toughness for the microcapsule/epoxy composite material were investigated by bending testing and DMA testing.

2. Experimental Section

2.1. Preparation of MUF Microcapsule

At room temperature, melamine (Analytical-grade, Sigma-Aldrich China), urea (Analytical-grade, Sigma-Aldrich China) and formaldehyde (37 wt.%, Analytical-grade, Tianjin Chemical Plant, China) in a molar ratio of 0.05:1:2 were added into three-necked round-bottomed flask fitted with a stirrer. After a fully mixing, pH of the mixed solution was adjusted to 8-9 bytriethanolamine (Analytical-grade,Merck China). The three-necked flask was suspended in a temperature-controlled water bath. The temperature

Content from this work may be used under the terms of the Creative Commons Attribution 3.0 licence. Any further distribution of this work must maintain attribution to the author(s) and the title of the work, journal citation and DOI.
Published under licence by IOP Publishing Ltd

of the system was raised to 70°C. and kept for 1 hour under stirring and refluxing until a transparent viscous MUF prepolymer was obtained. At this stage, the viscous MUF prepolymer was emulsified by Sodium dodecyl benzene sulfonate (SDBS) (0.5% wt%, 99% purity, Merck China). The emulsion was used to form shell material after cooling to room temperature. The MUF prepolymer and diglycidyl ether of bisphenol A epoxy E-51 (density is 1.20 g·ml-1 and epoxide equivalent weight is 196 g·mol-1,Shanghai Resin Plant, China) which was previously diluted by 17.5 wt% n-butyl glycidyl ether (BGE) (Analytical-grade, Sigma-Aldrich China) were together added into a three-necked round-bottomed flask fitted with stirrer in a molar ratio of 1:1. The O/W emulsion was obtained after stirring for 20-30 minutes at a agitation rate of 200-1000rpm. Various sizes of MUF microcapsules were prepared by the adjustment of agitation rates. The pH of the mixed chemicals was adjusted to 2–3 by drop-wise addition of dilute sulphuric acid (0.05 wt%, Analytical-grade, Sigma-Aldrich China). When the walls of microcapsules were fully formed after a curing reaction of 70°C. for 2 hours, the mixture was cured further at 60°C. for 2 hours under continuous agitation. Finally, MUF capsules were filtered and rinsed three times with ethanol and deionized water respectively and vacuum dried at 40°C. for 10 hours.

2.2. Preparation of Microcapsule /Epoxy Self-Healing Composite

Epoxy was added into suction flask and vacuum filtrated for 10 minutes in order to remove bubbles in the resin. BGE diluent, MC120D curing agent(Industrial product, Guangzhou Chuanjing Electronic Material Corporation, China) and MUF microcapsules with different sizes (Average diameters are 90μm,200μm and 550μm respectively) and various contents (2wt%, 5wt% and 10wt% respectively) were mixed with epoxy resin in beakers. The mixture was evenly blended until a uniform mixture was achieved. Diethylenetriamine (DETA) (Analytical-grade, Sigma-Aldrich China) of a curing agent for the epoxy matrix was blended into the mixture with a mass ratio for curing agent to epoxy resin of 15:100. The mixture was stirred for 5-10 minutes and vacuum filtrated for 10-30 minutes in order to remove bubbles. The blended resin was injected carefully into polytetrafluoroethylene mould (Mould dimension: 65×14×2mm) , following cured at a room temperature in vacuum drying oven for 24 hours. The specimens were taken out from the oven and demoulded, measured by vernier caliper and polished by rough and fine sandpapers to ensure that each batch of specimen has identical thicknesses, smooth edge surfaces and no visible defects.

MC120D chemicals used in this work is an imidazole compound, 6-(2-(2-methyl-1H-imidazol-1-yl) ethyl)-1, 3, 5-triazine-2, 4-diamine, white powder, thesoftening point is 95-115°C., particle diameter is less than 10μm. The stoichiometric ratio of curing agent to epoxy used is 20:100. The curing temperature is 100°C. for 40 minutes as the supplier suggests[1].

2.3. SEM Study

SEM (S-3400N(II), HITACHI, Japan) is used with an acceleration voltage of 15-20kV. The microcapsules and the fractured composite specimen were placed on conductive carbon tapes attached to a mounting piece, then sputter-coated with an approximate 10 nm conductive gold layer. MUF microcapsules ruptured using sharp razor blade were analyzed for measuring the average shell thickness for broken microcapsules varied at least 10 locations.

2.4. Testing Sample

At least ten MUF microcapsule/epoxy composite specimens were selected to form a batch of MUF microcapsule/epoxy self-healing sample. Nine specimens were divided into three groups including virgin sample (Type I), damage-controlled sample (Type II) and self-healing sample (Type III). The tenth specimen was used for DMA test. Because these specimens may occur post curing reaction during heat process that interfere with evaluation of the self-healing efficiency, the needed thermal preprocessing conditions were discussed in the relevant DSC measurement.

The Type II and the Type III samples were subjected to a controlled constant strains using three-point bending fixture with deformation of 90% bending fracture strains of the specimens which produce cracks within the materials based on the bending deformation vertical distance of the specimens in the fixture. The specimens were clamped at this bending deformation distance for 24 hours. After the cracks observed, the specimens were removed off and flattened under a loading, then tested by universal tensile machine for bending properties and DMA instrument for the dynamic mechanical property.

The Type III samples after subjection to constant strain damaging were heated in vacuum drying oven at 100°C. for 1 hour. After the self-healing process, bending properties of these specimens and the dynamic mechanical property were measured through bending tests and DMA test.

2.5. Bending Test

Bending properties for three types of samples were measured by universal tensile machine (model AGX, Shimadzu, Japan). The Type I samples were directly measured without any treatment. In this three-point bending test, bending span was 30 mm and bending speed was 5mm/min. Force-displacement curves were recorded for data analysis. The average value of three specimens data was taken as corresponding type bending strengths (I) and bending fracture strains (ε) calculated.

2.6. DMA Test

Storage modulus (E'), loss modulus (E'') and mechanical loss factor (Tanδ) of the tested specimen were measured via DMA temperature sweep experiment (Q800, TA Instruments, UK). DMA test was performed using a double cantilever beam, heating from 25°C.to 140°C at a heating rate of 5°C/min and an oscillator frequency of 1 Hz under nitrogen atmosphere.

3. Results and Discussions

3.1. MUF Microcapsule and Epoxy Composite

The applications of microcapsules in self-healing material are significantly influenced by stability, mechanical performance and interfacial property of the microcapsules. Many problems including low water-resistance, formaldehyde emission and short shell-life exist in previous used urea-formaldehyde (UF) microcapsules. UF resin modified with melamine can improve thermal stability, water resistance, durability, roughness and shell strength of the microcapsule, and simultaneously reduce its formaldehyde emission content[2]. The surface features, geometry and properties of microcapsules depend on the experimental core/shell ratio, reaction temperature, agitation speed, and surfactant, etc. The prepared MUF microcapsules have been optimized at various parameters of chemicals ratio and synthesis condition , etc, as described in our previous work[3]. Various diameters of the MUF microcapsules were obtained at different agitation rates, typical at 250rpm, 600rpm and 1000rpm respectively, and average sizes of MUF microcapsules were 550μm, 200μm and 90μm correspondingly. Surface morphology and shell thickness for MUF microcapsules are shown in Figure 1, indicating that size distribution of the microcapsules is a typical normal distribution and the spherical MUF microcapsules surfaces are rough and uneven. There is little adhesion each other and MUF resin accumulation among the microcapsules. The shell which was formed by MUF resin during the polymerization deposition is compact with thickness of 3.2μm. Additionally, there is no significant difference in the shell thickness among the several sizes microcapsules.

Core content of the microcapsule was obtained by extracting method using acetone. Microcapsules were ruptured adequately in agate mortar and socked in acetone for 24 hours, then treated by ultrasonic apparatus for 30 minutes and continuously socked in acetone for 24 hours to fully dissolve the core material. After a final filteration, the residual was the shell material, which was vacuum dried and weighed. The core contents of MUF microcapsules were obtained according to Equation. The core contents of MUF microcapsules with diameters of 550μm, 200μm and 90μm are 74.52%, 70.15% and 68.71% respectively.

$$\text{Microcapsule core content}\ \% = \frac{M_0 - M}{M_0} \times 100\% \qquad (1)$$

Where, M0(g) is weight of MUF microcapsules, M(g) is weight of the shell material.

As seen in Figure 2, the dispersity of MUF microcapsules in the epoxy material is relatively uniform. Good interfacial bonding between microcapsules and matrix is achieved due to the rough surface of MUF microcapsules increasing the contact area with epoxy substrate.

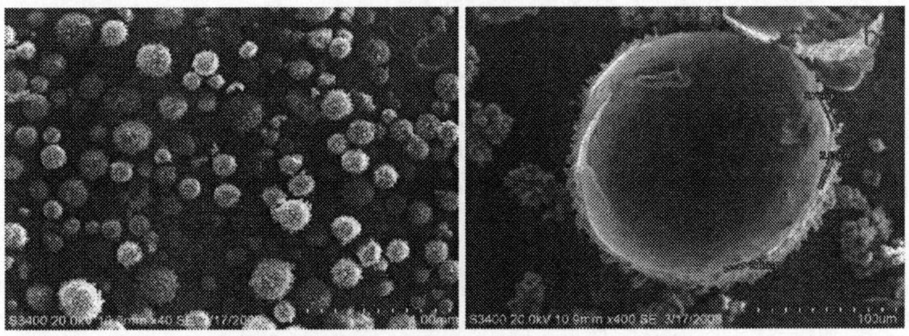

Figure 1. Morphology of MUF microcapsules. a) Surface characteristic; b) Shell thickness.

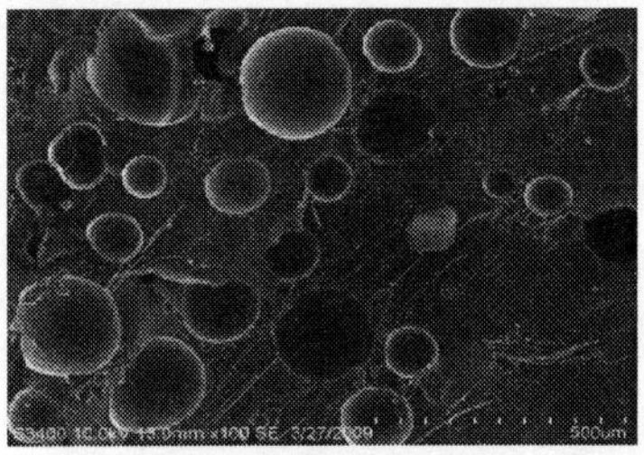

Figure 2. MUF microcapsule distribution in epoxy matrix (5wt% microcapsule content)

3.2. Bending Testing

Tensile performance and bending performance are important mechanical properties of material. Fracture elongation is a crucial parameter to evaluate fracture toughness. The previous results indicated that the microcapsules reduce tensile performance of polymer composite, but toughen the composite by 2-5% contents of microcapsule addition. With an increase of microcapsule content, tensile strengths of microcapsule/polymer self-healing materials are reduced while the elongations are increased. From 5% to 10% microcapsule content, the elongations decline further[4].In tensile testing, fracture location of the material is often unexpected due to imprecisely controlled damages introduced, thus affecting the testing accuracy. Three-point bending experiment is therefore used to make stress more concentrated on central area of material specimen along longitudinal direction where cracks are more likely to be produced rapidly, which facilitates the damaging process with an accurate control. As shown in Figure 3 (a), compared with the bending strength of epoxy specimen, those of microcapsule/epoxy specimens are drastically reduced. The microcapsules embedded in epoxy matrix may play a function of the material defects as stress concentration points inducing micro-cracks around the MUF microcapsules on the loading region where stress was introduced by three-point bending, decreasing mechanical strength of the material. When microcapsule content is 2wt%, the bending strengths of specimens whose MUF microcapsule diameters are 90μm, 200μm and 550μm are reduced to 90%, 81% and 53% value of the epoxy specimen respectively. The microcapsule size is a crucial factor affecting the descend range of bending strength for the material with this microcapsule content. The smaller microcapsules size, the better microcapsules dispersity exists in the matrix. The specific surface areas are bigger which increase the contact areas between microcapsules and epoxy substrate, thereby being capable of bonding the epoxy matrix more effectively and bearing bigger stress. When the content exceeds 2wt%, the difference is not so significantly affected by microcapsule size in the descend range of bending strength for the material. At the same condition of microcapsule mass fraction, although the larger size microcapsules lead to greater stress concentration degree on themselves, the smaller amounts microcapsules are contained in the matrix so fewer locations for stress concentration exist. Thus, the numbers of induced micro-cracks are relatively less exhibited. In contrast, more micro-cracks are produced by smaller size microcapsules throughout the specimen, but the degree of stress concentration for the material is relatively lower by the small size microcapsules. Therefore, microcapsule numbers and sizes of the epoxy material with the condition of high microcapsule mass fraction together affect the descending bending strength for the epoxy composite.

Bending fracture strain is the other important parameter to evaluate fracture toughness. The larger strain of bending fracture, the better toughness of material indicates. The effects of microcapsule content and size on bending fracture strain of the epoxy composites are seen in Figure 3(b). When the epoxy is subjected to an external force, matrix itself bears the load, so the lower toughness of material is shown. When the microcapsule content is 2wt%, the bending fracture strain of the specimen is visibly increased which demonstrates a toughening effect of MUF microcapsules on epoxy material. With the increase of MUF microcapsule content, bending fracture strains of the specimens embedded MUF microcapsules, whose diameters are 90μm and 200μm, are slightly increased at a low microcapsule content, reaching a maximum strains at 2wt% microcapsule content, then that of the specimens are gradually decreased at high microcapsule contents. Particularly, within whole range of microcapsule content, the bending fracture strain of epoxy specimen embedded microcapsule whose diameter is 90μm is larger than that of epoxy specimen with 200μm microcapsules, showing a significant toughening effect of microcapsules on epoxy specimen. In contrast, the epoxy specimen embedded 550μm microcapsules have obviously negative effect on both the bending fracture strain and the bending strength. Additionally, when microcapsule content increases to 10wt%, a large amount of MUF microcapsules are served as stress concentration points and defects within the material, the bending strength and fracture strain of the epoxy specimens embedded different size microcapsule especially 550μm microcapsule are deteriorated significantly. Therefore, the MUF microcapsule/epoxy composites embedded 2wt% and 5wt% microcapsule contents whose diameters are 90μm and 200μm were studied in the following.

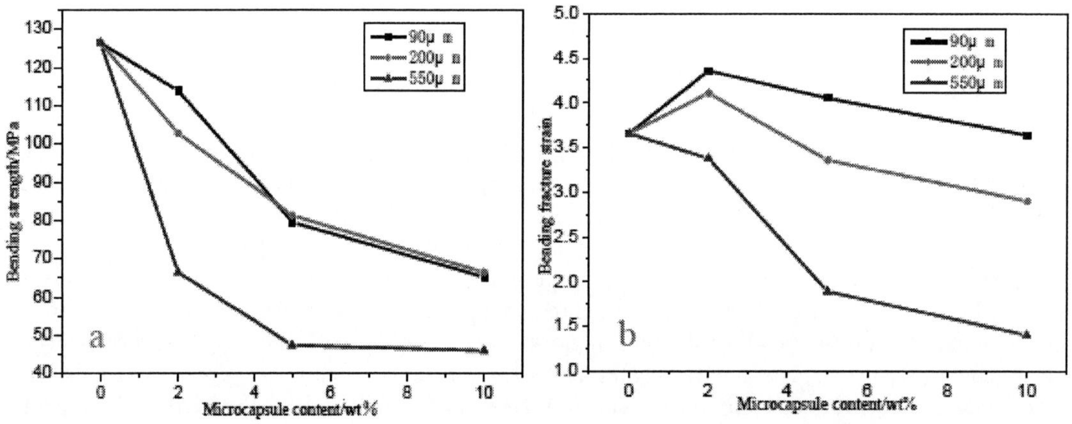

Figure 3. Effect of MUF microcapsules on bending properties for epoxy composites.
(a) Bending strength comparison; (b) Bending fracture strain comparison

3.3. Dynamic Mechanical Properties

Dynamic mechanical analysis is non-destructive to the measured material and sensitive to the mechanical relaxation transition for chain segments and molecular motion of polymeric material. The structural factors of dynamic modulus and mechanical loss factor of polymer with temperature under fixed frequency are simultaneously obtained to analyse microscopic movements within material. Storage modulus is a measure of material stiffness to characterize the ability to resist deformation of material. The higher storage modulus material, the stronger resist deformation ability has. Polymer mechanical performance is relative to the crosslink density of polymeric network. The higher crosslink

IOP Conf. Series: Materials Science and Engineering **275** (2017) 012017 doi:10.1088/1757-899X/275/1/012017

degree of polymeric network, the shorter length of chain segment between cross-link points and the more compact cross-link networks exist within polymer, forming high strength of cured substance. The temperature corresponding to the peak of Tanδ is defined as glass-transition temperature[5].

As shown in Figure 4(a), compared to epoxy specimen, the storage modulus of epoxy composite specimen declines because crosslink density and resist deformation ability for the composite is reduced due to the addition of MUF microcapsules. With an increase of microcapsule content, initial storage moduluses of epoxy composites with 2wt% and 5wt%microcapsule content are respectively decreased to 93% and 84% value of the epoxy, which is consistent with the changes of bending strength in the bending performance discussed before. The distance between molecular chains and free volume in crosslink network of epoxy resin are increased with the rise of MUF microcapsule content, thus decreasing the storage modulus and improving the epoxy material toughness. In Figure 4(b), two or more transition peaks are not appeared in Tanδ curves of the three specimens, which indicates a good compatibility and interfacial adhesion between MUF microcapsules and epoxy matrix. The glass-transition temperatures of the epoxy specimens with 0wt%, 2wt% and 5wt% microcapsule are 80.2°C, 77.4°C and 77.5°C respectively. The cross-linking density of the epoxy material is decreased by microcapsule addition, causing the Tg of the epoxy composite specimens embedded MUF microcapsules is slightly lower than that of the epoxy specimen. Besides that, there are few differences in both the peak values of Tanδ and the Tg for the epoxy composites specimens with 2wt% and 5wt% MUF microcapsules, indicating that compared to the epoxy specimen with 2% MUF microcapsule content, the addition of 5% MUF microcapsule content into the epoxy specimen is not capable of obviously reducing the thermostability for the material[6].

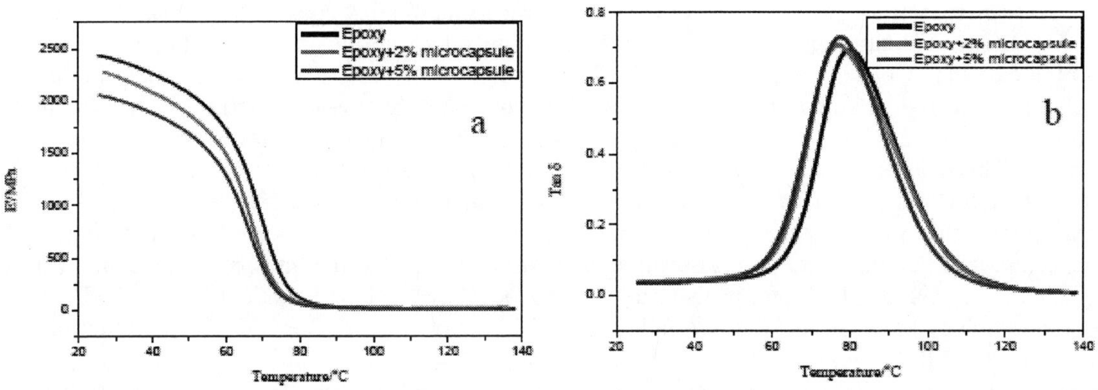

Figure 4. Effect of MUF microcapsule content on dynamic mechanical properties for epoxy composites.
(a)storage modulus (E'); (b)mechanical loss factor (Tanδ) .

3.4. The Toughening Effect of Microcapsule on Epoxy Composite

In the previous discussion, bending properties and dynamic mechanical performances indicated that the MUF microcapsule/epoxy composite materials were toughened by the addition of MUF microcapsules. Besides these data, morphology analysis on the fractured surface of the composites reveals this toughening effect.

Typical features of the epoxy composites with different MUF microcapsules size were shown in Figure 5. As seen in Figure 5(a) and 5(b), the smaller microcapsules size contained in epoxy specimen, the rougher fracture surfaces on the epoxy surrounded with microcapsules appear. Figure 5(c) shows a typical brittle fracture surface of the epoxy specimen embedded 550μm size microcapsules. In toughening materials, the distance between two toughening particles is important to affect the toughness. With a decrease of this distance between poly(acrylonitrile-butadiene-styrene) copolymer (ABS) particles within polycarbonate (PC) matrix, the stress fields around ABS particles occur overlap leading to the generations of more crazings and shear yielding, thus improving the material toughness[7]. For the case studied here, at a same level of microcapsule content, the smaller microcapsule size, the greater microcapsules spatial density and the shorter distance between microcapsules are within matrix, resulting in larger toughening areas which consume more surface energy. This observation also explains the increase of bending fracture strain for the epoxy composites by the small size microcapsules addition.

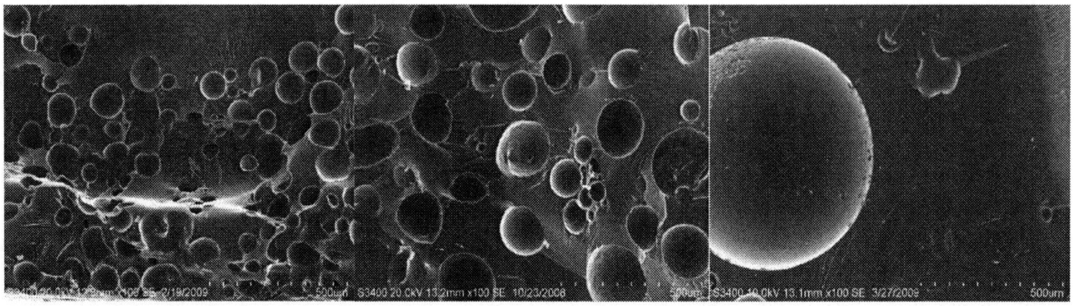

Figure 5. Effect of MUF microcapsules size on fracture morphology of epoxy composites (5wt% microcapsule content).
(a) specimen with 90μm microcapsules diameter; (b) specimen with 200μm microcapsules diameter; (c) specimen with 550μm microcapsules diameter.

Several types of toughening effects for MUF microcapsules/epoxy composites are shown in Figure 6. Caudiform structure appears around the surrounding microcapsule in Figure 6(a), which can be attributed to shear band within matrix initiated by the microcapsules as stress concentration points. A number of crazings are induced at the end of shear bands to absorb stressing energy[8]. When fracturing occurs in the material, deformation area may be enlarged by the microcapsules incorporated in epoxy matrix. This consumes an extra energy, which in fact toughens the epoxy material. Besides, a multilevel fracture path is exhibited within the matrix indicating a good fracture toughness can be realized by this rough fracture surface which is initiated by the microcapsules producing stress concentration, guiding cracks to pass through microcapsules at different spatial locations, further extending on multilevel fracture paths to delay the process of crack propagation[9]. Figure 6(b) shows the morphology of fracturing and debonding for MUF microcapsules. The crack is finally terminated by fractured MUF microcapsules, which reveals that addition of the microcapsules into epoxy is effectively prevent the crack propagation. A destruction of the interface between microcapsules and epoxy matrix consume considerable extra energy, which also toughens the epoxy materials.

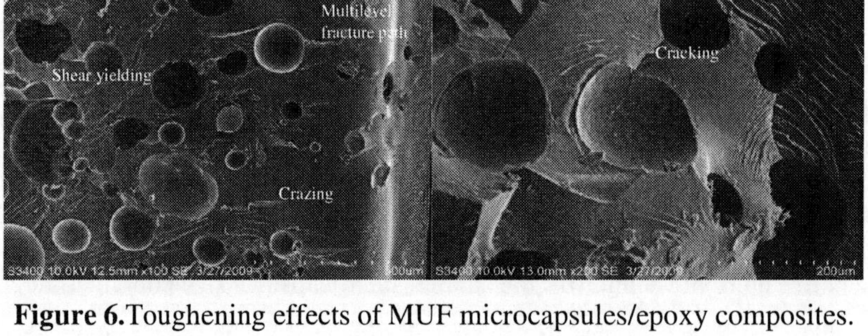

Figure 6. Toughening effects of MUF microcapsules/epoxy composites.
(a)multilevel fracture plane; (b)fracturing and debonding microcapsules

4. Conclusion

MUF microcapsules composed of a shell of MUF resin and a core of modified epoxy healing agent were adopted together with imidazole chemical for designing a self-healing MUF microcapsule/epoxy composite material. The spherical microcapsule shell is compact, rough and uneven with a thickness of 3.2μm, having normal size distribution. There is no adhesion and accumulation between MUF microcapsules. Core contents of the MUF microcapsules for the diameters of 550μm, 200μm and 90μm are 74.52%, 70.15% and 68.71% respectively. There is a good dispersion of MUF microcapsules in epoxy composite and a good interface between the microcapsule and epoxy matrix as designed. Both bending strength and storage modulus of the microcapsule/epoxy composite are considerably reduced with the increasing addition of the microcapsules, whereas the glass transition temperatures of microcapsule/epoxy composites are not visibly influenced. Significant toughening effects of MUF microcapsules on the epoxy composites are obtained at the conditions of different content and size of microcapsule especially at low microcapsule contents and small microcapsule sizes and observed several toughening morphologies including shear yielding, crazing effect, micro-cracking, multilevel fracture path, microcapsule fracture and debonding. Toughening effect of the shear yielding is the most effective.

5. Acknowledgment

Foundation: National Natural Science Foundation of China (51378315); Research & Development Fund of Shenzhen (JCYJ20130329114709152)

6. References

[1]. NI Zhuo, DU Xue-Xiao, Polymer Materials Science & Engineering, 25, 133-136, 2009.
[2]. A. Philbrook, C. J. Blake, N. Dunlop, C. J. Easton, M. A. Keniry, J. S. Simpson, Polymer, 46,2153-2156, 2005.
[3]. Z Ni, XX Du, F Xing, L Zhou, Journal of Shenzhen University Science & Engineering, 25, 351-357, 2008.
[4]. NI Zhuo, P Zhang, YL Lin, S Wang, LI Wei wen, Journal of Shenzhen University Science & Engineering, 27, 260-266, 2010.
[5]. W. Stark, Polymer Testing, 32, 231-239, 2013.
[6]. X.-M. Tong, T. Zhang, M.-Z. Yang, Q. Zhang, Colloids and Surfaces A: Physicochemical and Engineering Aspects, 371, 91-97, 2010.
[7]. M. Ishikawa, I. Chiba, Polymer, 31, 1232–1238, 1990.
[8]. C. Kaynak, O. Cagatay, Polymer Testing, 25, 296-305, 2006.
[9]. H. Jin, G. M. Miller, S. J. Pety, A. S. Griffin, D. S. Stradley, International Journal of Adhesion and Adhesives, 44, 157-165, 2013

Self-Repairing Mechanism of MUF/Epoxy Microcapsules for Epoxy Material

Zhuo Ni[1], Yuhao Lin[2] and Xiaobo Zhou[3]

[1]College of Chemistry and Chemical Engineering, Shenzhen University Shenzhen, China. e-mail: zhuoni@szu.edu.cn
[2]Lin. College of Chemistry and Chemical Engineering, Shenzhen University Shenzhen, China. e-mail:823394522@qq.com
[3]College of Chemistry and Chemical Engineering, Shenzhen University Shenzhen, China. e-mail: xuexiaodu@szu.edu.cn

Abstract: In this paper, a post curing reaction for the microcapsule/epoxy composite material and the conditions of thermal treatment for self-healing process were studied by differential scanning calorimetry (DSC). The condition of thermal treatment for post curing (60°C, 2 hours) was employed to fully cure the epoxy composite. Damage mechanism for the epoxy material was demonstrated via data simulation and three-point bending experiment for the stress distribution reveals that micro-cracks are more likely to be generated on the central region in stress concentration area of two constrained boundaries and the numbers of micro-cracks are reduced from the central area to the two ends of the material. Self-repairing performances of MUF microcapsule/epoxy composite materials were characterized using both destructive bending tests and non-destructive DMA measurements. Self-healing efficiencies of the composites embedded 2% and 5% microcapsule content measured by DMA are 101% and 104% respectively which are close to those results of 104% and 113% correspondingly measured by bending tests. Crack formation and development, core material releasing for MUF microcapsules and physiochemical process of the self-repairing were investigated by using OM, fluorescent technique and infrared microscope. These provide detailed evidences and important information on self-healing mechanism of the microcapsule/epoxy self-repairing material.

1. Introduction

At present, self-repairing properties including fracture toughness or varied mechanical properties of virgin, damaged and healed material are tested by these typical destructive methods at the cost of damaging the sample integrity [1]. In contrast, little work demonstrates a non-destructive test to reasonably investigate the self-healing performance of the materials. Dynamic mechanical performance of the same specimen exhibits a stable response undergoing several DMA non-destructive testings. The changes in dynamic modulus and mechanical loss factor with temperatures and frequency initiated by material molecules or structural modification are simultaneously provided, which is informative to understand the situation of segmental motion for the self-healing material after damaging and self-repairing process on a microscopic level.

Damage mechanism for the epoxy material was demonstrated via data simulation for the stress distribution in three-point bending [2-4]. The post curing reaction for the microcapsule/epoxy composite material, the conditions of thermal treatment to eliminate the interference of the post curing on the evaluation of the self-healing property and self-repairing cross-linking reaction were studied by differential scanning calorimetry (DSC). Self-healing process for MUF microcapsule/epoxy composites

Content from this work may be used under the terms of the Creative Commons Attribution 3.0 licence. Any further distribution of this work must maintain attribution to the author(s) and the title of the work, journal citation and DOI.
Published under licence by IOP Publishing Ltd

are applied at a high temperature. Damage degrees of these materials were controlled by using constant strain three-point bending fixture. Additionally, self-repairing performance of the MUF microcapsule/epoxy composite material was quantitatively characterized using both bending destructive tests and DMA lossless tests. Core material releasing for MUF microcapsules and physiochemical process of the self-repairing were investigated by using fluorescent technique and infrared microscope. These provide important information on the formation and propagation of micro-cracks in the composite, particularly on self-healing mechanism of the microcapsule/epoxy self-repairing material.

2. Experimental Section

2.1 Preparation of MUF Microcapsules
Refer to previous works, reference 5

2.2 Self-Repairing Efficiency Measurements

2.2.1 Self-Healing Efficiency Characterized by Bending Properties. Self-repairing effects can be evaluated by the variations of bending properties for the microcapsule/epoxy composite subjected to the damages and self-repairing process. The healing efficiency (self-healing), which is the ratio of regained strength for the self-healing material to losing strength for the damaged material.

*2.2.2 Self-Healing Efficiency Characterized by Dynamic Mechanical Properties.*Self-repairing effects can also be evaluated by the differences in DMA data for the microcapsule/epoxy composites subjected to damages and self-repairing process. The healing efficiency (P'self-healing), which is the ratio of regained storage modulus for the self-healing material to losing storage modulus for the damaged material.

2.3 Data Simulation for The Stress Distribution in Three-Point Bending
ABAQUS finite element software was adopted to compute numerical values of stress distribution in the epoxy composites. The dimension of calculation model was the same as that of tested specimen. A simply-supported edge was selected on a center span of 51mm for long edge and 4 mm vertical displacement was exerted on the center in this model. The data simulation was obtained by C3D8 module with an elastic modulus of 2GPa and a Poisson ratio of 0.38. The stress response at elastic stage was merely considered in order to simplify stress concentration analysis.

2.4 DSC Test
Differential scanning calorimetry (DSC-60, Shimadzu, Japan) calibrated by high purity indium and zinc was employed to study the epoxy post curing reaction and the condition of post curing treatment and self-healing thermal treatment for MUF microcapsule/epoxy composite. The specimens were heated from 25 to 220 °C at a heating rate of 10 °C /min under a nitrogen atmosphere at a flow rate of 50mL/min.

2.5 Testing Sample
At least ten MUF microcapsule/epoxy composite specimens were selected to form a batch of MUF microcapsule/epoxy self-healing sample. Nine specimens were divided into three groups including virgin sample (Type I), damage-controlled sample (Type II) and self-healing sample (Type III). The tenth specimen was used for DMA test. Because these specimens may occur post curing reaction during heat process that interfere with evaluation of the self-healing efficiency, the needed thermal preprocessing conditions were discussed in the relevant DSC measurement.

The Type II and the Type III samples were subjected to a controlled constant strains using three-point bending fixture with deformation of 90% bending fracture strains of the specimens which produce cracks within the materials based on the bending deformation vertical distance of the specimens in the fixture according to Equation 3. The specimens were clamped at this bending deformation distance for 24 hours. After the cracks observed, the specimens were removed off and flattened under a loading, then tested by universal tensile machine for bending properties and DMA instrument for the dynamic mechanical property.

The Type III samples after subjection to constant strain damaging were heated in vacuum drying oven at 100°C for 1 hour. After the self-healing process, bending properties of these specimens and the dynamic mechanical property were measured through bending tests and DMA test.

2.6 Bending Test

Bending properties for three types of samples were measured by universal tensile machine (model AGX, Shimadzu, Japan). The Type I samples were directly measured without any treatment. In this three-point bending test, bending span was 30 mm and bending speed was 5mm/min. Force-displacement curves were recorded for data analysis. The average value of three specimens data was taken as corresponding type bending strengths (I) and bending fracture strains (ε) [5].

2.7 DMA Test

Storage modulus (E'), loss modulus (E'') and mechanical loss factor (Tanδ) of the tested specimen were measured via DMA temperature sweep experiment (Q800, TA Instruments, UK). DMA test was performed using a double cantilever beam, heating from 25°C to 140°C at a heating rate of 5°C /min and an oscillator frequency of 1 Hz under nitrogen atmosphere[6].

2.8 Inverted Fluorescence Microscope Analysis

In order to study self-repairing mechanism of the MUFmicrocapsule/epoxy material, a batch of MUF microcapsules was particularly prepared by the addition of sodium fluorescein (F6377, BioReagent, Sigma-Aldrich China) into the epoxy healing agent. These special microcapsules were added into epoxy resin in similar ways to fabricate the composites previously to emphasize self-healing processing and mechanism. Crack propagation of the special specimen after subjected to the damage was observed using inverted fluorescence Microscope (IX70, Olympus, Japan) equipped with a 100W high-pressure mercury lamp as excitation source while a objective lens (40x, N.A. = 0.6, Olympus, Tokyo, Japan) and a blue optical filter were employed. An SIT camera(C2741-08, HAMAMATSU, Shizuoka, Japan) and a digital videocassette recorder (DVCAM, SONY, Tokyo, Japan) were also used for detection of an emission light resulting from the releasing of the healing agent mixed with the fluorescein, recording the rupture and release information of MUF microcapsules in the damaged specimen.

2.9 Infrared Microscope Analysis

Infrared microscope (Nicolet Continuμm, ThermoFisher, USA) was employed to analyze the filling materials on the fracture plane of the microcapsule/epoxy composite. Before measurement, cooling container of the microscope was filled with liquid nitrogen. The apparatus was stabilized for 1 hour in order to fully cool MCT detector. The optical gating of the infrared microscope was adjusted to 100μm×100μm. The image observation software attached to the apparatus was used to select the areas of visible fillings within the cracking of the composite specimen and to identify its chemistry. The IR spectrum was obtained in the wave number range 4000-450 cm-1 with a resolution of 4cm-1 from 28 scans.

3. Results and Discussions

3.1 Damage Mechanism for Epoxy Composites by Three-Point Bending

Micro-crack damage within material can be induced via a process of stretching. An effective course is needed for micro-crack propagation, while the pre-stretching damages are hardly controlled by tensile machine for a long time so that crack growth cannot be located using microscope technique. In contrast, constant strain applied in three-point bending fixture was employed to act on the self-healing material in a process of pre-bending, which could design various stages of the damage for detailed study.

Figure 1(a) shows the constant strain of three points bending fixture is applied in controlled damage for the epoxy composite specimens. Stress distribution on the composite material was analyzed in the condition of constant strain using three-point bending model. The data analysis in Figure 1(b) predicts the Mises stress distribution of equipotential line for the designed composite material and counter-force on vertical displacement is 0.448N. The Mises stress has a maximum value of 0.684 N/mm2 located at the central of the specimen, which is quickly attenuated from the center region along the z direction to the two ends because of small thickness of the specimen, so micro-cracks are more likely to be initiated on the center region of two constrained boundaries in stress concentration area. The OM photograph in Figure 1(c) shows that crack patterns for MUF microcapsule/epoxy self-healing material damaged by constant strain in three-point bending fixture. The region shown in Figure 1(c) is corresponding to the red frame of simulative specimen in Figure 1(b). Cracks are mainly observed in the center of the specimen where the specimen is subjected to the maximum vertical stress. With the gradual reduction of stress along the specimen edge, the numbers and the widths of micro-cracks are decreased from the central area to the two ends of the epoxy composite specimen. Crack initiation of the material subjected to bending by the fixture was located as expected, which could help explain the generation of cracks discussed previously. As shown in Figure 1(d) magnified by using monitoring software, the MUF microcapsule whose diameter is approximate 90μm of the epoxy composite specimen after subjection to constant strain damaging is entirely ruptured by a crack with 20-50μm width that causing adequate epoxy healing agent is capable of releasing into the crack surface. After self-healing process seen in Figure 1(e), the cracking and the MUF microcapsule in the epoxy matrix are entirely rebonded because of the cross-link reaction between the released epoxy healing agent and curing catalyst. However, those 200μm MUF microcapsules incorporated in the epoxy composite after damaging may not be fully ruptured by this width of crack causing less volume of epoxy healing agents flow into the crack plane directly reducing self-healing performance of the material. Therefore, the epoxy composite specimens embedded 90μm MUF microcapsules are merely applied in the following testings.

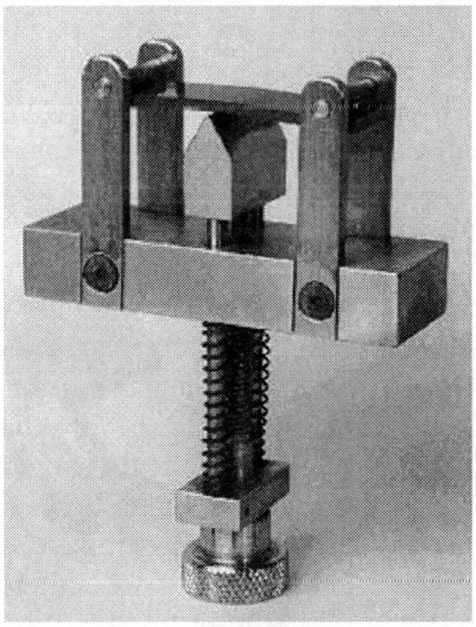

(a)

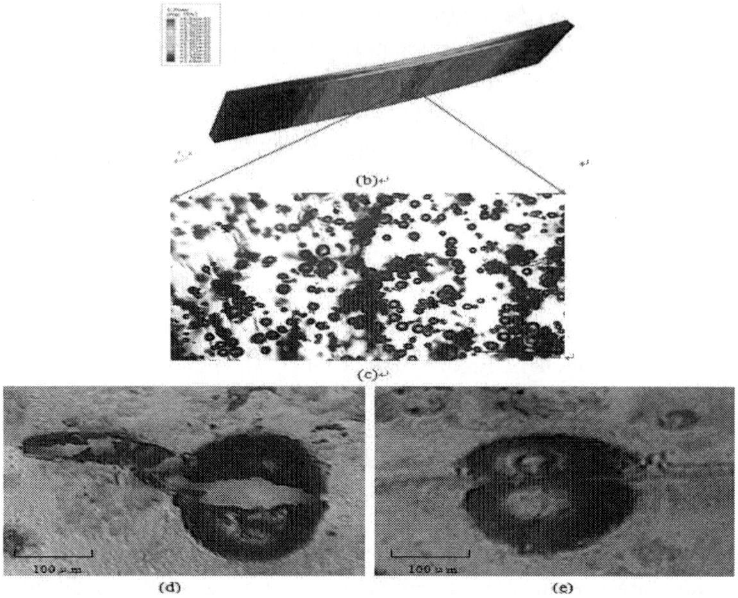

Figure 1. Data of three point bending testings for the MUF microcapsule/epoxy self-healing materials. (a) constant strain of damaging fixture photo; (b) Mises stress distribution simulation; (c) crack distribution after subjection to constant strain damaging. (d)MUF microcapsule morphology after subjection to constant strain damaging; (e) MUF microcapsule morphology after self-healing process;

3.2 Self-Healing Performance Test

3.2.1 Post Curing Reaction and The Condition of Self-Healing Process. The DSC curves of MUF microcapsule/epoxy composite are shown in Figure 2. The exothermic peak between curve A and baseline (dash line in this figure) over the temperature range above the Tg of epoxy material may demonstrate an existence of post-curing reaction within the epoxy matrix[7]. During the later stage of specimen curing, diffusions of epoxide active centers are difficult. The epoxy composite system transforms to glassy state and the cross-linking reaction is controlled by the chemical diffusion rather than reaction kinetics so that some unreacted active centers still remain within the epoxy matrix[8]. With the increase of testing temperature, chemical diffusions of residual active epoxides within the matrix are improved initiating the post curing reactions within the matrix between the residual epoxy active centers and unreacted curing agent DETA. Therefore, a small exothermic peak of post curing for the epoxy composite specimen cured at room temperature is observed in curve A. Because the post curing reaction may interferes with the self-healing efficiency evaluation for MUF microcapsule/epoxy composites, a post-curing thermal treatment (60°C, 2 hours) for the epoxy composite specimen was applied in order to fully cure the epoxy specimen used in following self-healing performance testings. The curve B exhibits there is no exothermic peak over the temperature range above the Tg of the MUF microcapsule/epoxy composite specimen after the post-curing heat treatment applied, which demonstrates the epoxy composite specimen was fully cured in this thermal treatment condition. The curve C indicates that multiple exothermic peaks of the specimen Type II are observed at approximate 100°C which illustrates significant curing reaction between the epoxy resin and the MC120D catalyst occurs at this temperature. In addition, a broad exothermic peak over the temperature range above the Tg of the epoxy composite demonstrates an effective cross-linking reaction between the healing agent released from ruptured MUF microcapsules and the latent imidazole curing agent embedded within the matrix occurs at the high temperature range, being capable of automatically repairing the cracks within the materials.

IOP Conf. Series: Materials Science and Engineering **275** (2017) 012018 doi:10.1088/1757-899X/275/1/012018

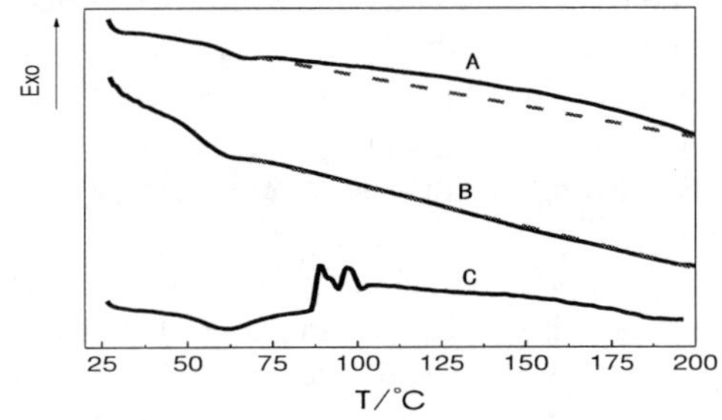

Figure 2. DSC curves of MUF microcapsule/epoxy composites.
(a) curing at room temperature; (b) curing at 60°C for 2 hours; (c) heat flow changes in self-healing process.

3.2.2 The Self-Healing Efficiencies Characterized by Bending Property. Bending properties for three types of the epoxy composites designed at different microcapsule contents are shown in Figure 3. Bending strength of the specimen Type I with 2wt% content microcapsule is reduced to 90.4% value of the epoxy specimen. Microcapsules as stress concentration points directly cause the generation of cracks in the specimen after the constant bending applied so that bending strength of the specimen Type II is decreased. Bending strength of the specimen Type III is recovered and the self-healing efficiency is 104%, while that of the epoxy specimen is not regained. In the condition of 5wt% microcapsule content, bending strength of the specimen is reduced to 64.3% value of the epoxy specimen and the self-healing efficiency of the specimen achieve 113%, which can be attributed to the higher microcapsule content in the self-healing material, the more volumes healing agents release from the ruptured microcapsules into the cracking so that the self-repairing ability of the material is effectively improved.

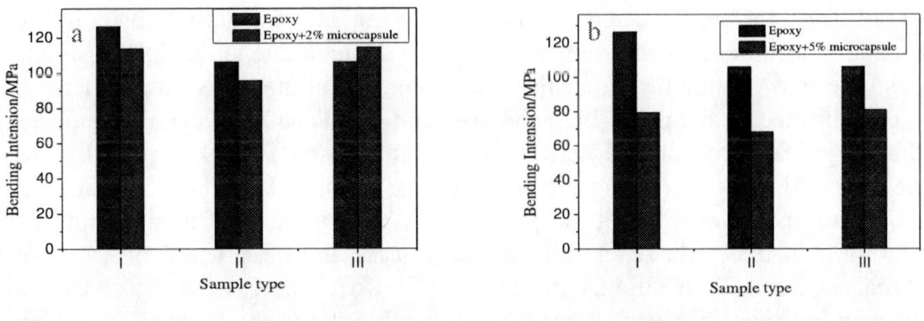

Figure 3 Bending strength comparisons of self-healing effects for epoxy composites.
(a)2wt% microcapsule content; (b)5wt% microcapsule content.

3.2.3 The Self-Healing Efficiencies Characterized by Dynamic Mechanical Property. Dynamic mechanical properties with the temperatures and the information of chain segment movement for three types of the epoxy composites can be simultaneously monitored[9]. As shown in Figure 4(a), value of initial storage modulus for epoxy specimen is 2.44GPa. That of the epoxy specimen subjected to damage by constant strain dramatically reduces to 1.82GPa and that of the epoxy specimen after self-healing is 1.84GPa. No obvious repairing effect is founded in the epoxy specimen. Figure 4(b) indicates that the value of initial storage modulus for MUF microcapsule/epoxy composite specimen with 2% microcapsule content is 2.27GPa. That of the epoxy composite specimen subjected to damage is decreased to 1.97GPa and that of the epoxy composite specimen after self-healing process is recovered to 2.27GPa. In Figure 4(c), the value of initial storage modulus for MUF microcapsule/epoxy composite specimen with 5% microcapsule content is 2.04GPa. That of specimen type II declined to

1.65GPa and that of specimen type III is considerably recovered to 2.06GPa, slightly exceeding original value of the material. These changes indicate that cross-linked network within epoxy matrix subjected to damage process is broken and molecular chains are fractured under loading, thus reducing the ability to resist deformation. After self-repairing, because epoxy composite is incorporated with MUF microcapsules contained epoxy healing agent and latent curing agent which are able to initiate cross-linked reaction in the self-healing process at a high temperature, cracks are rebonded by this chemical reactions and the broken cross-linked epoxy network is repaired in some extent, leading to a recovery for the material properties. The 3.12% self-healing efficiency of the epoxy specimen may result from sample error, experimental error and instrumental error, which need to be taken into consideration in evaluation of self-healing efficiency. The significant self-repairing in the MUF microcapsule/epoxy composite specimen with 2% and 5% microcapsule content achieve high self-healing efficiencies of 101% and 105% respectively.

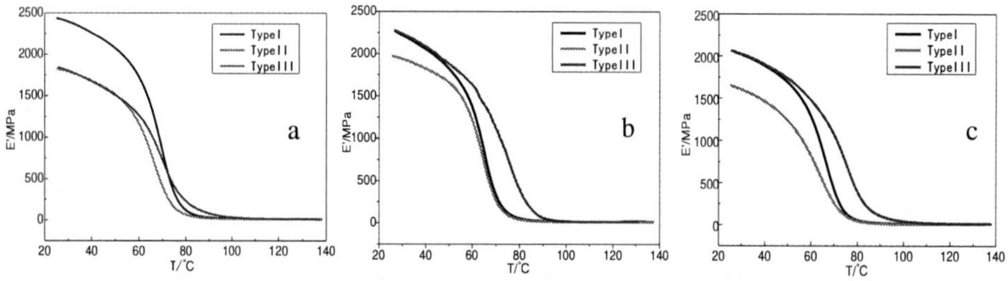

Figure 4 Storage modulus for three types of epoxy composite specimens.
(a) epoxy specimen; (b) MUF microcapsule/epoxy self-healing specimen (2% microcapsule content);
(c) MUF microcapsule/epoxy self-healing specimen (5% microcapsule content);

With regard to thermoset epoxy resin, peak value of Tanδ reflects relative degree of cross-linking density. Peak width of Tanδ is related to the relaxation transition for chain segment[10]. In Figure 5(a), the peak value of Tanδ for epoxy specimen is relatively high with a glass-transition temperature of 80.2°C. Relaxation transition of chain segment in epoxy is difficult due to its high cross-linking density which provides little free volume for molecular movement within the cross-linking network. After the epoxy specimen subjected to damages, both the area and the Tg of Tanδ peak are decreased, Tg is 78.7°C. At this stage, The cross-linked network within the matrix is broken partially and molecular chains may be fractured, thus decreasing the cross-linking density. In addition, because the length of chain segment is shortened, flexibility of molecular chains and mobility of motion units are promoted, resulting in this shift of the peak. After self-healing process, the peak width of Tanδ for the epoxy specimen is broadened and Tg is slightly increased to 79.8°C, whereas the peak value considerably decreases. The rearrangement of chain segment occurs within the epoxy material broadening the peak width. However, the damaged cross-linked network is not repaired so the Tg is not obviously increased. As shown in Figure 5(b), the peak value of Tanδ for MUF microcapsule/epoxy composite specimen with 2% microcapsule content is relatively high with a glass-transition temperature of 77.4°C. After damages applied in the epoxy composite specimen, both the shape and the corresponding temperature of Tanδ peak are reduced with the Tg of 75.9°C. The molecular chains in the composite are fractured during damage and the cross-linked network is partially broken, which reduces the cross-linking density. Flexibility of molecular chains and segment mobility are increased, leading to a movement toward a low temperature of the peak. The peak width of Tanδ for the epoxy composite specimen after self-healing process is broadened and the Tg is significantly increased to 85.3°C, while the peak value is considerably reduced. After self-repairing, it is speculated that non-uniformity of chain segments in the self-healing process at high temperature causes an improved mobility of segments so that peak width of the Tanδ is broadened. Compared with epoxy specimen, glass-transition temperature of MUF microcapsule/epoxy composite specimen is considerably increased by approximate 10°C that could result from the damaged material is more densified by the cross-linking reaction between healing agent and curing catalyst involved. Additionally, the cross-linked network is introduced with long chain segments from special curing agent molecular of latent imidazole curing catalyst MC120D. The length and the stiffness of the chain segment related to the cured epoxy resin in the cracking location and steric hindrance of internal rotation for the chain segments are increased. As a result, the glass-transition temperature of epoxy composite specimen is considerably raised and the dynamic mechanical performance is also enhanced. Additionally, the changes in the Tg and the peak value of Tanδ for MUF microcapsule/epoxy composite specimen with 5% microcapsule content are similar to that of the epoxy composite specimens with 2% microcapsule content shown in Figure 5(c). The Tg for three types of the epoxy composite specimens are 77.5°C, 73.3°C and 84.8°C respectively.

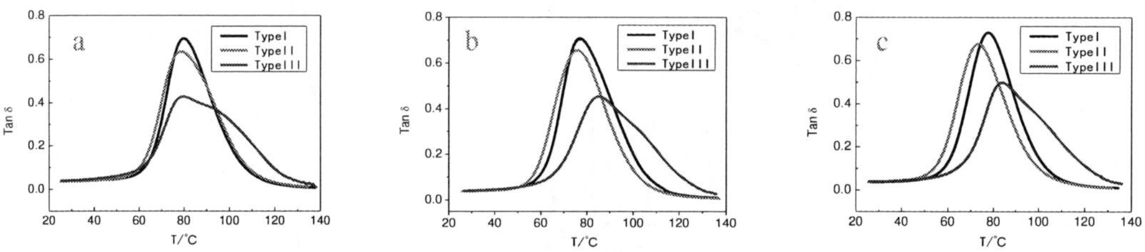

Figure 5. Mechanical loss factors for three types of epoxy composite specimens.
(a) epoxy specimen; (b) MUF microcapsule/epoxy self-healing specimen (2% microcapsule content); (c) MUF microcapsule/epoxy self-healing specimen (5% microcapsule content).

Healing efficiency of MUF microcapsule/epoxy material with 2% and 5% microcapsule content measured by DMA are 101% and 104%, which are close to those of the materials characterized by bending properties. Because bending testing is destructive, bending properties of three type specimen cannot be directly compared. A large numbers of experimental specimen having same composition and closing property must be used in order to have specimen repeatability and testing reliability. In contrast, DMA which is non-destructive, sensitive, repeatable and reliable provides a new method in the research of self-healing material.

3.3 Self-Repairing Mechanism

In this part, formation and propagation of the micro-cracks in MUF microcapsule/epoxy composite, reason and process of core healing agent releasing from microcapsule are observed using OM and particularly emphasized by inverted fluorescence microscope. The course of autonomic repairing micro-crack and the chemistry of the crack filling material are studied specifically by infrared microscope discussing the reaction mechanism between epoxy healing agent and latent imidazole curing catalyst.

Figure 6 shows a series OM image of a cracking development for MUF microcapsule/epoxy self-healing composite damaged by a constant strain. The rupture process of MUF microcapsule and the release process of healing agent contained sodium fluorescein can be observed using inverted fluorescence microscope directly. The crack of the specimen is marked so that the rupture details of MUF microcapsules can be observed. The cracking is produced from a bubble as an edge defect at location A. Bubbles are coated in epoxy material due to a high viscosity of epoxy resin, which are difficult to be entirely removed during the process of specimen preparation. When the crack is propagated through MUF microcapsule, the microcapsule is ruptured by cracking stress to release epoxy adhesives that wet crack plane owing to the capillary effect. Epoxy resin is an oligomer and its molecular chains have numerous oxygen and hydrogen active atoms. Viscosity of epoxy E-51 is relatively high at room temperature because of strong intermolecular forces such as hydrogen bond are existed, which restrict molecular motion. This could be a possible reason that those reported self-healing materials have a limited self-repairing efficiency. BGE is used to dilute the epoxy so that the epoxy viscosity is reduced considerably as shown in Figure 7 . When BGE content is increased to

10wt%, the viscosity of epoxy resin is considerably reduced. In addition, the micro-crack passes two microcapsules at location B and C in Figure 6 then extending continuously until the crack ruptures a microcapsule at location D. Modified epoxy healing agent in this microcapsule flows into the micro-crack and wets micro-crack surface then contacts curing agent previously incorporated in the matrix triggering a curing chemistry to produce a cured substance that is capable of repairing the cracks more effectively. It is observed that width of the crack is obviously decreased at location D and the crack is finally terminated at location E, which illustrate functions of both that the healing agent released from the ruptured microcapsule at location D bonds the crack surfaces and that the process prevents the crack propagation.

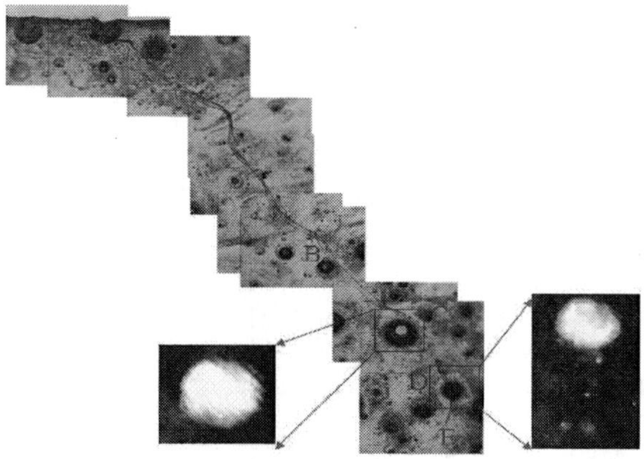

Figure 6. Crack propagation of MUF microcapsule/epoxy self-healing material. (40x lens of fluorescence microscope)

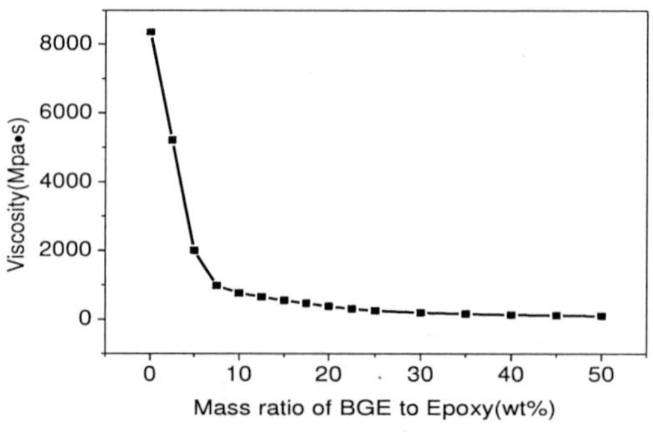

Figure 7. Viscosity of epoxy resin with BGE diluents.

Fracture surface for the healed region of the MUF microcapsule/epoxy self-healing composite was analyzed by SEM. Figure 8(a) shows that a cured layer material within matrix visibly repairs the micro-crack. In Figure 8(b), infrared spectrum of this filling material (spectrum a) in comparison with that of epoxy diluted by BGE and cured with MC120D curing agent (spectrum b) show shapes and locations for the characteristic peaks of epoxy resin, mainly include hydroxyl at 3482cm-1, benzene ring at 1610 and 1506cm-1 and aryl ether at 1252cm-1. The characteristic peak of epoxide in 772cm-1 is almost disappeared in two spectra and that of imidazole ring is seen at 1510 cm-1 in both spectra, which illustrate an expected chemical reaction occurred. These results prove that the cross-link reaction occurred between the epoxy shelled by MUF resin and imidazole chemical inside the matrix cracking which formed an effective bonding between cracked surfaces[11].

a

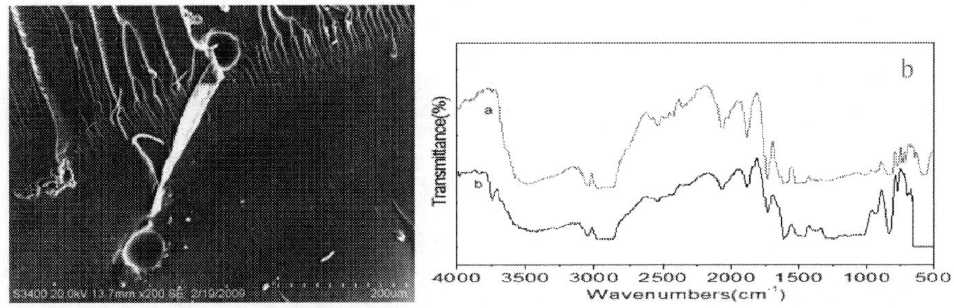

Figure 8 Characterizations for the filling material within the crack.
(a)SEM photo for the filling material; (b) Infrared spectra of the filling material and the reference material.

Molecular structure of the selected latent imidazole hardener contains a relatively long molecular chain positioned at the 3-substituded nitrogen, increasing the molecular steric hindrance so that the reaction between released epoxy adhesive and the hardener cannot be initiated at low temperature, whereas the latency is disappeared under a high temperature of heating discussed previously. BGE as reactive diluent has ether group and epoxide functional group that can join into the reaction of epoxy resin and imidazole hardener, together forming the cross-linking network. Because of charge polarization on epoxide group, it is formed that higher electron cloud density at the oxygen atom and lower electron cloud density at the carbon atom. The oxygen atom with negative charge on epoxide group of the epoxy resin or BGE initiates electrophilic addition for the reactive hydrogen atom of amine from the hardener, which results in ring-open polymerization (a) that the adduct (1:4) is generated by a ring-opened of epoxide group. Besides that, the lone pair electrons positioned at the 1-substituted nitrogen of imidazole ring attack the epoxide group, taking place the 1:1 addition reaction (b) and forming clathrate with coexistence of positive and negative ions. The negative ions of clathrate as active centers catalyze the epoxide ring open, triggering copolycondensation of epoxy resin, BGE and imidazole[12].

4. Conclusion

The conditions of thermal treatment for post curing (60°C, 2 hours) and cross-linking reaction (100°C, 1 hour) are important factors in the measurement of self-healing efficiency for MUF microcapsule/epoxy composite. The data simulation of stress distribution for epoxy composite reveals that micro-cracks are more likely to be initiated on the center region of stress concentration area for two constrained boundaries and the amounts of micro-cracks are decreased from the central area to the two ends of the material. Self-healing efficiencies of the epoxy self-healing composites embedded 2% and 5% microcapsule content measured by DMA are 101% and 104% respectively which are close to those results (104% and 113% correspondingly) measured by bending tests. This demonstrates DMA can be used to characterize self-healing efficiency of composite material. The self-repairing mechanism of the MUF microcapsule/epoxy composite materials is that numbers of micro-cracks within the composite after subjection to comstant strain damages are generated. Special prepared microcapsules are ruptured by cracking stress releasing adhesive with fluorescence substance, flowing along crack propagation direction into cracking gaps and wetting its surface. The infrared analysis of the cured filling material inside the cracking captured by SEM prove that the cross-link reaction between the epoxy shelled by MUF resin and imidazole chemical inside the matrix cracking occurred, forming an effective bonding between cracked surfaces.

5. Acknowledgment

Foundation: National Natural Science Foundation of China (51378315); Research & Development

Fund of Shenzhen (JCYJ20130329114709152)

6. References

[1]. J Lee, D Bhattacharyya, MQ Zhang, YC Yuan, Express Polymer Letters, 5, 246-253, 2011
[2]. LH Liang, XN Li, HY Liu, YB Wang, YG Wei, Surface & Coatings Technology, 285, 113-119, 2016.
[3]. H Ridha, PJ Thurner, Journal of the Mechanical Behavior of Biomedical Materials, 27, 94-106, 2013.
[4]. N Liu, D Liu, W Zhou, Applied Mathematics & Mechanics, 38, 99-110, 2017.
[5]. TR Wilshaw, CA Rau, AS Tetelman. Engineering Fracture Mechanics, 1, 191-196, 1968.
[6]. Z Ni, J Huang, D Gui, Journal of Shenzhen University Science & Engineering, 32, 458-464, 2015.
[7]. LC Kontaxis, E Kollia, A Kotrotsos, V Kostopoulos, GC Papanicolaou, Therma, 2014
[8]. PL Geiss, M Kopnarski, International Journal of Adhesion and Adhesives, 29, 478-487, 2009.
[9]. L Banks, B Ellis, Polymer, 23, 1466-1472, 1982.
[10]. F Aghadavoudi, H Golestanian, YT Beni, Polymer, 23, 1466-1472, 2016.
[11]. KR Hart, NR Sottos, SR White. Polymer, 67, 174-184, 2015.
[12]. S. K. Ooia, W. D. Cook, G. P. C. Simona, H. Suchb, Polymer, 41, 3639-3649, 2000.

Mix Proportion Design of Asphalt Concrete

Xianhu Wu[1], Lingling Gao[2] and Shoujun Du[3]

[1]Hebei University of Science and Technology, China. Email: 835671852@qq.com
[2]Hebei University of Science and Technology, China. Email:dutumu@163.com
[3]Hebei University of Science and Technology, China. Email: 34436371@qq.com

Abstract: Based on the gradation of AC and SMA, this paper designs a new type of anti slide mixture with two types of advantages. Chapter introduces the material selection, ratio of ore mixture ratio design calculation, and determine the optimal asphalt content test and proportioning design of asphalt concrete mix. This paper introduces the new technology of mix proportion.

1. Introduction

At present, there are a lot of large longitudinal slope pavement in mountainous highway, however, we hope that it has better anti sliding performance for the long and steep slope pavement. There are many factors that determine the performance of pavement skid resistance, among which asphalt concrete mix design is one of the key factors. In this paper, the advantages of AC-13 and SMA asphalt concrete are combined to design an economical and practical asphalt concrete mixture ratio.

2. Selection and Testing of Asphalt Concrete Raw Materials

The experiment of coarse aggregate by basalt, fine aggregate selection mechanism of sand filling, limestone slag, asphalt with SBS modified asphalt. Inspection results of raw materials are shown in Table 1, Table2, Table3, Table 4. The experiment of powder of 5 types of ores were sieved, the results shown in Table 1~5.

Table 1. Test results of basalt coarse aggregate

Index	Unit	Test result				Technical requirement
Stone specifications	mm	13.2mm	9.5mm	4.75mm	2.36mm	-
Crushed stone	%	14.1				≤26
Losangeles abrasion loss	%	12.8				≤28
Apparent relative density	g/cm^3	2.867	2.818	2.862	2.775	≥2.60
Gross volume relative density	g/cm^3	2.829	2.782	2.817	2.715	-
Water absorption	%	0.35	0.44	0.48	0.51	≤2.0
Needle particle content	%	5.7	7.6	3.9	3.4	≤15
Washed process<0.075mm Particle content	%	0.5	0.6	0.6	0.7	≤1

Content from this work may be used under the terms of the Creative Commons Attribution 3.0 licence. Any further distribution of this work must maintain attribution to the author(s) and the title of the work, journal citation and DOI.
Published under licence by IOP Publishing Ltd

Table 2. Test results of fine aggregate

Index	Unit	Test result					Technical requirement
Stone specifications	mm	1.18	0.6	0.3	0.15	0.075	-
Apparent relative density	g/cm3	2.688	2.699	2.705	2.695	2.703	≥2.50
Mud content (less than 0.075mm)	%	0.9	0.9	1.0	1.1	1.1	≤3
Sand equivalent	%	95.1	95.6	95.4	96	95.0	≥60

Table 3. The main technical index of mineral powder

Index	Unit	Test result	Technical requirement
Apparent relative density	g/cm^3	2.781	≥2.50
Water content	%	0.2	≤1
Hydrophilic coefficient	-	0.70	< 1
Sand equivalent	%	/	No caking aggregate

Table 4. Technical results of modified asphalt SBS

Test items	Unit	Test value	Specified value
Penetration (25°C, 100g,5s)	0.1mm	71.8	60~80
Penetration index	-	0.296	≥-0.4
Softening point (R & B)	°C	87.2	≥55
15°C Ductility	Cm	≥100	≥100
135°C Kinematic viscosity	Pa.s	2.23	2.0 ~ 3
Density	g/cm3	1.080	-
ΔElastic recovery 25°C	(%)	83	≥65

Table 5. Screening results of mineral materials

Screen size(mm)	Pass rate(%)				
	10~15	5~10	3~5	0~3	Powder
16	100	100	100	100	100
13.2	81.55	100	100	100	100
9.5	0.67	95.27	100	100	100
4.75	0.67	25.09	94.67	99.55	100
2.36	0.67	1.04	96.66	81.05	100
1.18	0.67	1.04	92.74	57.44	100
0.6	0.67	1.04	8.39	32.59	95.0
0.3	0.67	1.04	2.32	18.43	88.3
0.15	0.67	1.04	0.5	10.95	81.9
0.075	0.67	1.04	0.33	8.07	67.6

3. Target mix Design

3.1. selection Gradation

As the main body of the mixture, the gradation of aggregate plays an important role in the structure and performance of the mixture. The experiment was conducted based on AC grading, SMA grading, according to the grading range and performance index, mixture was given to study the anti slide pavement gradation advice, specific data is shown in Table 6.

Table 6. Design range of anti slide pavement material

Gradation types	The following screen (mm) quality percentage%									
	16	13.2	9.5	4.75	2.36	1.18	0.6	0.3	0.15	0.075
Grading limit	100	100	80	40	30	27	22	18	15	10
Gradation limit	100	90	55	25	20	16	14	9	8	6

Because of this experiment are not graded fiber stabilizer, so compared to ordinary SMA mixture, the contents of above 4.75mm decreased, the content of fine aggregate increased, gradation of the fine, but not on the formation of interlocking skeleton structure formation important influence. Due to the fine gradation, the aggregate clearance rate will be reduced, and the amount of asphalt will be reduced accordingly. The gradation given in this test is shown in Table 6. The position of the curve in the recommended range of AC, SMA and anti slide pavement is shown in Figure 1~3.

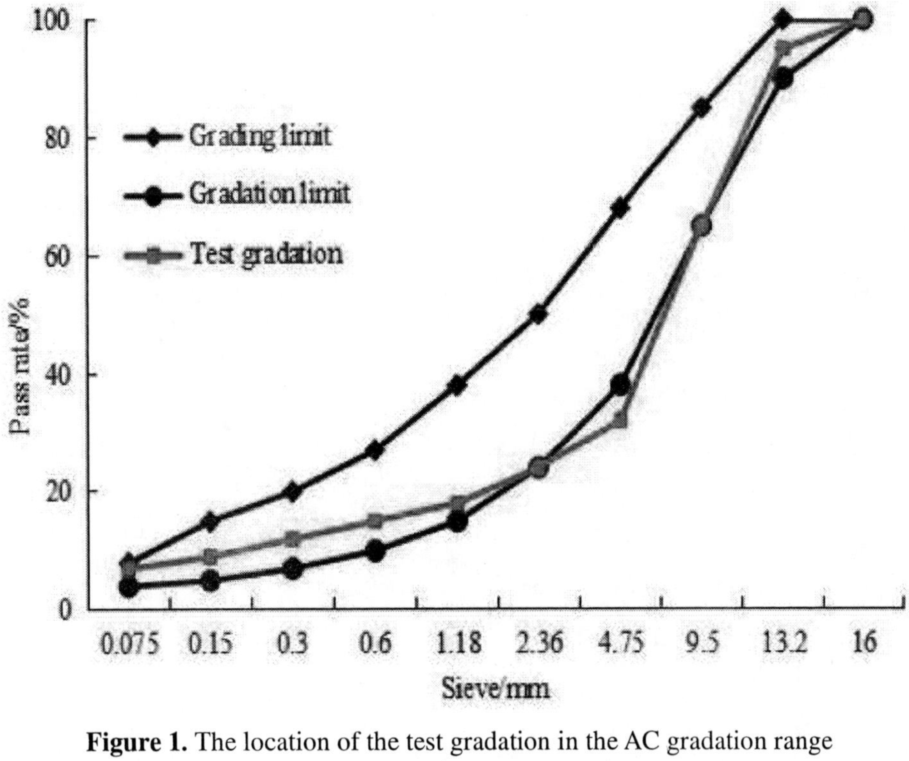

Figure 1. The location of the test gradation in the AC gradation range

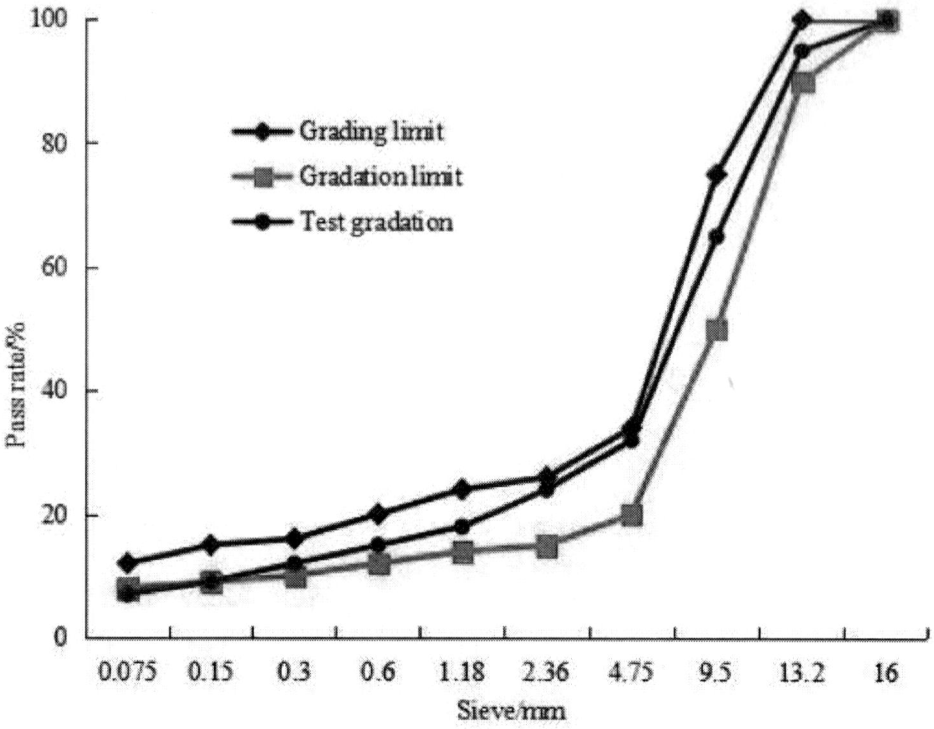

Figure 2. The location of the test gradation in the SMA gradation range

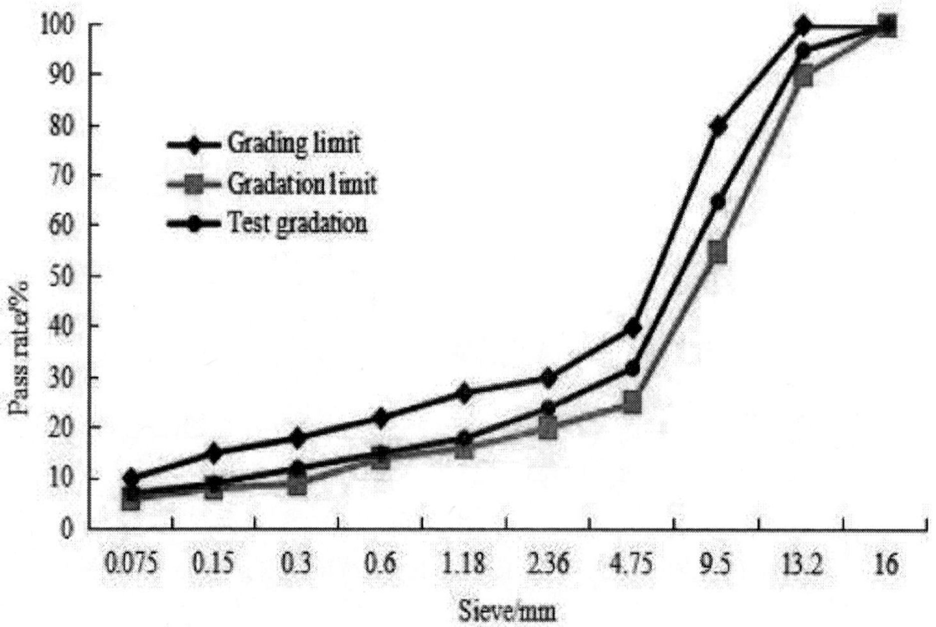

Figure 3. The location of the test gradation in the proposed gradation range

3.2 Determination of Optimum Asphalt Content by Marshall Test

The target mix were 4.9%, 5.3%, 5.7%, 6.1% Marshall test according to the asphalt stone, detection of different asphalt aggregate ratio of density, porosity, saturation, stability and flow value etc.. According to Marshall and other than the test index of different stone, determine a optimum asphalt content (asphalt aggregate ratio).

After the specimen is formed, the density is measured by the immersion balance method, and the specific data are shown in Table7.The volume parameters are calculated according to the density. Using Marshall stability meter to measure the stability and flow value and other mechanical indicators, technical requirements and test results are shown in Table 8, Table 9.

Table 7 Test specimen quality

Asphalt aggregate ratio(%)	Air quality m_a (g)	Water quality m_w (g)	Dry mass m_f (g)
4.9	1194.4	711.6	1196.3
5.3	1195.8	714.7	1197.7
5.7	1192.5	713.8	1194.5
6.1	1192.6	713.1	1194.0

Table 8. Technical requirements for mix proportion design of Marshall test

Test items	Nnit	Technical requirement Using modified asphalt
Marshall specimen size	mm	φ101.6mm×63.5mm
Marshall compaction times	-	Both sides hit 50 times
Void fraction VV	%	4
Mine clearance VMA (Not less)	%	15.0
Coarse aggregate clearance VCA_{mix} (Not greater)	-	VCA_{DRC}
Asphalt saturation VFA	%	60~70
Stability (Not less)	kN	6.0
Flow value	mm	-

Table 9. Marshall test parameters

Asphalt aggregate ratio (%)	Coarse aggregat e porosity VCR_{DRC} (%)	Clearance rate of coarse aggregate in mixture VCAmix (%)	r_f	VV (%)	VMA (%)	VFA (%)	MS (kN)	FL (mm)
4.9	35.4	35.0	2.465	5.5	15.4	64.2	11.57	5.9
5.3	35.4	35.4	2.476	4.6	15.0	69.5	12.63	6.0
5.7	35.4	35.1	2.481	3.9	14.9	73.3	12.45	6.3
6.1	35.4	34.9	2.480	3.4	14.9	77.0	11.83	5.7

Based on the test results in Table 8~9, draw the curve of the relative density, stability, flow value, air VV, VMA, VFA index and the asphalt content of the hair volume, respectively, see Figure 4~9.

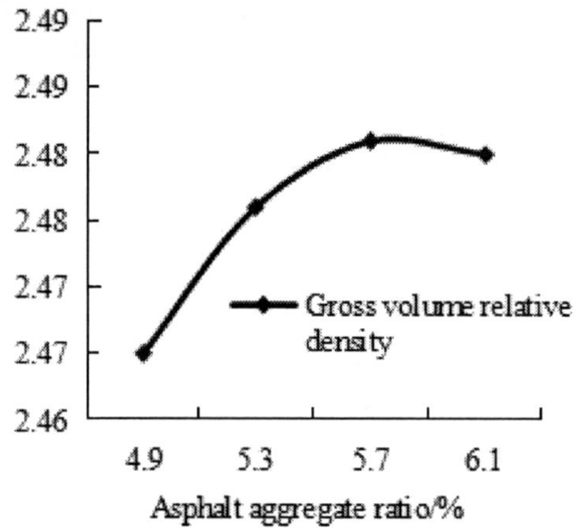

Figure 4. Relationship between density and aggregate ratio;

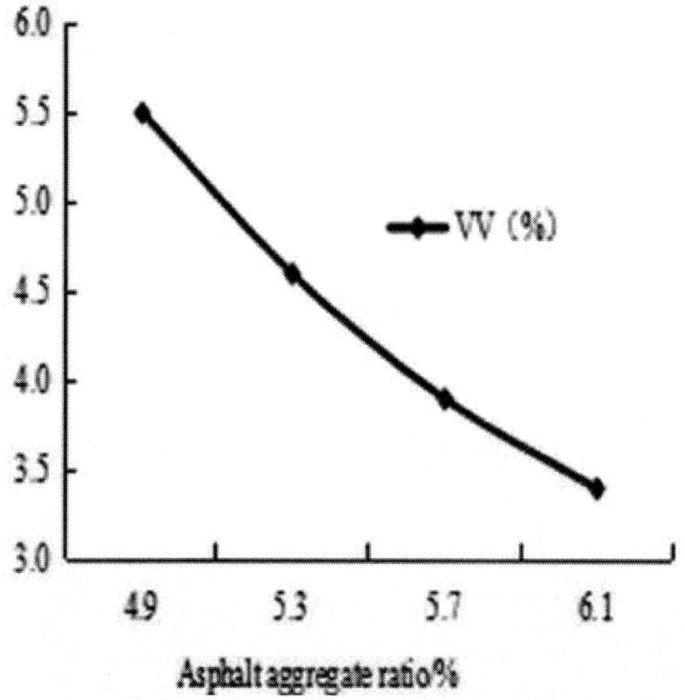

Figure 5. VV and asphalt aggregate ratio

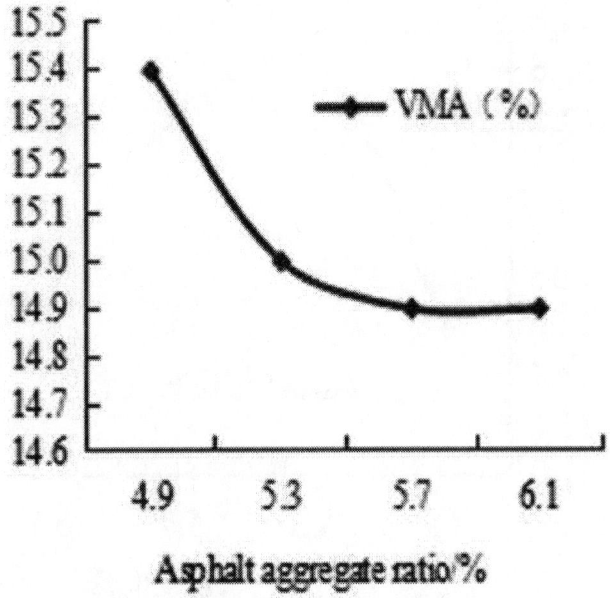

Figure 6. VMA and asphalt aggregate ratio;

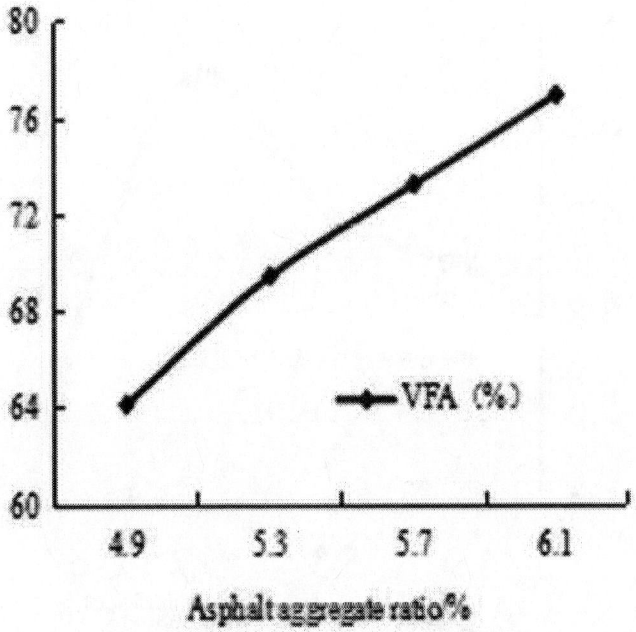

Figure 7. VFA and asphalt aggregate ratio

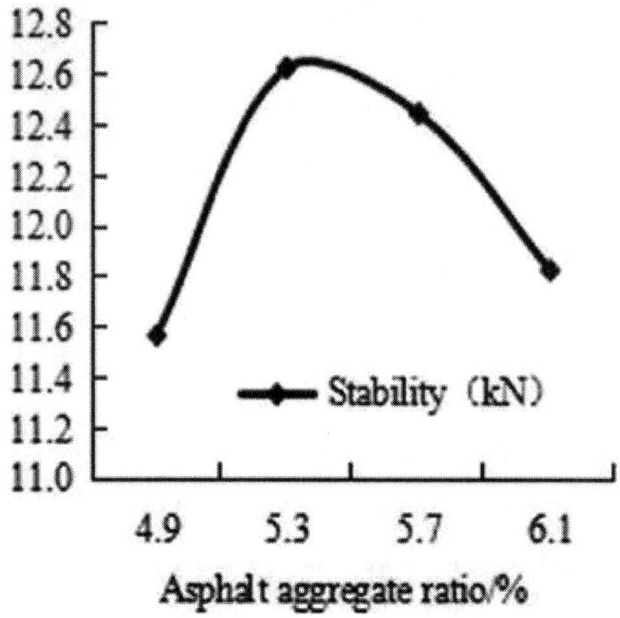

Figure 8. Stability and asphalt aggregate ratio;

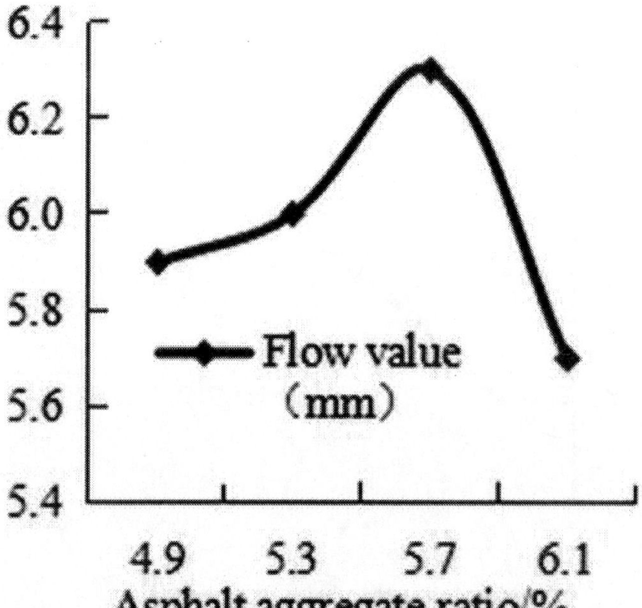

Figure 9. Diagram and flow value of asphalt aggregate ratio

The graph can be obtained by density maximum a1=5.4% and stability when the maximum a2=5.4%, void rate was 4%, a3=5.6%, the asphalt content range chosen not covered asphalt saturation range, the average value for 3 so as =5.47%.Since the Table 7 is the recommended indicators, not the strict requirements of the specification, determined, is of no practical significance.So as the optimum asphalt content calculation, the optimum asphalt aggregate ratio is 5.5% design gradation.

4 Mix Design Inspection

The Marshall mixture ratio design of asphalt mixture was carried out by using the volume design method of the test piece of Marshall, and the stability and flow value of the test were not the only index to accept or reject the mixture ratio design.The test results of the stability test is a key evaluation index, comprehensive rationality and residual stability test, leakage test, cantabro test to test the gradation and asphalt content.

The 4.75mm passing rate for the key rate, i.e. 25%, 32%, 40% for 3 groups of mix design, concrete gradation results and grading curve shown in Table 10 and Figure 10.

Table 10. Gradation composition

Gradation Types	Through the mesh quality percentage (mm)									
	16	13.2	9.5	4.75	2.36	1.18	0.6	0.3	0.15	0.075
1	100	95	60	25	20	16	14	9	8	6
2	100	95	65	32	24	18	15	12	9	7
3	100	95	75	40	30	25	22	16	13	9

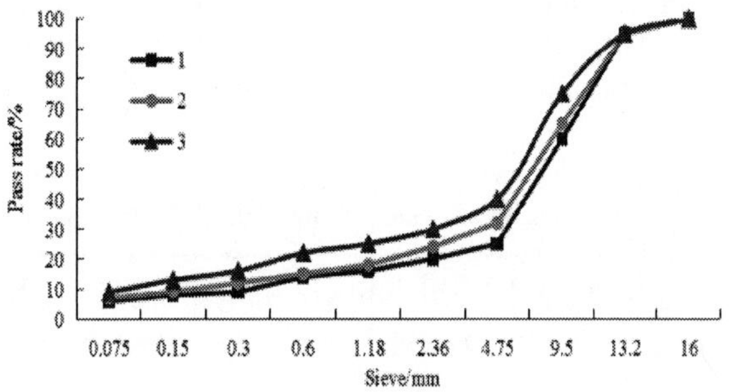

Figure. 10. Gradation curve

From Table 11 and Figure 10 shows that No.1 gradation of coarse, No.3 gradation is fine, so based on the optimum asphalt aggregate ratio on the corresponding changes in aggregate ratio were 5.2%, 5.8%, made Marshall specimen test stability and flow value is reasonable, the key is to verify the dynamic stability will how to change with the gradation. The 300mm * 300mm * 50mm size specimen was measured by a rut tester and its dynamic stability was measured at 60 degrees celsius. Test results are shown in Table 10.

Table 11 Experimental results

Number	Stability (kN)	Flow value(mm)	Dynamic stability(second/min)
1	10.39	4.29	4532>3000
2	11.70	5.34	3684>3000
3	9.88	4.96	2815<3000

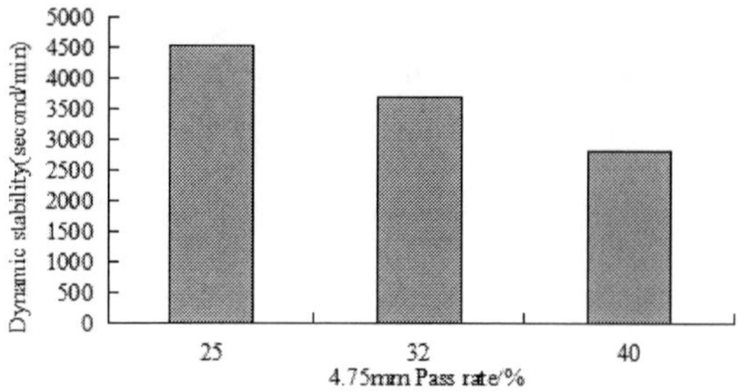

Figure 11. 4.75Pass rate and dynamic stability

It can be seen from Table 11 that there is no necessary relation between the Marshall stability and dynamic stability of the mixture. So when testing its performance, as long as the stability is greater than 6kN.From Figure 11 shows that 4.75mm pass rate is small, the dynamic stability is larger, indicating coarse aggregate content on the dynamic stability has obvious effect, the reason is that the content of coarse aggregate, > degree is bigger, the easier the formation of interlocking skeleton structure, make the mixture with high temperature stability good.

In addition to stability and dynamic stability, residual stability test, leakage test, cantabro test results page should meet the requirements, to number 2 as an example, to test the results as shown in Table 12, Table 13.

Table 12. Immersion Marshall test results

Specimen	Stability of 0.5h (kN)	Stability of 48h(kN)	Residual stability(%)
1	9.52	8.71	91>80
2	9.81	9.04	92>80
3	11.90	11.21	94>80

Table 13. Leakage and dispersion test results

Specimen	Leakage analysis	Scattering loss(%)
1	0.06<0.1	12.3<15
2	0.07<0.1	11.6<15
3	0.06<0.1	13.0<15

5. Production Mix Design

In the design of the target mix ratio, due to the screening of various specifications of the aggregate is too cumbersome and time-consuming, so according to the screening results of the various sets of production mix design. However, no matter how to change the proportion of the 5 kinds of aggregate, cannot match with the target mix ratio. Description of aggregate gradation used in the test is not very reasonable, so in the actual construction, should adjust the aggregate rate by itself, improve the levels of mineral aggregate gradation, the construction will be carried out through the development of 5 construction aggregate proportion, greatly improve work efficiency.

6. Conclusions

This paper through the research of AC, SMA grading, gives a mixture of anti slide the proposal and determine the target gradation range, design gradation, using the Marshall test method, to determine the optimum asphalt aggregate ratio, and verify the rationality of the physical and mechanical indexes. Through rutting test, Xie lueneburg asphalt leakage test, cantabro test performance test of asphalt mixture, pointed out that the dynamic stability and stability are not necessarily related, as well as the coarse aggregate effect on dynamic stability, the dynamic stability in a certain extent with the increase of coarse aggregate.

7. References

[1] JTGF422005. technical specification for construction of highway asphalt pavement S.
[2] Liang Qiyong. Design method of mix proportion of asphalt mixture [J]. Shanxi architecture, 2003, (4): 118-119.
[3] JTGF40-2004, technical specification for construction of highway asphalt pavement.
[4] AASHTO. Standard practice for Designing Stone Matrix Asphalt (SMA). 1998.
[5] Peng Bo. Influencing factors and variation law of pavement skid resistance. North China Communications, 2011, (8): 9-10.
[6] Kanhdal P. S,Koehler W S,The Alsphalt Handbook. The Alsphalt Institute Manual Series No4, 1989.

IOP Conf. Series: Materials Science and Engineering **275** (2017) 012020 doi:10.1088/1757-899X/275/1/012020

Research on Improving Low Rank Coal Caking Ability by Moderate Hydrogenation

Peng Huang

Beijing Coal Chemistry Research Institute, China Coal Research Institute, Beijing, China. Email:squallok@qq.com

Abstract: The hydrogenation test of low metamorphic coal was carried out by using a continuous hydrogen reactor at the temperature of 350 ~ 400°C and the initial hydrogen pressure of 3 ~ 6Mpa. The purpose of the experiment was to increase the caking property, and the heating time was controlled from 30 to 50min. The test results show that the mild hydrogenation test, no adhesion of low metamorphic coal can be transformed into a product having adhesion, oxygen elements in coal have good removal, the calorific value of the product has been improved significantly and coal particles during pyrolysis, swelling, catalyst, hydrogenation, structural changes and the combined effects of particles a new component formed between financial and is a major cause of coal caking enhancement and lithofacies change, coal blending test showed that the product can be used effectively in the coking industry.

1. Introduction

China had the most coke production in the world in 2016, about 450 million Tons coke were generated. Coking process under the current conditions, which required coking coal with good adhesion. Moderate metamorphic bituminous coal is a conventional coking coal, but in the various types of coal in China, moderate metamorphic bituminous coal reserves are low, and conventional coking coal reserves account for only 26% of the national coal reserves, of which more than 50% are distributed in Shanxi Province, the geographical distribution is extremely uneven. Compared with the low rank coal, the coking coalash and sulfur is relatively higher, and has a poor washability. Due to the influence of resource conditions, coking technology development and coke demand, the shortage of high quality coking coal will exist for a long time. [1,2]

Low rank coal is abundant in China, accounting for more than half of the total reserves. At the same time, characteristic of most low metamorphic bituminous coal with low ash, low sulfur, low phosphorus, but because there is no bond or bond is very weak, the current can not be used as the conventional coking coal use, even in the presence of strong caking coal, under the coke condition, the blending ratio is generally not more than 5%. [3] In the early study, some foreign experts have increased the coking characteristics of coal by thermal hydrogenation,, this study was conducted by using 4 kinds of thermal hydrogenation without nonagglomerating coal, including lignite, sub bituminous and sub bituminous coal and oxidized bituminous coal, the results show that the product of plasticity, free expansion coefficient has been compared to the raw coal some improvement, but still do not have adhesion. [4,5] therefore, the main goal of this paper is to increase the caking property of non caking coal by moderate solvent hydrogenation.

Content from this work may be used under the terms of the Creative Commons Attribution 3.0 licence. Any further distribution of this work must maintain attribution to the author(s) and the title of the work, journal citation and DOI.
Published under licence by IOP Publishing Ltd

2. Experimental

2.1. Material and Equipment

Table 1. Analysis of Yulin coal

wt/%	A_d	V_{daf}	V_{daf}	F_{Cd}	Qgr.d/ MJ/kg	C_{daf}	H_{daf}	O_{daf}	N_{daf}	S_{daf}
YulinCoal	4.1	44.16	44.16	51.86	30.79	80.67	5.36	12.54	1.01	0.4

Raw coal was selected which was long flame coal from Yulin mining area. The element analysis and industrial analysis of raw coal show table 1 (G was bituminous coal caking index). Table 1 shows that Yulin coal belongs to low metamorphic bituminous coal with high oxygen content and volatile content, and has no caking property. Broken about grinding to 0.16-0.18mm dry spare, hydrogenation solvent was anthracene oil. Hydrogenation experiments were carried out in continuous hydrogen type autoclave, autoclave by mechanical stirring, the stirring paddle makes the gas phase and liquid phase reactant mixing, stirring speed is 200r/min, using oil bath heating, the operating pressure is 0 ~ 35MPa, the operating temperature is 0 to 480°C, according to the temperature in the reactor, and regulation the control of oil bath, to achieve the purpose of indirect control of reaction temperature. The flow chart of the experimental device is shown in figure 1.

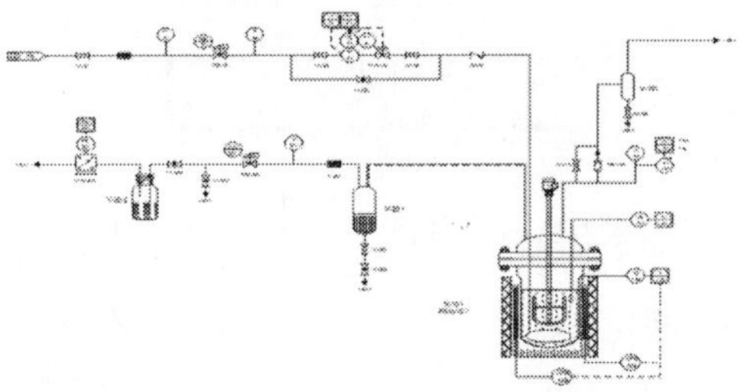

Figure 1. Schematic diagram of the experimental apparatus.

2.2. Procedures

The extraction flow chart is shown in figure 2. When the reaction temperature and pressure reached the set conditions, keep a certain amount of time, open the tank cover, get the products out, calculate reaction kettle liquid and solid volume, solid product into the vacuum oven; a part of the product was fractional extracted after drying; the reaction products all transferred to the quantitative filter tube, extracted by Soxhelt continuous extraction unit, followed by n-hexane, toluene and tetrahydrofuran , reflux extraction time for each solvent 48h, until the filtrate is clear.

IOP Conf. Series: Materials Science and Engineering **275** (2017) 012020 doi:10.1088/1757-899X/275/1/012020

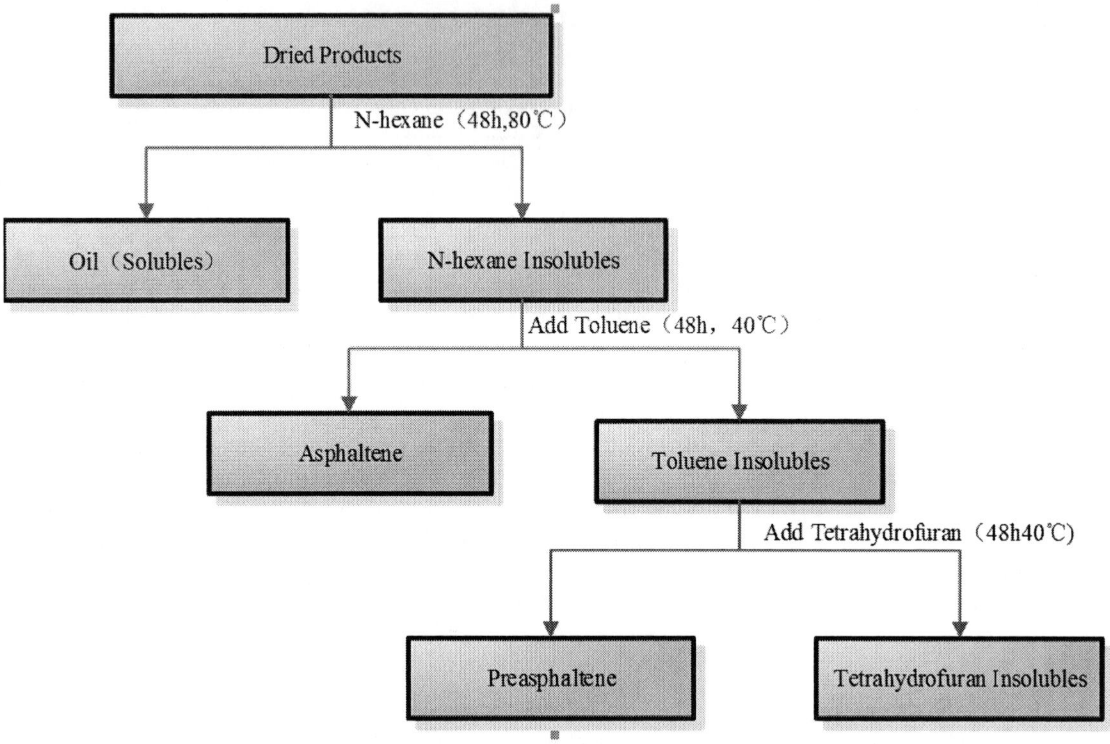

Figure 2. The products extraction

2.3. Material Balance Calculation

Gas Yield

$$\eta_g = \frac{m_g}{m_{C_{daf}}} \times 100\% \tag{1}$$

m_g :weight of the gas products

m_{Cdaf} :weight of the dry air free based coal.

H_2O yield

$$\eta_{H2O} = \frac{m_{Ocoal} - m_{Og}}{m_{Cdaf}} \times \frac{18}{16} \times 100\% \tag{2}$$

m_{Ocoal} :weight of oxygen element in the coal

m_{Og} : weight of oxygen element in the gas

Coal conversion rate

$$\sigma = (1 - \frac{m_{THFI} - m_{A}}{m_{Cdaf}}) \times 100\% \tag{3}$$

m_{THFI} : weight of residue can't dissolve in tetrahydrofuran

m_{A} :the weight of ash

asphaltene yield

$$\eta_a = \frac{m_a}{m_{Cdaf}} \times 100\% \tag{4}$$

m_a :weight of asphaltene (dissolve in toluene, can't dissolve in n-hexane)

preasphaltene yield

$$\eta_{pa} = \frac{m_{pa}}{m_{Cdaf}} \times 100\% \tag{5}$$

m_{pa} :weight of preasphaltene(dissolve in tetrahydrofuran , can't dissolve in toluene)

H_2. consumption

$$\varepsilon_H = \frac{m_{H0} - m_{Ht}}{m_{Cdaf}} \times 100\% \tag{6}$$

m_{H0} and m_{Ht} : weight of H_2 in the autoclave (before and after reaction).

Oil yield

$$\eta_{oil} = \sigma + \varepsilon_H - \eta_g - \eta_{H2O} - \eta_a - \eta_{pa} \tag{7}$$

In this experiment, 3 conditions were selected before optimization, 350°C was the initial reaction point; 380°C was the maximum temperature point of Yulin coal loss rate. 400°C, 6MPa was the best reaction condition. The reaction products were analyzed, the reaction conditions and the conversion rate under this condition were shown in table 2:

Table 2. Experimental conditions of coal hydrogenation in Yulin

Samples	Temperature/°C	Pressure/MPa	time/ min	σ/ %
1	350	4	30	35.47
2	380	4	50	37.16
3	400	6	50	54.14

3. Results and Discussion

3.1. Products Properties

The properties of Yulin coal after moderate hydrogenation modification are listed in Table 3(vacuum drying at 110C for 2 hours).

Table 3 Properties of Yulin coal after hydrogenation

wt/%	A_d	V_{daf}	V_{daf}	Qgr.d/MJ/kg	C_{daf}	H_{daf}	N_{daf}	S_{daf}	O_{daf}	R_{max}
1	7.39	45.35	45.35	35.36	82.65	4.99	2.55	0.93	1.49	0.96
2	6.61	44.31	44.31	36.66	86.15	4.63	1.42	0.87	1.32	0.98
3	3.77	41	41	36.74	90.93	5.13	1.79	0.98	1.17	0.98

After the reaction, the properties of the product were analyzed. The results showed that the modified product had adhesive property. The caking index(G) can reach about 70 at 350°C, and the caking index (G) can reach more than 90 at 400°C.

From the analysis of the properties of hydrogenation products in Table 4, it can be seen that the

IOP Conf. Series: Materials Science and Engineering **275** (2017) 012020 doi:10.1088/1757-899X/275/1/012020

content of carbon in mild hydrogenation products is higher than that of raw coal, and the structure of the product is more compact than that of the raw coal. From the angle of coal hydrocracking, during the reaction, the bridging reaction of oxygen bridge and side chain of low rank coal macromolecular structure was formed, and other forms such as H_2O, CO_x and H_2S were produced. On the other hand, the content of N in the mild hydrogenation product is much higher than that of the raw coal, which indicates that the N element in coal is difficult to be removed under the condition of hydro upgrading. From the structural analysis, the N element mainly exists in the aromatic structure.

From the results of industrial analysis in Table 3, it can be seen that the ash content has increased in comparison with the industrial analysis, which is due to the escape of some volatile components, and the heat value of the mild hydrogenation product is significantly higher than that of the raw coal. The reason is that in the process of mild hydrogenation, due to the fracture of the macromolecular side chain, the oxygen-containing functional groups and the intrinsic water in the low rank coal are removed, and the latent heat of vaporization of the water molecules during combustion is eliminated, and the calorific value is increased. [6-8]

The analysis of hydrogenation pyrolysis process, a large molecular fracture generated asphaltene, with activity of preasphaltene and other active components, with the increase of temperature and pressure, asphaltene substances increased, increasing coal pyrolysis, bond strong structural fracture and continue to generate free when the base fragments, these fragments are hydrogen stable liquid and gas production to small molecules, on the other hand, when the hydrogen pressure is conducive to inhibit the reaction process of the dehydrogenation, thus improving the yield of oil and gas, the higher the initial hydrogen pressure would be beneficial to solvent on molecular hydrogen solubility, increase the concentration of active hydrogen the solvent, solvent to dissolve hydrogen fast transfer to free radical fragments. The analysis of the composition of the products under different conditions is shown in figure 3.

From Figure 3 (condition 1,2, 3 respectively in the picture: 1, 2, 3) can be seen, generation of oils and gas molecules products increase, reaction to obtain solid product having adhesion for the purpose of increasing the conversion rate will bring a solid product yield loss.

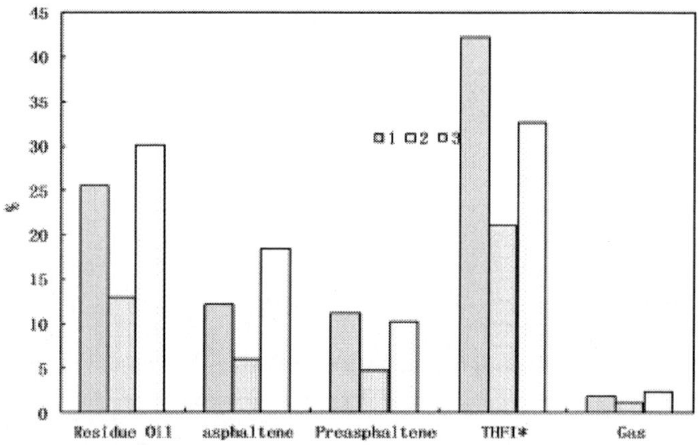

Figure 3. Hydrogenation products composition

3.2. Maceral Specification

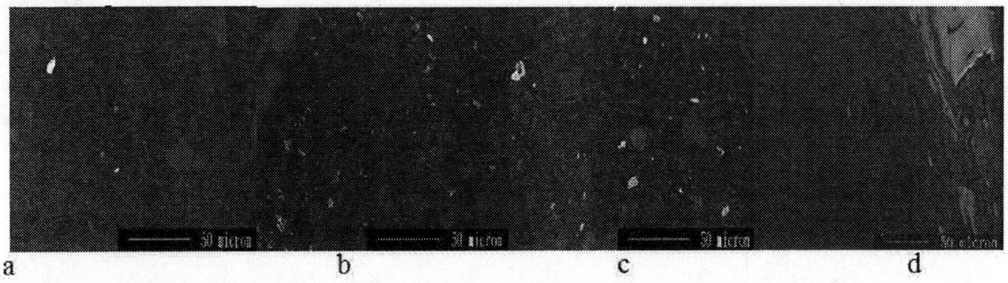

(a) 623K,Coal particles diameter increased, microlithotype of the modified coal change from monocomponent to vitrinertite;
(b) 653K, microlithotype of the modified coal formed by vitrinertite mainly;
(c) 673K, microlithotype of the modified coal formed by vitrinertite mainly, the particle of the Fe-catalyst and the particle of the Inertinite spreaded in vitrinite base;
(d) Raw coal maceral specification

Figure 4. Plasticized Coals Microphotograph (350°C ~400°C)

The change of petrographic characteristics during the modification of low metamorphic bituminous coal was studied by polarizing microscope, and the main factors affecting the caking property of modified coal were analyzed. The petrographic characteristics of the modified products at temperatures from 350 to 400°C are shown in figure 4.

From Figure 4 (a) can be seen before in the reaction temperature reaches 350°C, coal bond hasn't changed basically, vitrinite reflectance increased slightly with the rise of temperature, coal particle size increases, microlithotype by single component type into a micro mirror by inert coal; Figure 4 (b) can be seen, when the temperature reached 380°C, microlithotype in micro mirror inert coal; It can be seen from Figure 4 (c) that when the reaction temperature is 400°C, the micro coal type is dominated by micro mirror inert coal, and the catalyst particles and the inert group particles are dispersed in the vitrinite matrix. Figure 4 (d) is a micrograph of raw coal.

When the reaction temperature is higher than 380°C, the caking property of coal was obviously enhanced, and the characteristics of modified coal and coal rock changed significantly, The micro inert component is the main micro coal type in the mild hydrogenation product (microlithotype containing vitrinite and inertinite and exinite less than 95%, less than 5%). The structural changes and inter particle melting of coal particles under the combined action of pyrolysis, swelling, catalyst, hydrogenation and formation of new components are the main reasons leading to the increase of coal caking property and the change of petrographic characteristics. The caking property of modified coal mainly comes from the new soluble component at 350°C, and the increase of caking property caused by the change of coal structure needs higher reaction temperature.

3.3. Coal Blending of Yulin Coal

Choose the best modified products and coking performance of Shanxi 1/3 coking coal comparison, the use of these two samples and non caking coal blending. [9-12]At different ratios, the bond index between mild hydrogenation products and Shanxi coking coal and non caking coal is compared. The main coal quality characteristics of 1/3 Shanxi coking coal are shown in table 5.

Table 5 Characteristics of Puxian coal

wt/%	A_d	V_{daf}	C_{daf}	H_{daf}	N_{daf}	S_{daf}	O_{daf}	G	$\overline{R}^0_{max}$
Puxian coal	31.24	6.91	87.83	5.68	1.11	0.40	5.39	85	0.99

* $\overline{R}^0_{max}$: mean max reflectance of the coal

Overall, the modified product properties was between gas fat coal and coking coal.

The modified products and 1/3 coking coal were mixed according to the amount of 0%, 20%, 40%, 50%, 60%, 70%, 80%, 90% and 100%, respectively, and non caking coal (the percentage of total mixed coal). The caking index(G) of modified products was determined by GB/T5477 method, then draw the corresponding curve, coal blending curve as shown in figure 5.

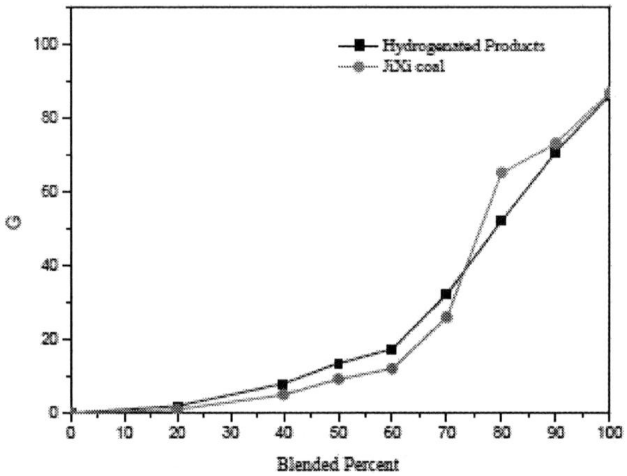

Figure 5. Blending Curve of Coal

We can see from Figure 5, the mixing quantity is between 66%~81%, change coal caking index curve and hydrogen production after coal blending coking coal faster; in the mixing quantity of 74% previous bond index of hydrogenation products than blending coking coal blending high coal blending effect is good; and after 74%, and hydrogen bond index products with the high coal blending coking coal.

4. Conclusion

(1) The cohesiveness of mild hydrogenation products has been obviously improved, and the expected purpose has been achieved. The deoxidation effect is obvious, and the calorific value has been improved obviously.

(2) During the mild hydrogenation process, the coal particles experienced pyrolysis, swelling, hydrogenation and inter particle melting, leading to the micro coal type of the modified coal mainly with micro mirror inert coal, and the inert group was often dispersed in the vitrinite matrix in fine particles.

(3) From the coal blending results, between mild hydrogenation product properties between gas fat coal and 1/3 coking coal.

5. Acknowledgements

This research was supported by the National Key R&D Plan (2016YFB0600305)

6. References

[1]Sun P, Wang B, Sun F, et al. Research on reservoir patterns of low-rank coal-bed methane in China [J]. Acta Petrolei Sinica, 2009, 30(5):648-653.

[2]Liu H L, Jing-Ming L I, Wang H Y, et al. Discussion on finding coalbed methane in low rank coal in China[J]. Journal of China Coal Society, 2006, 31 (1):50-53.

[3]Li X, Fu X, Liu A, et al. Methane Adsorption Characteristics and Adsorbed Gas Content of Low-Rank Coal in China[J]. Energy & Fuels, 2016, 30 (5) :51-54.

[4]Nandi B N, Ternan M, Belinko K. Conversion of non-coking coals to coking coals by thermal

hydrogenation [J]. Fuel, 1981, 60(4):347-354.
[5]Mukherjee D K, Sengupta A N, Choudhury D P, et al. Effect of hydrothermal treatment on caking propensity of coal [J]. Fuel, 1996, 75(4):477–482.
[6]K. Belinko, M T, Nandi B N. Mesophase Development during Thermal Hydrogenation of an Oxidized Bituminous Coal [J]. Acs Symposium, 1979, 141(90):96-105.
[7]Liu S Y, Lu H Z. Research of Low Rank Bituminous Coal Heat Treating and Hydrogenation Upgrading [J]. Advanced Materials Research, 2012, 347-353:3740-3743.
[8]Bai X F, Wang D S, Zhang Y H. The experimental research on petrological transformation of low rank coal during its enhancement of plastic characteristics by hydrogenation [J]. Journal of China Coal Society, 2009, 34(2):247-251.
[9]Shih J S, Frey H C. Coal blending optimization under uncertainty [J]. European Journal of Operational Research, 1995, 83(3):452-465.
[10]Vuthaluru H B, Brooke R J, Zhang D K, et al. Effects of moisture and coal blending on Hardgrove Grindability Index of Western Australian coal [J]. Fuel Processing Technology, 2003, 81(1):67-76.
[11]Vuthaluru H B, Zhang D K. Effect of coal blending on particle agglomeration and defluidisation during spouted-bed combustion of low-rank coals [J]. Fuel Processing Technology, 2001, 70(1):41-51.
[12]Nomura S, Arima T, Kato K. Coal blending theory for dry coal charging process [J]. Fuel, 2004, 83(13):1771-1776.

IOP Conf. Series: Materials Science and Engineering **275** (2017) 012021 doi:10.1088/1757-899X/275/1/012021

PIV Measurements of Gas Flow Fields from Burning End

Yifei Huang[1], Junzhang Wu[1], Jingsong Zeng[2,] *, Darong Tang[2], Liang Du[2]

[1]China Tobacco Guangdong Industrial Co.Ltd, Guangzhou 510385, China
[2]State Key Laboratory of Pulp and Paper Engineering, Plant micro/nano fiber research center, South China University of Technology, Guangzhou 510640, China

Abstract: To study the influence of cigarette gas on the environment, it is necessary to know the cigarette gas flow fields from burning end. By using PIV technique, in order to reveal velocity characteristics of gas flow fields, the velocities of cigarette gas flow fields was analyzed with different stepping motor frequencies corresponding to suction pressures, and the trend of velocity has been given with image fitting. The results shows that the velocities of the burning end increased with suction pressures; Between velocities of the burning end and suction pressures, the relations present polynomial rule; The cigarette gas diffusion in combustion process is faster than in the smoldering process.

1. Introduction

PIV technology is a non-contact flow field measurement technique based on cross-correlation analysis of flow field images. Many scholars use PIV to analyze the flow fields of gas and liquid at home and abroad. [1-2] Jinbang Hu et al. captured the gas flow by using PIV technology in the intake distributor; [3] Lingbing Bu et al. studied flow fields of the self-developed gas distributor by PIV technique; [4] Mizeraczyk J et al. explained the velocity flow fields of the cigarette gas in an electrostatic precipitator model by using PIV method [5] Kashyap Metal described turbulence in a circulating fluidized bed by using PIV techniques[6] This paper provided a valuable method for experimental study on flow fields of cigarette gas in burning end by using PIV techniques.

2. Experimental Setup

A self-made test platform measuring flow fields of cigarette gas was established, which included PIV system, cigarette, PLC control platform, stepping motor system and so on. As shown in Figure 1, the laser sheet irradiated the burning end of the cigarette from the vertical direction, CCD was used to shoot the laser sheet to snap the flow fields of cigarette gas. During the measurement, the stepping motor system was controlled by the PLC program to finish the different suction process with different frequencies of the stepper motor (i.e. "different suction pressure"). From the velocity flow fields, the flow track of cigarette gas and the related values were gained.

Content from this work may be used under the terms of the Creative Commons Attribution 3.0 licence. Any further distribution of this work must maintain attribution to the author(s) and the title of the work, journal citation and DOI.
Published under licence by IOP Publishing Ltd

IOP Conf. Series: Materials Science and Engineering **275** (2017) 012021 doi:10.1088/1757-899X/275/1/012021

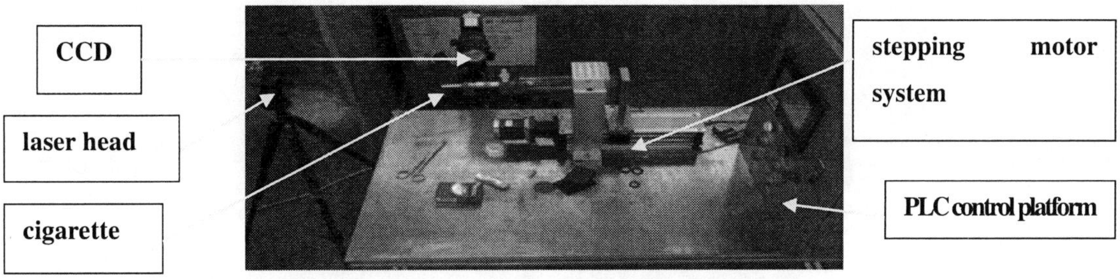

Figure 1. Scheme of experimental set up

3. Cigarette Gas Measurement of Burning End

1) Burning Process at Different Suction Pressures:

When the suction pressures are 0.01KPa, 0.05KPa, 0.13KPa, 0.21KPa, 0.29KPa and 0.38KPa respectively, the corresponding velocity flow fields of the burning end are as shown in Figure 2, Figure 3, Figure 4, Figure 5, Figure 6 and Figure 7.

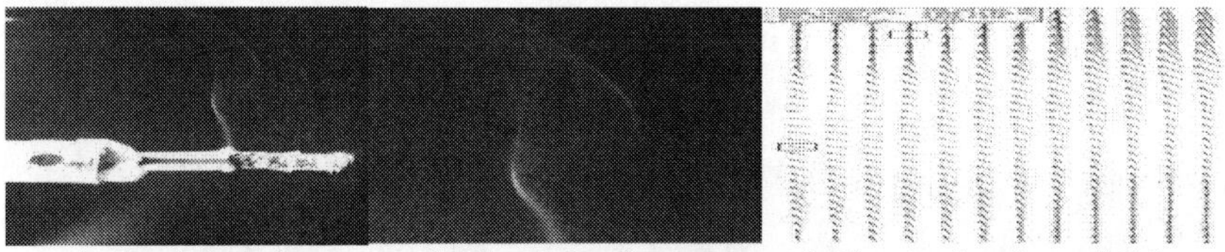

(a1) original image (a2) cutting image (a3) velocity flow field

Figure 2. PIV image for 0.01KPa suction pressure

(b1) original image (b2) cutting image (b3) velocity flow field

Figure3. PIV image for 0.05KPa suction pressure

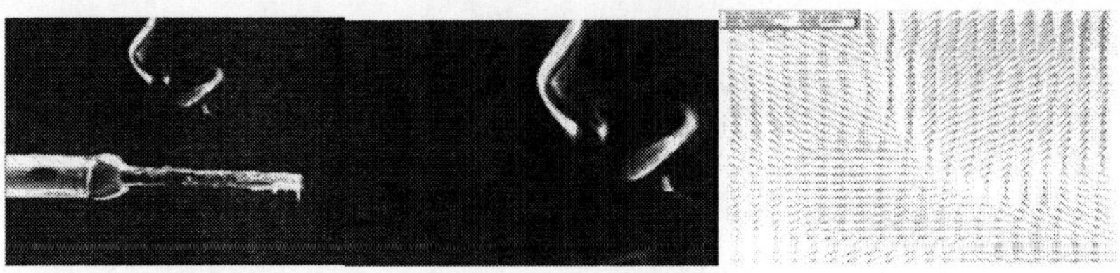

(c1) original image (c2) cutting image (c3) velocity flow field

Figure 4. PIV image for 0.13KPa suction pressure

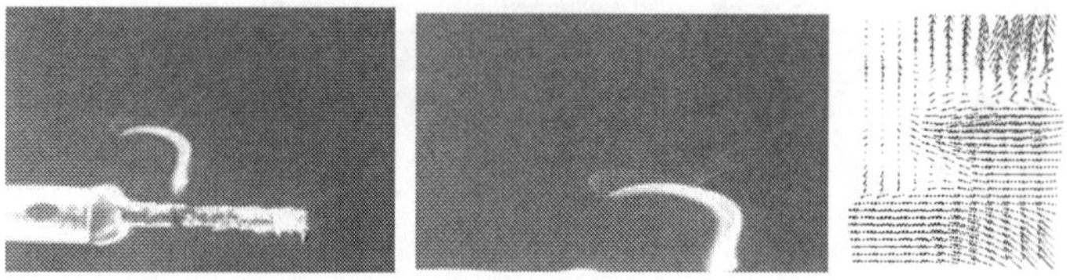

(d1) original image (d2) cutting image (d3) velocity flow field

Figure 5. PIV image for 0.21KPa suction pressure

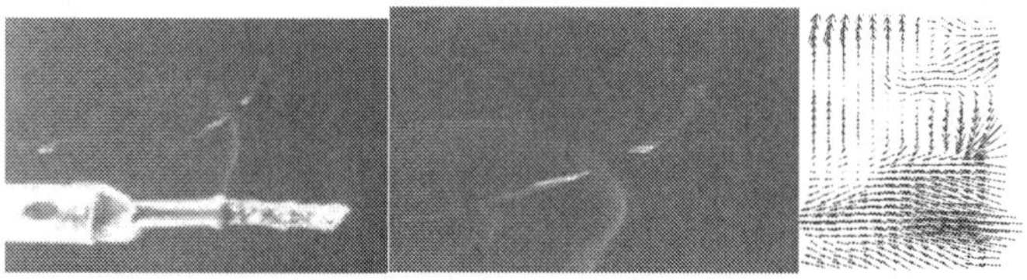

(e1) original image (e2) cutting image (e3) velocity flow field

Figure 6. PIV image for 0.29KPa suction pressure

(f1) original image (f2) cutting image (f3) velocity flow field

Figure 7. PIV image for0.38KPa suction pressure

2) Smoldering Process:

When the suction pressure is 0KPa, it is called smoldering process. The smoke flow field is as shown in Figure 8.

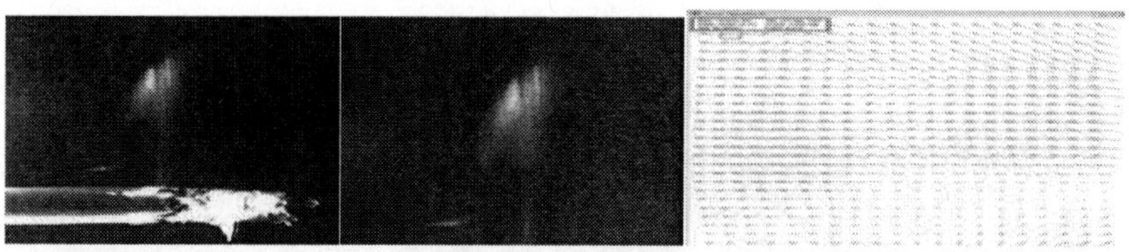

(g1) original image (g2) cutting image (g3) velocity flow field

Figure 8. PIV image for smoldering process

From the above experiments, the fitted curve between the velocities of the burning end and suction pressures is shown in Figure 9.

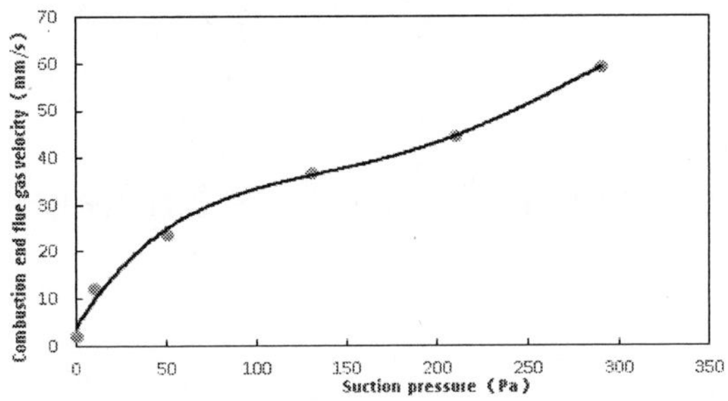

Figure 9. Velocities of the burning end with different suction pressures

As shown in Figure 9, the relationships between velocities of the burning end and suction pressures present curve properties, the fitting equation is:

$$y = (-2E-0.8)x^4 + (2E-0.5)x^3 - 0.0046x^2 + 0.6081x + 4.1182 \tag{1}$$

Correlation coefficient is $R^2 = 0.9944$. So the velocity flow fields of the burning end increase with the suction pressures.

4. Conclusions

The velocity flow fields of the burning end was measured by the smoke experiment machine, velocity characteristics of the flow fields are revealed with smoke particles directly as tracer particles, and the different flow fields are obtained under different suction pressures. The conclusions are as follows: the cigarette gas is mainly above the burning end of cigarette, and later gradually disappears due to the diffusion effect. It is obvious that the cigarette gas diffusion in combustion process is faster than in the smoldering process; the velocities of the burning end increased with the suction pressures.

5. References

[1] Zeng J, Chen K, Jun L I, et al. Effect of S-type discharger structure on flow field in dilution zone at bottom of tower [J]. Ciesc Journal, 2011, 62(3):685-691.

[2] Hardalupas Y, Sahu S, Taylor A M K P, et al. Simultaneous planar measurement of roplet velocity and size with gas phase velocities in a spray by combined ILIDS and PIV techniques [J]. Experiments in Fluids, 2010, 49 (2):417-434.

[3] Jin-Bang H U, Yan-Ping L I, Chen A X, et al. Effect of the venturi inlet configuration on gas flow in circulating fluidized bed desulfurization reactor-numerical simulation [J]. Chemical Engineering, 2005.

[4] Ling-Bing B U, Gao Y C, Ke-Bing L I, et al. PIV measurement of flow field for gas distributors [J]. Natural Gas Chemical Industry, 2012.

[5] Mizeraczyk J, Kocik M, Dekowski J, et al. Measurements of the velocity field of the flue gas flow in an electrostatic precipitator model using PIV method [J]. Journal of Electrostatics, 2001, s 51–52(1):272-277.

[6] Kashyap M, Chalermsinsuwan B, Gidaspow D. Measuring turbulence in a circulating fluidized bed using PIV techniques [J].PARTICUOLOGY, 2011, 09(6):572-588.

IOP Conf. Series: Materials Science and Engineering **275** (2017) 012022 doi:10.1088/1757-899X/275/1/012022

Research on a Sudden Explosion and its Environmental Impact

Maosheng Ye[1,*], Hui Ma[2] and Qingwei Ni [1]

[1]China Eastern Route Corporation of South-to-North Water Diversion, Beijing, China
[2]Beijing Municipal Commission of Urban Planning tongzhou District Branch, Beijing, China
E-mail: 78256510@qq.com

Abstract: A sudden blast was chosen as the studied topic. Also, one computer based virtual experimentation was used to estimate the dimensional impact of initial pollutant plume from blasts. Self-made method using Mathcad code was used to generate the output for the period of the first tenth of a second (1deci-second) to 1minute (60s) of the blast at the point source. It also depicted long-range air pollution travel within the first 1 to 10 minutes. In the case study, it assumed an average directional diffusivity of 1720 m^2s^{-1} which is about 25 per cent of the average generated speed of common explosives. The newly developed model revealed a plume cloud impact of $6.8\times10^7\mu gm^{-3}$ in the first 1millisecond (0.01s) which decayed suddenly to a value of $1.7\times10^7\mu gm^{-3}$ in the first 1decisecond (0.1s). The impact concentration at the point source by the end of the first second (1.0s) was $3.2\times10^5\mu gm^{-3}$ which implied a 99.5% sudden decay when compared to 0.01s concentration value at the emission point source. Computerized experiments observed that air pollutants release from explosives/blasts were dispersed into the atmosphere in the first few seconds by forceful injection instead of by gradual dispersion as is the case with normal air pollutants plume releases.

1. Introduction

Explosion is a hazard, and, besides the many services it provides to humankind, it may cause nuisance, damage to property and to people (injury and loss of life). Damage may be ranked in top-down severity order as: a) loss of life, i.e. mortality; b) loss of health, i.e. morbidity; c) loss of property and d) loss of activity.

Besides damage caused by the explosion process itself, there is also damage associated to the environment nearby. In 1997, Magurchara gas field caught fire and then caused huge damage of life and property over the whole area. The Magurchara gas field explosion damaged about 60 hectors of natural forest and 300 hectors of land were burnt. A large part of wild life (deer, birds, foxes, monkeys etc.) was destroyed or displaced to other places. Around 3000 people were affected because 31 hectors of tea garden was completely damaged during explosion (EPCT, 1997). This was a typical industrial accidental explosion which imitated follow-up negative environmental impact.

Also, the sudden explosions like volcanic eruptions will affect the environment, even human health. The 15 June 1991 eruption of Mt Pinatubo on the island of Luzon in the Republic of Philippines forced the evacuation of more than 200,000 people and caused the immediate deaths of more than 300, many from the collapse of homes due to the combination of heavy ashfall and rain from the nearly simultaneous passage of Typhoon Yunya(Hansen et al. 1992;McCormick et al. 1995). Actually there are three major areas of potential chronic human health impacts from volcanic activity which include respiratory problems, particularly silicosis (Buist et al. 1986; Baxter et al. 1999; Allen et al. 2000),

Content from this work may be used under the terms of the Creative Commons Attribution 3.0 licence. Any further distribution of this work must maintain attribution to the author(s) and the title of the work, journal citation and DOI.
Published under licence by IOP Publishing Ltd

psychological stress (e.g., Shore et al. 1986), and chemical impacts of gas or ash(e.g., Giammanco et al. 1998). So, the overall environment nearby and people's health can be highly affected by volcanic eruptions.

In the natural environment, magma contains dissolved gases that are released into the atmosphere during eruptions. At high pressures deep beneath the earth's surface, volcanic gases are dissolved in molten rock (Sigurdsson, 1999). But as magma rises toward the surface where the pressure is lower, gases held in the melt begin to form tiny bubbles. The increasing volume taken up by gas bubbles makes the magma less dense than the surrounding rock, which may allow the magma to continue its upward journey (Heiken et al. 1991). Closer to the surface, the bubbles are increasing in number and size, so the gas volume may exceed the melt volume in the magma and create foams (Settle et al. 1978). The rapidly expanding gas bubbles of the foam can lead to explosive eruptions in which the melt is fragmented into pieces of volcanic rock, known as tephra. If the molten rock is not fragmented by explosive activity, a lava flow will be generated (Durand et al. 2001). Together with the tephra and entrained air, volcanic gases can rise tens of kilometers into Earth's atmosphere during large explosive eruptions. Once airborne, the prevailing winds may blow the eruption cloud hundreds to thousands of kilometers from a volcano (Makhviladze et al. 1995).

Air pollution can be defined as an atmospheric condition in which substances are present at concentrations higher than their normal ambient levels due to certain natural or anthropogenic activities. The environmental impact of specific air pollutant depends on its ability to produce adverse effects/hazards on lives. In a broader sense, gaseous air pollutants include all the gases that are found in the atmosphere above their normal ambient levels. Here, airborne particulates include both solid and liquid particles have similar environmental impacts (Seinfeld, 1986; Hopke, 2009; DiGiovanni et al., 2006; Smodis, 2007; Smith et al., 2001).

Currently, the famous plume model mainly includes the model of Thomas (Thomas et al, 1963), the model of McCaffrey (McCaffrey, 1983), the model of Zukoski (Zukoski et al, 1981), the model of Heskestad(Heskestad,1984) etc. These models are mainly used to analyze the amount of smoke generated in the fire. They have different experimental basis and the applicable conditions, the results of the models for the same problem are often different. The plume flow values predicted by the model of Thomas and the model of McCaffrey were both a little big, which caused a lower calculated smoke temperature than the experimental values, so they are inapplicable to large power fires. The plume flow value predicted by the model of Zukoski was a little smaller and the calculated smoke temperature was bigger than the experimental value, so it is applicable to small power fires. The plume flow value predicted by the model of Heskestad was steady, and the result of the zone simulation agreed well with the experimental value, so it could be used to predict large power fires.

Another plume model is the Gaussian plume diffusion model which is widely used in the study of air pollution dispersion and the field of atmospheric environment quality prediction (Arystanbekova, 2004). It is a model used to describe the distribution of the mass concentration of atmospheric pollutants or toxicants in the atmospheric diffusion process. But this model did not involve the role of explosion on the pollutant diffusion. Besides, there was some error between the result from the model and the measured one. This model only applied to analyze the diffusion of continuous point source.

There are also some other air pollution models. TAPM is a PC-based, nestable, prognostic meteorological and air pollution model (with photochemistry) driven by a Graphical User Interface, and is a viable tool for year-long simulations (Hurley et al, 2005). Barsotti et al. (2008) presented a new modeling tool, named VOL-CALPUFF, that is able to simulate the transient and three-dimensional transport and deposition of volcanic ash under the action of realistic meteorological and volcanological conditions throughout eruption duration. Karamchandani et al. (2009) used a sub-grid scale modeling approach to simulate the transport and fate of toxic air pollutants. Molnar et al. (2010) studied about air pollution modeling using a Graphics Processing Unit with CUDA. Tominaga et al. (2011) have presented CFD modeling using RANS and LES of pollutant dispersion in a three-dimensional street canyon which is investigated by comparison with measurements. But these models can not reflect a sudden explosion and its environmental impact.

2. Theories

Here, an explosion can be defined as a sudden release of energy that produces a sudden volume expansion (dV) of the conveying material due to large changes in pressure dP within the shortest possible time (dt). This causes high pressure waves in the local medium in which they occur. These shock waves can either be subsonic or supersonic, natural or anthropogenic (Basarov et al., 1991, Slotnick, 2008). The intensity of an explosion (I_{exp}) therefore is proportional to the magnitude of $\frac{dP}{dt}$ and $\frac{dV}{dt}$

$$I_{\exp} \alpha \frac{dP}{dt} \tag{1}$$

$$I_{\exp} \alpha \frac{dV}{dt} \tag{2}$$

As we know, parts of the natural largest sources of sudden explosions are volcanic explosive eruptions. The sudden volcanic eruption occurs as a result of high pressure build-up in the volcano's magma chamber, which pushes up the magma with great force out of the chamber in the same way a warm soda drink will spray out vigorously out of its bottle when its CO_2 or H_2S bubbles have a higher pressure when the bottle is violently shaken and the lid opened suddenly (HVO, 2011). Major volcanic eruptions eject large amounts of particulate matter in form of volcano ashes. Also, gases in volcanic eruptions mainly include carbon diode, sulphur dioxide, hydrogen sulfide and other gases into the ambient air. Part of these pyroclastic materials gets transported to high enough altitudes and stays there for months or several years to impact on the global climate system (Neff, 1998). Volcanic pyroclastic have been observed at very high altitudes, 25,000m and above. This implies that materials from volcanic eruptions easily transcend the lowest portion of atmosphere which is the troposphere into the stratosphere (Groisman, 1998; Kane, 1998). The stratosphere is the region of the atmosphere where the ozone layer is found within the altitudinal range of 18,000m and 32000m.

Volcanic explosion can be in the form of sudden eruption of hot lava and thick smoke unto the surface, sudden release of hot spring of water and steam also known as Geysers, and violent vibration of the earth crust to a certain depth also known as the earth quake. The depth of the vibration determines the magnitude of the earth quake (Robock, 2003; Devine et al., 1984; Arya, 1999). These effects can be summarized by the following mathematical expression, which gives the discharge rate Q at which magma is ejected out of a volcano conduit of average radius R and length L as:

$$Q = \frac{\pi R^4 P}{8 \mu L} \tag{3}$$

Here, μ is the viscosity of the magma;
P is the pressure inside the magma chamber.

Under the explosive blast, a shock wave will be produced when the rate of combustion of the explosive substances generates a sharp pressure gradient. Chemical potential energy has been converted to kinetic and heat energy during the process of an explosion. For example, trinitrotoluene (TNT) is characterized by a large amount of chemical potential energy that is nearly instantaneously converted to kinetic and heat energy. The resulting generated speed in the medium is approximately $6900ms^{-1}$ for TNT. This results in very high kinetic energy of approximate value 4.7 kJg^{-1} or 4.7 $MJkg^{-1}$. According to Neff, (1998) the chemical reaction in an average TNT explosive is typically ninety per cent complete in between 10^{-9} and 10^{-6} second (1nanosecond to 1microsecond), and as the energy of the blast dissipates with increasing distance from the blast, the wave dissipates into a sound wave resulting in the loud blast.

In industrial circles, Environmental Impact Assessment (EIA) for all petroleum projects will allow the planners and implementers to understand the environmental impacts and con-sequences of their projects (NEAB, 1998; DoE, 2001). The EIA process is thought to consist of six main components: basics; impact identification; description of the affected environment; prediction and assessment of impacts; selection of proposed action and documentation in accordance with extant guidelines (Canter LW, 1997).

3. Methods

PTC Mathcad is the industry standard software for solving, analyzing, and sharing most vital engineering calculations. Its live mathematical notation, units intelligence, and powerful calculation capabilities, presented within an easy-to-use interface, allows engineers and design teams to capture and communicate their critical design and engineering knowledge. Here, it was used for the simulation of sudden air pollutants release.

For the scenarios setting, it considered different area spread. Further, it was defined as a sudden plume cloud release. It has to exclude a continuous smoke release situation. Momentum and kinetic energy variations are the prominent factors for detonations, whereas transport and dispersion processes are more relevant for the air pollution generated. The self-similarity of the solution at different scales was an important characteristic of the plume wave profile generated by the model.

In the study, the explosive blast waves were hypothesized as identical to volcanic eruption except for scaling in magnitude and duration. The activation of the model was made to be proportional to the mass and the initial velocity, which also implies the embedding of the initial kinetic energy of the blast into the model.

4. Results and Discussions

Empirically, air pollutants generated under a normal gradual circumstance get dispersed depending on the prevalent atmospheric conditions such as local wind speed, wind direction, humidity, at the time of release. The composition and chemistry of the atmosphere is of importance for several reasons, but primarily because of the interactions between the atmosphere and living organisms. Atmospheric physicists attempt to model Earth's atmosphere and the atmospheres of the other planets using fluid flow equations, chemical models, radiation balancing, and energy transfer processes in the atmosphere and underlying oceans.

Atmospheric stability, mixing height and atmospheric ventilation index are the primary influencing factors of the atmospheric conditions, which is categorized into five sections (A-E). A stable atmosphere is one that is strongly resistant to change, while atmospheric turbulence results in significant displacement of air parcels both in horizontal and vertical directions and vice versa. In order to model weather systems, atmospheric physicists employ elements of scattering theory, wave propagation models, cloud physics, statistical mechanics and spatial statistics, each of which incorporate high levels of mathematics and physics.

The numerical simulation revealed that air pollutants generated from sudden explosion/blast get dispersed very fast from emission point source due to the large forces and pressures which accompany the air pollutants at the moment of release.

In the simulation, it assumed an average directional diffusivity of 1720 m^2s^{-1}. It is about 25% of the average generated speed of common explosives. The outcome put forward a plume cloud impact of $6.8\times10^7\mu gm^{-3}$ at the emission point source in the first 1millisecond (0.01s) decayed suddenly to a value of $1.7\times10^7\mu gm^{-3}$ in the first 1decisecond (0.1s) (Figure1a). Further, by the end of the first one second (1.0s), the plume concentration at the emission point source was 3.2×10^5 μgm^{-3}. This result implied a 99.5% sudden dispersion of the matters released by the eruption within the shortest possible time. Please see Figures 1(a-f).

We can found that the instant dispersion of air pollutants and displacement of air parcel in the immediate vicinity of the emission point source is directly resulted in the extreme forces and pressures generated by the blast. This aided the propagation of the air pollutants emitted both horizontally and

IOP Conf. Series: Materials Science and Engineering **275** (2017) 012022 doi:10.1088/1757-899X/275/1/012022

vertically at shortest time frame. The model simulation revealed that the rate of dispersion of the air pollutants depended directly on the associated mass of the pollutants, the speed and the momentum generated by the explosive blast. Thus the kinetic energy associated with the blast played more significant role in the almost instant dispersion from the point source than the prevalent atmospheric conditions. It would have been the case for normal emissions. The simulation outcome revealed that within an infinitesimal fraction of a second of the explosion was very significant as the result has shown in Figures (2a and 2b). From these figures, the sharp variation of concentration impact at the source from 1.7×10^{7} μgm^{-3} in the first 1decisecond (0.1s) to $3.2\times10^{5}\mu gm^{-3}$ by the end of the first one second (1.0s) and to just $300\mu gm^{-3}$ by the end of the first one minute (60.0s) was an index to the rapid decay which might not be easy to observe under real physical conditions but only in virtual experimentation as performed in this study. Here, we can see the significance of numerical experimentation in the reveal of potential physical process.

Clearly, this can thus be suggested that air pollutants emitted from sudden explosion is more of forceful injection into the atmosphere than gradual dispersion in the very few seconds of the blast occurrence. And like air pollutants from volcanic eruption, it also has capacity to penetrate higher altitude of the atmosphere at shorter time than normally released air pollutants would do.

The obvious example can be found in the eruption of Mt. Pinatubo in June, 1991. It was found to have made significant chemical perturbation to the ozone layer and the resulting observed depletion in the ozone concentration for that period (Groisman, 1998). Likewise, Kane, (1998) noted the occurrence of a spectacular quassi biennial oscillation (QBO) wave between 1991 and 1993 with a range of about ±10% which is associated partly with the volcanic eruptions of Mt. Pinatubo in June 1991. Similarly, Morris et al., (2003) observed significant increase in stratospheric chlorine, which was linked with the El Chichon eruption in 1982.

Figure 3 revealed that the long range air pollutant dispersion from sudden explosion. Air pollutant has the ability to reach a distance of 1.2km within the first minute of the explosion, and could be propagated as far as 4km within the first ten to fifteen minutes of the release.

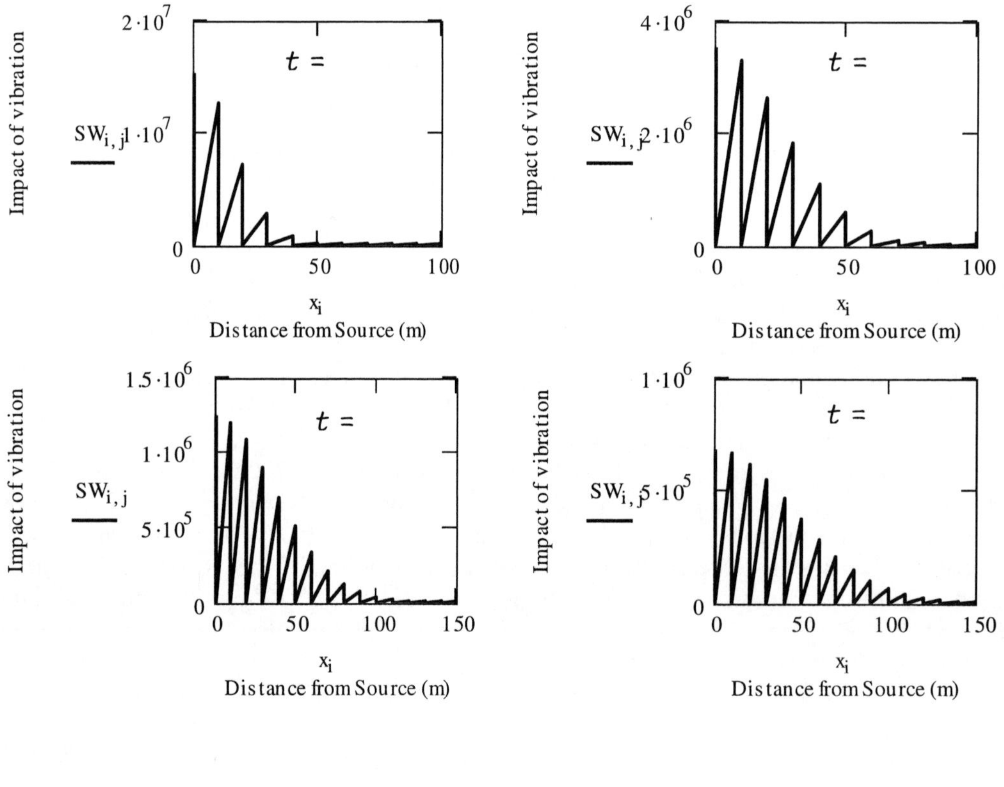

IOP Conf. Series: Materials Science and Engineering **275** (2017) 012022 doi:10.1088/1757-899X/275/1/012022

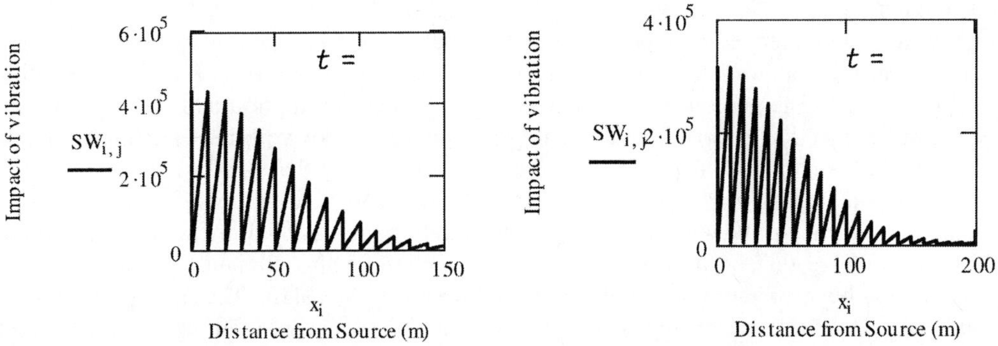

Figures 1(a-f): Shockwaves of air pollution concentration (***μgm***$^{-3}$) at the point source vicinity at various moments in the first one second of blast

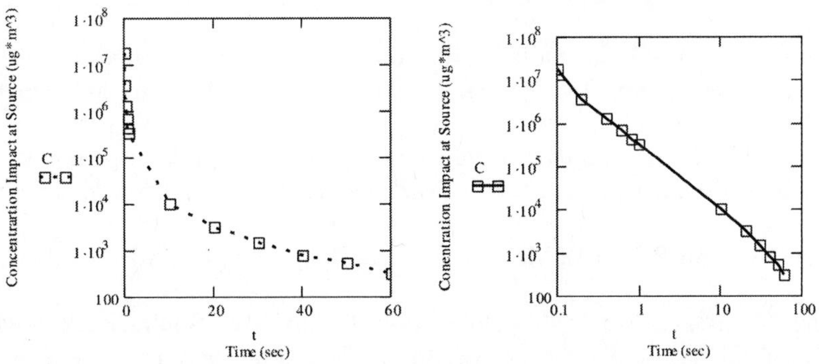

Figure 2(a-b): Variation of Concentration Impact at emission point source in the first one minute (60s) of blast

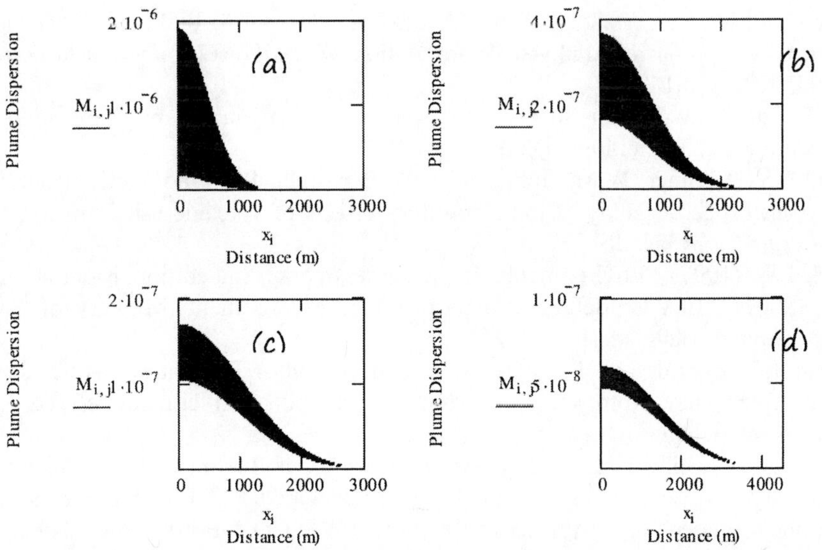

Figure 3(a-d): Long range Air pollution Dispersion (kgm^{-3}) in 1min, 3mins, 5mins and 10mins of blast respectively

5. Conclusions

This paper used computerized method to make a virtual experimentation to explain how the rate of dispersion of air pollutants changes in certain time and make a model to reveal what is relating with plume cloud impact. The gaseous products created by a volcanic eruption and by an explosive blast follow a similar injection pattern, but this study revealed that like volcanic eruptions, air pollution from sudden blast is more of forceful injection into the atmosphere than gradual dispersion in the first few seconds of the explosion. The length and time scales involved are orders of magnitude different. In the study, it obtained a sudden decay of the concentration dimensional impact at source point from $1.7\times10^{7}\mu gm^{-3}$ to $300\mu gm^{-3}$ in the first 60s. Sudden injection of air pollutants into the ambient air is enhanced by the huge pressure and momentum generated by the blast. The ability of air pollutants from sudden explosion to reach higher altitudes within a short period of time made it to have both immediate and long term adverse effect on the environment and the life form. It also implies a further complication of the already existing air pollution challenge confronting the environment.

6. References

[1] Allen, A.G., Baxter, P.J. and Ottley, C.J. (2000) Gas and particle emissions from Soufriere Hills Volcano,Montserrat, West Indies: characterization and health hazard assessment. Bull. *Volcano.* **62**, 8–19.

[2] Arystanbekova, N. K. (2004). Application of Gaussian plume models for air pollution simulation at instantaneous emissions. *Mathematics and Computers in Simulation*, **67**(4), 451-458.

[3] Arya P, (1999). Air Pollution, Meteorology and Dispersion**.** Oxford University Press, New York, 310pp.

[4] Bazarov SB, Bazhenova V, Bulat OV, Golub VV, ShulmeisteR AM, (1991). Three dimensional diffraction of shock wave_ In K_ Takayama_ editor, Shock Waves Proceedings of Sendai Japan Springer-Verlag, **1**: 251.

[5] Baxter, P.J., Baubron, J.C. and Countinho, R. (1999) Health hazards and disaster potential of groundgas emissions at Furnas volcano, Sao Miguel, Azores. *J. Volcanol. Geotherm. Res.* **92**, 95–106.

[6] Barsotti, S., Neri, A., & Scire, J. S. (2008). The VOL—CALPUFF model for atmospheric ash dispersal: 1. Approach and physical formulation. *Journal of Geophysical Research: Solid Earth* (1978–2012), 113(B3).

[7] BUET, (2003). Workshop in Oil and Gas Sector", organized by Bangladesh University of Engineering and Technology, Dhaka.

[8] Buist, A.S., Vollmer, W.M., Johnson, L.R., Bernstein, R.S. and McCamanat, L.E., (1986) A four-yearprospective study of the respiratory effects of volcanic ash from Mt. St. Helens. *Am. Rev. Respir.Dis.* **133**, 526–534.

[9] Canter LW, (1997). "Environmental impact assessment", 2nd edition, Mcgraw-hill Inc., p. 640.

[10] DoE, (2001). "EIA Guidelines", Department of Environment, Ministry of Environment and Forest, Govt. of Bangladesh.

[11] Devine, J.D., Sigurdsson, H., Davis, A.N., Self, S, (1984). Estimates of sulfur and chlorine yield to the Atmosphere from volcanic eruptions and potential climatic effects. *J. Geophysical Research* **89**: 6309.

[12] DiGiovanni F, Fellin P, (2006). Trans-boundary air pollution, in Environmental Monitoring , [Eds.Hilary I. Inyang, John L. Daniels], in Encyclopedia of Life Support Systems (EOLSS), Developed under the Auspices of the UNESCO, Eolss Publishers, Oxford ,UK, [http://www.eolss.net]

[13] Durand, M., & Grattan, J. (2001). Effects of volcanic air pollution on health. *The Lancet*, **357**(9251), 164.

[14] EPCT, (1997). "Draft Report - Environmental Impact Assessment in Oil and Gas Sector", survey work conducted by Engineering Planning Consultancy Team, Sylhet.

[15] Fire Research Station, & Thomas, P. H. (1963). *Investigations into the flow of hot gases in roof venting*. HM Stationery Office.
[16] Groisman, PY, (1992). Possible regional climate consequences of the Pinatubo eruption: an empirical approach. *Geophysical Research Letters* **19**(15):1603.
[17] Giammanco, S., Ottaviani, M., Valenza, M., Veschetti, E., Principio, E., Giammanco, G. and Pignato, S. (1998) Major and trace elements geochemistry in the ground waters of a volcanic area: Mount Etna (Sicily, Italy). *Water Res*. **32**, 19–30.
[18] Kato N, Akimoto H, (1992). Anthropogenic Emissions of SO2 and NOx in Asia: Emission Inventories. *Atmospheric Environment*, **26**A: 2997-3017.
[19] Kane RP, (1998). Ozone depletion, related UV-B increase and increase skin cancer incidences. *International J. Climatology* **18**: 457.
[20] Karamchandani, P., Lohman, K., & Seigneur, C. (2009). Using a sub-grid scale modeling approach to simulate the transport and fate of toxic air pollutants. *Environmental fluid mechanics*, **9**(1), 59-71.
[21] Hawaii Volcano Observatory, (HVO), (2011). Volcanic Air Pollution- A Hazard in Hawaii. http://hvo.wr.usgs.gov/).
[22] Hansen, J., Lacis, A., Ruedy, R., & Sato, M. (1992). Potential climate impact of Mount Pinatubo eruption. *Geophysical Research Letters*, **19**(2), 215-218.
[23] Heiken, G., & Wohletz, K. E. N. N. E. T. H. (1991). Fragmentation processes in explosive volcanic eruptions. *Sedimentation in volcanic settings*, **45**:86.
[24] Hopke PK (2009). Contemporary threats and air pollution. *Atmospheric Environ* **43**: 87.
[25] Hurley, P. J., Physick, W. L., & Luhar, A. K. (2005). TAPM: a practical approach to prognostic meteorological and air pollution modelling. *Environmental Modelling & Software*, **20**(6), 737-752.
[26] Heskestad, G. (1984). Engineering relations for fire plumes. *Fire Safety Journal*, **7**(1), 25-32.
[27] Leopold LB, Clarke FE, Manshaw BB, Balsley JR, (1971). A Procedure for Evaluating Environmental Impacts, U.S. Geological Survey Circular No. 645, Government Printing Office, Washington, D.C.
[28] Makhviladze, G. M., Roberts, J. P., & Yakush, S. E. (1995). Modelling of atmospheric pollution by explosions. *Environmental Software*, **10**(2), 117-127.
[29] McAdam M, (2008). Bangladesh Travel Guide, Lonely Planet, 6th Edition ISBN: 9781741045475, p. 153.
[30] McCaffrey, B. J. (1983). Momentum implications for buoyant diffusion flames. *Combustion and Flame*, **52**, 149-167.
[31] Morris G, Barbara G, Newman PA, Aikin A, Heaps W, Crum F, Larko D, Todaro RM, (2003). The Chemical, Thermal, and Dynamical Structure of Earth's Atmosphere http://hyperion.gsfc.nasa.gov/ code 916.
[32] Molnar Jr, F., Szakaly, T., Meszaros, R., & Lagzi, I. (2010). Air pollution modelling using a graphics processing unit with CUDA. *Computer Physics Communications*, **181**(1), 105-112.
[33] Neff M, (1998). A Visual Model for Blast Waves and Fracture. PhD Thesis, University of Toronto pp11.
[34] NEAB, (1998). "EIA Manual", National Environmental Association of Bangladesh.
[35] McCormick, M. P., Thomason, L. W., & Trepte, C. R. (1995). Atmospheric effects of the Mt Pinatubo eruption. *Nature*, **373**(6513), 399-404.
[36] Robock A, (2003). Volcanoes: Role in climate in Encyclopedia of Atmospheric Sciences, J. Holton, J. A. Curry, and J. Pyle, Eds., (Academic Press, London), 10.1006/rwas.2002.0169: 2494.
[37] Seinfeld JH, (1986). Atmospheric chemistry and physics of air pollution. John Wiley and Sons, New York, Toronto, pp77-91,
[38] Slotnick J, (2008). Explosive Forces of Improvised Explosive Devices, http://www.securitydiver.com/aic/stories/aritcle-114.html (accessed August 2008).

[39] Smith SJ, Pitcher H, Wigley TML, (2001). Global and regional anthropogenic sulfur dioxide emissions. *Global and Planetary Change* **29**: 99.

[40] Smodis B, (2007). Investigation of trace element atmospheric pollution by nuclear analytical techniques at a global scale: harmonized approaches supported by the IAEA. *Journal of Environmental Management* **85**:121.

[41] Shore, J.H., Tatum, E.L. and Vollmer, W.M. (1986) Psychiatric reactions to disaster: the Mt. St. Helensexperience. *Am. J. Psychiatry* **143**, 590–598.

[42] Sigurdsson, H., Houghton, B., Rymer, H., Stix, J., & McNutt, S. (1999). *Encyclopedia of volcanoes*. Access Online via Elsevier.

[43] Settle, M. (1978). Volcanic eruption clouds and the thermal power output of explosive eruptions. *Journal of Volcanology and Geothermal Research*, **3**(3), 309-324.

[44] Tominaga, Y., & Stathopoulos, T. (2011). CFD modeling of pollution dispersion in a street canyon: Comparison between LES and RANS. *Journal of Wind Engineering and Industrial Aerodynamics*, **99**(4), 340-348.

[45] Zukoski, E. E., Kubota, T., & Cetegen, B. (1981). Entrainment in fire plumes. *Fire Safety Journal*, **3**(3), 107-121.

IOP Conf. Series: Materials Science and Engineering **275** (2017) 012023 doi:10.1088/1757-899X/275/1/012023

The Method of Calculating the Shock Effect of Falling Rock Research

Kexuan Guo[1*], Hongkai Chen[1, 2] and Tao Chen[1]

[1]College of Hydraulic & environmental Engineering, China Three Gorges University, Yichang,Hubei,443002.China

[2]Institute of geotechnical engineering,Chonging Jiaotong University,400074,China

[*]Email:374550280@qq.com

Abstract: The paper study on the process of rockfall falling, consider the air below the rockfall will be compressed, calculate the force of the compressed air to the rockfall; Set up theory mode and divide the process into n parts, using the theory of Aerodynamics, Conservation of energy theorem and Air moving theory to derive the method of calculate the rockfall impacts; The results have certain reference, it can be used in the theory study of disaster reduction and technical of rockfall.

1. Introduction

China is a mountainous country with many high and steep slopes, dangerous rocks and rockfall are common forms of geological disasters rockfall. Disaster is characterized by randomness, sudden and unpredictable. Once it happens, it will cause serious casualties and economic losses. Such as August 1983,Xiangyu line Bashan a state highway section of the road slope affected by heavy rain caused rockfall disaster, a volume of about 12.6 rock fall from the height of about 80 meters, and a bridge was destroyed causing traffic interruption about 23 hours, the direct economic losses up to 22.1 million. So far, there are a large number of domestic and foreign scholars have carried out extensive and in-depth research of rockfall disaster [1], but they were committing on the research of the formation mechanism evolution [2] and fall to the impact of failure mechanism [3], neglecting the research of rockfall movement process and the impact factors for rockfall force was not considered perfect, the compressed air resistance when the dangerous rock fall. Based on the research status [4], this paper deduces the calculation method of impact velocity of dangerous rockfall by combined the viewpoint of energy with aerodynamics. The research results to improve the calculation accuracy of rockfall impact force [5],

Further prediction of rockfall disaster, and to carry out disaster prevention work has a positive guiding significance.

2. Derivation the Method of Calculate the Speed of Rockfall

Dangerous rockfall has many characteristics, which are high speed and the huge destructive. Scholars at home or abroad often ignore the air impact but pay too much attention to free-faller when they research this type disasters [9]. Making inevitable error when we apply these conclusion to the prevention and repair on dangerous rockfall. In this paper, establishing a rockfall theory model, using aerodynamic theory, principle of conservation of energy and air dynamic theory aims to analyse the compression process, combined with the air leakage coefficient, to derived the calculation methods of rockfall fall speed.

Content from this work may be used under the terms of the Creative Commons Attribution 3.0 licence. Any further distribution of this work must maintain attribution to the author(s) and the title of the work, journal citation and DOI.
Published under licence by IOP Publishing Ltd

2.1. Establish Model

The theory model of air compressed by dangerous rockfall [6].Shown in figure 1.

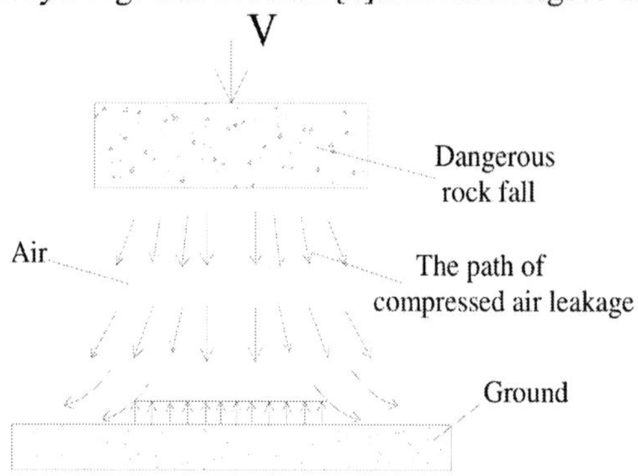

Figure1. The mechanical model of air be compressed by rockfall

Added a rookfall fall height of H, divided the H into n parts, Constant n is large enough, each part $dh = H / n$. The process of rookfall fall is continuous. After a part of dh relies on a part. The whole process is shown in figure 2.The first wave of A_1-A_1 propagation speed is C_0, same as the sound and shown in figure2-b. After the first dh, rookfall to continue the second dh, a new compression wave was create. As shown in figure2-c.Propagation speed affected by transmission medium density and pressure, the wave of A_2-A_2 propagate in compressed air by the wave of A_1-A_1, propagation speed $C_1 > C_0$. In a similar way, and the third (figure 2-d, 2-e) even countless(figure 2-f)wave will be produced to compress air. Out of rookfall fall in the ith ($i \in (1.2 \cdots n)$) part to analyze [7].

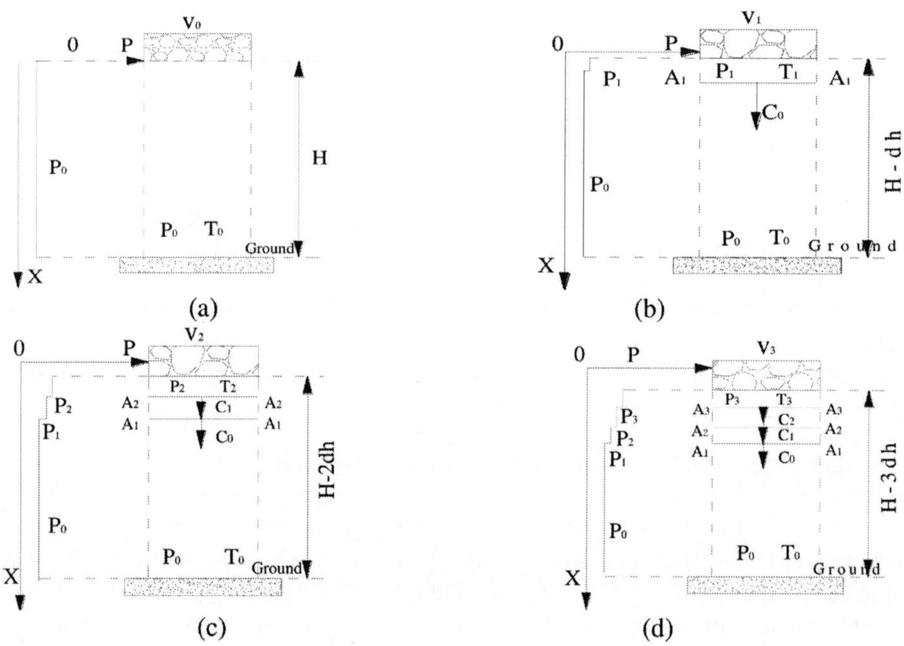

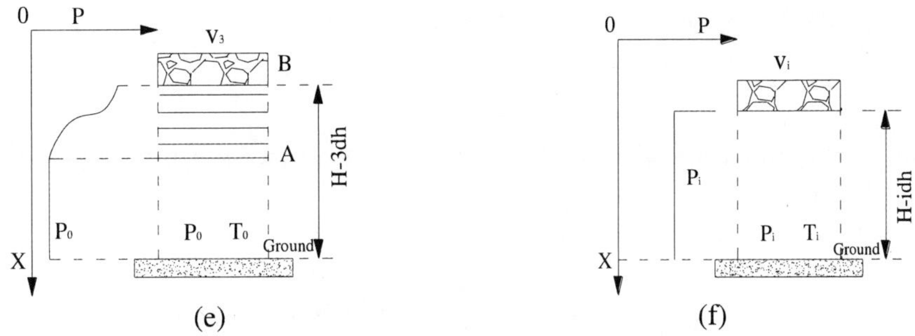

Figure 2. The process diagram of air be compressed

2.2. *Derivation the Method of Calculate the Speed of Rockfall*

Analysis the first dh of rookfall falling.

First law of thermodynamics: Change of internal energy=The energy of action+heat transfer

The gas state equation form air dynamic theory

$$E=\frac{\gamma}{2}kNT=\frac{\gamma}{2}k\frac{M}{m}T \tag{2.1}$$

γ is degree of freedom,air generally take 5; k is ideal gas constant,generally take $8.314J/mol$; M_i/m can calculate the amount of gas materia.

Ignore the leaked air internal energy change.Combined with the energy conservation can get the formula

$$\frac{1}{2}Mv^2{}_0+\frac{1}{2}M_0C^2{}_0+Mgdh+M_0\frac{dL_1}{2}=\frac{1}{2}Mv^2{}_1+\frac{1}{2}M_1C^2{}_1+\frac{\gamma}{2}k\frac{M_1}{m}(\Delta T) \tag{2.2}$$

$v_0=0m/s$ said the rookfall original speed, $C_0=340m/s$ is wave propagation velocity in the air, ΔT is changed temperature of the compressed air

$$M_0=\pi R^2\rho_{air}C_0t_1 \tag{2.3}$$

M_0 is the quality of air below the rookfall in initial stage; R is the equivalent radius of rookfall

$$dh=v_0t_1+\frac{1}{2}a_1t_1{}^2=\frac{1}{2}gt_1{}^2 \tag{2.4}$$

a_1 is the acceleration of the rockfall falling in first dh,take the free acceleration

$$t_1=\sqrt{\frac{2dh}{g}} \tag{2.5}$$

t_1 is the time of the rockfall falling in first dh

$$dL_1=C_0t_1 \tag{2.6}$$

dL_1 is the length of the compression wave forward in first dh, $\frac{dL_1}{2}$ is the height of compressed air gravity drop

$$v_1=v_0+gt_1 \tag{2.7}$$

v_1 is the the velocity of the rockfall when finished first dh

$$M_1 = (1-\alpha_1)M_0 \tag{2.8}$$

M_1 is the quality of air below the rookfall when finished first dh; α_1 is the leakage coefficien of air when finished first dh;Air leakage coefficient can be obtained through experiment.
Take formula form (2.3) to (2.8) into (2.2)

$$\frac{1}{2}\pi R^2 \rho_{air} C_0^{\ 3}\sqrt{\frac{2dh}{g}} + \pi R^2 \rho_{air} C_0^{\ 3}\frac{dh}{g} = \frac{1}{2}(1-\alpha_1)M_0 C_1^2 + \frac{\gamma}{2}k\frac{(1-\alpha_1)M_0}{m}(\Delta T) \tag{2.9}$$

By the gas state equation

$$PV = \frac{M}{m}kT \tag{2.10}$$

$$P_1 = \frac{M_1}{mV_1}kT_1 \tag{2.11}$$

$$T_1 = T_0 + \Delta T \tag{2.12}$$

$$\rho_1 = \frac{M_1}{V_1} = \frac{(1-\alpha_1)M_0}{\pi R^2 (C_0 t_1 - dh)} \tag{2.13}$$

According to the wave propagation velocity formula which is included in aerodynamics

$$C = \sqrt{\frac{rp}{\rho}} \tag{2.14}$$

r is the adiabatic coefficient, air usually take1.402; p is the pressure of the propagation medium, ρ is the density of the propagation medium; C associated with the density and the pressure of the air.

The rookfall finished first dh is shown in figure 2-d.Simultaneous equation can get the air temperature changes, then to derived the next movement dh;

That i $(0 \le i \le n)$ is the serial number of the current dh in all n parts dh; M_i is the quality of air below the rookfall when finished ith dh; V_i is the volume of air below the rookfall when finished ith dh; a_i is the acceleration of the rookfall falling in ith dh; P_i is the intensity of pressure of air to rookfall when finished ith dh; v_i is the velocity of the rookfall when finished ith dh; t_i is the time of the rookfall falling in ith dh; C_i is the wave propagate velocity produce in the period of ith dh which dpropagate in compressed air by i-1th wave, ρ_i is the density of the propagation medium when finished ith dh, d_{Li} is the length of the wave propagate in ith dh, R is the rockfall equivalent radius. Analysis the ith dh of rookfall falling. Strong energy balance equation and the change of each parameter conditions are as follows

$$\frac{1}{2}Mv^2_{i-1} + \frac{1}{2}M_{i-1}C^2_{i-1} + Mgdh + M_{i-1}\frac{d_{Li}}{2} = \frac{1}{2}Mv^2_i + \frac{1}{2}M_i C^2_i + \frac{\gamma}{2}k\frac{M_i}{m}(\Delta T) \tag{2.15}$$

$$a_i = g - \frac{F_i}{M} = g - \frac{sp_{i-1}}{M} = g - \frac{\pi R^2 p_{i-1}}{M} \tag{2.16}$$

$$dh = v_{i-1}t_i + \frac{1}{2}a_i t_i^2 \tag{2.17}$$

$$t_i = -\frac{v_{i-1}}{a_i} + \sqrt{\left(\frac{v_{i-1}}{a_i}\right)^2 + \frac{2dh}{a_i}} \tag{2.18}$$

The speed of rockfall falling

$$v_i = v_{i-1} + a_i t_i = v_{i-1} + \left(g - \frac{\pi R^2 p_{i-1}}{M}\right) \bullet \left(-\frac{v_{i-1} \cdot Mg}{Mg - \pi R^2 p_{i-1}} + \sqrt{\frac{v_{i-1} \cdot Mg + 2dhMg}{Mg - \pi R^2 p_{i-1}}}\right) \tag{2.19}$$

The length of the wave propagate

$$d_{Li} = C_{i-1} t_i \tag{2.20}$$

The volume of air below the rookfall when finished i-1th dh

$$V_{i-1} = \pi R^2 d_{Li} = \pi R^2 C_{i-1} t_i \tag{2.21}$$

The volume of air below the rookfall when finished ith dh

$$V_i = \pi R^2 (d_{Li} - dh) = \pi R^2 (C_{i-1} t_i - dh) \tag{2.22}$$

The process of rookfall fall is the process of air compressed.There is some air escaping in the compression process, so the M of air will changed before and after the compression.

The quality of the air before it be compressed is M_{i-1}

$$M_{i-1} = \rho_{i-1} V_i = \pi R^2 \rho_{i-1} C_{i-1} t_i \tag{2.23}$$

The quality of the compressed air is M_i

$$M_i = (1 - \alpha_i) M_0 \tag{2.24}$$

The density of the compressed air is ρ_i

$$\rho_i = \frac{M_i}{V_i} = \frac{(1-\alpha_i) M_0}{\pi R^2 (C_{i-1} t_i - dh)} \tag{2.25}$$

$$T_i = T_{i-1} + \Delta T_i \tag{2.26}$$

$$P_i = \frac{M_i}{m V_i} k T_i \tag{2.27}$$

In the same way,take various parameters into type (2-16) to type (2-28).Simultaneous solution is obtained ΔT_i. When the air is continue compressed by the back wave,the propagation speed will increase.In the end,the wave before and after will be added together to promote with a common velocity.Finally rockfall and air motion with the same speed.Assumption after the sth dh $(s < n, s \in i)$.Rockfall and air motion with the same velocity.That is;

$$\frac{1}{2} M v_s^2 + \frac{1}{2} M_s C_s^2 + Mgdh + M_s \frac{C_s t_{s+1}}{2} = \frac{1}{2}(M + M_{s+1}) v_{s+1}^2 + \frac{\gamma}{2} k \frac{M_{s+1}}{m} (\Delta T) \tag{2.28}$$

Compression wave will contact with the ground with dangerous rookfall at the same time. Assumes that the compression wave all absorbed by the ground in the final.

$$dL_i = C_{i-1} t_i > H - idh = (n-i) dh \quad (s \le i \le n) \tag{2.29}$$

IOP Conf. Series: Materials Science and Engineering **275** (2017) 012023 doi:10.1088/1757-899X/275/1/012023

Said the rookfall motion a dh length and compression wave also spread to the ground. Since then the pressure change is not a curve,but there is a saltation(figure 2-f),and each parameter changes

$$dL_i = (n-i)dh \tag{2.30}$$

Among them $(s \le i \le n)$, the formulae of below all in the same way.

The volume of the air before it be compressed

$$V_{i-1} = \pi R^2 \left[(n-i+1)dh\right] = \pi R^2 (n-i+1)dh \tag{2.31}$$

The volume of the compressed air

$$V_i = \pi R^2 \left[(n-i)dh\right] = \pi R^2 (n-i)dh \tag{2.32}$$

The quality of the air before it be compressed

$$M_{i-1} = \rho_{i-1}\pi R^2 (n-i+1)dh \tag{2. 33}$$

The quality of the compressed air

$$M_i = (1-\alpha_i)M_0 \tag{2.34}$$

The pressure change of air be compressed when finished ith dh

$$P_i = \frac{M_i}{m} \cdot \frac{kT_i}{V_i} \tag{2.35}$$

The calculation method is same as above. The air below the rockfall pressure gradually increase when rookfall fall, and the acceleration of rockfall will decrease. Until the adverse effect of compressed air for rookfall and the gravity of the rockfall are the same, and reach the limit state, Acceleration is reduced to 0.After, rockfall will fall with a constant velocity.
Assumption when $i = m$

$$P_m = \frac{Mg}{\pi R^2} = \frac{M_i}{m} \cdot \frac{kT_i}{V_i} \tag{2.36}$$

when $i > m$ accelerated velocity $a_i = 0$

$$\frac{1}{2}Mv^2_{i-1} + \frac{1}{2}M_{i-1}C^2_{i-1} + Mgdh + M_{i-1}\frac{C_{i-1}t_i}{2} = \frac{1}{2}(M+M_i)v^2_i + \frac{\gamma}{2}k\frac{M_i}{m}(\Delta T_i)$$

$$(2.37)\ \frac{1}{2}(M+M_{i-1})v^2_{i-1} + Mgdh + M_{i-1}\frac{dh}{2} = \frac{1}{2}(M+M_i)v^2_i + \frac{\gamma}{2}k\frac{M_i}{m}(\Delta T_i) \tag{2.38}$$

The rookfall falling speed is

$$v_i = v_m \tag{2.39}$$

The above content is the calculation method of the rookfall falling speed be effected by the air compression. The whole calculation process is relatively complicated, The calculation process can use the VB program.

3. The Method to Calculate the Impact Force of Rockfall

The scholar Ye Siqiaoproposed a new calculation method which is based on a variety of classical methods [8].

$$F_{n\max} = k\left[\frac{Mv_{bn}\left(1+e_n\right)}{\Delta t} + Mgcox\alpha\right] \tag{3.1}$$

$$\Delta t = \frac{1}{100}\left(0.097Mg + 2.21h + \frac{0.045}{H} + 1.2\right) \tag{3.2}$$

e_n is the normal recovery coefficient $e_n = v_{bn} / v_{an}$. v_{bn} is the normal component of the velocity along the surface of the slope before the rockfall impact the slope surface(m/s), v_{an} is the normal component of the velocity along the surface of the slope after the rockfall impact the slope surface(m/s).
Combine this calculate method of the rockfall impact force with calculate method of the rockfall impact speed drived in this paper, and apply to engineering example. The results show that the affect of compressive effect is not negligible. Compared with other papers, the results were different because the starting point and the theory used of the study is different, but the difference is not huge. Further research is needed on the accuracy of the two, but there is no doubt that air compressibility has a certain effect on the impact of rockfall.

4. Conclusion

(1)In the process of rockfall falling the rock will compress the lower the air. During the process of compression, the part of the air leak and the other part density and pressure of the air are increase, the air pressure increase will form a certain air resistance to the rockfall mass. The air resistance directly affect the impact velocity of rockfall.

(2)The gas-shrinking process is equivalent to the propagation process of compressive wave, and the ideal hypothesis of rockfall fall height is divided into n parts. Combining aerodynamic theory, energy conservation theorem and aerodynamic theory (gas equation of state), the method of calculating the impact speed of rockfall is carried out by microanalysis.

5. References

[1] Chen H K,Dong P and Tang H M.Dangerous rock collapse disaster research present situation and the trend[J]. *Journal of chongqing normal university(Natural science edition)* , 2015, **32**(6): 53- 60.

[2] Buzzi O, Giacomini A and Spadari M. Laboratory investigation on high values of restitution coefficients[J]. *Rock mechanics and rock engineering,* 2012, **45**(1): 35-43.

[3] Mitchell W A, McSaveney M J, Zondervan A, Kim K, Dunning S A and Taylor P J.The Key long Serial rock avalanche, NW Indian Himalayas geomorphology and palaeoseismic implications[J]. *Landslides*, 2007,(4):245254

[4] Hu C. High and steep slope instability mechanism and the law of collapse rockfall motion [D]. *Shandong University,*2014.

[5] Binal A and Ercanoglu M. Assessment of rockfall potential in the Kula (Manisa, Turkey) geopark region [J].*Environmental Earth Sciences,* 2010, **61**(7): 1361-1373.

[6] Alfredsson K S and Stigh U. Stability of beam-like fracture mechanics specimens [J]. *Engineering Fracture Mechanics,* 2012, **89**: 98-113.

[7] Wang L,Li D F,Ye J ,Xu H X and Yu X L. Compressed air brake vehicle engines cycle characteristics[J]. *Journal of Zhejiang University(Engineering science)*,2014,**48**(1):56-62.

[8] Ye S Q,Chen H K and Tang H M. Rockfall impact force calculation method [J]. *China Railway Science,*2010, **31**(6):56-62.

IOP Conf. Series: Materials Science and Engineering **275** (2017) 012024 doi:10.1088/1757-899X/275/1/012024

Research on the Calculation Method of Internal Force of Combined Energy Dissipation Shed-tunnel Roof

Fiao Liao, Deyu Meng, Zhiyong Yang, He Wang and Ranjie Wu

Institute of Geotechnical Engineering, Chongqing Jiaotong University, Chongqing 400074, China.
E-mail:1185249623@qq.com. Tel: 15123146089

Abstract: The paper studied on the internal force calculation process of roof of the combined energy dissipation shed-tunnel through using the theory of thin plates, the theory of structural dynamics and coordinated deformation theory, the paper taked the top slab as one-way slab, used the heavy series theory to study the equation, and got the internal force analytical solution under the action of a rockfall impact.The research results can provide reference for shed-tunnel structure design.

1. Introduction

Dangerous rock collapse is a common geological disaster in mountain and hilly area. It has the characteristics of randomness and suddenness. There are countless traffic accidents caused by the collapse of the crisis every year. At present, passive protection techniques such as shed-tunnel are adopted in the prevention and control of dangerous rock collapse, which make use of plastic deformation of cushion soil to absorb and dissipate the impact energy of rockfall. A lot of researches have been carried out on the shed tunnel at home and abroad. The damping plate structure is proposed by Mougin et al, which can achieve energy dissipation by deflection deformation or local damage [1]. Delhomme et al have found that under the impact of low strength, reinforced concrete slabs would produce severe cracking in the impact area and the edge of the Influence area would produce warping deformation [2][3]. OLSSON studied the plastic effect of the bearing plate under impact loading and the dynamic response of the bearing plate under the impact of sphere [4]. He Siming uses energy method to analyze the shock resistance mechanism of a new type of energy dissipation shed tunnel [5]. Yang Lu has studied the dynamic response of the rolling stone impact load to shed-tunnel structure under different speeds and angles by using the finite element numerical simulation and obtained the relationship between the displacement and the rockfall velocity and the impact angle [6]. Yang Lu has studied the dynamic response of the rolling stone impact load to shed-tunnel structure under different speeds and angles by using the finite element numerical simulation and obtained the relationship between the displacement and the rockfall velocity and the impact angle [7].

In summary, the analysis of the force structure of the rockfall shed tunnel under the impact load mainly depends on the numerical simulation and model test, and the relevant analytic study are less. In particular, there is less research on shed hole with buffer energy dissipation bearing. In view of this, this paper uses the plate theory, structural mechanics and deformation coordination theory to analyze the force characteristics of the shed tunnel with buffer energy dissipation bearing.

2. The Structure Characteristics of Combined Energy Dissipation Shed Tunnel

The object of this study is the energy dissipation shed tunnel structure proposed by Professor Chen

Content from this work may be used under the terms of the Creative Commons Attribution 3.0 licence. Any further distribution of this work must maintain attribution to the author(s) and the title of the work, journal citation and DOI.
Published under licence by IOP Publishing Ltd

Hongkai [8] (ZL201420649539.5), As shown in Figure 1. The structure from the bottom as follows: two longitudinal beams fixed longitudinally on both sides of the road on the energy dissipating bearing (1),a transverse beam (2) fixed transversely on two longitudinal beams, a protective roof (3) secured to a beam and a energy dissipation bearing is arranged between the connecting plate and the support.The energy dissipation support is arranged at the top of each column, and the crossbeam and the longitudinal beam of the shed hole are combined with the support column, the composite support structure of the rock falling energy dissipation shed cavity is formed by these structures.

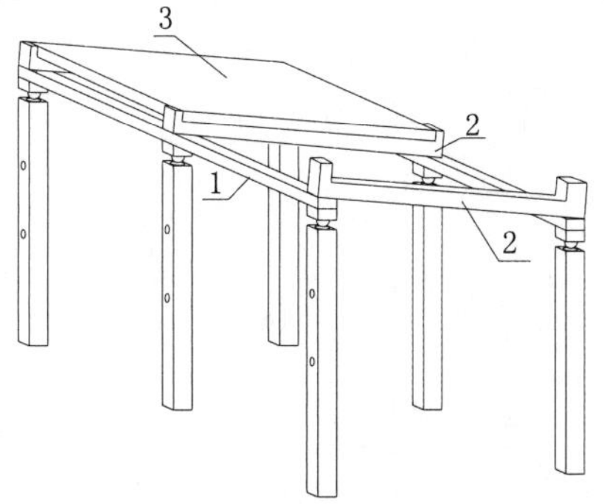

Figure 1. Schematic diagram of the combined energy dissipation shed tunne

3. Calculation of Roof Force in Shed Tunnel

Because the roof and the beam of the combined shed are simply supported and overlapped, the roof can be simplified as a simply supported plate, as shown in figure 2, The overlapping end of unidirectional simply supported plate is b. The width of the free end is a, and the thickness of the plate is t. According to the theory of elasticity, when the ratio of the thickness of the plate t to the minimum characteristic size b in the plate is 1/80~1/5, it is called a thin plate. Referring to the general code for design of highway bridges and culverts (JTG D60-2015), it is known that the roof of tunnel shed belongs to thin plate type.

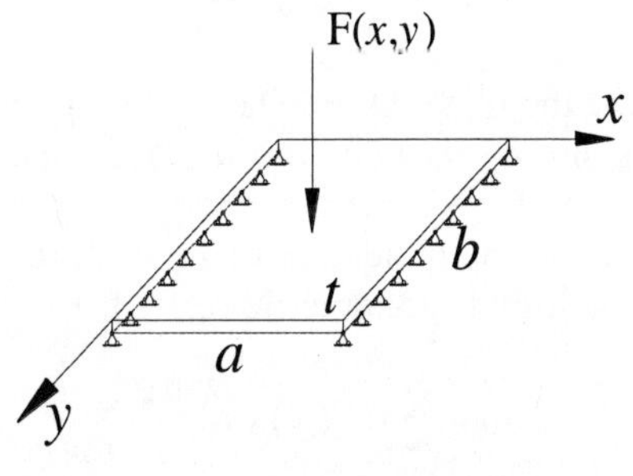

Figure 2. Schematic diagram of the simplified roof slabs

Shed cave is mainly to withstand the impact of falling rocks, when the rockfall hits the roof of the shed tunnel, the impact position is random, and the impact force changes dynamically with time.It is assumed that the impact of rockfall is at one point of the roof and the maximum impact load is F (x, y), as shown in figure 2.

According to the differential equation of the elastic surface, the expression of the major stress components in the thin plate can be derived:

IOP Conf. Series: Materials Science and Engineering **275** (2017) 012024 doi:10.1088/1757-899X/275/1/012024

$$\begin{cases} M_x = -D(\frac{\partial^2 \omega}{\partial x^2} + \nu \frac{\partial^2 \omega}{\partial y^2}) \\ M_y = -D(\frac{\partial^2 \omega}{\partial y^2} + \nu \frac{\partial^2 \omega}{\partial x^2}) \\ M_{xy} = M_{yx} = -D(1-\nu)\frac{\partial^2 \omega}{\partial x \partial y} \\ F_{sx} = -D\frac{\partial}{\partial x}\nabla^2 \omega \\ F_{sy} = -D\frac{\partial}{\partial y}\nabla^2 \omega \end{cases} \tag{1}$$

Furthermore, the secondary stress component can be obtained according to the equilibrium equation and the boundary condition of the simply supported plate:

$$\begin{cases} \tau_{xz} = -\frac{Et^2}{2(1-\nu^2)}(\frac{t^2}{4} - z^2)\frac{\partial}{\partial x}(\nabla^2 \omega) \\ \tau_{xz} = -\frac{Et^2}{2(1-\nu^2)}(\frac{t^2}{4} - z^2)\frac{\partial}{\partial x}(\nabla^2 \omega) \\ \tau_{xz} = -\frac{Et^2}{2(1-\nu^2)}(\frac{t^2}{4} - z^2)\frac{\partial}{\partial x}(\nabla^2 \omega) \\ \tau_{xz} = -\frac{Et^2}{2(1-\nu^2)}(\frac{t^2}{4} - z^2)\frac{\partial}{\partial y}(\nabla^2 \omega) \end{cases} \tag{2}$$

The deflection of a roof under normal concentrated load F(x,y)can be calculated by the following formula:

$$\nabla^4 \omega = \frac{F(x,y)}{D} \tag{3}$$

type:*D* is the bending rigidity of the plate, D=Et3/12(1-v2), t is the thickness of the plate, and V ,E are the Poisson's ratio and the modulus of elasticity of the plate respectively.

Taking the single span of the roof of the combined energy dissipation shed as the object of study, it is simplified as simply supported slab.The deflection of Levi solution to shed roof is solved, then the deflection expressions as single rewriting trigonometric series [9]:

$$\omega = \sum_{m=1}^{\infty} Y_m(y)\sin\frac{m\pi x}{a} \tag{4}$$

In the above expression, Y_m (y) is an undetermined function, and *M* is any positive integer.

It can be seen from formula that (4) the deflection single trigonometric series formula should satisfy the following boundary conditions:

$$\begin{cases} \omega|_{x=0} = 0, \frac{\partial^2 \omega}{\partial x^2}|_{x=0} = 0 \\ \omega|_{x=a} = 0, \frac{\partial^2 \omega}{\partial x^2}|_{x=a} = 0 \end{cases} \tag{5}$$

To bring (4) into (3):

$$\sum_{m=1}^{\infty}\left[Y_m^{(4)}-2(\frac{m\pi}{a})Y_m^{"}+\left(\frac{m\pi}{a}\right)^4 Y_m\right]\sin\frac{m\pi x}{a}=\frac{F(x,y)}{D} \tag{6}$$

Further, the sides of the formula (6) are multiplied by sin (M PI x/a) at the same time, then the above formula of X from 0 to a integral.By using orthogonality of trigonometric functions, we can obtain:

$$Y_m^{(4)}-2(\frac{m\pi}{a})Y_m^{"}+\left(\frac{m\pi}{a}\right)^4 Y_m=\frac{2}{a}\int_0^a\frac{F(x,y)}{D}\sin\frac{m\pi x}{a}dx \tag{7}$$

The upper order is a non-homogeneous linear differential equation of order four, assuming the general solution $Y_m(y)=Y_m^0(y)+Y_m^*(y)$.Among them,$Y_m^0(y)$ is the general solution of homogeneous equation, and $Y_m^*(y)$ is the special solution of non-homogeneous equation.

Equation (7) the general solution of the homogeneous equation on the left $Y_m^0(y)$ as follow:

$$Y_m^0(y)=A_m\text{ch}\frac{m\pi y}{a}+B_m\frac{m\pi y}{a}\text{sh}\frac{m\pi y}{a}+C_m\text{sh}\frac{m\pi y}{a}+D_m\frac{m\pi y}{a}\text{ch}\frac{m\pi y}{a} \tag{8}$$

Therefore, when we obtain the analytic solution of Ym (y), we can deduce the deflection equation of the thin plate,then we can get the deflection of the roof of the shed roof under the maximum impact force, so that we can get the internal force of the roof of the roof of the shed under the impact force *F* (*x*, *y*).

For maximum impact load *F* (*x*, *y*), do the following assumption:

$$F(x,y)=p\delta(x-x_0,y-y_0) \tag{9}$$

$$\delta(x-x_0,y-y_0)=0,\left(x\neq x_0,y\neq y_0\right) \tag{10}$$

At this point, the expression on the right side of equation (7) is:

$$\frac{2}{a}\int_0^a\frac{F(x,y)}{D}\sin\frac{m\pi x}{a}dx=\frac{2}{Da}p\delta(y-y_0)\sin\frac{m\pi x_0}{a} \tag{11}$$

Transform the formula (11) into series, as folllow:

$$\frac{2p}{Da}\sum_{n=-\infty}^{\infty}\frac{1}{T}e^{tn\omega(y-y_0)}\cdot\sin\frac{m\pi x_0}{a} \tag{12}$$

Among them, T is the period of Fourier variation. For the particular solution of equation (8), the following assumptions can be made:

$$Y_m^*(y)=G\sum_{m=1}^{\infty}Z_{mn}e^{\frac{n\pi(y-y_0)}{b}} \tag{13}$$

Among them: $G=\frac{p}{Dab}\cdot\sin\frac{m\pi x_0}{a}$

The formula (13) is obtained by substitution (7):

$$Y_m^{(4)}-2(\frac{m\pi}{a})Y_m^{"}+\left(\frac{m\pi}{a}\right)^4 Y_m=G\sum_{n=-\infty}^{\infty}Z_{mn}e\frac{tn\pi(y-y_0)}{b}\left[(\frac{m\pi}{a})^2+(\frac{n\pi}{b})^2\right]^2 \tag{14}$$

So there: $Z_{mn}=\left[\left(\frac{m\pi}{a}\right)^2+\left(\frac{n\pi}{b}\right)^2\right]^{-2}$

$$Y_m^*(y)=G\sum_{m=1}^{\infty}\left[\left(\frac{m\pi}{a}\right)^2+\left(\frac{n\pi}{b}\right)^2\right]^{-2}e^{\frac{n\pi(y-y_0)}{b}} \tag{15}$$

So the general solution Ym (y) takes the form of:

$$Y_m(y)=\left(A_m\text{ch}\frac{m\pi y}{a}+B_m\frac{m\pi y}{a}\text{sh}\frac{m\pi y}{a}+C_m\text{sh}\frac{m\pi y}{a}+D_m\frac{m\pi y}{a}\text{ch}\frac{m\pi y}{a}\right)+Y_m^*(y) \tag{16}$$

In the formula: $Y_m^*(y)=G\sum_{m=1}^{\infty}\left[\left(\frac{m\pi}{a}\right)^2+\left(\frac{n\pi}{b}\right)^2\right]^{-2}e^{\frac{n\pi(y-y_0)}{b}}$

According to the analysis of the structural characteristics of the energy dissipation rock shed cave, the boundary condition of the roof of the shed cave can be expressed as:

$$\begin{cases}\left(M_y\right)_{y=b}=0,\left(\frac{\partial^2\omega}{\partial y^2}+\nu\frac{\partial^2\omega}{\partial x^2}\right)\Big|_{y=b}=0\\ \left(V_y\right)_{y=b}=0,\left[\frac{\partial^3\omega}{\partial y^3}+(2-\nu)\frac{\partial^3\omega}{\partial y\partial x^2}\right]\Big|_{y=b}=0\\ \left(M_y\right)_{y=0}=0,\left(\frac{\partial^2\omega}{\partial y^2}+\nu\frac{\partial^2\omega}{\partial x^2}\right)\Big|_{y=0}=0\\ \left(V_y\right)_{y=0}=0,\left[\frac{\partial^3\omega}{\partial y^3}+(2-\nu)\frac{\partial^3\omega}{\partial y\partial x^2}\right]\Big|_{y=0}=0\end{cases} \tag{17}$$

Taking the boundary conditions into the type (16), we can get A_m, B_m, C_m, D_m , and then the deflection expression of the roof can be obtained:

$$\omega=\sum_{m=1}^{\infty}\left[\left(A_m\text{ch}\frac{m\pi y}{a}+B_m\frac{m\pi y}{a}\text{sh}\frac{m\pi y}{a}+C_m\text{sh}\frac{m\pi y}{a}+D_m\frac{m\pi y}{a}\text{ch}\frac{m\pi y}{a}\right)+Y_m^*(y)\right]\sin\frac{m\pi x}{a} \tag{18}$$

In the formula (18) substituting formula (1), the internal forces of the roof under the concentrated load $F(x, y)$ can be obtained.

4. Example Analysis

Xiang Yu line Laohekou East to Ankang section is located in the low mountain of Eastern Qinling Mountains, in which taking a shed hole as an example to calculate the roof internal force of combined energy dissipation shed-tunnel.There are 3 steep cliffs, high 35~50m, almost vertical, there are perlious rock, broken stone and stone particle at the foot of cliff slope, and the size is about 0.2~0.35m. It is a limestone area with a bulk density of 28kN·m-3.When the perlious rock collapses, it falls from 18m and hits the roof of the shed. The position of the roof is vertical and horizontal, and the incidence is about 45 degrees, C30 concrete is used in shed tunnel structure, whose elastic modulus is 31.5GPa, and the Poisson's ratio of concrete plate is 0.2, the spring stiffness coefficient of energy dissipation bearing is 105kN/m.The dimensions of the shed holes are shown in next Figure.

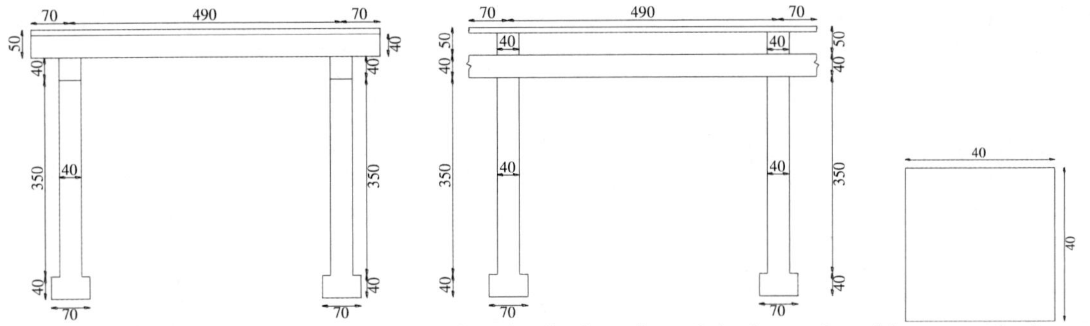

(a) cross section of shed tunnel (b) longitudinal section of shed tunnel (c) rectangular beam

Figure 3. Geometrical dimensions of shed holes in Xiang Yu railway line (unit: cm)

The maximum impact force of rockfall is calculated in JHPC algorithm [9], and the formula is as follows:

$$P=2.108M^{\frac{2}{3}}\lambda^{\frac{2}{5}}H^{\frac{3}{5}} \tag{19}$$

Type: P the maximum rockfall impact force, *kN*; M rock mass, *t*;λa lame coefficient, generally take 1000 $kN\cdot m^{-2}$; H rockfall free height, *m*.

Using the above calculation method, the maximum rockfall impact force is 34.145 kN. According to the calculation method of the internal force corresponding to the above energy dissipation shed tunnel structure, the MATLAB software is used to solve the problem, and the shear force and the bending moment of the roof of the shed tunnel under the maximum impact force are obtained.

Due to the impact of rockfall, the roof is in the middle of the roof of the shed, the internal forces of the longitudinal cross section of the plate are shown in the following figure. The maximum shear value is located at the support of the beam at 4.4kN, and the maximum value of the bending moment is 3.56 kN·m in the midspan.

(a) shear force(unit: kN) (b) bending moment(unit: kN·m)

Figure 4. The internal force diagram of the roof of the energy dissipation roof

5. Conclusion

(1)This paper theoretically analyzed the loading performance of the composite energy dissipation shed tunnel, and obtained the general series expression of roof disturbance of shed under the impact of concentrated load. Make it in accordance with the requirements and conditions through certain mathematical transformation, and then bring into the corresponding boundary condition so as to obtain the general perturbation expression of the roof of shed under the impact of concentrated load which has certain reference significance for the design of the combined energy dissipation shed.

(2)The calculation method of the roof disturbance not only applied to the dynamic response of the plate under concentrated load, but also can be used to obtain the disturbance expression of the roof of the shed tunnel under the linear load by a slight change.

(3)This paper considered only a cross and simplified the shear stress to boundary stress when analyzing the stress of the shed tunnel, which made the calculation results with some errors. Therefore, the following research should consider the influence of adjacent cross plates, in order to obtain a more accurate theoretical results.

6. Acknowledgements
This study in the paper was funded by The Chongqing Jiaotong University Innovation Fund (No. 20150111, CYB15107, CYB15110)

7. Reference
[1] Mougin J, Perrotin P, Mommessin M, et al. Rock fall impact on reinforced concrete slab: an Experimental approach [J].*International Journal of Impact Engineering*, 2005, **31(2)**:169-183.
[2] Delhomme F, Mommessin M, Mougin J P, et al. Simulation of a block impacting a reinforced concrete slab with a finite element model and a mass-spring system[J].*Engineering Structures*, 2007, **29(11)**:2844-2852.
[3] Delhomme F, Mommessin M, Mougin J P, et al. Behavior of a structurally dissipating rock-shed:experimental analysis and study of punching effects[J].*International Journal of Solids and Structures*,2005,**42(14)**:4204-4219.
[4] OLSSON R. Impact response of orthotropic composite plates predicted from a one-parameter differential equation [J].*AIAA Journal,* 1992, **30(6)**:1587-1596.
[5] He SM, WU Y. Study on the Action Mechanism of a New Type of Energy-saving Rolling Shore [J].*Journal of Rock Mechanics and Engineering*, 2010, **29(5)**:926-932.
[6] Zhang QL, Wang QC, Wu Q, etc.Study on impact resistance of sheds with different structural types [J].*Journal of Vibration and Shock*, 2015, **12(3)**:72-76.
[7] Yang L, Li SM, Wu ZM, etc.Study on impact dynamics of rolling stone on shed structure [J]. *Journal of Traffic and Transportation Engineering*, 2012, **12 (1)**: 25-30.
[8] Chen Hk, Tang HM, Wang LF.A kind of energy dissipation shed and its energy dissipation bearing: China, 201010242005 [P].2012-02-19.
[9] Ye SQ, Chen Hk, Tang HM.Method for calculating the impact force of falling stone [J], *China Railway Science*, 2010, **31(6)**:56-62.

IOP Conf. Series: Materials Science and Engineering **275** (2017) 012025 doi:10.1088/1757-899X/275/1/012025

Study on Excitation-triggered Damage Mechanism in Perilous Rock

Hongkai Chen[1,*] **and Shengjuan Wang**

[1]Institute of Geotechnical Engineering, Chongqing Jiaotong University, Chongqing 400074, China

E-mail: chkcxf@163.com.Tel:023-62982694

Abstract: Chain collapse is easy to happen for perilous rock aggregate locating on steep high slope, and one of the key scientific problems is the damage mechanism of perilous rock under excitation action at perilous rock rupture. This paper studies excitation-triggered damage mechanism in perilous rock by wave mechanics, which gives three conclusions. Firstly, when only the normal incidence attenuation spread of excitation wave is considered, while the energy loss is ignored for excitation wave to spread in perilous rock aggregate, the paper establishes one method to calculate peak velocity when excitation wave passes through boundary between any two perilous rock blocks in perilous rock aggregate. Secondly, following by Sweden and Canmet criteria, the paper provides one wave velocity criterion for excitation-triggered damage in the aggregate. Thirdly, assuming double parameters of volume strain of cracks or fissures in rock meet the Weibull distribution, one method to estimate micro-fissure in excitation-triggered damage zone in perilous rock aggregate is established. The studies solve the mechanical description problem for excitation-triggered damage in perilous rock, which is valuable in studies on profoundly rupture mechanism.

1. Introduction

As one of global far-ranging geological disasters, collapse seriously threatens highways and railways in mountainous areas, mines, urbans and residents. Hitherto, foreign scholars have studied collapse after perilous rock rupture, such as Strom and Korup considered that both perilous rock rupture and rockfall movement are the kinetic process for rock slope evolution [1]. At present, researches gradually focus on rupture of perilous rock at source of collapse, for example, Xuqiang et al. discovered there was key block for Jiweishan avalanche in Chongqing [2]. Based on grey theory and catastrophe theory, Tian Qingyan and Fu Helin established a new model to predict collapse time of rock slope [3]. In view of geostress and rockcell depth, development of perilous rock was discussed by Zhang Yongxing et al [4]. A catastrophe model to determine landslide was established in elastic mechanics by Wang Zhiqiang et al [5]. Frayssines and Hantz discussed failure of limestone slope [6]. One method to determine Poisson's ratio and elastic modulus of rock at end of dominant fissure in perilous rock had been established by Chen et al [7]. Especially, excitation effect at perilous rock failure is discovered by Chen et al [8], and the effect is significant in degrading stability of perilous rock. This paper is focus on excitation-triggered damage mechanism in perilous rock, which must be facilitation to mechanical rupture of perilous rock.

2. Method to Identify Peak Vibration Velocity Excitation-triggered at Rupture of Perilous Rock

Cases identify that perilous rocks exist in cliff as perilous rock aggregate. In an instant at rupture of any one perilous rock block in the aggregate, excitation wave is produced and passed to surrounding

Content from this work may be used under the terms of the Creative Commons Attribution 3.0 licence. Any further distribution of this work must maintain attribution to the author(s) and the title of the work, journal citation and DOI.
Published under licence by IOP Publishing Ltd

perilous rock block certainly, and this process belongs to vibration process, but also a energy consumption process. Surfaces among perilous rock blocks are likened to mass rock joint, and perilous rock aggregation can be likened to joint rock mass. Joints belong to interface of wave impedance. When excitation wave passes to interface the wave will produce reflection and transmission of wave (Fig.1), whose transmission path alter and arouse reassignment of wave field energy. Only the situation of excitation wave's normal incidence is considered in this paper, and takes energy loss of excitation wave in the interface into account.

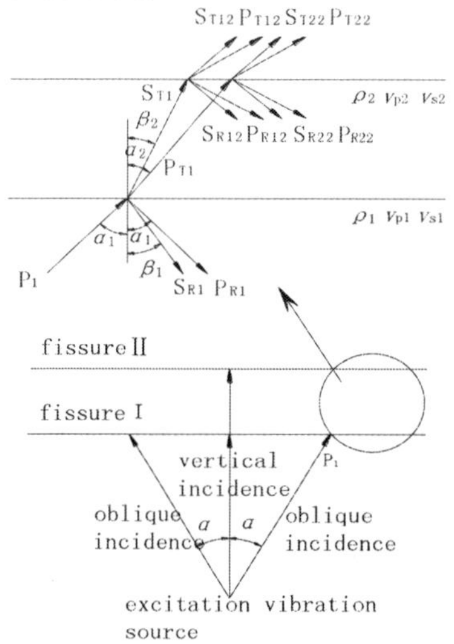

Figure1. Transmission and reflection model for excitation wave to pass through boundary between perilous rock blocks in perilous rock aggregate

Based on equation Zoeppritz, wave will produce four kinds of wave, reflection P-wave, reflection S-wave, transmission P-wave, transmission S-wave to pass the interface (Fig.1). where, SR1, PR1, ST1, PT1 are reflection P-wave, reflection S-wave, transmission P-wave, transmission S-wave, SR12, PR12, ST12, PT12 are reflection P-wave, reflection S-wave, transmission P-wave produced by ST1, SR22, PR22, ST22, PT22 are new reflection P-wave, reflection S-wave, transmission P-wave, transmission S-wave produced by PT1, respectively. This equation expresses the relation between incidence P-wave and S-wave, transmission coefficient and angle of incidence, density and velocity, and determined energy distribution of excitation wave. The equation is as follows,

$$\begin{bmatrix} \sin\alpha_1 & \cos\beta_1 & -\sin\alpha_2 & \cos\beta_2 \\ -\cos\alpha_1 & \sin\beta_1 & -\cos\alpha_2 & -\sin\beta_2 \\ \sin 2\alpha_1 & \frac{v_{P1}}{v_{S1}}\cos 2\beta_1 & \frac{\rho_2 v_{S2}^2 v_{P1}}{\rho_1 v_{S1}^2 v_{P2}}\sin 2\alpha_1 & -\frac{\rho_2 v_{P1} v_{S2}}{\rho_1 v_{S1}^2}\cos 2\beta_2 \\ \cos 2\beta_1 & -\frac{v_{S1}}{v_{P1}}\sin 2\beta_1 & -\frac{\rho_2 v_{P2}}{\rho_1 v_{P1}}\cos 2\beta_1 & -\frac{\rho_2 v_{S2}}{\rho_1 v_{P1}}\sin 2\beta_2 \end{bmatrix} \begin{bmatrix} R_P \\ R_S \\ T_P \\ T_S \end{bmatrix} = \begin{bmatrix} -\sin\alpha_1 \\ -\cos\alpha_1 \\ \sin 2\alpha_1 \\ -\cos 2\beta_1 \end{bmatrix} \tag{1}$$

in which, α_1 , β_1, α_2, β_2 are normal angles of reflection P-wave, reflection S-wave, transmission P-wave, transmission S-wave on joint plane; R_P and R_S are respectively reflection coefficient of P-wave and S-wave; T_P and T_S respectively are transmission coefficient of P-wave and S-wave; ρ_1 and ρ_2 are reflection plane density on two sides of medium, kg/m^3; v_{P1} and $v_{S1}$are velocity of incidence P-wave and S-wave, m/s; v_{P2} and $v_{S2}$are velocity of reflection P-wave and S-wave, m/s. In the paper, only normal incidence transmission problem of excitation wave is considered, namely $\alpha_1=0$, and normal incidence wave coefficient is following by Equation (1),

$$T_P = \frac{2v_{p1}\rho_1}{v_{p2}\rho_2 + v_{p1}\rho_1} \tag{2}$$

Excitation wave energy depends on its square of the amplitude by wave theory. Due to the P-wave in vertical incidence doesn't produce reflection S-wave and transmission S-wave, so parameters are zeros such as reflection P-wave and transmission P-wave's reflection and transmission angle. For the study of energy distribution characteristics of excitation wave transmission in the interface, transmission coefficient E_k was defined by according to equal interface normal energy flux density energy principles.

$$E_k = \frac{Z_2 A_2^2}{Z_1 A_1^2} \tag{3}$$

Where, $Z_2=V_{p2}\rho_2$,; A_2 is amplitude of transmission wave; $Z_1=V_{p1}\rho_1$,; A_1 is amplitude of incidence wave.

Excitation wave transmission coefficient can express to the ratio of transmission wave and incidence wave, so

$$\frac{A_2}{A_1} = \frac{2v_{p1}\rho_1}{v_{p2}\rho_2 + v_{p1}\rho_1} \tag{4}$$

Put Eq.(4) into Eq.(3), excitation wave energy transmission coefficient expression is follows as

$$E_k = \frac{4Z_1 Z_2}{(Z_2 + Z_1)^2} \tag{5}$$

For example, E_k of marble is 30.8281 by Ju Yang's studies, and E_k of granite is 30.7088, elastic strain energy expression $\overline{W}$ [8] produced sudden failure of perilous rock, i.e.,

$$\overline{W} = \int_0^{l_2} \frac{M^2(x)}{2EI} dx + \frac{\gamma^2 l_1^2 l_2 bh}{2G} + \frac{\overline{\sigma}^2 ebl_2}{2E} \tag{6}$$

Where, h and b are height and width of perilous rock respectively, (m); $A=bh$, (m^2) ; E is elastic modulus, (kPa); I is inertia moment of cross section for perilous rock, (m^4); M is bending moment, (kN·m); l_1 is length for perilous rock 1#, (m); l_2 is length for perilous rock 2#, (m); e is full water depth of dominant fissure, (m); $\overline{\sigma}$ is fissure water of dominant fissure, (kPa); G is shear modulus, (kPa); γ is unit weight, (kN/m^3).

After excitation wave transmit the interface, formula W_T for transmitted wave energy is as followings.

$$W_T = E_k \overline{W} = E_k \left[\int_0^{l_2} \frac{M^2(x)}{2EI} dx + \frac{\gamma^2 l_1^2 l_2 bh}{2G} + \frac{\overline{\sigma}^2 ebl_2}{2E} \right] \tag{7}$$

Where, units of $\overline{W}$ and W_T are KJ; the others of the variable physical meaning is the same as above.

Further peak vibration velocity formula was deduced by rest energy from excitation wave transmitted the interfaces.

$$V_T = \sqrt{\frac{2vE_k}{EAT} \left[\int_0^{l_2} \frac{M^2(x)}{2EI} dx + \frac{\gamma^2 l_1^2 l_2 bh}{2G} + \frac{\overline{\sigma}^2 ebl_2}{2E} \right]} \tag{8}$$

where, V_T is peak vibration velocity of excitation wave to cross the interface, (m/s); v is propagation velocity of excitation wave, (m/s); The others is the same as above.

Similarly, peak vibration velocity for perilous rock triggered by excitation wave to cross interface between anyone of perilous rock blocks, and with the damage criterion excitation damage scope from failure of perilous rock can be determined preliminarily.

3. Damage Criterion for Perilous Rock under Excitation Effect

Excitation effect from rupture of perilous rock is likened to vibration load produced by miniature blasting in structural plane of jointed rock mass. Peak vibration velocity is available to confirm damage scope to adjacent perilous rock under excitation effect. For acquiring damage threshold of perilous rock on the basis of particle peak vibration velocity, this paper used two kinds of damage criterion[12], Sweden criterion and Canmet criterion, which respectively confirmed maximum and minimum peak vibration velocity for perilous rock.

Sweden criterion adopted upper limit threshold under the condition of rock stretch to determine maximum peak vibration velocity that rock can bear:

$$V_{\max} = \frac{\sigma_T v}{E} \tag{9}$$

in which, V_{max} is the maximum peak vibration velocity that rock can bear before rock broke, m/s. σ_T is uniaxial tension strength, kPa. The rest of the variable meaning is the same as above.

Canmet criterion obtained minimum peak vibration velocity that rock can bear on the basis of lower limit threshold value of damage degree under weak blasting, and its expression is as follows,

$$V_{\min} = \frac{0.021\sigma_c}{v\rho} \tag{10}$$

in which, V_{min} is the minimum peak vibration velocity that protogenesis crack of perilous rock expanded, m/s. σ_c is uniaxial tension strength, kPa. The rest of the variable meaning is the same as above.

It is thus clear that damage criterion of unstable under excitation effect is,

For $V_T < V_{min}$, excitation effect from rupture of perilous rock to damage and break effect to adjacent perilous rock can be neglected;

For $V_{min} \leq V_T < V_{max}$, excitation effect from rupture of perilous rock has damage effect to adjacent perilous rock;

For $V_{max} \leq V_T$, excitation effect from rupture of perilous rock has destructive effect to adjacent perilous rock.

4. Microcosmic Crack Estimation Method in the Damage Scope by Excitation Effect from Rupture of Perilous Rock

Perilous rock is formed by brittle rock blocks, such as sandstone marlstone, limestone, which develops a lot of microcrack and microfissure. Excitation wave produced by abrupt failure of perilous rock inevitably arouses microcrack and microfissure to be nucleus, grow up and expand. Futher, mechanical property decayed, even perilous rock block broke. According to studies of Grady and Kipp, crack density is the ratio of rock volume for influenced crack scope and total rock volume, and numbers of fissure and crack activate obey two-parameter Weibul1 distribution of volume tensile strain:

$$C_d = \beta N a^3 \tag{11}$$

where, C_d is crack density; β is coefficient, approximately equals to 1; N is the number of microfissure or microcrack excited; a is characteristic length of microfissure or microcrack under actiong of excitation, can be calculated by the following formula,

$$a = \frac{1}{2}\left(\frac{\sqrt{20}K_{\mathrm{IC}}}{\rho v \theta_{\max}}\right)^{2/3} \tag{12}$$

where, K_{IC} is fracture toughness for intact rock of perilous rock, (kN.m$^{-3/2}$). ρ is rock density, (kg/m^3), v is velocity of excitation wave, (m/s), θ_{max} is maximum volume tension strain rate, (s^{-1}).

Damage variable D was defined by elastic modulus of medium as followings:

$$D = 1 - \frac{\overline{E}}{E} \tag{13}$$

where, E is initial elastic modulus for rock, (MPa), $\overline{E}$ is effective elastic modulus after perilous rock damaged, (MPa), Yu Shouwen and Feng Siqiao [14] suggested the following formula to calculate the parameter.

$$\frac{\overline{E}}{E} = \left[1 + \frac{16}{45}\frac{(1-\mu^2)(10-3\mu)}{2-\mu}C_d\right]^{-1} \tag{14}$$

in which, μ is poisson ratio of perilous rock.

Combining Eq. (12), Eq. (13) and Eq.(14), and β=1, expression of damage variable D becomes,

$$D = 1 - \left[1 + \frac{8}{9}\frac{(1-\mu^2)(10-3\mu)}{2-\mu}N\left(\frac{K_{IC}}{\rho v\theta_{\max}}\right)^2\right]^{-1} \tag{15}$$

In addition, based on $V_{\max}$ that rock can bear before rock broke and $V_{\min}$ expanded by original fissure damage variable expression was defined as follows:

$$D = \frac{V_T - V_{\min}}{V_{\max} - V_{\min}} \tag{16}$$

Vibration velocity equation[9] of perilous rock under excitation was obtained as formula (17).

$$C = \sqrt{\frac{2v}{EAT}\left[\int_0^{l_2}\frac{M^2(x)}{2EI}dx + \frac{\gamma^2 l_1^2 l_2 bh}{2G} + \frac{\overline{\sigma}^2 ebl_2}{2E}\right]} \tag{17}$$

Where, C is vibration velocity of perilous rock under excitation effect: the rest of physical meaning of variable is the same as above.

Combining Eq. (15), Eq. (16) and Eq.(17),

$$D = \frac{\sqrt{\frac{2v}{EAT}\left[\int_0^{l_2}\frac{M^2(x)}{2EI}dx + \frac{\gamma^2 l_1^2 l_2 bh}{2G} + \frac{\overline{\sigma}^2 ebl_2}{2E}\right]} - \frac{21\sigma_c}{1000v\rho}}{\frac{\sigma_T v}{E} - \frac{21\sigma_c}{1000v\rho}} \tag{18}$$

Numbers of microfissure of left perilous rocks likely excited under excitation can be estimated by formula (19).

$$N = \frac{\sqrt{\frac{2v}{EAT}\left[\int_0^{l_2}\frac{M^2(x)}{2EI}dx + \frac{\gamma^2 l_1^2 l_2 bh}{2G} + \frac{\overline{\sigma}^2 ebl_2}{2E}\right]} - \frac{21\sigma_c}{1000v\rho}}{\frac{8}{9}\left(\frac{\sigma_T v}{E} - \sqrt{\frac{2v}{EAT}\left[\int_0^{l_2}\frac{M^2(x)}{2EI}dx + \frac{\gamma^2 l_1^2 l_2 bh}{2G} + \frac{\overline{\sigma}^2 ebl_2}{2E}\right]}\right)\frac{(1-\mu^2)(10-3\mu)}{2-\mu}\left(\frac{K_{IC}}{\rho v\theta_{\max}}\right)^2} \tag{19}$$

5. Case Analysis

Cliff of YangCha River is located in south of QiJiang in Chongqing of Three Gorges Reservoir, lies in core of anticline, is composed of limestone and mudstone, and presents remarkable geomorphology of footstep state. Dispersion of cliff is between 34 m and 36 m. Stratum is composed of interactive limestone and mudstone, and thickness of lower mudstone is about 20 m with 4.3 m limestone placing in between. About 5 m high and 3 m deep rock cell develops in Mudstone outcropped place above the limestone, which is a relieved eroded platform with about 25 m wide, and SongZao mining area railway across the platform. Bottom of perilous rock zone is weak foundation bed composed of mudstone. Due to mudstone's weak resistance to weather, concave rock cell usually develops in cliff. Upper perilous rock is mainly close vertical limestone mass, and its physics and mechanical parameters are listed in table 1. Geometric dimensions of perilous rocks W10 is 5m(height)×5m(width)×4m(length), and length of dominant fissure is 2.5 m.

V_{min} for Yangcha River perilous rock is 2.9 mm/s, and V_{max} is 68.1 mm/s combining with Eq. (9) and Eq. (10). Peak vibration velocity of perilous rock deduced by rupture of perilous rock is 27.7 mm/s, and it is greater than V_{min}, which clearly states that excitation wave from rupture of perilous rock could damage adjacent perilous rock. With the fractured rock mass physical and mechanical parameters in the joint surface rather hard to obtain, this paper takes energy transmission coefficient 0.6 crossing jointed rock mass referring to SHPB test conducted by Ju Yang etc, and put 0.6 into Eq. (8) peak vibration velocity of perilous rock W11 is 2.9 mm/s; Microcrack numbers of each perilous rock can be obtained by Eq. (19) in the damage scope. For example, excitation wave excited 38 microcracks with excitation vibration source crossing the interface between perilous rock W10 and W11.

Table 1. Physical and mechanical parameters of perilous rocks in YangCha River

category	Unit weight (kN/m^3)			Elastic modulus (10^4MPa)	Tension strength of rock (MPa)	Tension strength of rock mass (MPa)	Maximum volume tension strain rate (/s)	Poisson ratio	Fracture toughness ($kPa\sqrt{m}$)
	dry	natural	saturation						
limestone	/	24.8	25.2	0.52	2.60	0.52	0.03	0.21	1600
mudstone	25	25.4	/	0.31	/	/	/	0.32	1300

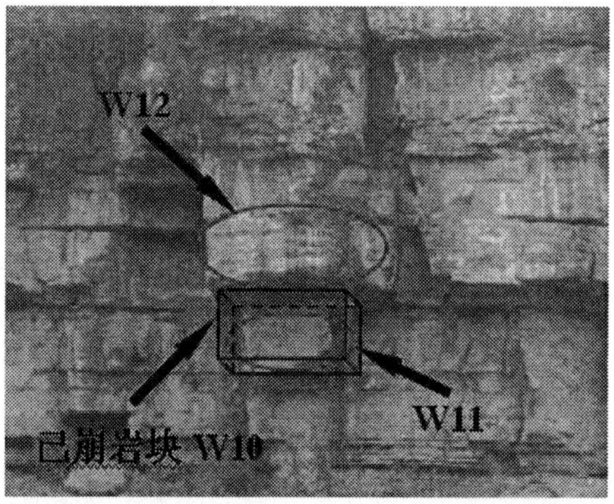

Figure 2. The W10, W11 and W12 perilous rock in the middle reaches of Yangcha river basin

6. Conclusion

Firstly, pay attention to the truth of excitation effect at rupture of perilous rock in the aggregate, the authors establish one method to calculate peak excitation velocity when excitation wave pass interface between perilous rock blocks by wave mechanics.

Secondly, based on the rule of Sweden and the rule of Canmet, wave velocity damage criteria for perilous rock excitation effect was proposed.

Thirdly, Abide by the assumption of bi-parameters Weibull distribution function for volume tension strain of cracks in rock, crack estimation method in the damage scope by excitation effect from rupture of perilous rock is established.

7. Acknowledgements

This paper was funded by the National Natural Science Foundation of China (51678097, 51378521), the innovative team building programs of Chongqing University in 2016 (CXTDG201602012), and the Par-Eu Scholars Program of Chongqing(201309).

8. Reference

[1] Strom, A.L. and Korup, O. Extremely large rockslides and rock avalanches in the Tien Shan mountains 2006. *Kyrgyzstan. Landslides*, **3(2)**:125-136.

[2] XU Qiang, Huang Runqiu, Yin Yueping, et al. The Jiweishan landslide of June 5, 2009 in Wulong Chongqing: characteristics and failure mechanism 2009. *Journal of engineering geology*, **17(4)**:433-444.

[3] Tian Qinyan, Fu Helin. Failure time prediction of slope collapse of block rockmass based on gray and catastrophic theories 2009. *Journal of south china university of technology (natural science edition)*, **37(12)**:122-126.

[4] Zhang Yongxing, Lu Li, Zhang Siping, et al. Development and failure principle of differential weathering overhanging rock 2010. *Journal of Civil, Architectural & Environmental engineering*, **32(2)**:1-6.

[5] Wang Zhiqiang, Wu Minying, Pan Yue. Fold catastrophe model of slope destabilization and its starting velocity 2009. *Journal of China University of Mining & Technology*, **38 (2)**:175-181.

[6] Frayssines M. and Hantz, D. Modelling and back-analysing failures in steep limestone cliffs 2009. *International Journal of Rock Mechanics and Mining Sciences*, **46(7)**:1115-1123.

[7] Chen Hongkai, Tang Hongmei, Ye Siqiao. Damage model of control fissure in perilous rock 2006. *Applied Mathematics and Mechanics*, **27 (7)**:967-974.

[8] Chen Hongkai, Tang Hongmei, Wang Zhi, et al. Study on frequency domain characteristics of excitation signals at rupture of perilous rock 2014. *Journal of Vibration and shock*, **33(19)**:64-68.

IOP Conf. Series: Materials Science and Engineering **275** (2017) 012026 doi:10.1088/1757-899X/275/1/012026

Failure Analysis of a Service Tube

Zhongdong Xie, Weiguo Cai, Zhenxing Li, YiMing Guan, Baocheng Zhang, XiaoTong Yang

Department of Machinery and Power Engineering, Dalian Ocean University, Dalian, Liaoning

Abstract: One tube was cracked used in the primary reformer furnace in a fertilizer plant for two and half years. In order to find out the causes of cracking, the methods for chemical composition analysis, macro- and microstructure analysis, penetrant testing, weld analysis, crack and surface damage analysis, mechanics property analysis, high temperature endurance performance analysis, stress and wall thickness calculation were adopted. The integrated assessment results showed that the carbon content of the tube was in the lower limit of the standard range; the tube effective wall thickness was too small; local overheating leads to tube cracking in use process.

1. Introduction

The converter furnace tube used in a fertilizer plant for 2.5years ruptured. Many kinds of analytical methods such as chemical composition, macro- and microstructure, mechanical properties and other checks are adopted in order to find out failure reasons.

2. Experimental Materials

The whole part of cracking tube was removed, dissected, and sampled. The segment of physical fracture of the tube is shown in Figure1.The crack in the outer wall of tube is longer than that in the inner wall, about 218mm and 203mm, respectively.

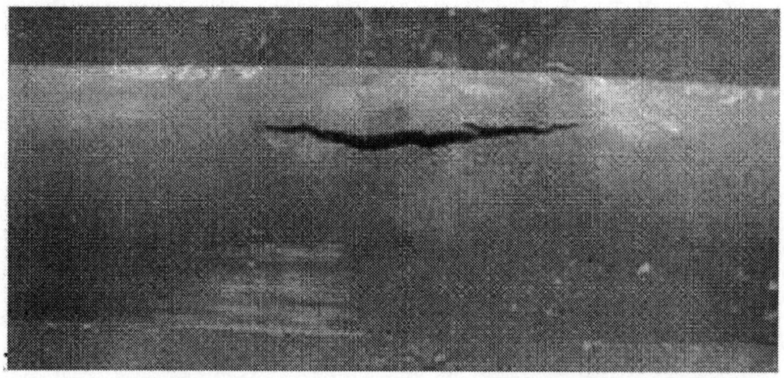

Figure 1. The object of cracking tube

3. Experiment Result and Analysis

3.1. Chemical Composition Analysis

In the cracking tube, three samples were taken at vicinity and wide berth ends of the fracture,

Content from this work may be used under the terms of the Creative Commons Attribution 3.0 licence. Any further distribution of this work must maintain attribution to the author(s) and the title of the work, journal citation and DOI.
Published under licence by IOP Publishing Ltd

respectively.

The results of Chemical composition analysis meet basically the technical contract requirement and KHR35CT standard provided by Kubota. But there were three issues to be explanted:

(1) Cr content of the sample was close to the requirements of 24% with a minor difference, But the type of material, general control level is of 23~27% at home and abroad, and on behalf of 25%.The tube property and structure are all normal within the control level. So the Cr content was not as a problem, and the accuracy for different testing equipment is not the same. By KHR35CT standard, an allowable error is of 0.25%.

(2) The wall thickness of tube were relatively small, So the carbon content of tube should be taken at upper limit level of KHR35CT standard so as to increase high temperature strength, and ensure the operation safety, While carbon content of the tubes is lower, the possibility of explosion would raise up.

(3) The influence of alloy elements as Cr,Ni, Nb and trace elements on the tube property should be further researched.

Overall, the results of chemical composition meet the technical specification and KHR35CT standards, but lower carbon content in design of wall thickness for tube must be drawn attention.

3.2. Penetrant Testing

Diameter measurement of all parts of the tube sample were examined by dye perpendicular on the wall thickness of creeping expansion section. Each section of tube were made within the margin of manufacturing error, in addition to explosion vent with a little bulging. The results showed that the tube wall thickness was uniform.

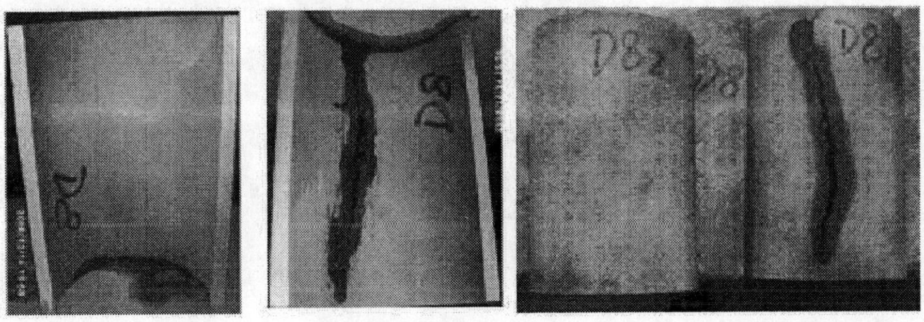

Figure 2. Results of dye penetrant testing on the surfaces both inside and outside of sample

The dye penetrant testing were used in both inside and outside surfaces of the sample in which no other defects were found in addition to crack. The inner dissected surface and the annular section were inspected also by dye penetrant testing, the results have been shown in Figure2. The main results showed that there were many long cracks in inner surface, these cracks extended from inside to outside surfaces, and a bit small cracks were found in the front end and vicinity of the fracture.

3.3. Macro- and Microstructure Analysis

The macrostructure consists of the columnar crystals in outer surface and the equiaxed crystals in inner layer, which was normal. From up to down, the proportion of columnar crystal and equiaxed is different, and the changes in the proportion would not influence the service life of tube by a quantity of studies[1,2], There was no casting defect found in all of the three section tubes.

The microstructure of the tube can reflecused its used temperature condition and time. From upper to lower of the tube, the microstructure was shown in Figure3. The microstructure of upper tube was the complete cast eutectic structure with a lot of fine secondary carbide, the results showed that the tube had been used at low temperature. The part was basically in good state on the above. The microstructure nearby the cracking had changed obviously with eutectic carbide as a chain-like intermittent network, and the number of secondary carbide decreased, Eutectic carbide particles began to grow up and a little of rod-like carbide occurred. Both of them were accessed at high temperature. A sample was taken in opposite of the crack. Its secondary carbide particles decreased, gathered, and grew up, and the rod-like carbides increased. The microstructure was deteriorated. The microstructure near voids was composed of original chiain-breaking carbide and the amount of the secondary carbide aggregating to the big blocks. It was clearly caused by over high temperature. According to the microstructure analysis, the temperature would be more than 900 °c.

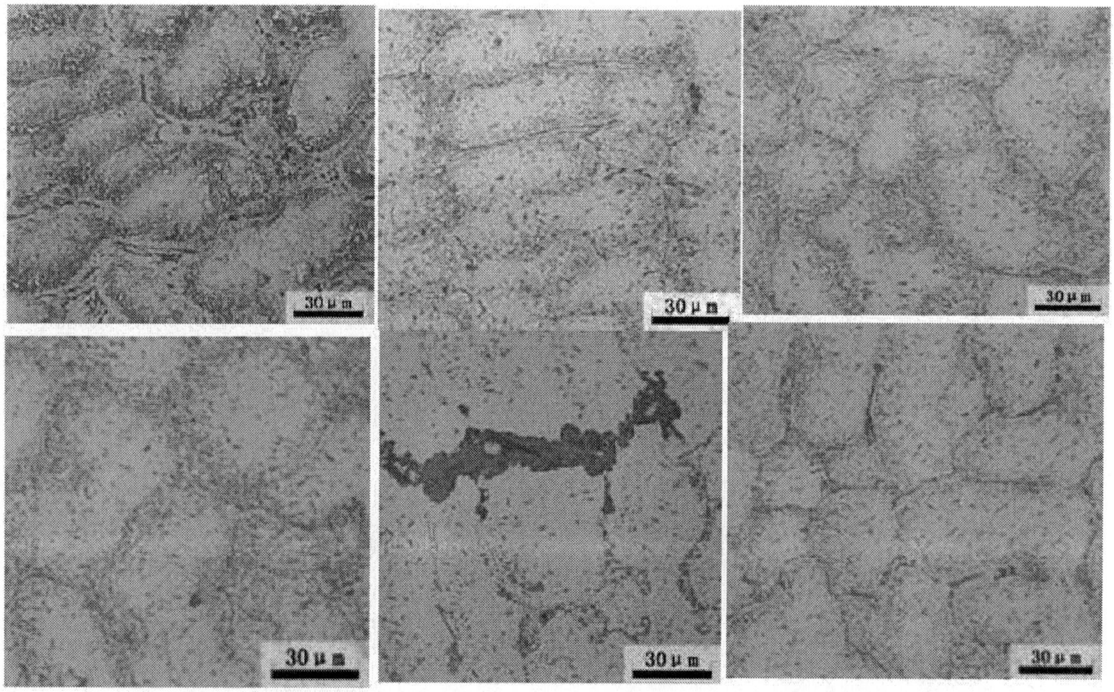

Figure 3. The microstructures of the crack and its upper part, near part, and lower part

By above analysis of various parts, the microstructure of the upper parts was normal. In the lower parts, where exists different degree of overheating structure, the microstructure was deteriorated obviously near crack. So the tube cracking was caused by local overheating

3.4. Crack Analysis

The microstructures on the internal and external surfaces are shown in Figure 4.

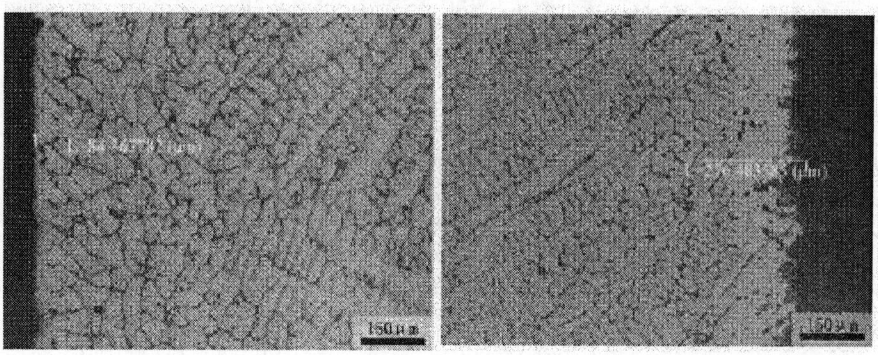

Figure 4. Damages of the internal and external surfaces

The damages of inner and outer surfaces were relatively minor, exception of the cracking, The damage in the outer surface was less than 0.5mm, and that about 0.1mm in the inner surface. This is normal oxidation corrosion in the use of furnace tube and it is not directly related to tube cracking.

By x-ray diffraction analysis on inner and outer surfaces, there were Cr and Fe oxides shown in Figure5. Compared with other used tubes, the diffraction effects is basically the same.

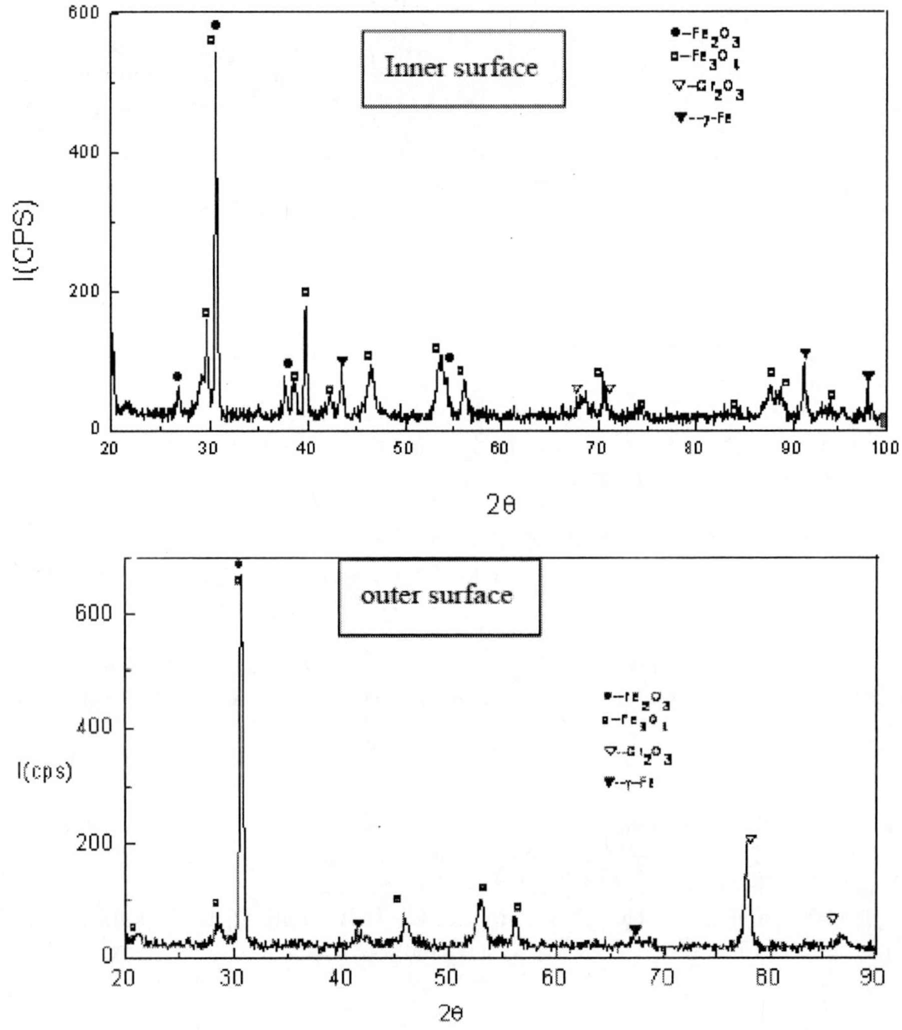

Figure 5. X-ray diffraction analysis results of inner and outer surface

The observation and analysis on the dissection segments of the main crack were done in various parts. The edge of the main crack were made into a several dozens of samples for micro analysis, there was so many micro analysis pictures, not list all in the paper, but these results were consistent. After the observation it can be looked at the oxidation corrosion layer in internal and external surfaces, and there was no oxidation corrosion layer in the fracture part, The result showed that the main crack had been formed in a short time, not developed from those cracks extending to outside surface. Around the primary fracture and the secondary crack tips, where no original structure defects has been found, and

IOP Conf. Series: Materials Science and Engineering **275** (2017) 012026 doi:10.1088/1757-899X/275/1/012026

no inclusions existed also.

By the crack analysis, there were some creep voids existed in the front end of crack and in the vicinity of micro-cracks with a tendency of being gradually connected to micro-cracks. It was a typical structure characteristic of high temperature creep cracks caused by high temperature and stress [3,4]. All of them has no the original structure defects. The morphologies of the secondary micro-cracks in front end and creep voids were shown at Figure 6. The results showed that these creep cracks are the products of high temperature

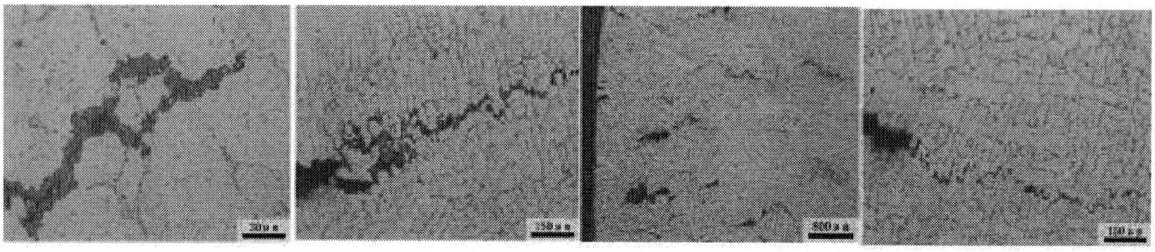

Figure 6. Creep voids and micro-cracks

4. Conclusion

(1) The effective wall thickness of the original tube was too small, it was the root cause for the explosion of furnace tube.

(2) The microstructure of the tube in the lower part (cracks and nearby) were significant over temperature. High temperature creep rupture was verified by the crack analysis, It was the direct reason for the tube cracking at a local overheating condition

(3) According to the chemical composition analysis, the carbon content of tube was in the lower limit of the standard level; the possibility of tube bursting would increase.

(4) All of the above analysis, there was no manufacturing defect found in the tube.

5. References

[1]. Crispin Hales, Kelley J. Stevens, Phillip L. Daniel, Mehrooz Zamanzadehand Albert D. Owens, Boiler feed water pipe failure by flow-assisted chelant corrosion, Engineering Failure Analysis,Volume 9, Issue 2, 2002, 235-243

[2]. Wu Liansheng, Failure Analysis of Petrochemical equipment, Physical and Chemical Analysis Testing, 2000, 36(2), 64-69

[3]. Hong Zhang and Jun Zhuang, Research, development and industrial application of heat pipe technology in China, Applied Thermal Engineering,Volume 23, Issue 9, 2003, 1067-1083, 12TH INTERNATIONAL HEAT PIPE CONFERENCE, 2002

[4]. F. Z. Shen Y. Ding D. M. Hou Z. P. Ning Y. Lü T. B. Dong A. R. Tuohuti. FAILURES AND THE LIFE OF FURNACE TUBES, Acta Metallurgica Sinica (English Letters), 2004, 8

Analysis and Life Evaluation of CS_2 Reaction Tube

Zhenxin Li, Xiaofei Ding, Zhongdong Xie, Xiaotong Yang and Yiming Guan

College of Mechanical and Power Engineer in Dalian Ocean University, Dalian 116023, China

Abstract: The chemical composition analysis, macro and microstructure analysis, mechanical properties and other methods were used to analyze the state and life of CS2 reaction tube for a period of time. The results showed that there is no sulfur corrosion ditch in the furnace tube, and the other one has a uniform corrosion state except for the severe local corrosion. The mechanical properties are good at room temperature. The number and appearance of σ phase are the main reason for making material brittle, welding property degradation and causing crack. The approximate residual life of the furnace tube is also evaluated by various means analysis and life calculation. It is recommended that the tube be replaced with a higher grade nickel-chromium alloy instead of HK40.

1. Introduction

The main components of the reactor tube material is the nickel-chromium, the mass fraction of 20%and25% of nickel-chromium alloy steel has been used for CS_2 reaction tube materials which can withstand high temperature of 1050°C. With the updating and optimization of petrochemical processing technology, the alloys and can withstand high temperature 1100°C [1]. The reaction tube is made by centrifugal casting method and Smooth inner surface machining process is to reduce the reaction of coke [2]. The CS_2 reaction radiant tubes were used for 18 months in which the three of them were severe failure. The three tubes were analyzed respectively, the one is under the eastern side of the 8th layers of a furnace (aslo called tube 1), another one is in 2th layer of bottom of the west side (called tube2) and the 15th layer one (called tube3).

2. Test Results and Discussion

The spectral analyzers, metallographic microscope and other equipment are used to research the cracked tube section. The comprehensive anslysis will be done to find out the failure reasons.

2.1. Chemical Composition

The chemical composition analysis in Table 1 shows that the carbon mass fraction and common elements Si, Mn, P and S satisfy standard requirements [3].

Table 1. Chemical composition analysis results (w%)

sample	C	Si	Mn	S	p	Cr	Ni	Mo	Fe
tube 1	0.45	1.02	1.11	0.0264	0.00677	26.32	33.93	0.0506	other
tube 2	0.438	0.941	0.892	0.0213	0.0113	25.70	20.93	0.0468	other
tube 3	0.413	0.976	0.987	0.0171	0.016	24.53	20.53	0.0381	other
standard	0.35-0.45	≤1.5	≤1.5	≤0.03	≤0.03	24/28	34/37	≤0.5	other

Content from this work may be used under the terms of the Creative Commons Attribution 3.0 licence. Any further distribution of this work must maintain attribution to the author(s) and the title of the work, journal citation and DOI.
Published under licence by IOP Publishing Ltd

2.2. Marcostructure Inspection

The result of Penetrant inspection is shown in Figure 1 a, b, c respectively. Figure (a) is the inner surface of the tube No. 1 for selective corrosion and appears pit. The inner surface of pipe 2 is basically uniform corrosion in Figure (b). Figure (c) is the No. 3 crack pipe ring.

The past many cases show that a more serious sulfur corrosion groove was found in the inspection and assessment process for CS_2 reaction furnace tube serviced for many years. But the analyzed furnace pipes were not found to have sulfur corrosion groove.

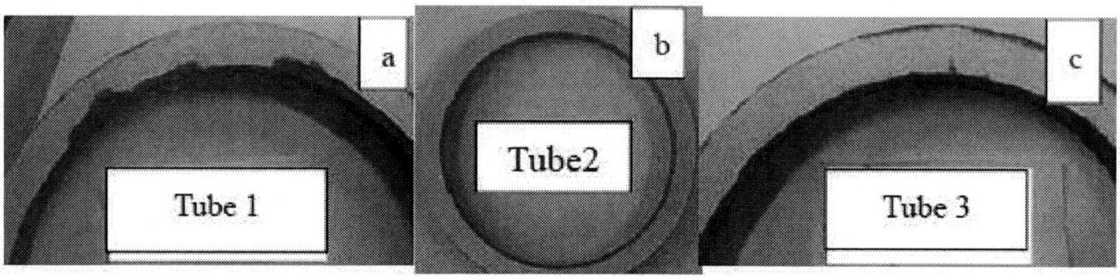

Figure 1. The penetration inspection results of three tube

The diameter and wall thickness of furnace tube are measured. The standard size of the furnace tube is$\Phi 168\times 14$. Measurement results showed that the outer diameter size within the manufacturing error range. The diameter size of three furnace tubes is slightly larger than the standard size. The varying degrees of corrosion results in uneven wall thickness.

2.3. Microstructure

The microstructure the matrix of the three furnace tube shows in Figure 2a, b, c, respectively, a eutectic carbides state remains intact in the microstructure, the secondary carbides is small and diffused, and there is a small amount of acicular σ phase. The state of carbides in tube 1 and tube 3 is basically the same, the number of needle-like σ phases to tube 3 is relatively small. Microstructure analysis for three tube used for a period of time shows that the eutectic carbides change little, which still maintain a better state and the secondary carbides are fine and dispersed which are in good condition. Precipitation of a large number of needles σ phase which is a hard and brittle leads to lower plasticity. This is the root cause of reduced weldabilit.

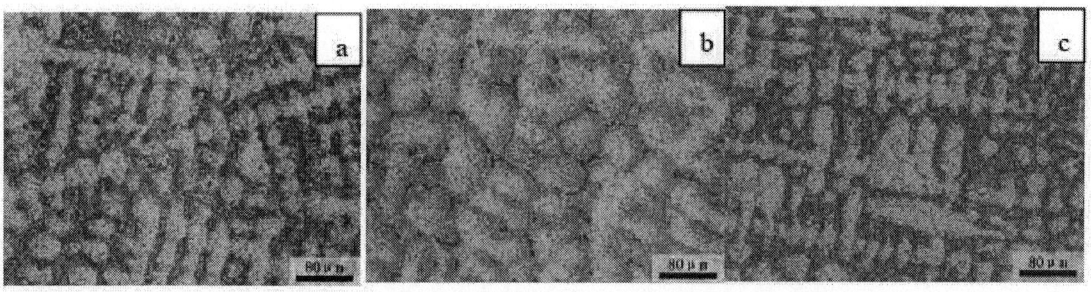

Figure 2. Microstructure of the matrix of three tube oxalic acid electrochemical corrosion

2.4. Damage of Internal and External Surface of Furnace Tube

The external surface corrosion is slight by detection for the three tube ring .The damaged layer within 0.5mm, including the transition layer. In Figure 3, a, b and c are the outer surface microstructure of the tubes 1, 2 and 3, respectively.

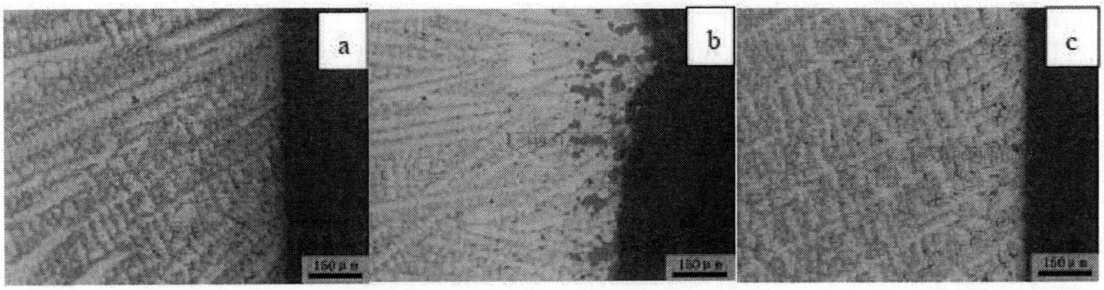

Figure 3. The microstructure of the outer surface of the three furnace tubes Oxalic acid electrolytic corrosion

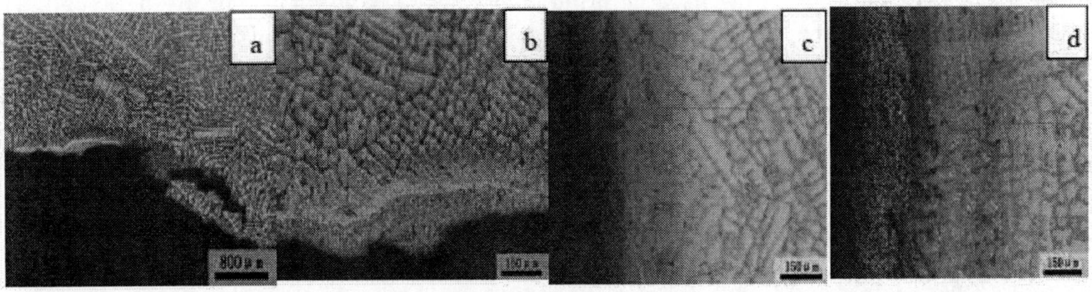

Figure 4. The microstructure of the inner surface of the furnace tube Oxalic acid electrolytic corrosion

The local severely corroded part of tube1 is in Figure4 (a), the more uniform corrosion sites in tube1 was shown in Figure 4 (b).The corrosion of the inner surface of tube 2 and tube 3 are shown in Figure 3c and Figured respectively. The corrosion of inner surface of tube2 and tube3 is uniform.

2.5. Mechanical Property Analysis and Life Assessment

The mechanical properties of three kinds of furnace tubes at room temperature and high temperature were tested. The results show that the strength of the material is very good, Plasticity decreased. The results meet technical contact requirement.

In this detection, the corrosion of the furnace tube1 is the most serious, so the furnace tube will be used to evaluate the remaining life. With the more serious corrosion, its effective wall thickness will be reduced, which will cause the furnace to bear the stress increased, the corresponding P value decreases, the residual life of the tube will also be reduced, In this paper, the medium diameter formula and L-M curve are used to evaluate the life [4].

The design of the furnace tube diameter 168mm, Wall thickness 14mm, The design pressure is 0.9MPa and the applicable temperature is 820 ~ 860°C. Furnace tube measured temperature between 706 ~ 841 °C. For safety, the temperature used for life prediction is 860°C. Measured maximum corrosion layer thickness of 6.4mm, used time is the 15 month (40,000hrs), the corrosion rate calculated by uniform corrosion is 0.356mm per hour. According to HK40 alloy L-M curve of $P = T * (15 + \log tr) * 10^{-3}$ formula, P value for 40,000 hours is 22.21, L-M curve corresponding to its lower limit stress is 18.7MPa [5], the 80% of the stress is 15.0MPa as shown in Figure 5. According to the relevant provisions of GB150-2011 and JB4732-1995, the furnace tube in this paper belongs to the internal pressure cylinder equipment, the main stress can be obtained by the following formula $\sigma = \frac{Po \cdot \Phi m}{2W}$ (1-1).

P_0: tube design pressure, MPa;

Φ_M: the value of the tube diameter minus the effective wall thickness, mm

W: Effective wall thickness mm

The measured minimum wall thickness is 14-6.4=7.6mm. The outer surface corrosion layer is 0.5mm and that of the inner surface is 1mm. The minimum effective wall thickness for calculation is 6.1mm. The safest minimum wall thickness for 40,000 hours can be calculated by the formula (1-1) is 3.88mm. Safe and effective wall thickness of only 6.1-3.88 = 2.22mm, according to the corrosion rate of 0.356mm / month, the tube safe use of time only 6 months.

3. Conclusion

None of the three tubes was found to have sulfur corrosion ditches. Good mechanical properties, the strength index is good and is higher than the standard, the plastic properties of materials decreased. The eutectic carbides remained in a good condition by microstructure analysis. The secondary carbides are small and divergent and have a large amount of needle-like σ phases precipitated. Needle-like σ-phase precipitation leads to the material brittle and cracks and welding property decreased. The inside of the furnace tube contacts the reaction gas and the outside contacts the combustion air. Usually furnace tube are beard high temperature, internal pressure, weight, other stress, as well as repeated low-cycle fatigue caused by the parking[6]. In order to improve its high temperature corrosion resistance and high temperature strength should be replaced with a higher level of nickel-chromium alloy.

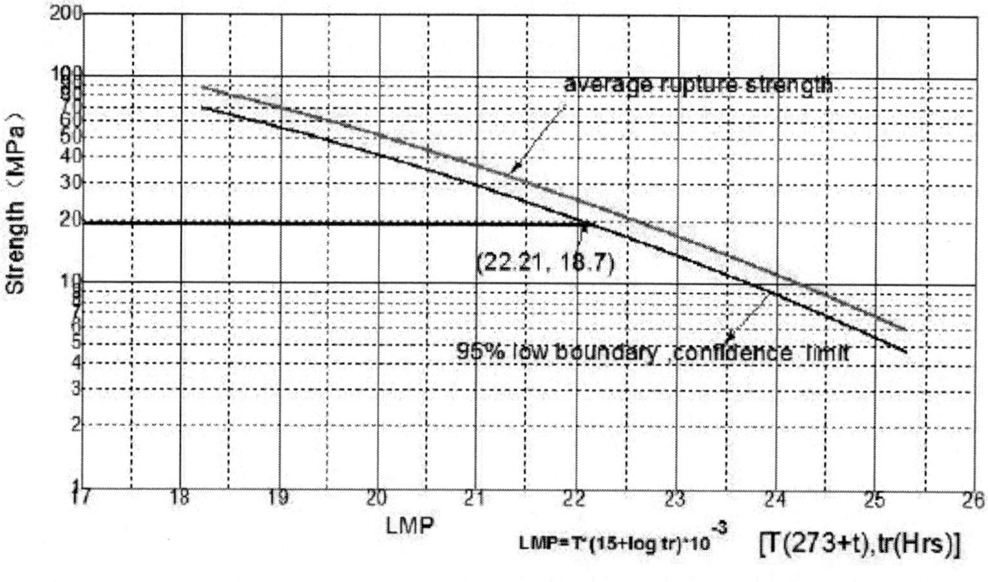

Figure 5. The L-M curve of HK40 alloy and evaluated point of remaining life

4. References

[1] Long Guo-wen, Lu Yong-ming. Carbonization Analysis and Life Prediction of Domestic Refining Ethylene Cracking Furnace [J]. Petroleum Machinery, 1999, 27 (7): 13-14.

[2] ZHANG Jun, LI Lian-bo, SHI Zhi-qiang. Analysis of Corrosion Failure of N80 Oil Well Water Injection Pipe [J]. Corrosion and Protection, 2008 (4): 6-228.

[3] zeng zhengming. Mechanical Engineering Materials Handbook [M].Beijing, Machinery Industry Press, 2003. 345-411.

[4] Journal of Lanzhou Petrochemical Vocational and Technical College, 1998, 26 (2): 46-49. [Article in Chinese] Lanzhou: Journal of Lanzhou Petrochemical College of Technology, 1998, 26(2): 46-49.

IOP Conf. Series: Materials Science and Engineering **275** (2017) 012027 doi:10.1088/1757-899X/275/1/012027

[5]Yang, R. C., et al. "Determination and application of larson-miller parameter for heat resistant steel 12CrlMoV and 15CrMo." Acta Metallurgica Sinica (English Letters) 17.4 (2009): 471-476.
[6] Xu Zili, Huang Jirong et al. Prediction of residual life of HK40 alloy furnace tube [J]. "Mechanical Engineering Materials", 1984, Issue 5: 11-15.

IOP Conf. Series: Materials Science and Engineering **275** (2017) 012028 doi:10.1088/1757-899X/275/1/012028

The Design and Analysis of the Hydraulic-pressure Seal of the Engine Box

Zhenya Chen[1,2], Xingquan Shen[1,2], Zhijie Xin[1,2],TingtingGuo[1,2] and Kewei Liao[1,2]

[1]School of Mechanical Engineering, North University of China, Taiyuan 030051, China;
[2]Research Center of Deep-hole Machining Engineering Technology, Taiyuan 030051, China

Abstract: According to the sealing requirements of engine casing, using NX software to establish three-dimensional solid model of the engine box. Designing two seals suppress schemes basing on analyzing the characteristics of the case structure, one of seal is using two pins on one side to localize, the other is using cylinder to top tight and fasten, Clarifying the reasons for the using the former scheme have a lower cost. At the same time analysesing of the forces and deformation of the former scheme using finite element analysis software and the NX software, results proved that the pressure scheme can meet the actual needs of the program. It illustrated the composition of the basic principles of manual pressure and hydraulic system, verifed the feasibility of the seal program using experiment, providing reference for the experimental program of hydrostatic pressure in the future.

1. Introduction

Engine is a kind of can taking other forms of energy into mechanical energy of the machine, and tank of the engine is the base of the transmission parts, with gears and bearing parts in it, equivalent to muscle cavity supporting the heart movement. In the processes of energy conversion, engine tends to produce high pressure and heat ,transferring a larger force and torque at the same time, so the cabinet design requires sufficient strength and stiffness; At the same time, in the process of working, the function of gears, the work by cylinders and the rotation of the bearings require lubrication and clean environment to ensure efficient operation system, so box body needs to be sealed well .The hydraulic pressure test of the engine box is an important means of verifying the qualification of sealing, as well as the compactness of the material.

The holes in the engine box are divided into two types: one is the thread hole and the other is the non-threaded hole. Because of its connection, screw holes are tightly sealed by thread through sealing tape .Non-threaded hole, can be sealed by sealing pad on the cylinder to cling to the tank. Connecting the water injection mouth with pump by hose and inject water, the pump the air from the cylinder to loosen seal and release water, by the test bench around the pool should be built to realize the recycling of water. Operation process requires 0.1 Mpa for water pressure which should be maintained for 15 min; if the pressure reading is constant and deformation of tank could be neglected, meanwhile the injection process should be quickly, then the case is on test,; in addition, the exhaust should be reasonable and thorough, fast clip with quick pressure relief is also in need, etc.

2. The analysis and Design of the Tank Structure

There are 25 hole should be sealed on the tank. In this project, the structure design of the cylinder

Content from this work may be used under the terms of the Creative Commons Attribution 3.0 licence. Any further distribution of this work must maintain attribution to the author(s) and the title of the work, journal citation and DOI.
Published under licence by IOP Publishing Ltd

tightening device is marked. The construction of the cabinet is shown in Figure 1.

(1)Front view: there are 6 holes. Two of them (R116) are on the same plane, whose boss height is 5mm.the radius of the other four with uniform plateau and boss height (5mm) are R120, R118, R65, R50.Exhaust vent is arranged on the upper right of the view; the acreage for sealing of front view is 921.1cm2; the pressure of the outlet should be 921.1kg, according to its hydraulic pressure of 0.1Mpa.

(2)The left view: there are 2 holes.one of them is a hole with radius of 165mm.The acreage for sealing of front view is 283.3cm^2.the pressure of the outlet should be 283.3kg.What's more, there is a screw hole of M30x1.5,whose shape are uniform viewed from left and right.

(3)The back view: 7 holes. R117.5 and R31.3(be covered),both of them are in the same plane; the other five are in another plane, with a] threaded hole (M27x2), and another threaded hole (M32), R151.1, as well as R141.The area need to be jam is 99.2cm2, while the pressure is 998.2kg. There is a] threaded hole which is not shown in the picture.

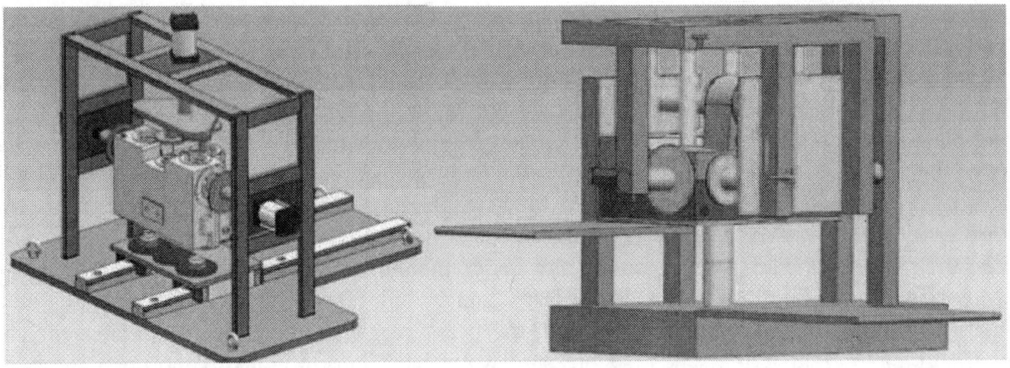

Figure 1.

3. The Plan of Pressure Loading

The hole-bracing sealing scheme requires that the supporting poles meet high positioning accuracy and individual freedom movement, which makes it difficult to design, thus the hole-pressing sealing scheme should be adapt.

Based on the characters of the tank, two programs for sealing are put forward:

Pro.1 is to sealing with two pins on one side, shown in Figure 2, which contains three air cylinder, a suit of bed-stand support, a pair of guide rail and support circuit. The over-all structure has been illustrated in picture. Whole process is divided into 5 parts: work locating, workpiece transmitting, cylinder closing and tightening, cylinder withdraw and output.

Pro.1 is made up by two pins on one side to orientate, making use of bottom and mechanical cooperate of short pin and hole. To achieve bottom sealing by its mass, elastic sealing rubber mat is distributed on the supporting surface.

Pro.2, seals suppress schemes of cylinder, includes two conveyer belts, six cylinders, a suit of bed-stand support, as well as support circuit and motor, as is shown in Figure 3.To achieve the function, the device finishes six steps in the test: placing and transferring, identifying the situation, locating the work, closing and tightening, dropping out of cylinder, exportation of products.

Compared with pro.1, pro. 2 is tedious and money-cost. So pro.1 is recommended, and following is analysis.

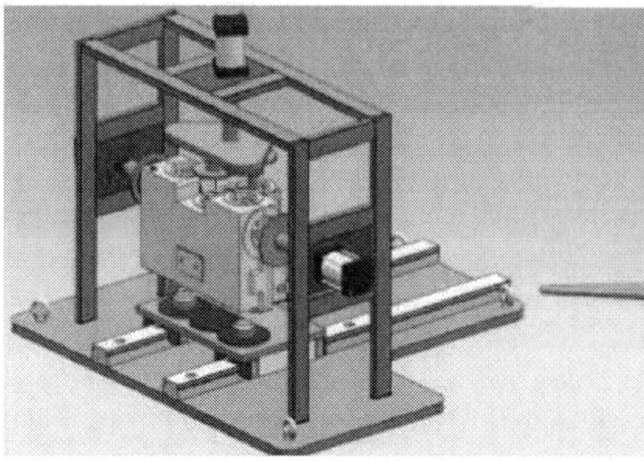

Figure 2. The structure of two pins on one side

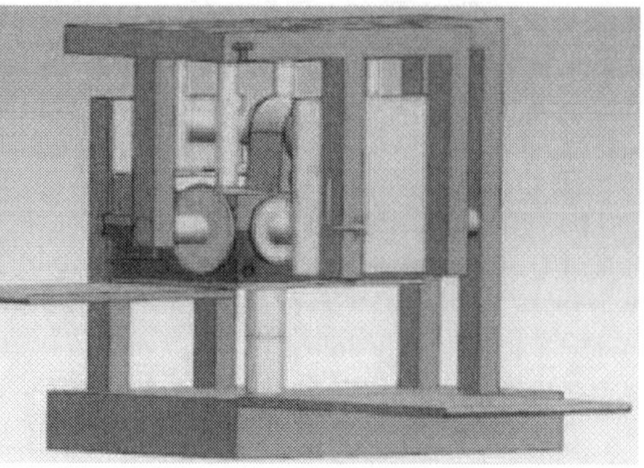

Figure 3. Structure of cylinder sealing

4. Principle and Project of Hydraumatic System

The principle of system of hydraumatic pumping test is shown in pic.7.the pressure for tank depends on P1, if the tank is sized. Start the hydraulic pump. Make the electromagnetic commutation valve 1 charge, clamp the cylinder action, will be clamped by the specimen. The pressure oil enters the left cavity of the cylinder and the piston rod goes right. The water in the tank opens a one-way valve to the left cavity of the water cylinder. When the switch is touched, the signal is sent for the electromagnetism valve 1.Then, the pressure oil is transferred to the right cavity of the cylinder, pushing the piston to the left, the pressure water inside the tank, along the pipe to the test piece. If the test valve is not leaking, the contact pressure gauge will show the pressure in the tank and automatically press for one minute. After that, the pressure gauge sends an electrical signal, which opens the unloading valve and sends the pressure water back into the tank. Then press the button to power the reverse valve 2, loosen the cylinder and remove the workpiece. The cycle turns over and over. If the valve is leaking, open the unloading valve immediately and let the pressure water return to the tank, as shown in figure 4.

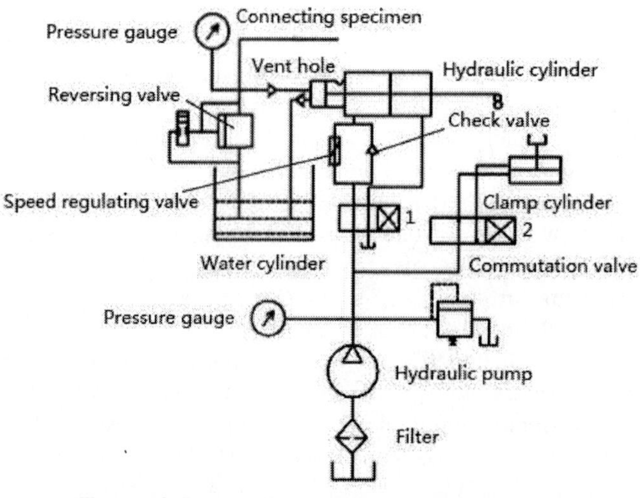

Figure 4. Hydraulic System Principle Diagram

5. Suppressing the Analysis of the Box Body

Engine cases are usually made of gray cast iron, and gray cast iron has good casting performance and vibration damping performance. But its density is larger caused in the weight of per unit volume is larger, in order to satisfy the use requirement, to realize the lightweight design, this experiment adopts cast aluminum as the material of body, whose density of 2700 kg/m3, thus the body weight is 212 kg,

IOP Conf. Series: Materials Science and Engineering **275** (2017) 012028 doi:10.1088/1757-899X/275/1/012028

and at the same time the weight of support is 89 kg, so the total weight is 301 kg.

Using the ANSYS finite element software to simulate, we got the simulation parameters for aluminum alloy, whose density is 2700 kg/m3, 7 e10pa with young's modulus, and Poisson's ratio is 0.33.All constraints are imposed on the casing wall. Putting the pressure (0.45 Mpa) on the side face of box, the simulation results are as follows:

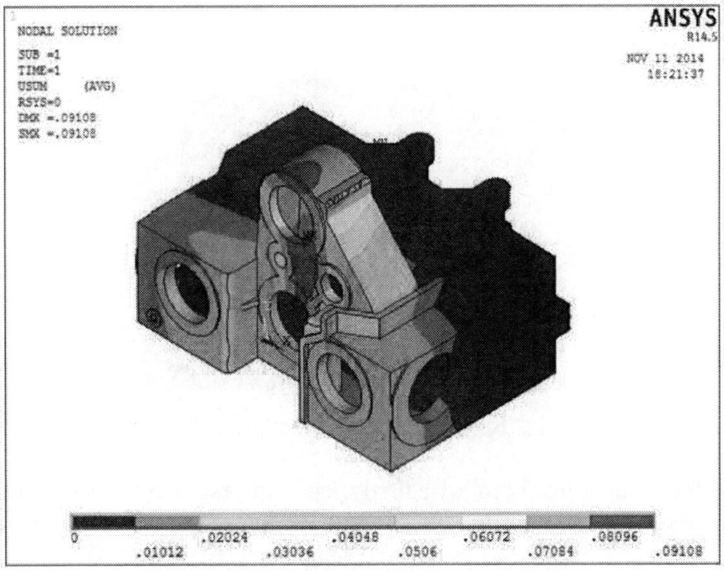

Figure 5. Displacement diagram of box

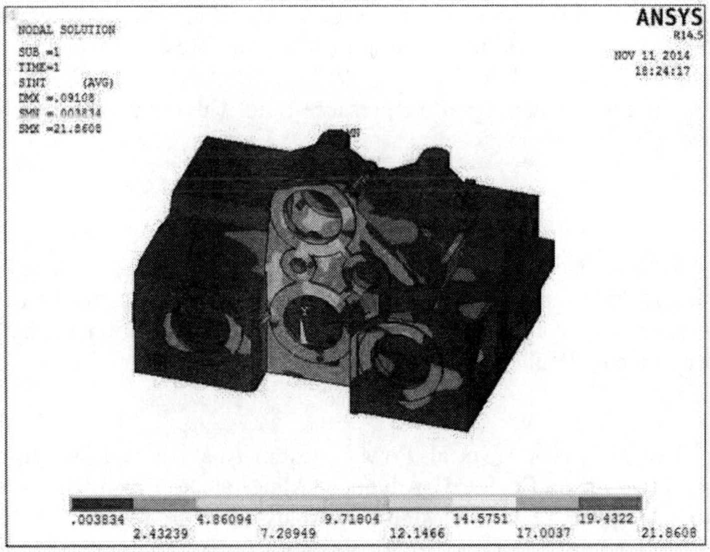

Figure 6. Stress diagram of box

The maximum displacement of the box is 0.09108 mm when the external pressure is 0.45 MPa, and the maximum displacement occurs in the outer edge of the ear, got from figure 5.

The maximum stress of the box in figure 6 is 21.8608 MPa, the maximum stress occurs in the fascia joint.

Taking use of the finite element analysis software NX to simulate the features, the simulation results are as follows: in the figure 11, when the outer pressure of 0.7 MPa, the maximal displacement of box body is 0.140 mm, the largest displacement occurs in the bearing outer ears.

Conclusion: the results of ANSYS and NX simulation results show that the pressure hole sealing method is very small and does not have a large effect on the seal.

6. Pressure Test of the Tank

The manual test pump is shown in figure 7.

Figure 7. Manual test pump

The process of the manual test is follow: firstly, open the tap and the valve, casing to the engine with water. When water spilled from the overflow valve, close the overflow valve and water-filling valve;

Start the main motor metering pump and keep pump running, closing circulating valve and open the hydraulic valve as well as pressure valves, then the press rises up .When the pressure rises to 0.45 MPa, servo motor offers small flow rate until pressure to 0.5 MPa, then hold the pressure on .

The circulating valve keeps working by pump to retain the pressure, when the engine body at 0.5 MP and send a signal.

After the test pressure is finished, open the pressure-relief valve firstly and then open the overflow valve.

7. Conclusion

Pressure test could guarantee the airtight of the box .This paper has designed two kinds of box seal testing scheme, the latter adopts positioning method called "two pins on a plane" .The finite element analysis and experiments validation prove that the second method of engine box functions very well.

The hydraulic system can complete the automatic pressure regulation of the system, which has guiding significance to the hydraulic pressure test.

8. Reference

[1] Cheng L J, Zhang J P, Huo W, et al. Power System Analysis and Design of the Six-Cylinder Hydraulic Confined Piston Engine[J]. Advanced Materials Research, 2013, 765-767:57-61.

[2] Wang Z, Zhao J, Zhang Z, et al. Design and Analysis of the Limiting Valve Group of Rotary Pressure for Full Hydraulic Rotary Drill[J]. Machine Tool & Hydraulics, 2016.

[3] Sun M, Zhang Z Y, Kong W J. Design of Sealing Structure and Analysis of the Flow Field in Micro Internal Combustion Swing Engine[J]. Applied Mechanics & Materials, 2014, 670-671(2):930-935.

[4] Liu F, Cai X. Design and Analysis of New Type of Four Equal Pressure Seal Sealing Fixture [J]. New Technology & New Process, 2011.

[5] Shen Z, Li L, Wang P. Analysis on the Effect of External Pressure Function on Internal Pressure Seal of Special Thread Connection[J]. Technology Supervision in Petroleum Industry, 2012.

[6] Zhang S, Zhao Z F, Zhao C, et al. Design Approach and Dimensionless Analysis of a Differential Driving Hydraulic Free Piston Engine[C]// SAE 2016 Commercial Vehicle Engineering Congress. 2016.

Investigation of Comfort Properties of Knitted Denim

Abdul R Akbar[1, 2], Siwei Su[1], Junaid Khalid[2], Yingjie Cai[1, 3] and Lina Lin[1, 2, 3 *]

[1]Hubei Provincial Engineering Laboratory for Clean Production and High Value Utilization of Bio-based Textile Materials, Wuhan Textile University, Wuhan 430073, China
[2]College of Textile Science and Engineering, Wuhan Textile University, Wuhan, 430200, China
[3]Engineering Research Centre for Clean Production of Textile Dyeing and Printing, Ministry of Education, Wuhan Textile University, Wuhan, 430073, China
Email: linalin@wtu.edu.cn

Abstract: Knitted denim was designed by using cross terry structure on circular knitting machine. Knitted denim looks like a denim fabric which has visual appearance like woven denim. Two type of cross terry structure 2/1 and 3/1 were used which gives twill effect with 2 and 3 floats respectively. Four types of materials, cotton, polyester, flax and polypropylene were used. With four materials and two structural combinations 8 samples were produced. Comfort properties of knitted denim including moisture management, air permeability, thermal, and bursting strength were tested. For checking the inherent anti-microbial property of materials anti-microbial test was also applied. Samples containing flax and polyester were found with best results and not even a single sample was found anti-microbial.

1. Introduction

Knitting is fabric construction technique in which a continuous yarn is changed into the vertical or horizontal intermeshed loops of yarn either with machine or by hand. According to the movement of yarn during knitting process, it is categorized into two main classes i.e. weft knitting and warp knitting [1]. Weft knitting is mostly used for wearing purposes. Basic weft knit structures is single jersey, rib, interlock and purl. A fabric known as terry is a derivative of single jersey. The knitted denim has a basic structure of terry [2, 3].

Denim is the kind of a fabric, and a traditional blue denim is warp faced woven cotton fabric in 3/1 or a 2/1 twill constructions. Warp is dyed with Indigo dye while weft yarn is remaining undyed. In twill weave the weft yarn interlaces with more than one yarn [4]. Denim is a complete lifestyle statement. On the other hand, for knit denim, cross terry structure is used. While in knitting the knit yarn is dyed and tuck is remained as undyed. In knitting one feeder knits and other feeder loops in this way diagonal lines are formed [5]. Usually knit denim is used for multi purposes as trouser, shoes, hats, socks, shirts, collars and cuffs, curtains as well as fabric's elements [6]. The statistics tells as that denim accounts about 3% of the total world fabric production. An American has eight pairs of jeans, while the Europeans having almost five to six pairs of jeans [7, 8].

The fibre we are going to use to make these knitted denim are flax, cotton, polypropylene and polyester. Knitted denim can be combined with some functional fibres like Tencel, bamboo fibre, spandex, soybean fibre and other fibres to create innovative products with enhanced properties and appearance [9]. The process of manufacturing knitted denim involving ball warping, re-beaming, indigo dyeing, re-winding to cheese/cone form and finally knitted in single jersey circular knitting machine

Content from this work may be used under the terms of the Creative Commons Attribution 3.0 licence. Any further distribution of this work must maintain attribution to the author(s) and the title of the work, journal citation and DOI.
Published under licence by IOP Publishing Ltd

[10]. Knitted denim can also be dyed and converted to soft cones by re-winding and knitted in a circular knitting machine [11].

Knitted denim fabric provides good wrinkle resistance, improved stretch, extensibility to body fit ease, long lasting neat appearance and provide high level of comfort in the garments, as it is highly pliable, breathable, difficult to tear, highly crease resistant, highly air permeable and limb fabric [12]. Also the knitted denim are piece dyed as previously the woven denim fabric are generally made from dyed yarns. Denim knit fabric can be produce by float plated technology, thread fleece and interlocked plated jacquard. In float plated technology 70% of indigo yarn and 30% cotton yarn is used and the output is almost 220 GSM fabric weight. In thread fleece technique, 50% indigo yarn and 50% cotton yarn is used, resulting in almost 130 GSM. Moreover, In interlocked plated jacquard technique 70% indigo yarn and 30% cotton yarn is used to get Greige GSM of 180 [13]. There are two types of knitted denim with respect to structural constructions, which are knit denim and tuck denim. Tuck denim was produced by two ways tuck denim on 4 tracks and Tuck denim on 3 tracks [14].

The human body produces heat during excessive workout, to eliminate this excessive heat a natural cooling system of body triggers and the sweat is produced from the glands on skin [15, 16]. Management of liquid water is mostly referred as the fabric's "moisture management" and the management of water vapours is termed as the fabric's "breathability" [17]. The management of fabrics moisture management is seen by adsorption, absorption/diffusion, wicking, desorption and evaporation which were tested in this research [18].

This project emphasises on making of knitted denim by using different types of yarn combinations. The main objective is to make knitted denim more cost effective have properties like good moisture management, good comfort level and anti-microbial properties.

2. Experimental

2.1. Materials

Materials used for the production of knitted denim were combination of natural and synthetic yarn. Natural fibre yarns were 30 Denier cotton and 27 Denier flax; synthetic yarns were 275 Denier polyester and 275 Denier polypropylene. To get the denim effect, the dyed yarn was used. Indigo dyes on natural yarns (cotton and flax) were used. While the synthetics yarns remain undyed because synthetics yarn is used on technical front side (commercially back side).

2.2. Knitting

As far as the machine is concerned circular weft knitting single jersey machine, Fukuhara Japan 1992, 90 feeders was used for the production of knitted denim. Stitch length of 0.32 cm for knit and stitch length of 0.15 to 0.16cm for loop. Method used for the production of knitted denim is precisely described below. For getting knitting denim effect cross-terry structure with 3 and 4 tracks repeat were used, and the experiment design is shown in table 1.

Table 1. Design of experiment.

Sr.	Structure	Technical Face	Technical Back	Sr.	Structure	Technical Face	Technical Back
1	2/1 Terry	Cotton	Polyester	5	3/1 Terry	Flax	Polyester
2	2/1 Terry	Cotton	Polypropylene	6	3/1 Terry	Cotton	Polypropylene
3	2/1 Terry	Flax	Polyester	7	3/1 Terry	Cotton	Polyester
4	2/1 Terry	Flax	Polypropylene	8	3/1 Terry	Flax	Polypropylene

2.3. Finishing

Finishing processes are applied to the samples because when samples are removed from the machine they possess a lot of shrinkage. When samples are washed and relaxed they come to their normal relaxed state. Also removal of shrinkage is very necessary because when fabric is converted into the garment if it contains shrinkage then dimensions of final garment will not stay in the required position. Its required shape will not achieve. Samples were washed in room temperature water to remove the dust and other stains got by the fabric during knitting process. All the samples were tumble dried for one cycle.

2.4. Test

ASTM D 3786 (textile fabrics diaphragm bursting strength tester) was used to check the bursting strength of knitted or woven fabric. Moisture management test (AATTC 195) method was used for the inspection of fabric for liquid moisture management. There are five grades scale (1-5) in assessing the moisture management according to the OMMC value. The grades of the indexes are: 1(0-0.2) = very poor, 2(0.2-0.4) = poor, 3(0.4-0.6) = good, 4(0.6-0.8) = very good, 5(>0.8) = excellent. ISO 11092 test method was used for checking the thermal properties of woven or knitted fabrics. ASTM D 737 test method measured the rate of air flow perpendicular to the surface of textile material at prescribed pressure. AATCC TM100-2004 was used for checking the anti-microbial activity of fabrics.

Culture of *Staphylococcus aureus* and *Escherichia coil* were cultivated in two flasks separately and the flask was incubated separately at 37°C for 18 hrs. After incubation growth of bacteria on the petri plates were counted and percentage reduction in bacteria for each sample against *Staphylococcus aureus* and *Escherichia coil* was calculated.

3. Results and Discussion

3.1. Thermal Property

The thermal properties of samples are showed in figure 1. It is obvious that the sample made from cotton and polypropylene (2/1) shows the highest thermal insulation among the samples with the value of 0.0964 m^2k/W, because the thermal properties of polypropylene is very good, thus it added with flax gives the highest thermal insulation amongst other samples. While the sample made from flax/polyester has the least heat insulation with the value of 0.0725 m^2k/W. As polyester is poor resistant of heat, the samples which have polyester show comparatively low thermal insulation.

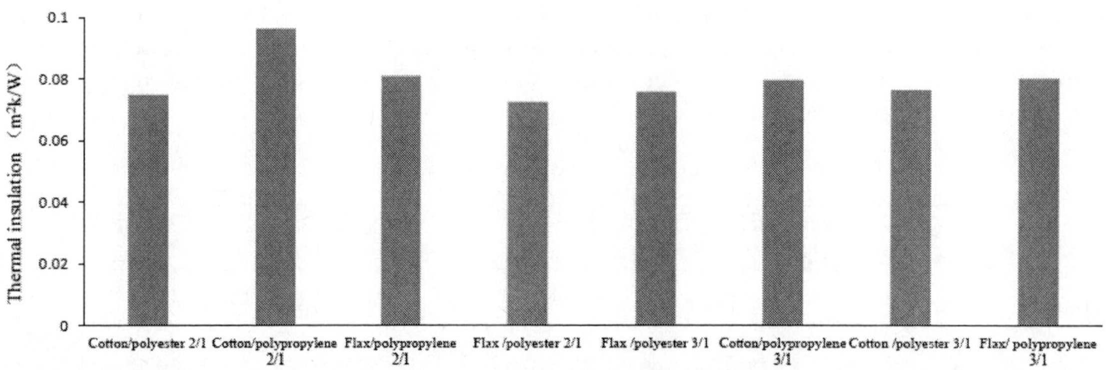

Figure 1. Thermal properties of the samples.

3.2. Air Permeability

In figure 2, the knitting sample made from flax/polyester shows the highest air permeability value of 964 mm/sec for the face side of the fabric and 556.33 mm/sec for the back side of the fabric, indicating the fabric has passed more air through it as compared to other. It can be said that the fabric was more permeable and had higher pore amounts than other fabrics. On the other hand the lowest value of air permeability was shown by cotton/polypropylene 3/1 with the value of 399 mm/sec for the front of the fabric and 396.33 mm/sec for the back of the fabric. There are many aspects, which effect the air permeability of the fabric, e.g. construction, fabric type, thickness, bulk density, and air porosity in the yarn. Seen like the combination of cotton and polypropylene has more bulk density and smaller air porosity in the interlooped yarns of the knitted structure.

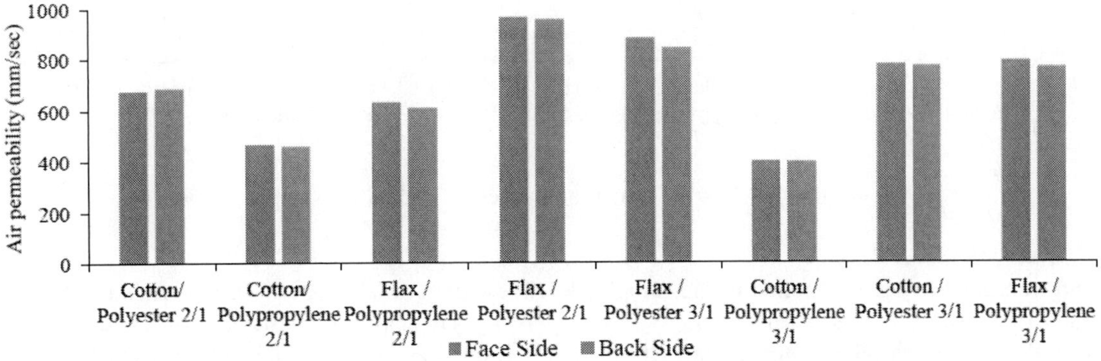

Figure 2. Air permeability of front and back of the fabric samples.

3.3. Moisture Management

This test was done on the moisture management tester machine. A solution of sodium chloride was dropped on the fabric surface and the time given to the solution to drop on the fabric was 20 sec. After that the spreading radius and speed or time taken by the sample to spreading was recorded and overall moisture management content was calculated. Three test samples were taken from each fabric and the average values for overall moisture management capacity (OMMC) were calculated, and the results are presented in figure 3. Moisture management results show that the flax/polyester 2/1 has the highest value of overall moisture management capacity of 0.878, which indicates that this sample had manage the moisture transmission in fabric in better way than others. It hints that this property was assistance to rapidly remove the sweat from the body by transmitting outside of the fabric. The flax/polyester 3/1 also shows comparatively higher value of moisture management than other samples. Whereas the cotton/polyester 2/1 shows the least value of 0.0698, which was possibly attributed to the flax and polyester. It can be seen from the graph where there is flax in the samples the values of moisture management are relatively higher. Flax fibre has naturally higher wicking ability, it can gain upto 20% water before feel damp, which contributes a better moisture management with flax contents.

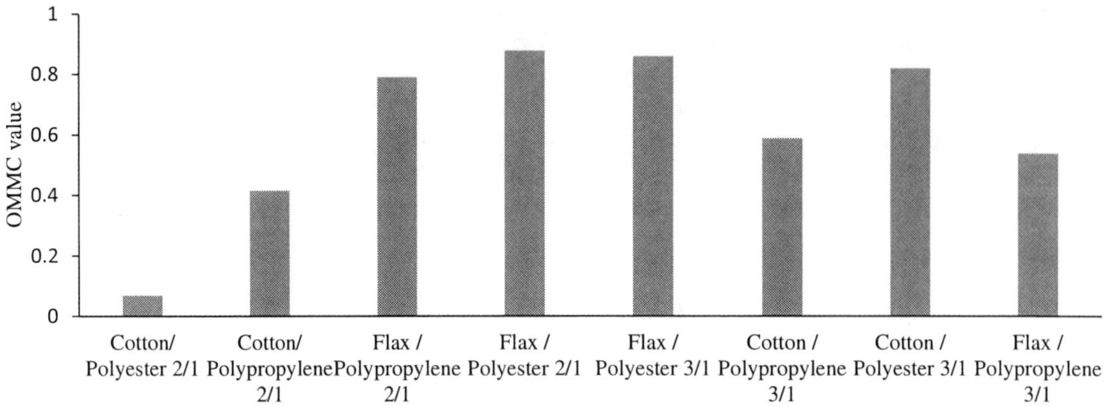

Figure 3. Moisture management test of the fabric samples.

3.4. Anti-Microbial Property

Anti-microbial test were applied to the samples to check the natural property of materials to resistant attack of microorganisms. The following results (figure 4) were achieved by the test. It is concluded that from these pictures taken after the activity of bacteria on these samples, the material used in these samples were not naturally anti-microbial because bacteria had grown on them. None of the fabric sample showed any resistance to bacterial growth showing their inability to resist microorganism.

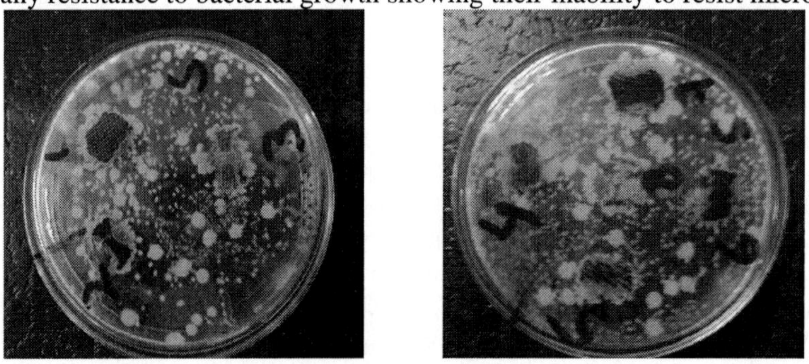

Figure 4. Anti-microbial test of the fabric sample.

4. Conclusion

The flax/polyester demin fabric produced by the 2/1 twill effect showed the best results in moisture management and air permeability but showed average value in thermal test results. Antibacterial/anti-microbial activity was not observed at any fabric. Flax/polyester 3/1 and Polyester/Cotton 3/1 also showed better results. They also exhibited better properties than other sample. Samples with flax and polyester showed comparatively better results. Polypropylene with cotton gives medium results of thermal, moisture management and air permeability.

5. Acknowledgement

This work was financially supported by the China National Textile & Apparel Council (2013 "Textile Vision" Applied Basic Research, 2013-153); Hubei Province Science and Technology Support Program (Grant No. 2013BAA043) and the Collaborative Innovation Plan of Hubei Province for Key Technology of Eco-Ramie Industry (2014-8).

6. References

[1] Miller, D 2015 *Consump. Mark. Cult.* **18** 298

[2] Didar S A 2015 *Res. J. Engg. Sci.* **4** 9
[3] Lee, I Y, Jeong G E, Kim S R, Bengelsdorff C and Kim S D 2015 *Color. Technol.* **131** 192
[4] Wei B, Xu F, Azhar S W, Li W, Lou L, Liu W and Qiu Y 2015 *J. Ind. Text.* **44** 798
[5] Nazir A, Hussain A T, Ahmad F and Faheem S 2005 *Autex. Res. J.* **14** 39
[6] Maqsood, M, Nawab Y, Umar J, Umair M and Shaker K 2017 *J. Text. I.* **108** 522
[7] Hu J 2005 *J. TEXT. I.* **75** 57
[8] Shaker K, Umair M, Ashraf W and Nawab Y 2016 *Physical Sciences Reviews* **1** 7
[9] Kyosev Y and Renkens W 2007 *J. Fam. Ecol. Consum. Sci.* **35** 39
[10] ASTM D737 1996 American Society for Testing and Materials, West Conshohocken, PA, USA
[11] Patil, L, Patil T C, Raichurkar P and Dorugade V A 2016 *International Journal on Textile Engineering and Processes* **2** 20
[12] Spencer D J 2001 *Knitting technology: a comprehensive handbook and practical guide*, (Cambridge: CRC Press)
[13] Ray S C 2012 *Fundamentals and advances in knitting technology* (New Dehli: Wood Head Publishing Limited)
[14] Çelik H I, Topalbekitoglu M and Dülger L 2015 *Fibres Text. East. Eur.* **23** 85
[15] Haque M, Sharif A 2014 *Int. J. Innovative Sci. Mod. Eng.* **6** 24
[16] Delkumburewatte G B 2011 *Advances in knitting technology* (Leicester: Elsevier Ltd)
[17] Spencer D J 2012 *Knitting technology*, (Leicester: Wood Head Publishing Ltd)
[18] Gazzah, M, Jaouachi B, Schacher L, Adolphe D C and Sakli F 2015 *Int. J. Cloth. Sci. Tech.* **27** 922

IOP Conf. Series: Materials Science and Engineering 275 (2017) 012030 doi:10.1088/1757-899X/275/1/012030

Stability and Control of the Convergence Point for Two-strand Yarn Spinning

Wenyan Liu[1], Yanping Yu[2], Feng Li[1] and Yajun Zhao[1]

[1]China Construction Science & Technology (Beijing) CO.TLD. Longhe Hi-tech Park Tiangao Road No.86, Anci District, Langfang City, Hebei Province, China
[2]College of Textiles, Donghua University, Shanghai 201620, China
Email: w.y.liu@163.com

Abstract: In this paper, stability of the convergence point in Siro-spinning composite area and control of tension compensator (TC) to the unstable motion of the convergence point in Sirofil spinning were kinetically studied. The control effect was compared with corresponding experiments. It is concluded that the adoption of TC can greatly improve the stability of the system, apparently reduce the vibration of the convergence point, and simultaneously reduce the hairiness of Sirofil composite yarn and improve the property of abrasion-resistance enormously.

1. Introduction

Siro-spinning, one kind of ring spinning, is now widely employed in the textile industry. It is known that Sirospun yarns are conducted on a conventional ring frame by feeding two rovings, drafted simultaneously, into the apron zone at a predetermined separation. Emerging from the nip point of the front rollers, the two strands are twisted together to form the two-strand yarn. It is called Sirofil spinning if men substitute one filament yarn for one of the two rovings in the spinning process [1] (also see Figure 1). Mechanics and mathematics approaches have been involved in solving the problems in such a spinning process. Quasistatic models by He et al. [2], Yu et al.[5] and Liu [6], and dynamic models by He et al. [3] [4] for two-strand yarn spinning to analyze instability phenomenon of Siro-spinning have been established. We also studied the control of TC to the properties of Sirofil composite yarn from the static point of view [7], and now we study the control of TC to the unstable movement of the convergence point in the view of dynamics.

In Sirospun spinning, the component and modulus of the two rovings are the same. The twisting is stable and balanced, and the structure of resultant yarn is regular. But in Sirofil spinning, due to the modulus difference between roving and filament yarn, the two components are distributed asymmetrically over the Sirofil composite area, and the convergence point is thus unstable, leading to irregularity of yarn structure, i.e., staple yarn partially wrapping filament yarn. A tension compensator is introduced to control the irregular vibration of the convergence point, so as to improve the unevenness and the abrasion property of Sirofil composite yarn [7]. In this paper, from the dynamical point of view, we study the movement of convergence point before and after the control of TC, and the effect of TC on the stability of Sirofil composite area is apparently shown. It is analytically and qualitatively seen that our control results are consistent with the experimental data shown by [7].

Content from this work may be used under the terms of the Creative Commons Attribution 3.0 licence. Any further distribution of this work must maintain attribution to the author(s) and the title of the work, journal citation and DOI.
Published under licence by IOP Publishing Ltd

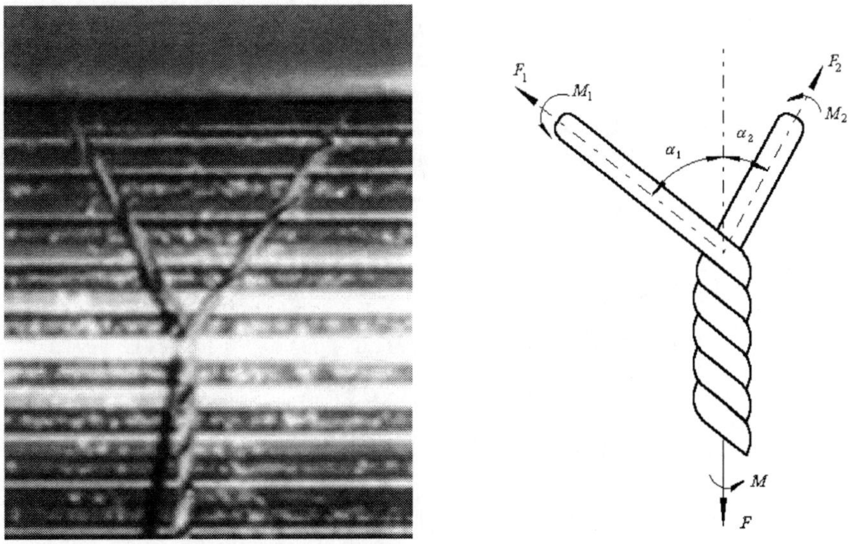

Figure 1. An asymmetric two-strand yarn spinning

2. Stability of the Convergence Point in the Symmetric Two-strand Yarn Spinning

We first study the symmetric two-strand yarn spinning. The system is assumed initially in a stable condition. Due to some perturbations, the convergence point (equilibrium position *O* in Figure 2) will move to an instantaneous position (*O*'). The distance *x* and *y* are measured from the equilibrium position.

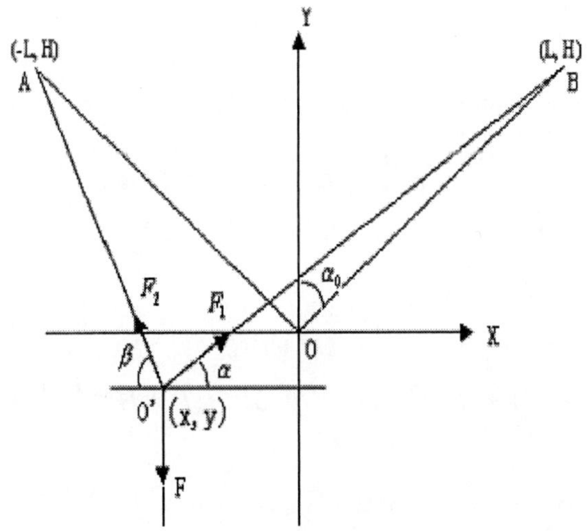

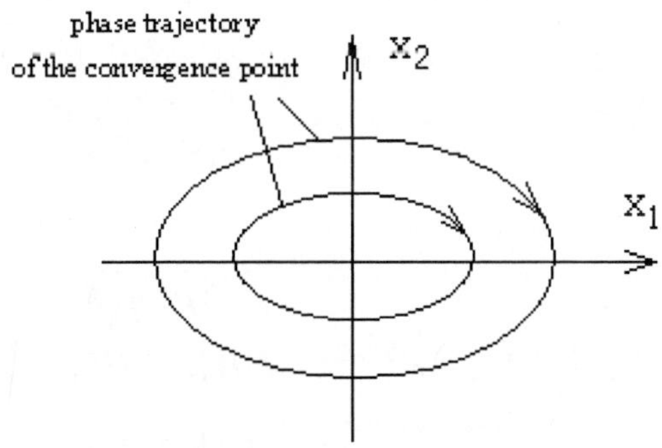

Figure 2. The dynamical illustration for Sirospun spinning;
Figure 3. Schematic diagram of horizontal motion of the convergence point for Sirospun spinning

For this case, let the ends of the two strands above convergence point be fixed a distance $2L$ apart, and let the equilibrium position be H below. F is the force in the two-strand yarn below the convergence point, and F_1, F_2 are the forces in the two strands above the convergence point. The angles α and β are also defined in Figure 2. It is easy to see that the equations of motion in *x*- and *y*-directions are

IOP Conf. Series: Materials Science and Engineering **275** (2017) 012030 doi:10.1088/1757-899X/275/1/012030

$$\begin{aligned} &M\ddot{x}+F_2\cos\beta-F_1\cos\alpha=0 \\ &M\ddot{y}+F-F_1\sin\alpha-F_2\sin\beta=0 \end{aligned} \tag{1}$$

Where M is the total mass of the control volume illuminated by [3][4].

For the small amplitude vibration of the convergence point, x and y are much smaller than $\sqrt{L^2+H^2}$. So applying the binomial theorem to expand the square-root terms, we have

$$\cos\alpha=\frac{L-x}{\sqrt{(L-x)^2+(H-y)^2}}\approx\frac{L}{\sqrt{L^2+H^2}}(1-\frac{x}{L}+\frac{Lx+Hy}{L^2+H^2})+\frac{1}{2}\left[-3L(L^2+H^2)^{-3/2}+3L^3(L^2+H^2)^{-5/2}\right]x^2$$

$$+\frac{1}{2}\left[-L(L^2+H^2)^{-3/2}+3LH^2(L^2+H^2)^{-5/2}\right]y^2+\left[-H(L^2+H^2)^{-3/2}+3L^2H(L^2+H^2)^{-5/2}\right]xy, \tag{2}$$

$$\sin\alpha=\frac{H-y}{\sqrt{(L-x)^2+(H-y)^2}}\approx\frac{H}{\sqrt{L^2+H^2}}(1-\frac{y}{H}+\frac{Lx+Hy}{L^2+H^2})+\frac{1}{2}\left[-H(L^2+H^2)^{-3/2}+3L^2H(L^2+H^2)^{-5/2}\right]x^2$$

$$+\frac{1}{2}\left[-3H(L^2+H^2)^{-3/2}+3H^3(L^2+H^2)^{-5/2}\right]y^2+\left[-L(L^2+H^2)^{-3/2}+3LH^2(L^2+H^2)^{-5/2}\right]xy, \tag{3}$$

$$\cos\beta=\frac{L+x}{\sqrt{(L+x)^2+(H-y)^2}}\approx\frac{L}{\sqrt{L^2+H^2}}(1+\frac{x}{L}+\frac{-Lx+Hy}{L^2+H^2})+\frac{1}{2}\left[-3L(L^2+H^2)^{-3/2}+3L^3(L^2+H^2)^{-5/2}\right]x^2$$

$$+\frac{1}{2}\left[-L(L^2+H^2)^{-3/2}+3LH^2(L^2+H^2)^{-5/2}\right]y^2+\left[H(L^2+H^2)^{-3/2}-3L^2H(L^2+H^2)^{-5/2}\right]xy, \tag{4}$$

$$\sin\beta=\frac{H-y}{\sqrt{(L+x)^2+(H-y)^2}}\approx\frac{H}{\sqrt{L^2+H^2}}(1-\frac{y}{H}+\frac{-Lx+Hy}{L^2+H^2})+\frac{1}{2}\left[-H(L^2+H^2)^{-3/2}+3L^2H(L^2+H^2)^{-5/2}\right]x^2$$

$$+\frac{1}{2}\left[-3H(L^2+H^2)^{-3/2}+3H^3(L^2+H^2)^{-5/2}\right]y^2+\left[L(L^2+H^2)^{-3/2}-3LH^2(L^2+H^2)^{-5/2}\right]xy. \tag{5}$$

Substitution of Equations (2)-(5) into Equations (1), approximately regarding that $F_1=F_2=f$, $F=2f\cos\alpha_0$, $\cos\alpha_0=H/\sqrt{L^2+H^2}$, $F=2fH/\sqrt{L^2+H^2}$, gives

$$\begin{aligned} &\ddot{x}+\omega_x^2x-axy=0 \\ &\ddot{y}+\omega_y^2y-bx^2-cy^2=0 \end{aligned} \tag{6}$$

where $\omega_x^2=2fH^2/\left[M(L^2+H^2)^{3/2}\right], \omega_y^2=2fL^2/\left[M(L^2+H^2)^{3/2}\right]$, $a=2f\left[-H(L^2+H^2)^{-3/2}+3L^2H(L^2+H^2)^{-5/2}\right]/M$, $b=f\left[-H(L^2+H^2)^{-3/2}+3L^2H(L^2+H^2)^{-5/2}\right]/M$, $c=f\left[-3H(L^2+H^2)^{-3/2}+3H^3(L^2+H^2)^{-5/2}\right]/M$.

The equivalence autonomous nonlinear system like $\dot{x}=X(x)$ of Equations (6) is

$$\begin{aligned} &\dot{x}_1=x_2 \\ &\dot{x}_2=-\omega_x^2x_1+ax_1y_1 \\ &\dot{y}_1=y_2 \\ &\dot{y}_2=-\omega_y^2y_1+bx_1^2+cy_1^2 \end{aligned} \tag{7}$$

In terms of Taylor expansion, it approximates the following first-order linear system

$$\dot{x}=Ax \tag{8}$$

Where $$A=\left.\frac{\partial X(x)}{\partial x}\right|_{x=0}=\begin{bmatrix}\frac{\partial X_1(x)}{\partial x_1}\cdots\frac{\partial X_1(x)}{\partial x_n}\\ \cdots\ \cdots\ \cdots\\ \frac{\partial X_n(x)}{\partial x_1}\cdots\frac{\partial X_n(x)}{\partial x_n}\end{bmatrix}_{x=0}=\begin{bmatrix}0&1&0&0\\ -\omega_x^2&0&0&0\\ 0&0&0&1\\ 0&0&-\omega_y^2&0\end{bmatrix}$$

is the Jacobi matrix of the system, i.e., the first-order linear system is

$$\begin{aligned}\dot{x}_1&=x_2\\ \dot{x}_2&=-\omega_x^2x_1\\ \dot{y}_1&=y_2\\ \dot{y}_2&=-\omega_y^2y_1\end{aligned}\tag{9}$$

In the phase space of $(x_1,x_2)=(x_1,\dot{x}_1)$ and $(y_1,y_2)=(y_1,\dot{y}_1)$, the phase trajectories of Equations (9) are the integrals of the following expressions

$$\frac{dx_2}{dx_1}=-\omega_x^2\frac{x_1}{x_2},\ \frac{dy_2}{dy_1}=-\omega_y^2\frac{y_1}{y_2}$$

i.e., $$\frac{1}{2}x_2^2+\frac{1}{2}\omega_x^2x_1^2=H_x,\quad \frac{1}{2}y_2^2+\frac{1}{2}\omega_y^2y_1^2=H_y\tag{10}$$

Where the integral constants H_x and H_y determined by the initial state, are respectively conservative, representing the horizontal and vertical energy of the system. $x_2^2/2$ and $y_2^2/2$ are the horizontal and vertical kinetic energy of the system, while $\omega_x^2x_1^2/2$ and $\omega_y^2y_1^2/2$ are the potential energy of the two directions. In the phase space, Equations (10) are respectively a set of closed elliptical curves around the center $(0,0)$ as shown ones in Figure 3. The motion of the convergence point is stable, though not asymptotically stable.

3. Stability and Control of the Convergence Point in the Asymmetric Two-strand Yarn Spinning

3.1 Stability of the Convergence Point in the Asymmetric Two-strand Yarn Spinning

For the asymmetric case, i.e. Sirofil spinning, the dynamics of the system is quite different. The schematic diagram is shown as Figure 4.

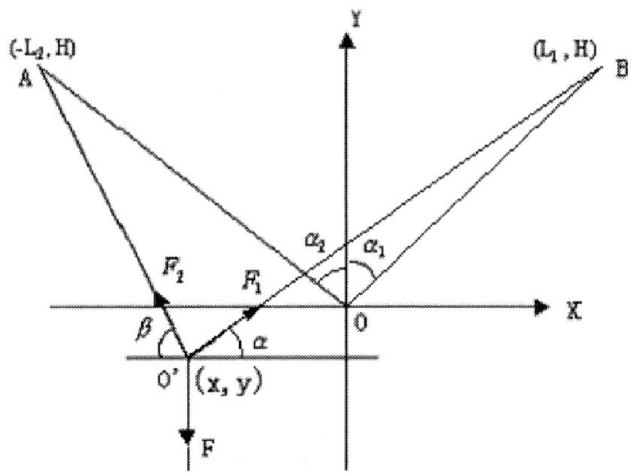

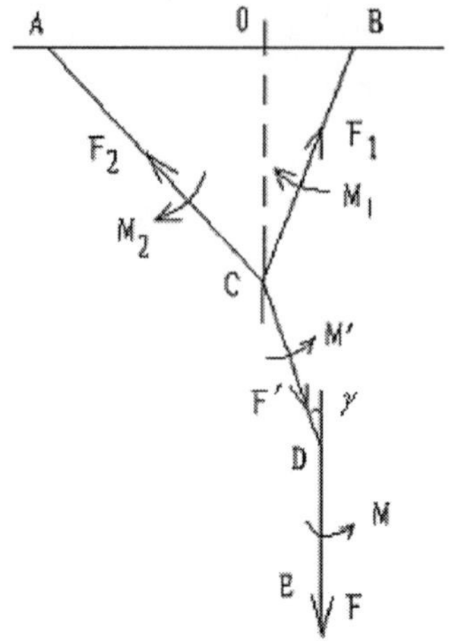

Figure 4. The asymmetric illustration of two-strand spun;
Figure 5. The tension compensator control diagram of Sirofil spinning

The equations of motion are established as the symmetric ones, i.e. Equations (1). Also applying the binomial theorem to expand the square-root terms, substitution of the corresponding $\cos\alpha$, $\sin\alpha, \cos\beta, \sin\beta$ into Equations (1) gives

$$\begin{aligned} &\ddot{x}+\omega_x^2 x+dy+a_1x^2+\frac{1}{2}by^2+cxy=0 \\ &\ddot{y}+\omega_y^2 y+dx+\frac{1}{2}cx^2+a_2y^2+bxy=0 \end{aligned} \tag{11}$$

where $\omega_x^2=\left(F_1\cos^3\alpha_1+F_2\cos^3\alpha_2\right)/MH$, $\omega_y^2=\left(F_1\sin^2\alpha_1\cos\alpha_1+F_2\sin^2\alpha_2\cos\alpha_2\right)/MH$, $d=\left(F_2\sin\alpha_2\cos^2\alpha_2-F_1\sin\alpha_1\cos^2\alpha_1\right)/MH$, $a_1=\left(3F_1\sin\alpha_1\cos^4\alpha_1-3F_2\sin\alpha_2\cos^4\alpha_2\right)/2MH^2$, $a_2=\left(3F_1\sin^2\alpha_1\cos^3\alpha_1+3F_2\sin^2\alpha_2\cos^3\alpha_2\right)/2MH^2$, $b=\left[-F_1\sin\alpha_1\cos^2\alpha_1(3\cos^2\alpha_1-1)+F_2\sin\alpha_2\cos^2\alpha_2(3\cos^2\alpha_2-1)\right]/MH^2$, $c=\left[-F_1\cos^3\alpha_1(3\sin^2\alpha_1-1)-F_2\cos^3\alpha_2(3\sin^2\alpha_2-1)\right]/MH^2$, and $\cos\alpha_1=H/\sqrt{L_1^2+H^2}$, $\sin\alpha_1=L_1/\sqrt{L_1^2+H^2}$, $\cos\alpha_2=H/\sqrt{L_2^2+H^2}$, $\sin\alpha_2=L_2/\sqrt{L_2^2+H^2}$.

The equivalence autonomous nonlinear system like $\dot{x}=X(x)$ of Equations (11) is

$$\begin{aligned} &\dot{x}_1=x_2 \\ &\dot{x}_2=-\omega_x^2x_1-dy_1-a_1x_1^2-\frac{1}{2}by_1^2-cx_1y_1 \\ &\dot{y}_1=y_2 \\ &\dot{y}_2=-\omega_y^2y_1-dx_1-\frac{1}{2}cx_1^2-a_2y_1^2-bx_1y_1 \end{aligned} \tag{12}$$

In terms of Taylor expansion, it approximates the following first-order linear system

$$\dot{x}=Ax \tag{13}$$

Where

$$A=\left.\frac{\partial X(x)}{\partial x}\right|_{x=0}=\begin{bmatrix}\frac{\partial X_1(x)}{\partial x_1}\cdots\frac{\partial X_1(x)}{\partial x_n}\\ \cdots\ \cdots\ \cdots\\ \frac{\partial X_n(x)}{\partial x_1}\cdots\frac{\partial X_n(x)}{\partial x_n}\end{bmatrix}_{x=0}=\begin{bmatrix}0 & 1 & 0 & 0\\ -\omega_x^2 & 0 & -d & 0\\ 0 & 0 & 0 & 1\\ -d & 0 & -\omega_y^2 & 0\end{bmatrix}$$

is the Jacobi matrix of the system, its characteristic equation is

$$|A-\lambda E|=\begin{vmatrix}-\lambda & 1 & 0 & 0\\ -\omega_x^2 & -\lambda & -d & 0\\ 0 & 0 & -\lambda & 1\\ -d & 0 & -\omega_y^2 & -\lambda\end{vmatrix}=\lambda^4+(\omega_x^2+\omega_y^2)\lambda^2+\omega_x^2\omega_y^2-d^2=0 \quad (14)$$

So the eigenvalues can be solved by

$$\lambda^2=\frac{-(\omega_x^2+\omega_y^2)\pm\sqrt{(\omega_x^2+\omega_y^2)^2-4(\omega_x^2\omega_y^2-d^2)}}{2} \quad (15)$$

If $\omega_x^2\omega_y^2<d^2,\lambda_{1,2}^2<0,\lambda_{3,4}^2>0$, one of the four eigenvalues λ is positive, thus the system is unstable; If $\omega_x^2\omega_y^2>d^2,\lambda_{1,2}^2<0,\lambda_{3,4}^2<0$, the eigenvalues are two pairs of pure imaginary, and the equilibrium point is called center. The movement of the convergence point is stable, though not asymptotically stable. Thus, it is known that bifurcation occurs when $\omega_x^2\omega_y^2=d^2$.

3.2 Control of the convergence point in the asymmetric two-strand yarn spinning

We then discuss the movement of the convergence point after tension compensator (TC) control. As shown in Figure 5, C is the convergence point, D is the point where the TC applies.
Analyzing the undergoing force at points C and D respectively, the corresponding equations of motion of convergence point C are

$$\begin{aligned}&M\ddot{x}+F_2\cos\beta-F_1\cos\alpha-F\tan\gamma=0\\&M\ddot{y}+F-F_1\sin\alpha-F_2\sin\beta=0\end{aligned} \quad (16)$$

Where γ is the 'β' arranged in the experiments [7] as shown in Figure 5. Expanding the Equations (16), the equivalence autonomous nonlinear expressions for the controlled system in terms of Taylor expansion approximate the same as Equations (13), but with different values of $\omega_x,\omega_y,a_1,a_2,b,c,d$ for the varying of F_1,F_2 when the tension compensator applied.

As we have known that the convergence point may occur large vibration and be unstable (i.e. $\omega_x^2\omega_y^2<d^2$) in the asymmetric two-strand yarn spinning, which do not appear in the symmetric case. Applying tension compensator (TC) to the system, and adjusting the γ, i.e. correspondingly the F_1,F_2 to some value may make the system stable, i.e. $\omega_x^2\omega_y^2>d^2$. So the system may turn out to be stable, and the effect of TC control can be apparently seen.

In the experiments [7], the properties of the hairiness and abrasion-resistance of Sirofil composite yarn with and without Tension Compensator control are shown in Table 1 and Fig.2-3, including hairiness quantity versus length or yarn configuration experiments with the same filament pre-tension and with different filament pre-tension, and shown in Table 2-5, including Two-sample average analysis, before and after yarn steaming of the abrasion-resistance experiments with and without TC. In general, the experiments concluded that the adoption of TC greatly reduce the hairiness of Sirofil composite yarn, and enormously improve the property of abrasion-resistance, which are consistent very well with the present theoretic derivation and analysis.

4. Conclusions

Following our work in the static aspect [7], we sequentially and kinetically disclosed the mechanism of TC control to Sirofil spinning. It is once more concluded that the TC control can increase the stability of the convergence point in Sirofil composite area, and correspondingly improve the Sirofil yarn properties.

5. References

[1] Yu Y P 2005 *Study on Performance and Mechanic Analysis of Compound Triangle Area for Siro-series Spinning Technology* (China: Donghua University Doctoral Dissertation)
[2] He J H, Yu Y P, Pan N, Cai X C, Yu J Y and Wang S Y 2005 Quasistatic Model of Two-strand Yarn Spinning *Mechanics Research Communications* 32(2) 197-200
[3] He J H, Yu Y P, Yu J Y, Li W R, Wang S Y and Pan N 2005 A Linear Dynamic Model for Two-strand Yarn Spinning *Textile Research Journal* 75(1) 87-90
[4] He J H, Yu Y P, Yu J Y, Li W R, Wang S Y and Pan N 2005 A Nonlinear Dynamic Model for Two-strand Yarn Spinning *Textile Research Journal* 75(2) 181-184
[5] Yu Y P, Liu W Y, Cheng L D, He J H, Yu J Y and Wang S Y 2005 A New Static Model for Sirofil Spinning *Int. Conf. on Advanced Fibers and Polymer Materials Shanghai, China* p 1277-1280
[6] Liu W Y 2010 Comment on He, J.H. et al., Quasistatic Model for Two-strand Yarn Spinning *Mechanics Research Communications* 37(1) 129
[7] Liu W Y, Yu Y P, He J H and Wang S Y 2007 Effect of Tension Compensator on Sirofil Yarn Properties *Textile Research Journal* 77(4) 195-199

Theoretical Studies of the Electron Paramagnetic Resonance Parameters and Local Structure for VO^{2+} in Oxyfluoroborate Glasses

Huaming Zhang[1, 2, a], Xiaopeng Yu[1, 2] and Wenbo Xiao[1, 2]

[1]Jiangxi Engineering Laboratory for Optoelectronics Testing Technology, Nanchang Hangkong University, Nanchang 330063, China
[2]Key Laboratory of Nondestructive Testing, Ministry of Education, Nanchang Hangkong University, Nanchang 330063, China
[a]Corresponding author: zhm2071@163.com

Abstract: The electron paramagnetic resonance parameters (*g* factors $g_{\parallel}$, $g_{\perp}$ and hyperfine structure constants $A_{\parallel}$, $A_{\perp}$) of a tetragonal V^{4+} center in oxyfluoroborate glasses ($20Li_2O$-$10Li_2F_2$-$70B_2O_3$) are theoretically investigated by using the perturbation formulas for a $3d^1$ ion in tetragonally compressed octahedra. The calculated results are in good agreement with the experimental data. Local structure parameters of $[VO_6]^{8-}$ clusters are obtained from the calculation (*i.e.*, $R_{\parallel} \approx 1.74$ Å and $R_{\perp} \approx 1.985$ Å for the metal-ligand distances parallel and perpendicular to the C_4 axis, respectively). It is shown that the local structure around the V^{4+} ion possesses a compressed tetragonal distortion along C_4 axis. The signs of the hyperfine structure constants $A_{\parallel}$ and $A_{\perp}$ for V^{4+} centers in oxyfluoroborate glasses were also suggested in the discussion.

1. Introduction

In recent years glasses doped with transition-metal (TM) ions have attracted a great deal of attraction because of their potential applications in the development of new tunable solid-state lasers, electro-optic switches, solar energy convectors, and fiber optic communication devices [1-6]. Usually, these properties are believed to be sensitive to the internal electric and defect structures produced by impurities. Since electron paramagnetic resonance (EPR) spectroscopy is a sensitive technique for the study local structures of transition-metal (TM) ions in solids. Many studies on EPR spectra of single TM ions in borate glasses have been made of recent years [7-10]. Amongst these TM ions, vanadyl ion (VO^{2+}) is the most stable and widely used EPR probe for probing the site symmetry of the central metal ion and the nature of the bonding with the ligands and this has also been used as an impurity for understanding the defect, structural, phase transition and orientation properties of the host lattice [1, 11, 12]. Because of the strong V^{4+}-O^{2-} bonding in VO^{2+} ion, most of the VO^{2+} complexes in the crystals posses tetragonally compressed (along theV^{4+}-O^{2-} bonding) octahedral or square pyramid symmetry (C_{4v}). This point has been confirmed by the optical and EPR spectral experiments for VO^{2+} complexes in many crystals [7, 8, 11]. For example, the EPR experiments were carried out for VO^{2+} doped oxyfluoroborate glasses ($20Li_2O$-$10Li_2F_2$-$70B_2O_3$), and the EPR parameters (*i.e.*, g factors $g_{\parallel} = 1.9370$, $g_{\perp}= 1.9802$ and hyperfine structure constants $A_{\parallel} = 176$, $A_{\perp}= 57\times10^{-4}$ cm^{-1}) were also measured [13]. As is known, an octahedral complex with a tetragonal compression would give $g_{\parallel} < g_{\perp} < g_s$ and $|A_{\parallel}| > |A_{\perp}|$, where g_s is the free-spin *g* value of 2.0023. The observed values of the EPR parameters agree

Content from this work may be used under the terms of the Creative Commons Attribution 3.0 licence. Any further distribution of this work must maintain attribution to the author(s) and the title of the work, journal citation and DOI.
Published under licence by IOP Publishing Ltd

with the relation. That is to say, VO^{2+} complexes in oxyfluoroborate glasses present in tetragonally-compressed octahedral.

However, until now, no satisfactory interpretation to the above experimental results has been made, and information about defect structure for the impurity V^{4+} center has not been obtained. Considering that microscopic mechanisms of the EPR parameters and information of impurity local structures can be helpful to understand the properties of this material with TM dopants, further studies on the above EPR parameters are of scientific and technical significance. In this work, the third-order perturbation formulas of these EPR parameters for a $3d^9$ ion in tetragonally compressed octahedra are applied. From the calculations, the experimental EPR parameters are explained and some useful information of defect structures for the impurity V^{4+} center in oxyfluoroborate glasses can be acquired.

2. Calculation

When VO^{2+} doped into oxyfluoroborate glasses, the V^{4+} ions lie in the interspace of the three-dimensional lattice and are surrounded by six oxygen atoms. For a V^{4+} ($3d^1$) ion in a tetragonally distorted octahedron, the energy level ^2D of the free $3d^1$ ion in a octahedral crystal field will split into ^2E_g and $^2T_{2g}$ levels [14]. In the C_{4v} crystal field, the $^2T_{2g}$ level is further split into $^2B_{2g}$ and ^2E_g levels, and the ^2E_g level into $^2B_{1g}$ and $^2A_{1g}$ levels [14]. According to the crystal-field theory and the experimental EPR parameters, one can conclude that the ground state for V^{4+} in oxyfluoroborate glasses is ^2B_2g. Utilizing the perturbation method, the third-order perturbation formulas of the EPR parameters for $3d^1$ ions in tetragonal symmetry with the ground state $^2B_{2g}$ can be derived from the perturbation theory as [15]:

$$g_{\parallel} = g_s - \frac{8k\varsigma}{E_2} - \frac{(k+g_s)\varsigma_d^2}{E_2^2} + \frac{4k\varsigma^2}{E_1E_2}$$

$$g_{\perp} = g_s - \frac{2k\varsigma}{E_1} + \frac{(k-g_s)\varsigma^2}{E_1^2} - \frac{2g_s\varsigma^2}{E_2^2}$$

$$A_{\parallel} = P[-\kappa - \frac{4N}{7} + (g_{\parallel} - g_s) + \frac{3(g_{\perp} - g_s)}{7}]$$

$$A_{\perp} = P[-\kappa + \frac{2N}{7} + \frac{11(g_{\perp} - g_s)}{14} \qquad (1)$$

In the above formulas, k is the orbital reduction factor, which is equivalent to the covalency factor N, a characteristic of the covalency between the central ion and the ligands. ζ and P are, respectively, the spin-orbit coupling coefficient and the dipolar hyperfine structure parameter for the $3d^9$ ion in crystals. κ is the isotropic core polarization constant. E_1 and E_2 are the energy separations between the excited $^2B_{2g}$ and ^2E_g and the ground $^2B_{1g}$ states under tetragonal crystal fields. Based on the crystal-field theory, they can be calculated from the energy matrices for a $3d^1$ ion under tetragonal symmetry in terms of the cubic field parameter D_q and the tetragonal field parameters D_s and D_t and can be expressed as:

$$E_1 = E(^2E) - E(^2B_2) = -3D_s + 5D_t$$

$$E_2 = E(^2B_1) - E(^2B_2) = 10D_q \qquad (2)$$

From the superposition model [16], the cubic and tetragonal field parameters can be determined as:

$$D_q = \frac{4}{3}\bar{A}_4(R_0),$$

$$D_s = \frac{4}{7}\bar{A}_2(R_0)[(\frac{R_0}{R_{\parallel}})^{t_2} - (\frac{R_0}{R_{\perp}})^{t_2}],$$

$$D_t = \frac{8}{21}\bar{A}_2(R_0)[(\frac{R_0}{R_\parallel})^{t_4} - (\frac{R_0}{R_\perp})^{t_4}]. \quad (3)$$

Here, $t_2 \approx 3$ and $t_4 \approx 5$ are the power-law exponents [14, 15, 17]. $R_\parallel$ is the impurity-ligand distance along the C_4 axis, $R_\perp$ is the bonding length between V^{4+} and the original planar oxygen ions. R_0 ($\approx R_\perp \approx$ 0.1985 nm [18]) is the reference distance. $\bar{A}_2(R_0)$ and $\bar{A}_4(R_0)$ are the intrinsic parameters with the reference distance R_0. For $3d^n$ ions in octahedra, the ratio $\bar{A}_2(R_0)/\bar{A}_4(R_0)$ is in the range of 9 ~ 12 in many crystals [14, 15, 17], and we take the average value $\bar{A}_2(R_0)/\bar{A}_4(R_0) \approx 10.5$ here. Thus, the EPR parameters is connected with the cubic and tetragonal field parameters and hence with the local structure of the studied systems.

Taking into account the covalency effect (characterized by the covalence reduction factor N), the spin–orbit coupling parameter ζ and dipolar hyperfine constant P can be given as [15]

$$\varsigma = N^2 \varsigma_0, \qquad P = N^2 P_0 \quad (4)$$

Hence, the spin-orbit coupling coefficient ζ and the dipolar hyperfine structure parameter P can be acquired for the studied systems by using the free-ion data ζ_0 ($\approx$ 248 cm^{-1} [19]) and P_0 ($\approx 172 \times 10^{-4}$ cm^{-1} [20]) for V^{4+} ion. From the optical absorption spectra for VO^{2+} doped in $C_3H_7NO_2$ powder with the same $[VO_6]^{8-}$ octahedral clusters, the cubic field parameter $D_q \approx 1466$ cm^{-1} [16] can be obtained and adopted in this work. The core polarization constant κ lies between 0.6 and 1 for VO^{2+} (or V^{4+}) in various oxide glasses and is sensitive to the even small deformations of the metal electron orbitals [14, 15, 17]. Here, we take $\kappa \approx 0.71$.

Thus, only the impurity-ligand distance $R_\parallel$ and the covalency factors N (*i.e.*, the orbital reduction factor k) are unknown in the formulas of the EPR parameters. Substituting these parameters into Eq.(1) and fitting the calculated EPR parameters to the experimental values, one can obtain:

$$R_\parallel \approx 0.174nm \qquad and \qquad N \approx 0.86 \quad (5)$$

The corresponding theoretical results of the spin Hamiltonian parameters are compared with the observed values in Table 1.

Table 1. The calculated and experimental spin-Hamiltoian parameters g factors and the hyperfine structure constants (in $10^{-4} cm^{-1}$) for VO^{2+} in oxyfluoroborate glasses ($20Li_2O$-$10Li_2F_2$-$70B_2O_3$).

Glasses		$g_\parallel$	$g_\perp$	$A_\parallel$	$A_\perp$
$20Li_2O$-$10Li_2F_2$-$70B_2O_3$	Cal.	1.9370	1.9698	-177.1	-58.6
	Expt. [13]	1.9370	1.9700	176	61

3. Discussion

From Table 1, one can be seen that the calculated EPR parameters (g factors $g_\parallel$, $g_\perp$ and the hyperfine structure constants $A_\parallel$, $A_\perp$) for VO^{2+} in oxyfluoroborate glasses based on the adopted parameters in Eq. (6) are in reasonable agreement with the experimental values. This indicates that the study method and the local structure parameters in Eq. (6) obtained in this work can be regarded as reasonable.

1) By fiting the calculated values to the experimental results, one can obtained that the impurity-ligand distance $R_\parallel$ ($\approx$ 1.74 Å) along the C_4 axis for V^{4+} center in oxyfluoroborate glasses, this means that the $[VO_6]^{8-}$ octahedron around V^{4+} ion is significantly compressed ($\Delta R = R_\perp - R_\parallel \approx 0.245$ Å) along the C_4 axis, which is consistent with the experimental EPR results (i.e. $g_\parallel < g_\perp < g_e$ and $|A_\parallel| > |A_\perp|$) under the ground state B_{2g}. The compressed distortion could be explained by Jahn–Teller effect. Because in compression, the ground state is orbit singlet, while in elongation, the ground state is orbit doublet, it can be expected that the latter is unstable. Similar local distortions were also found for the

$3d^1$ (VO^{2+}) ion in various oxides glasses. For example, $\Delta R \approx$ (0.395 Å) was found for VO^{2+} in Zn (antipyrine)$_2$(NO_3)$_2$ [21]. Therefore, the local distortion obtained in this study can be regard as reasonable.

2) The sign of the hyperfine structure constants is very difficult to ascertain [14, 15]. Thus, the experimental results of constants *A* reported in many crystals are actually the absolute values and they are often written as positive values [22-26]. Muncaster and Parke [27] have proved that the signs of $A_{\parallel}$ and $A_{\perp}$ should be negative for VO^{2+} ions in hydrated salts and glasses. In this paper, we found, that the signs of for VO^{2+} in oxyfluoroborate glasses also should be negative, thus, the signs of the $A_{\parallel}$ and $A_{\perp}$ obtained in this work can be regarded as suitable.

3) The validity of the covalency factors *N* adopted can be roughly illustrated from the empirical formula: $N^2 = 1 - h(L)k(M)$ [28], where the parameters $h(O^{2-}) \approx 1.0$ [28] is the characteristic of the ligand, and *k*(M) is the characteristic of the central metal ions. From the values $k(V^{2+}) \approx 0.1$ and $k(V^{3+}) \approx 0.15$ [28], one can reasonably obtain $k(V^{4+}) \approx 0.20$ here by extrapolation. Thus, we have *N* ($\approx$ 0.86). This value is the same as *N* ($\approx$ 0.86) adopted in this work.

4. Conclusion and

The experimental EPR parameters for V^{4+} center in oxyfluoroborate glasses are theoretically investigated from the perturbation formulas for a $3d^1$ ion in tetragonally compressed octahedra. From the calculations, the bond length are $R_{\parallel} \approx 0.174$ *nm* for $[VO_6]^{8-}$ octahedral clusters. The negative signs of the calculated $A_{\parallel}$ and $A_{\perp}$ are also suggested in this work.

5. Acknowledgement

This work was financially supported by Youth Science Foundation of Jiangxi Province (grant No. 20132BA212002) and Chinese Natural Science Foundation (grants 11304145, 11764028).

6. References

[1] A. Agarwal, V. P. Seth, P. S. Gahlot, S. Khasa, and P. Chand 2003 J. Phys. Chem. Solids **64** 2281.
[2] S. P. Singh, R. P. S. Chakradhar, J. L. Rao, and B. Karmakar 2013 J. Magn. Magn. Mater. **346** 21.
[3] S. Y. Marzouk, N. A. Elalaily, F. M. Ezz-Eldin, and W. M. Abd-Allah 2006 Physica B **382** 340.
[4] C. Lin, L. Li, S. Dai, C. Liu, Z. Zhao, C. Bocker, and C. Rüssel 2016 J. Phys. Chem. C **120** 4556.
[5] S.Y. El-Zaiat, M. Medhat, M. F. Omar and M. A. Shirif 2016 Opt. Commun. **370** 176.
[6] Samdani, G. Ramadevudu, M. N. Chary and M. Shareefuddin 2017 Mater. Chem. Phys. **186** 382.
[7] E. Bozkurt, B. Karabulut, İ. Kartal, 2009 Spectrochim. Acta A **73** 163-167.
[8] L.D. Bogomolova, T. F. Dolgolenko, V. A. Tachkin, and V.N Lazukin 1969 J. Magn. Res. **15** 283.
[9] G. Srinivas, J. S. Kumar, Md. Shareefuddin, M. N. Chary and R. Sayanna 2016 Glass Phys. Chem. **42** 141.
[10] A Yadav, S Khasa, A Hooda, MS Dahiya, A. Agarwal, P. Chand 2016 Spectrochim. Acta A **157** 129.
[11] P. G. Prakash, and J. L. Rao 2005 Spectrochim. Acta A **61** 2595.
[12] H. M. Zhang, X. Wan and Z. M. Zhang 2014 Physica B **407** 2201.
[13] M. V. Ramana, P. S. Lakshmi, S. Rahman, K. Sivakumar, and G. S. Sastry 1991Mat. Sci. Eng. **10** 25.
[14] H. M. Zhang, W. B. Xiao, and X. Wan 2014 Physica B **449** 225.
[15] C. Y. Li and X. M. Zheng 2014 Acta Phys. Pol. A **125** 73.
[16] D. J. Newman, and B. Ng 1989 Rep. Prog. Phys. **52** 699.
[17] J. Z. Lin, S. Y. Wu, Q. Fu, H. M. Zhang 2006 Radiat. Eff. Defects Solids **161** 571.

[18] G. Meunier, M. Bertaud 1974 J. Acta Crystallogr. B **30** 2834.
[19] J. S. Griffith. *The Theory of Transition-Metal Ions* (Cambridge University Press, London 1964.)
[20] B. R. McGarvey 1967 J. Phys. Chem. **71** 51.
[21] L. H. Xie, G. P. Zhao, M. Qiu 2010 J. Magn. Magn. Mater. **332** 2184.
[22] M. Q. Kuang, S. Y. Wu, G. L. Li, X. F. Hu 2015 Mol. Phys. **113** 698.
[23] H. M. Zhang, J. H. Duan, W. B. Xiao and X. Wan 2015 J. Non-Cryst. Solids **425** 173.
[24] W. L. Feng 2007 Phil. Mag. Lett. **87** 663.
[25] C. Y. Li, S. F. Liu, J. X. Fu 2016 Zeitschrift für Naturforschung A **71** 396.
[26] P. Zhou, L. Li 2015 Zeitschrift für Naturforschung A **113** 159.
[27] R. Muncaster, and S. Parke 1977 J. Non-Cryst. Solids **24** 399.
[28] K. H. Karlsson and T. Perander 1973 Chem. Scripta **3** 201.

IOP Conf. Series: Materials Science and Engineering **275** (2017) 012032 doi:10.1088/1757-899X/275/1/012032

Microstructure, Thermal, Mechanical, and Dielectric Properties of BaO–CaO–Al_2O_3–B_2O_3–SiO_2 Glass-Ceramics

Bo Li*[1,2,3], Haibo Bian[1] and Yi Fang[1]

[1]School of Microelectronics and Solid-State Electronics, University of Electronic Science and Technology of China, Chengdu 610054, China
[2]State Key Laboratory of Electronic Thin Films and Integrated Devices, Chengdu 610054, China
[3]National Engineering Research Center of Electromagnetic Radiation Control Materials, Chengdu 610054, China
Email: boli@uestc.edu.cn

Abstract: BaO–CaO–Al_2O_3–B_2O_3–SiO_2 (BCABS) glass-ceramics were prepared via the method of controlled crystallization. The effect of CaO modification on the microstructure, phase evolution, as well as thermal, mechanical, and dielectric properties was investigated. XRD identified that quartz is the major crystal phase; cristobalite and bazirite are the minor crystal phases. Moreover, the increase of CaO could inhibit the phase transformation from quartz to cristobalite, but excessive CaO would increase the porosity of the ceramics. Additionally, with increasing the amount of CaO, the thermal expansion curve tends to be linear, and subsequently the CTE value decreases gradually, which is attributed to the decrease of cristobalite with high CTE and the formation of $CaSiO_3$ with low CTE. The results indicated that a moderate amount of CaO helps attaining excellent mechanical, thermal, and dielectric properties, that is, the specimen with 9 wt% CaO sintered at 950 °C has a high CTE value (11.5 × 10^{-6}/°C), a high flexural strength (165.7 MPa), and good dielectric properties ($\varepsilon_r = 6.2$, $\tan\delta = 1.8 \times 10^{-4}$, $\rho = 4.6 \times 10^{11}$ Ω·cm).

1. Introduction

Low temperature co-fired ceramics (LTCC) is a well known technology, which has been attracted widely for fabrication of multilayer substrates and the application of semiconductor devices packaging, e.g. ceramic ball grid arrays (CBGA) for flip chip packages [1,2]. In order to meet the requirement for LTCC technology, the desired packaging materials need to possess low dielectric constant and loss ($\varepsilon_r < 7$, $\tan\delta < 1 \times 10^{-3}$), high mechanical strength (≥150 MPa), and a high coefficient of thermal expansion (CTE > 10 × 10^{-6}/°C) [3,4]. Therefore, the thermal match, to realize the high reliability [5], can be achieved between the packaging substrate and the printed circuit board (PCB, CTE = 12–16 × 10^{-6}/°C).

Recently, Li et al. [6] reported that BaO–Al_2O_3–B_2O_3–SiO_2 (BABS) glass-ceramics own good dielectric properties and high values of CTE. It was found that the thermal property play an important role in the packaging materials for LTCC applications. Especially, for the BABS glass-ceramic, the occurrence of phase transformation of α–cristobalite/β–cristobalite will lead to the nonlinear variation of thermal expansion [7], which will result in the mechanical failure by the thermal mismatch. In fact, some researchers successfully synthesized the composites with high CTE by using low softening-point glass as the flux agent and quartz powders as the filler [8,9]. Nevertheless, the doping modifications

Content from this work may be used under the terms of the Creative Commons Attribution 3.0 licence. Any further distribution of this work must maintain attribution to the author(s) and the title of the work, journal citation and DOI.
Published under licence by IOP Publishing Ltd

and the crystalline behaviors of these composites have not been extensively investigated yet. As we know, the calcium borosilicate glass-ceramics not only have a moderate CTE value, but also exhibit a linear trend of the thermal expansion [10]. To improve the mechanical and thermal properties of such material, that CaO partly replacing BaO was applied to research the doping behavior as well as the property change of BCABS glass-ceramics.

In the present work, we focus on the influence of CaO modification on the microstructures and the crystalline phases, as well as the mechanical, thermal, and dielectric properties of BCABS glass-ceramics. More importantly, the aim of this paper is to improve the thermal property and the mechanical property, and finally to achieve a novel glass-ceramic BaO–CaO–Al_2O_3–B_2O_3–SiO_2 (BCABS) as a candidate for CBGA packages.

2. Experimental Procedures

BCABS glass-ceramics were synthesized by the solid-state method, and the starting materials are analytical reagent $Ba(OH)_2 \cdot 8H_2O$, $Ca(OH)_2$, $Al(OH)_3$, H_3BO_3, and H_3SiO_3. The chemical compositions of this study were listed in Table 1. The powders were mixed adequately using the deionized water via ball-milling for 7 h. After been dried at 100 °C, these powders were preheated at 700 °C for 3 h in air. Then the powders were pressed into disks (Φ 23.6 mm × 1 mm) and bars (50 mm × 4 mm × 3 mm) at a compressive stress of 25 MPa. Thereinto, 5 wt% acrylic emulsion was used for the previous prilling process. The samples subjected to the controlled crystallization were sintered constantly from room temperature to 400 °C at a heating rate of 2 °C/min for 3 h to remove the organic binder, then heated up to 950 °C and annealed for 1 h at this temperature.

Table 1. Compositions of BaO–CaO–Al_2O_3–B_2O_3–SiO_2 glass-ceramics (wt%)

Sample	BaO	CaO	Al_2O_3	B_2O_3	SiO_2
C1	32	3	5	12	48
C2	30	5	5	12	48
C3	28	7	5	12	48
C4	26	9	5	12	48
C5	24	11	5	12	48
C6	22	13	5	12	48

X–ray diffractometer (XRD, PANALYTICAL PW3040/60, Netherlands) was employed to identify the crystalline phases of sintered samples, using Cu Kα radiation, in the 2θ range from 20° to 80°. Microstructures of cross–section were characterized by a scanning electron microscopy (SEM, JEOL 6490LV, Japan). The coefficient of thermal expansion of glass-ceramics was measured within the temperature range from 25 to 600 °C at a heating rate of 5 °C/min using a thermal dilatometer (NETZSCH 401 PC, Germany). The bulk density of sintered samples was detected by the precision density balance (AND GF300D, Japan) using Archimedes method. The flexural strength was determined by an electromechanical universal testing machine (MTS CMT6104, China). Dielectric constant and loss of the disk samples with silver electrode were examined at 1 MHz using a precision LCR meter (AGILENT 4284A, USA).

3. Results and Discussions

XRD analysis was used to investigate the effect of CaO addition on the crystallization behavior of BCABS glass-ceramics, and the result of samples sintered at 950 °C for 1 h is presented in figure 1. XRD patterns detected that the major crystal phase is quartz (JCPDS 70-3755), and the minor phases are cristobalite (JCPDS 77-1317) and bazirite (JCPDS 29-0214). Apparently, the bazirite was crystallized from BaO, ZrO_2, and SiO_2. It was found that the intensity of X-ray peaks for each phase has a great relation with the increase of CaO content. Moreover, a new crystal phase wollastonite (JCPDS 75–1396) appears in sample C6 (13 wt% CaO). In the ternary phase system, CaO–B_2O_3–SiO_2, wollastonite ($CaSiO_3$) could be formed as a major phase at 800–900 °C [11]. Thus, excessive CaO could react with SiO_2 to generate the $CaSiO_3$ phase.

The further calculation was figured out by the whole pattern fitting (WPF) method, and the crystal content and the crystallinity of BCABS glass-ceramics were listed in Table 2. The error of calculation was controlled in the range from 1.35% to 3%. It is clearly seen that with the increase of CaO content, the crystallinity of the BCABS glass-ceramic firstly increased from 33.29% to 39.61%, and then decreased to 35.18%. This suggests that only a proper amount of CaO could facilitate the crystallization process of BCABS glass-ceramics. Moreover, the crystal content exhibited a strong dependence as a function of the CaO content. With increasing the amount of CaO, the content of quartz remarkably increased from 86.7 to 94.1 wt%, but that of cristobalite rapidly decreased from 8.5 to 2.0 wt%. The bazirite, in general, behaved the trend of gradual decline. Actually, the cristobalite was not only transformed from the quartz, but also precipitated from the silica glass. The result indicated that the addition of CaO could effectively restrain the transformation from quartz to cristobalite. Also, sample C6 revealed that the excessive CaO will lead to the formation of the new crystal phase $CaSiO_3$.

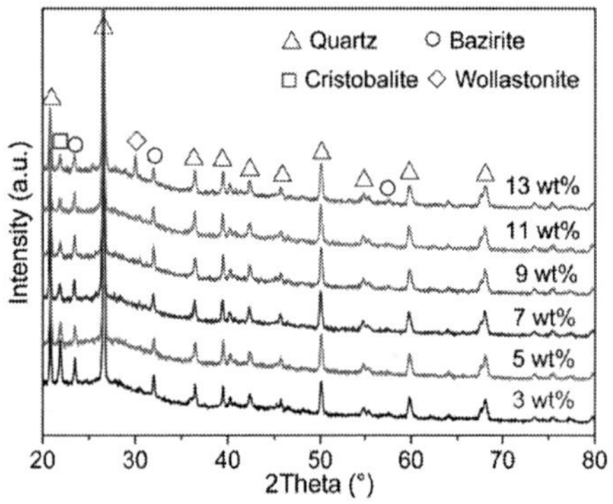

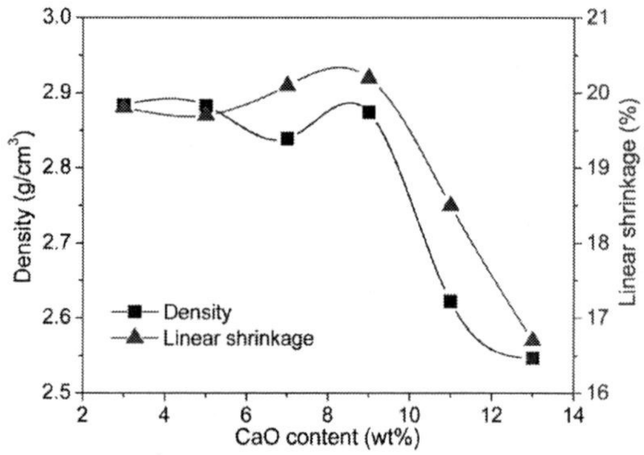

Figure 1. XRD patterns of BCABS glass-ceramics with different CaO content sintered at 950 °C.
Figure 2. Bulk density and linear shrinkage of BCABS glass-ceramics as a function of CaO content.

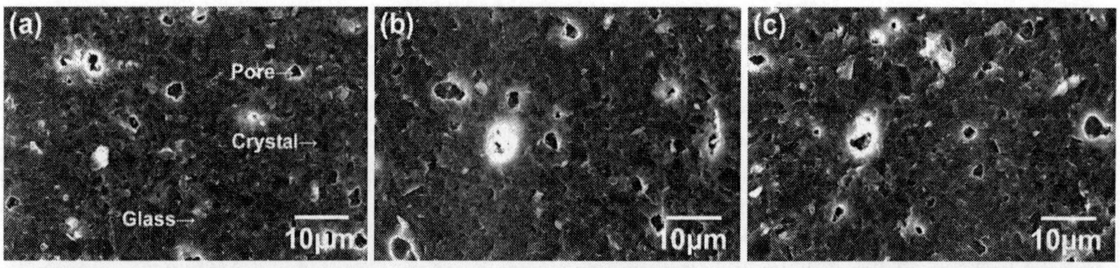

Figure 3. Cross-section SEM (SE mode) images of BCABS glass-ceramics with various CaO content sintered at 950 °C (a) 3 wt%, (b) 5 wt%, (c) 7 wt%, (d) 9 wt%, (e) 11 wt%, (f) 13 wt% CaO.

Table 2. Phase content and crystallinity calculated from XRD

Sample	Phase content (wt%)				Crystallinity (%)
	Quartz	Cristobalite	Bazirite	Wollastonite	
C1	86.7	8.5	4.8	–	33.3
C2	91.3	4.5	4.2	–	33.9
C3	92.9	3.4	3.7	–	34.7
C4	93.6	2.1	4.3	–	39.6
C5	94.1	2.0	4.0	–	35.3
C6	87.3	3.1	2.5	7.1	35.2

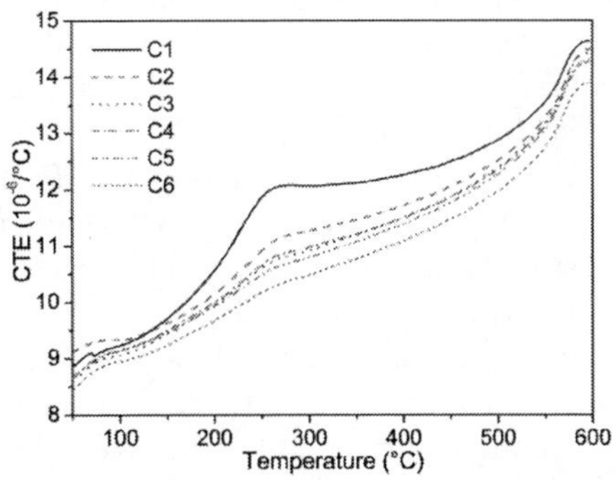

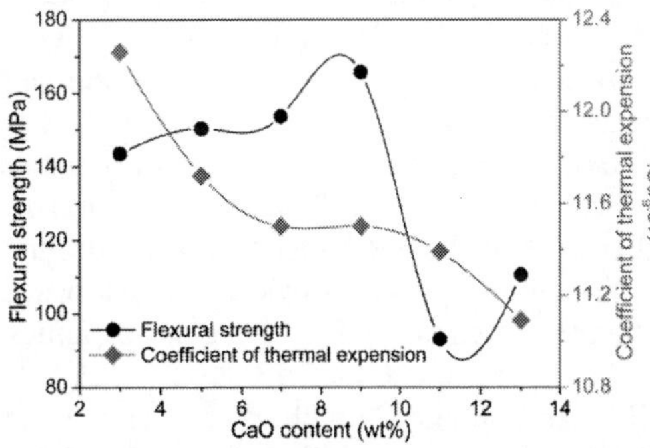

Figure 4. CTE (25 - 600 °C) of BCABS glass-ceramics as a function of CaO content.
Figure 5. Flexural strength and coefficient of thermal expansion (at 400 °C) of BCABS glass-ceramics versus CaO content.

In order to study the influence of CaO contents on the sintering behavior in detail, the bulk density and the linear shrinkage at 950 °C was shown in figure 2. For the specimens with ≤9 wt% CaO, the linear shrinkage slightly increased from 19.8% to 20.02%, while the bulk density almost kept in the range of 2.84–2.89 g/cm^3. However, both shrinkage and density decreased suddenly when the CaO content was more than 9 wt%. For specimen C6 (13 wt% CaO), the bulk density and the linear shrinkage reached the minimum value of 2.547 g/cm^3 and 16.7%, respectively. In fact, the crystal phase transformation from quartz to cristobalite is accompanied by a volume increase of approximate 17% [12]. As a result, the density and the shrinkage decreased markedly after extra CaO being added. Therefore, this result also proved that excessive CaO cannot promote the densification process of such system.

Figure 3 shows the cross-section SEM microphotographs (SE mode) of BCABS specimens with different CaO contents sintered at 950 °C. It is observed that the BCABS glass-ceramics is mainly composed of the crystal phase (dark gray particles) and the glass phase (light gray matrix). A few pores are randomly distributed in the microstructure. To a certain degree, the number of grains increased first and then decreased as the CaO addition increased, which is accordance with the crystallinity in Table 2. However, the number of pores in specimens C5-C6 (11-13 wt% CaO) is clearly more than that in specimens C1-C4 (3-9 wt% CaO) This finding indicates that BCABS samples with > 9 wt% CaO completed their sintering at higher temperatures, which also lead to the occurrence of their insufficient crystallization. Hence, the SEM image for Sample C4 (9 wt% CaO) demonstrated that an appropriate amount of CaO in BCABS will advantage the formation of the best microstructure.

To avoid the thermal mismatch between substrates and PCB, it is very important to investigate the thermal properties of BCABS glass-ceramics. Figure 4 illustrates the temperature dependence of CTE values from 25 to 600 °C for samples with various amount of CaO. Clearly, the nonlinear variation emerged on these curves for samples C1-C2, and the changing slope was observed around 220–270 °C [7]. In a previous study [12], BCABS glass-ceramic was found as a high expansion system because of the major crystal phases with high CTE value i.e. quartz (13 × 10^{-6}/°C) and cristobalite (50 × 10^{-6}/°C). Accordingly, this phenomenon can be explained reasonably through the phase transformation: α–cristobalite/β–cristobalite, which is accompanied by a dramatic expansion in volume. Nevertheless, with further increase of CaO content, the thermal expansion curves of samples became increasingly linear (samples C3-C6). This means that CaO could suppress the nonlinearity of the thermal expansion for the BCABS glass-ceramics, and thus decrease the thermal stress produced by the great volume change originated from the transformation of cristobalite, which will guarantee it to meet the requirements of IC package, such as thermal shock resistance and mechanical strength. Thus, it is believed that the increase of CaO content can effectively improve the thermal property of BCABS glass-ceramics. At the same time, the CTE values at 400 °C as a function of CaO content was shown in figure 5. Apparently, CTE values versus CaO content presents the tendency of gradual decrease from 12.26 to 11.09 ppm/°C. As we know, the value of CTE is directly related to the proportion of crystal phases and residual glass in specimens. As mentioned above, CaO additions can suppress the occurrence of phase transition from quartz to cristobalite. Consequently, the decrease in CTE of glass-ceramics was certainly caused by the reduction in the proportion of cristobalite with higher CTE, which also verified the XRD result. In addition, the lowest CTE of sample C6 could be attributed to the formation of new phase $CaSiO_3$ having a lower CTE value of 6.5 ppm/°C [13,14].

The flexural strength of BCABS glass-ceramics was measured and the result was also drawn in figure 5. For specimens C1-C3 (3-7 wt% CaO), the flexural strength was enhanced from 143.5 to 153.6 MPa. And sample C4 with 9 wt% CaO obtained a maximum strength of 165.7 MPa. As shown in Table 2, the highest crystallinity of sample C4 may be one of the most important reasons for the highest flexural strength. As is well known, the fluctuation of volume generated by the rapid and reversible structural transformation of cristobalite is disadvantage to the mechanical property. XRD results say that the addition of CaO restrained the precipitation of the cristobalite phase [15,16]. Hence, the improvement of flexural strength is ascribed to the decrease in the mass fraction of cristobalite. However, when further increasing CaO content (11 wt%), specimen C5 has the lowest flexural strength of 93 MPa. It was considered that the appearance of pores resulted in the remarkable decline of the mechanical property. The similar trend was also observed for the bulk density and the linear shrinkage in figure 2. Generally, the mechanical property is correlated with the crystal phases in the materials. It was notable that the flexural strength of sample C6 (13% CaO) rebounded to 110.6 MPa, which is contributed to the presence of $CaSiO_3$ phase.

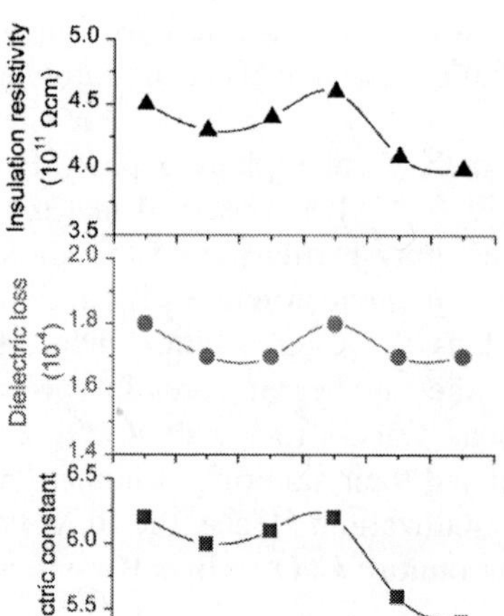

Figure 6. Dielectric properties of BCABS glass-ceramics as a function of CaO content.

Figure 6 shows the dielectric properties at 1 MHz, including dielectric constant, dielectric loss, and insulation resistivity, for BCABS glass-ceramics versus the amount of CaO. The dielectric constant remained stable around 6.0–6.2 when the CaO content within 3–9 wt%, whereas it decreases to 5.4 for CaO contents > 11–13 wt% due to the porous microstructure. Accordingly, the variation of dielectric constant is attributed to the emergence of the remaining pores in the microstructure of BCABS. Besides, independent of CaO content, the dielectric loss always maintained at the same magnitude of 10^{-4}. Obviously, the insulation resistivity first enhanced and then declined, which is coincident with the tendency of bulk density. Generally, sample C4 (9 wt% CaO) possesses the good dielectric properties: $\varepsilon_r = 6.2$, $\tan\delta = 1.8 \times 10^{-4}$, $\rho = 4.6 \times 10^{11}$ Ω·cm, meeting the requirement of the substrate materials.

4. Conclusions

$BaO–CaO–Al_2O_3–B_2O_3–SiO_2$ (BCABS) glass-ceramics were prepared for the LTCC application: CBGA package. The research indicated that the amount of CaO addition significantly affected the sinterability, structure, thermal, mechanical, and dielectric properties of BCABS system. The crystal phases Quartz (major) as well as Cristobalite and Bazirite (minor) were detected by XRD. Due to the phase transformation of quartz to cristobalite restrained by the CaO addition efficiently, the thermal property, especially the linearity, was improved evidently. However, an excess of CaO deteriorated the bulk density, bending strength, and dielectric properties, which was mainly attributed to the formation of more pores during the sintering process. 9 wt% CaO doped BCABS glass-ceramics sintered at 950 °C achieved high flexural strength (165.7 MPa), high CTE value (11.50×10^{-6}/°C), low dielectric constant ($\varepsilon_r = 6.2$), low dielectric loss (1.8×10^{-6}).

5. References

[1] Sebastian M T, Jantunen H. 2008 *Int. Mater. Rev.* **53** 57
[2] Imanaka Y. 2005 *Multilayered low temperature cofired ceramics (LTCC) technology*. (Springer Science & Business Media)
[3] Eberstein M, Glitzky C, Gemeinert M, Rabe T, Schiller W A, Modes C. 2009 *Int. J. Appl. Ceram. Technol.* **6** 1
[4] Makarovič K, Meden A, Hrovat M, Holc J, Benčan A, Dakskobler A, Kosec M. 2012 *J. Am. Ceram. Soc.* **95** 760
[5] Lo C L, Duh J G, Chiou B S, Li W H. 2002 *J. Am. Ceram. Soc.* **85** 2230
[6] Li B, Long Q, Duan D. 2016 *J. Mater. Sci. Mater. Electron.* **27** 2824
[7] San O, Özgür C. 2009 *J. Eur. Ceram. Soc.* **29** 2945
[8] Lim E S, Kim B S, Lee J H, Kim J J. 2006 *J. Non–cryst. Solids.* **352** 821
[9] Cui X, He Y, Liang Z, Zhang H, Zhou J. 2010 *Ceram. Int.* **36** 1473
[10] Chen X, Zhang W, Bai S, Du Y. 2013 *Ceram. Int.*, **39** 6355
[11] Chiang C C, Wang S F, Wang Y R, Wei W C. 2008 *Ceram. Int.* **34** 599
[12] MacMillan P W. 1979 *Glass–Ceramics* (Academic Press)
[13] Chang C R, Jean J H. 1999 *J. Am. Ceram. Soc.* **82** 1725
[14] Lei C D, Jean J H. 2005 *J. Am. Ceram. Soc.* **88** 599
[15] Hsiang H I, Yung S W, Wang C C. 2014 *Mater. Res. Bull.* **60** 730
[16] Wu C W, Jean J H. 2016 *J. Am. Ceram. Soc.* **99** 2664

Fabrication of PLA Filaments and its Printable Performance

Wenjie Liu, Jianping Zhou*, Yuming Ma, Jie Wang and Jie Xu

School of Material Science and Engineering, Nanchang Hangkong University, Nanchang 330063, China

*Corresponding author. *Email*: zf161162@163.com

Abstract: Fused deposition modeling (FDM) is a typical 3D printing technology and preparation of qualified filaments is the basis. In order to prepare polylactic acid (PLA) filaments suitable for personalized FDM 3D printing, this article investigated the effect of factors such as extrusion temperature and screw speed on the diameter, surface roughness and ultimate tensile stress of the obtained PLA filaments. The optimal process parameters for fabrication of qualified filaments were determined. Further, the printable performance of the obtained PLA filaments for 3D objects was preliminarily explored.

1. Introduction

Being able to construct complex three-dimensional (3D) objects by depositing materials layer by layer, additive manufacturing (AM), in particular, 3D printing has become a fast developing manufacturing technique nowadays.[1-2] Particularly, fused deposition modeling (FDM), which relies upon the extrusion of a thermoplastic monofilament through a heated nozzle, has developed to be one of the prevalent 3D printing techniques.[3-4] Due to its low cost, high reliability and simple operation, FDM 3D printers have correspondingly gained increasing popularity in both industries and home users, and has been started to be utilized in many areas ranging from manufacture, biomedicine to education. [5-7]

As the filaments serve as the premise for successful application of 3D printing, the fabrication of qualified filaments has therefore attracted numerous concerns from industry and academia. So far, the most commonly used polymers for filaments preparation are acrylonitrile butadiene styrene (ABS) and polylactic acid (PLA). Being a biodegradable, bio-absorbable and renewable thermoplastic polyester, together with its excellent mechanical strength and process ability, PLA filaments exhibit a more promising prospect when compared to ABS. For example, Gaal *et al.* [8] fabricated integrated microfluidic devices with PLA filaments via FDM 3D printing. Wang et al. [9] used PLA filament to print bone tissue scaffold in cold atmospheric plasma. Melocchi *et al.* [10] explored the feasibility of FDM 3D printing in the manufacturing of capsular model devices for oral pulsatile release by PLA filaments. Li *et al.* [11] manufactured continuous carbon fiber reinforced PLA composite by the rapid prototyping approach of 3D printing. Tiersch *et al.* [12] explored PLA filaments printed cryobiology device.

Although PLA filaments have been applied to FDM process successfully, there are still very few reports about the detailed extrusion process of PLA filaments. Especially, personalized 3D printing technique requires only a small amount of new filament material. Thus, it is very meaningful to explore the fabrication of PLA filament with a desktop extruder. In this work, we use a household extruder to prepare PLA filaments and examine the printable performance of the obtained qualified filament. It is believed the results could provide valuable information for the 3D printing community,

Content from this work may be used under the terms of the Creative Commons Attribution 3.0 licence. Any further distribution of this work must maintain attribution to the author(s) and the title of the work, journal citation and DOI.
Published under licence by IOP Publishing Ltd

IOP Conf. Series: Materials Science and Engineering **275** (2017) 012033 doi:10.1088/1757-899X/275/1/012033

especially for those who want to try novel filaments other than ABS and PLA to print new 3D structures.

2. Experimental

2.1. Fabrication of PLA Filaments

PLA (4032D type, America Nature Works) with an average particle size of 3.5 mm was bought commercially. To fabricate the filaments, PLA particles were first crushed into powders with a WL-A type grinder (Chengde testing machine factory, China). The powders were subsequently dried at 60 °C for 8 h in vacuum and were fed into a B-type desktop extruder (Wellzoom, China) to prepare filaments. The effect of extrusion temperature, ranging from 185°C to 200°C at an interval of 5°C, and screw rate, ranging from 2 to 6 rpm at an interval of 1 rpm, on the quality of the extruded filaments was investigated.

2.2. 3D Printing of Qualified PLA Filaments

To examine the printable performance of the obtained filaments, a cuboid plate with dimensions of 20 mm (L) × 20 mm (W) × 5 mm (H) was printed with a desktop 3D printer (Wellzoom, China, B type). The typical printing parameters were shown in table 1.

Table 1. Primary parameters for the 3D printer

Parameters	Values
Extruder temperature T_e (°C)	220
Build plate temperature T_b (°C)	60
Speed while extruding V_e (mm/s)	60
Speed without extruding V_t (mm/s)	120
Gap between nozzle tip & build layer d (mm)	0.1
Filling ratio F (%)	100

2.3. Materials Testing and Characterization

Diameter measurement. The diameter of the monofilament was measured by a digital vernier calipers (MNT, China). Five monofilaments were tested and ten measurements were performed for each monofilament. The measurement was performed every 0.5m along the filament and the results were averaged.

Ultimate tensile stress measurement. All tests are conducted at room temperature (T=20 °C) with a Sansi stretcher. The stretching speed is 2mm/s. The calculation of ultimate tensile stress σ is according to the equation (1):

$$\sigma = F/S \qquad (1)$$

where F is the ultimate tensile force as measured by a force meter, S is the fractured cross section area of the filament. During the test, the stretcher is moving upward at a speed of 2mm/s until the filament is broken. The maximum F is then automatically recorded and the fractured cross section area S of the filament is measured manually. The results of five samples were averaged. For reference, the ultimate tensile stress of commercial sample of PLA filament was tested to be 53.9±3 MPa.

3. Results and Discussion

3.1. Effect of Extrusion Temperature and Screw Speed on the Filaments' Diameter

The uniformity of the filaments' diameter is very important for its applications. The diameter data of the obtained filaments that fabricated under various processing conditions are shown in Table 2. As can be seen, when the screw speed keeps constant ranging from 2rpm to 6rpm, the filaments' diameter gradually decreases with the increase of the extrusion temperature. This conclusion is also clearly confirmed in Figure 1. It is well known that the melt viscosity reduces as the temperature increases, resulting in better melt flow and lower die swelling effect. However, we also observed that PLA powders cannot be completely plasticized when the extrusion temperature reduces to lower than 190°C, and thus no satisfied filaments can be obtained under this situation. Meanwhile, the data in Table 2 also show that, when the extrusion temperature keeps constant, the filaments' diameter keeps growing as the screw speed increases. This is because the melt volume extruded out of the die increases as the screw speed increases, causing a pronounced die swelling of the melt and a larger diameter in the subsequent cooling process.

Table 2. Diameter data of the filaments obtained under various extrusion temperature and screw speed

	185°C	190°C	195°C	200°C
2rpm	1.756	1.680	1.620	1.448
3rpm	1.760	1.721	1.680	1.575
4rpm	—	1.791	1.727	1.665
5rpm	—	1.831	1.810	1.747
6rpm	—	1.946	1.888	1.857

*(—) No satisfied filaments obtained

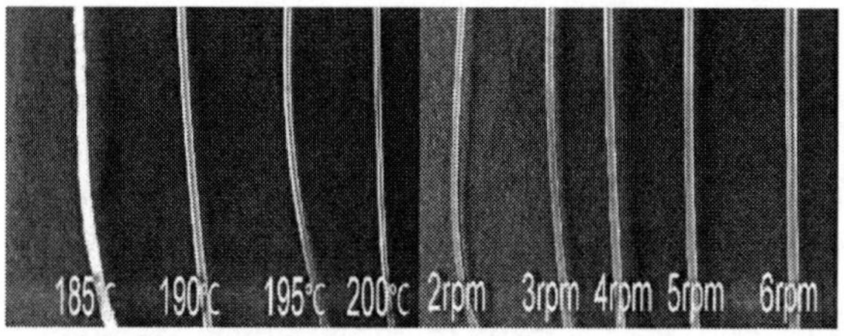

a. fabricated at 2rpm b. fabricated under 200°C

Figure 1. Photography of filaments fabricated under constant extrusion speed (a), or constant extrusion temperature (b).

3.2. Effect of Extruding Temperature and Screw Speed on Ultimate Tensile Stress

Ultimate tensile strength is very important for the applications of the filaments. If the filament is too fragile, it is liable to fracture during the depositing process, which is very detrimental to the continuity of printing. As shown in Table 3, the ultimate tensile stress of the filaments is substantially higher than 53.9±3 MPa when the extrusion temperature is 190°C or 195 °C. The value is slightly lower for the filaments obtained under 200°C, while it is very small for those fabricated under 185 °C. The poor plasticization or defects in the filaments obtained under 185°C or 200 °C may account for the lower mechanical property.

Table 3. Ultimate tensile stress of the filaments obtained under various extrusion temperature and screw speed

	185°C	190°C	195°C	200°C
2rpm	6.0	58.2	54.8	51.7
3rpm	5.4	57.5	55.1	53.3
4rpm	—	58.8	58.7	55.6
5rpm	—	58.4	60.4	56.6
6rpm	—	54.2	55.9	53.4

*(—) No satisfied filaments obtained

3.3. Printable Performance of the Qualified PLA Filaments

3.3.1. Effect of the Build Layer on Parts' Quality. Build layer is a key parameter to control the quality of the printed parts. Generally, the part's quality becomes better as the builder layer decreases. By varying the build layer *d* from 0.1 to 0.3 mm with 0.1 mm increment and keeping other printing parameters constant (as given in Table1), a series of samples were printed and the dimensional accuracy, surface roughness, density, tensile strength of the samples were tested respectively.

As shown in Table 4, the experimental results clearly indicate that the relative dimensional error of the printed parts increases with increase of the build layer, this is because the more the printing layers, the higher the printing accuracy. Also, the parts' surface roughness increases with increase of build layer (as shown in Fig. 2), and the parts' surface traces become more obvious with increase of the build layer, this is because the greater gap between the nozzle tip, the greater the printing roughness. Additionally, the parts' density increases with decrease of build layer, this is because the smaller gap between nozzle tip, the greater the printing compactness. As to the parts' tensile strength, it increases with decrease of the build layer, which can be due to the fact that the more the printing layers, the more the bond between layers, and the greater the cohesive force. Based on the data shown in table 4, it can be drawn that the quality of printing parts is relatively good when the build layer is 0.1 mm.

Table 4. Quality of parts printed with different build layer

d (mm)	0.1	0.2	0.3
Relative error ε	0.011	0.023	0.042
Surface roughness R_a(μm)	5.2	9.1	11.38
Density D(g/cm^3)	1.16	1.14	1.13
Tensile strength σ(MPa)	51.7	50.5	48.1

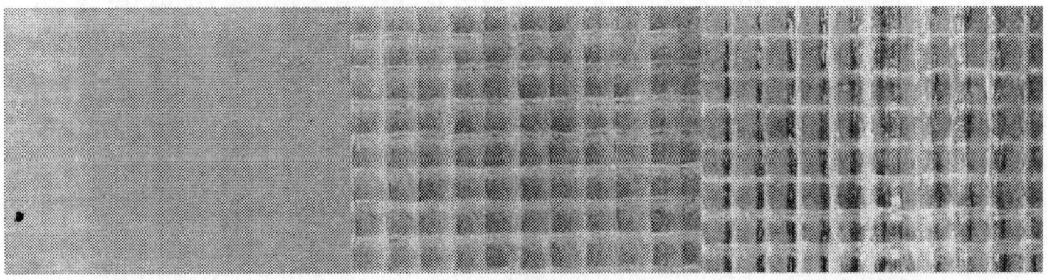

(a) *d*=0.1mm (b) *d*= 0.2mm (c) *d*= 0.3mm

Figure 2. Surface photography of parts that printed with various build layer. The extruder temperature is 220 °C, and the speed while extruding is 60 mm/s.

3.3.2. Effect of Extruder Temperature on Parts' Quality. Extruder temperature also plays an important role in controlling the printed parts' quality as it significantly affects the plasticization of the polymer filament. By adjusting the extruder temperature T_e from 200 to 230 °C with 10 °C increment and keeping the other printing parameters constant as given in Table 1, a series of samples were printed and the dimensional accuracy, surface roughness, density, tensile strength of the samples were tested respectively.

The data in Table 5 indicate that the part's relative dimensional error and surface roughness decreases with increase of extruder temperature when it is changing from 200 to 220*°C*. The latter is also clearly confirmed by Figure 3. As is well known that the liquidity of PLA increase as extruder temperature goes higher, leading to better spreading and rebonding between PLA droplets when extruded out through the nozzle. However, it should be noted that when the temperature is above 220 °C, the printed part's dimensional accuracy and surface roughness is hard to control with higher liquidity material. The parts' density observed in Table 5 increases with increase of the extruder temperature, which can be ascribed to the fact that the liquidity of PLA increase as extruder temperature goes higher, leading to more material being deposited on the cuboid model. The parts' tensile strength also increases with increase of extruder temperature. However, when it is above 220 °C, the thermal degradation of PLA filament may start to occur, causing the deterioration of the material's properties. It can be drawn from data of table 5 that the quality of the printed parts is relatively good when the extruder temperature is 220 °C.

Table 5. Quality of parts from various extruder temperature

T_e (°C)	200	210	220	230
Relative error ε	0.032	0.018	0.011	0.031
Surface roughness R_a(μm)	9.1	8.6	5.2	8.6
Density D(g/cm^3)	1.08	1.12	1.16	1.22
Tensile strength σ(MPa)	41.3	45.2	51.7	48.6

(a) T_e= 200°C (b) T_e=210°C (c) T_e= 220°C (d) T_e= 230°C

Figure 3. Surface photography of parts printed under different extruder temperature. The build layer is 0.1 mm, and the speed while extruding is 60 mm/s.

3.3.3. Effect of Speed while extruding on Parts' Quality. Speed while extruding V_e is another factor that accounts much for the parts quality. Accordingly, by adjusting V_e from 30 to 120 mm/s with 30 mm/s increment and keeping other printing parameters constant as given in Table 1, cuboid samples were printed and related properties were measured.

The experimental results in Table 6 show that ε and R_a increases with increase of V_e when it is above 60mm/s. It is worth noting that the printed part's dimensional accuracy and surface roughness is hard to control with faster speed while extruding. However, we also observed that, when V_e is only 30mm/s, the relative error value is somewhat larger than that when V_e is 60mm/s. This may because 30mm/s is fairly low that leads to more material deposited when fabricating the part. As shown in Fig. 4, the parts' surface traces also confirm the results of part's surface roughness. However, the parts' density and tensile strength are found to increase with decrease of the speed while extruding, this is because that a faster speed will lead to more internal defects of parts. Based on the data in Table 6, it can be drawn the quality of printed parts is relatively good when speed while extruding is 60 mm/s.

Table 6. Parts' quality of samples printed under various V_e

V_e (mm/s)	30	60	90	120
Relative error ε	0.018	0.011	0.024	0.052
Surface roughness R_a(μm)	8.7	5.2	6.7	7.9
Density D(g/cm^3)	1.18	1.16	1.05	0.85
Tensile strength σ(MPa)	53.3	51.7	47.2	45.5

(a) V_e =30 mm/s (b) V_e = 60 mm/s (c) V_e = 90 mm/s (d) V_e =120 mm/s

Figure 4. Surface photography of parts that fabricated under various speed while extruding. The build layer is 0.1 mm and extruder temperature is 220 °C.

4. Conclusion

Preparation of PLA filaments for FDM 3D printing was systematically investigated with a desktop extruder specially designed for household users. The effects of extrusion temperature and screw speed on the filaments' diameter and ultimate tensile stress were systematically investigated. It was found that qualified filaments could be obtained under conditions of extrusion temperature 190-195°C and screw speed 2-5 rpm. Furthermore, the printable performance of the obtained filaments was explored with a desktop 3D printer. It was found that when build layer is 0.1mm, extruder temperature is 220°C and speed while extruding is 60mm/s, best printing effect with a relative dimensional error of 0.011, surface roughness of 5.2 μm, density of 1.16 g/cm^3 and tensile strength of 51.7 MPa could be achieved. The finding in this work will be valuable for 3D printing community, especially helpful for those 3D printing lovers and small enterprises.

5. Acknowledgement

Financial support from National Natural Science Foundation of China (Grant Nos. 51363017 and Nos. 51303078) is gratefully acknowledged.

6. References

[1] Shirazi S F S, Gharehkhani S, Mehrali M, Yarmand H. *Sci. Technol. Adv. Mater.* 2015, 16: 33502-08
[2] Farahani R D, Chizari K, Therriault D. *Nanoscale,* 2014, 6:10470-85
[3] D-Y Chang, B-H Huang. *Int. J. Adv. Manuf. Technol.* 2011, 53:1027-37
[4] Krisztián Kun. *Procedia Engineering*, 2016, 149: 203-11
[5] Bikas H, Stavropoulos P, Chryssolouris G. *Int. J. Adv. Manuf. Technol.* 2015,1711-14
[6] Mannoor M S, Jiang Z, James T, Kong Y L. *Nano Lett.* 2013, 13:2634-39
[7] Tien Chi Huang, Chun Yu Lin. *Telematics and Informatics*, 2017, 34(2): 604-13
[8] Gabriel Gaal, Melissa Mendes, Tiago P Almeida. *Sensors and Actuators B,* 2017, 242: 35-40
[9] Mian Wang, Pelagie Favi, Xiaoqian Cheng. *Acta Biomaterialia,* 2016, 46:256-65
[10] Alice Melocchi, Federico Parietti. Alessandra Maroni. *Int. J. Pharm.*, 2016, 509(1-2): 255-63
[11] Nanya Li, Yingguang Li, Shuting Liu. *J. Mater. Proc. Tech.*, 2016, 238: 218-25
[12] Terrence R Tiersch, William T Monroe. *Cryobiology*, 2016, 73(3):396-98

IOP Conf. Series: Materials Science and Engineering **275** (2017) 012034 doi:10.1088/1757-899X/275/1/012034

Construction of a Dry Ash Dam with Soilbags and Slope Stability Analysis

Hui Li[1], Yingjun Song[2,*], Jiaorong Gao[2], Longhua Li[1], Yuqi Zhou[2] and Hui Qi[1]

[1]Sichuan Electric Power Design & Consulting Company, Chengdu, Sichuan, 610016, China
[2]College of Water Conservancy and Hydropower Engineering, Hohai University, Nanjing, 210098, China
*E-mail: yingjun@hhu.edu.cn

Abstract: In thermal power plants, it is necessary to build ash dams to store fly ash, which is the by-product after the combustion of coals. To solve the problem of lacking rockfill materials in Africa, A new technology of constructing ash dams using solibags filled with local sands is proposed and the method of analyzing its slope stability is suggested. The design of the ash dam using soilbags in Lamb Thermal Power Plant of Kenya is introduced in detail. The slope stability of the soilbags-constructed ash dam was analyzed by adopting the suggested method. The results show that the soilbags filled with ash or sands have high compressive strength, and the primary dam constructed with soilbags can effectively retain the backfill ash and the stacking dam reinforced with soilbags can stand stable even with the slope of 1:1.5.

1. Introduction

With the increasing construction quantity and scale of the thermal power plants, the emissions of fly ash are also increasing. As the main industrial waste of the thermal power plants, the fly ash can bring about a serious pollution to the environment if it was discharged directly without appropriate treatment, such as producing dust, polluting the atmosphere and silting the river. Besides, the toxic chemicals in the fly ash is harmful to humans and living creatures [1], [2]. However, it was shown by researchers [3], [4] that, the comprehensive utilization of fly ash has a very broad prospect. Due to its advantages such as large quantity, light weight, high strength and good permeability [5], the fly ash can be used as a basic material [6] when preparing building materials including bricks, blocks, cement and concrete, etc. Moreover, it can not only be made into fertilizers and soil amendments [7] but also be used for waste water treatment and flue gas desulfurization [8], [9]. Nevertheless, the quantity of fly ash in utilization is still far less than the total quantity discharged by thermal power plant, so it is necessary to solve the storage problems of the remaining large amount of fly ash.

At present, the most world-widely used storage method is to construct an ash dam. And ash dam is an important part of the thermal power plant [10]. It is composed of the primary dam and the stacking dam. In engineering, the key method of constructing ash dams is by installments.Figure1 shows the schematic diagram of a conventional ash dam and the construction stages. Firstly, the initial dam is built with soils and stones. After the ash bank behind of the initial dam is compacted by fly ash, the fill dam will be built layer by layer until reaching the designed dam height. Meanwhile, the local soils and stones will be used to build the slope of the fill dam. According to the “Design Specification of the Dry Ash-storage Field of the Thermal Power Plant (DL/T5488-2014)” of the People's Republic of China, there should be enough construction materials near the ash field. The initial dam is usually the rockfill dam

Content from this work may be used under the terms of the Creative Commons Attribution 3.0 licence. Any further distribution of this work must maintain attribution to the author(s) and the title of the work, journal citation and DOI.
Published under licence by IOP Publishing Ltd

or the masonry dam, using rockfill materials with compressive strength greater than 30MPa, weathering coefficient greater than 0.75 and rock softening coefficient greater than 0.8. The slope of the fill dam is built with soils and stones with weak permeability. The height of each layer of the fill dam should be 5~10 meters, and the gradient of the slope should be 1:3~1:4. However, for the thermal power plants where the traffic conditions are poor and the local materials are not suitable for dam construction, such as the Kenyan Ram Thermal Power Plant in Africa, the cost of purchasing and transporting dam construction materials is very high, which results in a high investment by using the traditional materials such as rockfills to construct the ash dams. Therefore, a new method to construct the ash dam by using the soilbags is proposed in this paper to make full use of local construction materials and reduce the dam construction costs.

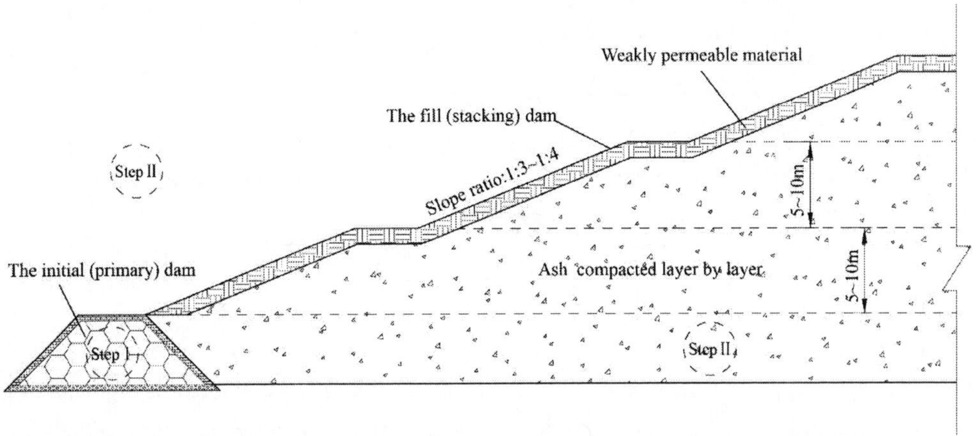

Figure 1. Schematic view of Conventional Ash dam construction Scheme

Recently, Matsuoka and Liu studied the strengthening mechanism and engineering characteristics of the solibags [11]-[14]. They found that soilbags had a high bearing capacity, good integrity, high ability in resisting deformation, make use of local materials and low construction cost. What's more, it had enough durability if the soilbags were buried in the soil or protected by the concrete pavement from the sunlight. Owning to these advantage, soilbags have been widely used in many geotechnical engineering, such as road foundation [15], slope protection [16] and dam reinforcement [17]. In this paper, the soilbags filled with local sands are firstly stacked on the slopes of the ash dam, which not only solve the problems that there is not enough rockfill materials to construct the ash dam but also take full advantage of the local materials.

Therefore, in this paper, based on the principle of soilbags technology, a construction method of the ash dam located in the Kenyan Ram Thermal Power Plant in Africa using soilbags filled with local sands is proposed for the first time. Furthermore, the slope stability of the ash dam is studied by using the limit equilibrium method.

2. Reinforcement Principle of Soilbags

As a kind of geotextile fabric, soilbags are filled with soil or soil-like materials. For a long time soilbags are widely used to raise embankments during floods and to construct temporary structures after disasters as well as other temporary projects. They are seldom applied in the permanent buildings such as the slope reinforcement, which is mainly because of the lack of understanding of the characteristics of solibags and the incomplete appearance of woven bags under the long-term irradiation of ultraviolet light. Recently, Matsuoka and Liu [18] found that soilbags had a very high compressive strength through experimental tests and theoretical analysis. The bearing capacity of a soilbag with size of 40 cm×40 cm×10 cm (length × width × height) is more than 20 tons, which is equivalent to 1/5 ~ 1/10 of the concrete block strength.

Figure 2 is a schematic illustration of the reinforcement mechanism of soilbags. The overall strength of soilbags is closely related to the friction strength between soil particles in the soilbags, which is developed due to the interparticle normal contact force N. Under the upper vertical pressure, the compression deformation of the soil causes the extension of the bag perimeter, which leads to the tensile force T along the bag. The tensile force T along the bag enhances the contacts between the soil particles inside the bag, resulting in an increase in the normal contact forces N and the frictional forces F between the soil particles (F=μN, where μ is the friction coefficient). Therefore, the overall strength of soilbags is improved.

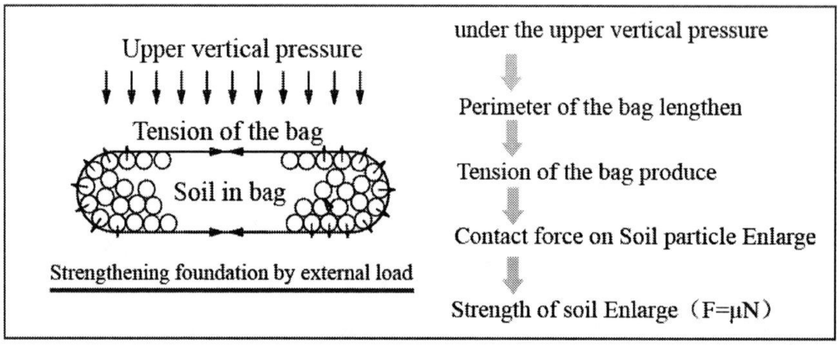

Figure 2. Schematic view of reinforcement mechanism of soilbags

Figure 3(a) shows a two-dimensional soilbag subjected to the principal stresses σ_{1f} and σ_{3f}. The material inside the soilbag is assumed to be frictional and granular. At constant volume condition, under the actions of σ_{1f} and σ_{3f}, the total perimeter of the bag usually increases because the solibag becomes flatter, as shown in Figure 4. Subsequently, the bag becomes flat and a tensile force T is developed high tensile forces. The tensile force T produces an additional stress on the particles inside of the bag with components $\sigma_{01}=2T/B$ and $\sigma_{03}=2T/H$, in which B and H are the width and height of the solibag, respectively. As illustrated in Figure 3(b), the stresses acting on the particles are thus the combination of the external stresses and the stress caused by the tensile force T. At failure, the major principal stress σ_{1f} can be calculated by:

$$\sigma_{1f}+\frac{2T}{B}=K_p\left(\sigma_{3f}+\frac{2T}{H}\right) \tag{1}$$

Therefore

$$\sigma_{1f}=\sigma_{3f}K_p+\frac{2T}{B}(\frac{B}{H}K_p-1) \tag{2}$$

in which $K_p=(1+\sin\phi)/(1-\sin\phi)$ is the lateral earth pressure ratio at passive state. For cohesive-frictional materials, the following relationship exists between the major and minor principal stresses at failure:

$$\sigma_{1f}=\sigma_{3f}K_p+2c\sqrt{K_p} \tag{3}$$

Where the apparent cohesion c is thereby expressed as [11]:

$$c=\frac{2T}{B\sqrt{K_p}}(\frac{B}{H}K_p-1) \tag{4}$$

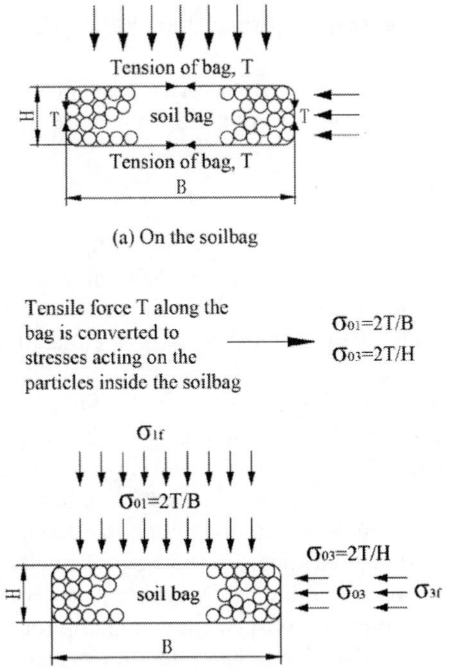

(b) On the partical inside the soilbag: the tensile force along the bag is converted to the stresses acting on the pariticles inside the soilbag

Figure 3. Force analysis on solibag and the particles inside the soilbag

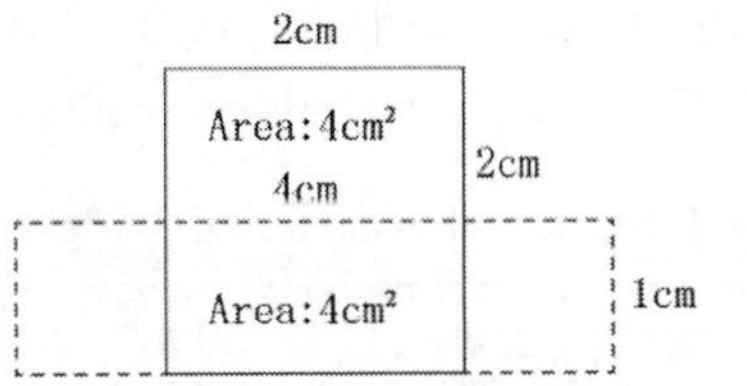

Under the same area of 4 cm2, the perimeter is extended from 8 cm to 10 cm when a 2 cm × 2 cm square becomes a 4 cm × 1 cm rectangle. As a result, a tensile force is produced along the bag

Figure 4. Tensile force is produced along the bag when a solibag becomes flat

3. Design of an Ash Dam with Soilbags

3.1. Geological Condition of the Project

The Lamb Thermal Power Plant is located in the north side of a Navy Terminal, which is in the northwest area of Manda Island, Kenya, with total installed capacity of 1080MW. Through geological prospecting, the region of the Thermal Power Plant is covered by Quaternary tectonics within depth of 50 meters, and the lithology is dominated by sand and clay. Sand I is mainly distributed in the depth of 0 ~ 15 meters, forming the bearing stratum of the foundation of the structure. The basic physical and mechanical parameters of sand I are: unit weight $\gamma = 17.2kN/m^3$, cohesive strength $C = 3.2kPa$, internal friction angle $\varphi = 35.4°$, and elastic modulus $E_0 = 18.2MPa$. The clay and sand II ,which are distributed in the depth of 12 ~ 30 meters and 30~ 50 meters respectively, constitute the substratum of the structure foundation. The basic properties of clay are $\gamma = 18.5kN/m^3$, $C = 60.3kPa$, $\varphi = 12.6$ ° and compression modulus Es = 9.75MPa while the properties of sand II are $\gamma = 21.1kN / m^3$, C = 5.1kPa, $\varphi = 42.3$ ° and $E_0 = 25.6$MPa.

According to geological data, there are not appropriate dam construction materials on the site of the power plant. Therefore, the solibags are used in this project and are filled with local materials to pile on the slope of the dam.

3.2. Properties of Soilbags

The soilbags used in the project are woven polypropylene (PP) bags filled with local sand. The size of the soilbag is 120 cm long, 60 cm wide and 30 cm high. The mechanical properties of the soilbags are as follows: the unit mass is 150 g/m^2; the radial and latitudinal tensile loads are greater than or equal to 30 kN/m and 25 kN/m, respectively; the radial and latitudinal elongations are both less than or equal to 25%.

As the Lamb Thermal Power Plant is located near the equator where ultraviolet rays are strong, the durability of soilbags is particular important to the long-term behavior of the ash dam. Therefore, the accelerated aging tests were carried out on the soilbags mixing with 1% UV resistant materials. The test equipment was the Type-Ci5000 artificial weathering test chamber made by the ATLAS company of United States. The light source is the water-cooled xenon arc lamp (5kW ~ 14kW) with irradiance 0.5W/m^2 • mm. The temperature of the blackboard is (65 ± 3) °C and the relative humidity is (65 ± 5) %. When conducting the tests, the soilbags with 1% UV resistant materials were put in the test chamber and were exposure under the ultraviolet rays of the xenon arc lamp for 1200h. The tests would be ended until the retention rate was no more than 50%. The retention rates of tensile strength and elongation were recorded at the time of 0h, 300h, 600h, 900h and 1200h, respectively.

Table 1 shows the results of accelerated aging tests. The test results showed that after explosion under the ultraviolet ray of the xenon arc lamp for 1200h, the retention rate of tensile strength of tested PP bags was 94% while the retention rate of elongation was 84%. The aging performance index of PP bags mixing with 1% UV resistant material was higher than the required value in the "Technical standard for applications of geosynthetics (GB/T17690-1999)" of the People's Republic of China. Therefore, the PP bags mixing with 1% UV resistant materials were used in this project.

Table 1. Results of accelerated aging tests

Irradiation time of Xenon arc lamp	0h	300h	600h	900h	1200h
Retention rate of tensile strength,%	100	106	96	94	94
Retention rate of elongation,%	100	99	91	82	84

The physical and mechanical properties of fly ash of the Lamb Thermal Power Plant in Kenya were analyzed by laboratory tests, as shown in Table 2. According to equation (3) , the maximum additional cohesion of solibags filled with ash and sand under the ultimate load was 165 kPa and181 kPa, respectively. According to the laboratory friction tests, the friction angle of the interface between soilbags and ash slopes was 16 degrees [11]. Because of the embedding function between two layers of the soilbags, the bottom horizontal friction coefficient of the soilbag slopes is as large as 0.7.

Table 2. Properties of the ash

Properties	Natural water content w (%)	Natural unit weight γ (kN/m^3)	Dry unit weight γ_d (kN/m^3)	Cohesion c (kPa)	Internal friction angle φ(°)
Ash	16.1	10.4	10.5	4.3	31.6

3.3. Design Scheme

Due to the lack of water in Kenya, choosing the way of dam construction with dry ash can achieve the objective of water saving. Moreover, compared with the wet ash dam, the investment of dry ash dam is lower, especially in the plain area. Figure 5 shows the dam body cross-section drawing of dry ash dam using soilbags. The initial dam of dry ash dam is piled up by sand to the final height of 3m. The initial dam is made up of soilbags filled with local sand, and the slope ratios are 1:1. Meanwhile, the crest was equipped with concrete pavement. The fill dam is constructed by using upstream method. The slope of the fill dam is piled up by two rows of soilbags filled with sand, with slope ratio 1:1.5. There are three layers of the fill dam. The height of each layer is 10m and the total height of the dam is 33m. Meanwhile, 8m berm in width is located in the slope per 10m height.

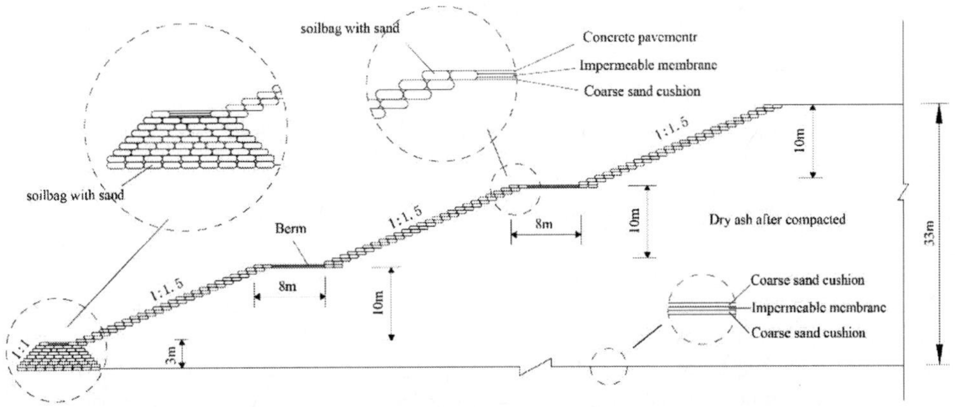

Figure 5. The cross-section of the ash dam built with soilbags

The dam body is quadrilateral distribution and the dam area is about 5.97×10^5 m^2. The total reservoir capacity is about 1.5×10^7m^3. The ash dam is constructed by layer buried method. The soilbags are packed and transported by large machines. The dry ash in the ash field and the soilbags in the slopes are finally compacted by using the vibrating rollers.

The soilbags filled with sand, which are placed in a certain arrangement on the ash slopes, can not only restrict the deformation of the ash behind the soilbags effectively, but also increase the stability of the ash dam slope. The main reasons are as follows: 1) Reinforcement effect. Because of the compaction effect in the process of construction, the compression deformation of the soil causes the extension along the bag perimeter and the tensile strength in the soilbags, which is equivalent to the effect of an additional cohesion and leads to the improvement of the overall strength of the soilbags. For the assembly of soilbags, because of the embedding and friction effect between two layers of soilbags[18], the overall strength is further improved, which is equivalent to a reinforced body; 2) Pressing slope effect. According to equation (3), the soilbags has high strength, which is similar to the rock., the stability of the ash dam slope will increase as the slope is pressed by stacking the soilbags on the slope.

4. Slope Stability Analysis of the Stacking Dam

The soilbags filled with sand have high compression strength, which is similar to the rock. When the soilbags are stacked on the slope of the ash dam, it is equivalent to a loading pressing on the ash slope. Therefore, the solibags can be considered as a pressing slope load when calculating the slope stability of the ash dam with soilbags. Due to the high compression strength of a single soilbag, the individual bag is hardly to be damaged. However, for the combination of soilbags, the interlayers between soilbags are weak surfaces and are easy to slide. Moreover, when calculating the stability of the ash dam slope piled up by soilbags, sliding failure will not occur in the contact surface between the pressing slope of soilbags and the ash behind it, that is to consider the soilbags and the ash slope as a whole when analyzing the slope stability of the ash dam. The sliding surface of the slope with soilbags is shown in

Figure 6.

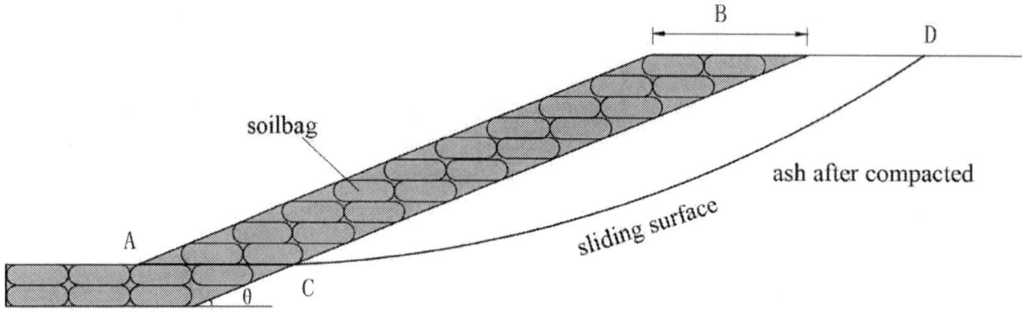

Figure 6. The sliding surface of slope with soilbags

Figure. 7 shows the diagram of the forces on the slope of the ash dam with soilbags. According to the force equilibrium conditions on the pressing slope of soilbags, the sliding resistance force T_1 and the vertical stress N_1 on the bottom of the pressing slope can be calculated by

$$T_1 = P_1 \sin(\theta - \phi_1) \qquad N_1 = W_1 - P_1 \cos(\theta - \phi_1) \tag{5}$$

where W_1 is the gravity load of the pressing slope piled up by soilbags, P_1 is the force of the pressing slope applied on the ash dam slope, θ is the slope angle, and φ_1 is the interlayer friction angle between the pressing slope of soilbags and the ash dam slope.

Assuming that F_S is the factor of safety of the slope, μ is the friction coefficient between different interlayers of soilbags in the pressing slope, the relationship between F_S and μ can be obtained by

$$T_1 = \frac{\mu \cdot N_1}{F_S} \tag{6}$$

According to the equation (5) and equation (6), the force of the pressing slope applied on the ash dam slope P_1 can be calculated by

$$P_1 = \frac{\mu \cdot W_1}{F_S \cdot \sin(\theta - \phi_1) + \mu \cdot \cos(\theta - \phi_1)} \tag{7}$$

In this paper, it is assumed that P_1 is uniformly distributed on the ash slope, so the magnitude of the uniform load p can be calculated by $p=P_1 sin\theta/H$, where H is the height of the pressing slope of soilbags.

As shown in Figure 7, the ash slope behind the pressing slope of soilbags is discretized into slices and the simplified Bishop method is used to determine the factor of safety F_S. According to the definition of factor of safety and the Mohr-Coulomb criterion, for slice i, the following equation can be obtained

$$T_i = \frac{1}{F_s} T_{fi} = \frac{(c_{total})_i l_i + N_i \tan(f_{total})_i}{F_s} \tag{8}$$

Where c_{total}, φ_{total} are shear strength indexes of the ash, T_{fi} is the shear force of slice i, l_i is the bottom length of slice i, $l_i = \Delta x_i / \cos\alpha_i$.

The simplified Bishop [19] method satisfies vertical force equilibrium for each slice and overall moment equilibrium about the center of the circular trial surface. Since horizontal forces are not considered at each slice, the interslice shear forces are assumed to be zero. For each slice, it can be obtained that

$$W_i + p\Delta x_i \frac{\cos(\theta - \phi_1)}{\cos\theta} + \Delta Z_i - T_i \sin\alpha_i - N_i \cos\alpha_i = 0 \tag{9}$$

$$\sum W_i R \sin\alpha_i - P_1 y - \sum T_i R = 0 \tag{10}$$

If the same factor of safety is assumed for all the slices, the safety factor F_s of the sliding surface can be calculated by combining equations. (8-10)

$$F_S = \sum \frac{1}{m_i}\left[(c_{total})_i \Delta x_i + (W_i + p\Delta x_i \frac{\cos(\theta - \phi_1)}{\cos\theta})\tan\left(\phi_{total}\right)_i\right]\left(W_i R \sin\alpha_i - \frac{P_1 y}{R}\right)^{-1} \tag{11}$$

in which $m_i = \cos\alpha_i + \frac{\tan(\phi_{total})}{F_s}\sin\alpha_i$. The cohesion c_i and internal friction angle φ_i are related to the normal stress on the bottom of slice i, $\sigma_n = N / l_i$. Since the m_i and P_1 in the right term of equation (11) are functions of the factor of safety, equation (11) can be solved by iterative method.

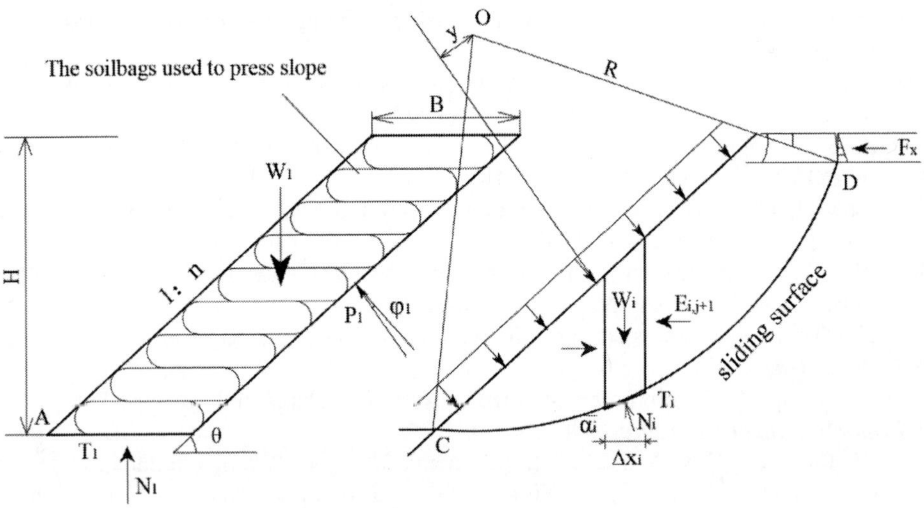

Figure 7. Diagram of the forces on the slope of the ash dam with soilbags

According to the calculation equations, the slope stability analysis of the ash dam with soilbags shown in Figure 2 is carried out. The results of calculation show that the slope factor of safety with soilbag is 1.441 while the slope factor of safety of the ash slope without soilbags is 1.113, with an increase of 29.4%. Therefore, the method of construction of an ash dam with soilbags was effective.

5. Conclusion

In this paper, a new technology to construct an ash dam with solibags filled with local sands is proposed and the method of analyzing the slope stability of the ash dam with solibags is suggested. The design of the ash dam with soilbags in Ram Thermal Power Plant of Kenya is introduced in detail. By using the suggested method, the slope stability of the soilbags-constructed ash dam is analyzed. The conclusions are summarized as follows:

(1) The soilbags filled with ash or sands have a high compressive strength. The durability of the PP bags mixing with 1% UV can be guaranteed near the equator in Kenya based on the results of the accelerated aging tests.

(2) The ash dam built with soilbags is composed of the primary dam and the stacking dam. The soilbags are used to build the primary dam and reinforce the slope of the stacking dam.

(3) The primary dam constructed with soilbags can effectively retain the backfill ash and the stacking dam reinforced with soilbags can stand stable even with the slope of 1:1.5.

6. Reference

[1] Yuanfeng, Wu, et al. Current situation of comprehensive utilization of fly ash. *Clean Coal Technology*. **19.6** (2013).
[2] McMurphy, Lisa M., and A. Lane Rayburn. Nuclear alterations of maize plants grown in soil contaminated with coal fly ash. *Archives of Environmental Contamination and Toxicology*. **25.4**, 520-524(1993).
[3] Clarke L B. Applications for coal-use residues: an international overview. *Studies in Environmental Science*. **60**, 673-686(1994).
[4] Ferreira C, Ribeiro A, Ottosen L. Possible applications for municipal solid waste fly ash. *Journal of hazardous materials*. **96.2**, 201-216(2003).
[5] Sarkar A, Rano R, Udaybhanu G, et al. A comprehensive characterisation of fly ash from a thermal power plant in Eastern India. *Fuel Processing Technology*. **87.3**, 259-277(2006).
[6] McCarthy M J, Dhir R K. Towards maximising the use of fly ash as a binder. *Fuel*, **78.2**, 121-132(1999).
[7] Menzies N W, Aitken R L. Evaluation of fly ash as a component of potting substrates. *Scientia Horticulturae*. **67.1-2**, 87-99(1996).
[8] Viraraghavan T, Dronamraju M M. Utilization of coal ash in water pollution control. *International journal of environmental studies*. **40.1**, 79-85(1992).
[9] Horiuchi S, Kawaguchi M, Yasuhara K. Effective use of fly ash slurry as fill material. *Journal of Hazardous Materials*, **76.2**, 301-337(2000).
[10] Li S, Zhao J, Chen Y, et al. Optimization research for ash dam cross section. *Advances in Engineering Software*. **36.2**, 109-1132005.
[11] Matsuoka, Hajime, and Sihong Liu 2005 *A new earth reinforcement method using soilbags*. (The Netherlands: A.A.Balkema Publishers-Taylor&Francis).
[12] Liu S H, Matsuoka H. A new earth reinforcement method by soilbags. *Rock and Soil Mechanics*. **28.8**, 1665-1670(2007).
[13] Liu S H, Wang Y S. Reinforcement mechanism of soilbags and its applications. *Journal of Hydraulic Engineering*. **5**, 001(2007).
[14] Liu S H, Bai F Q, Wang Y S, et al. Experimental Study on Wetting-Induced Deformation and Strength Characteristics of Soilbag Filled with Expansive Soil. *South-to-North Water Transfers and Water Science & Technology*. **7.6**, 54-27(2009).
[15] Liu S.H. 2017 *Principle and application of soilbags*. (China: Science Press).
[16] Liu S, Lu Y, Weng L, et al. Field study of treatment for expansive soil/rock channel slope with soilbags. *Geotextiles and Geomembranes*. **43.4**, 283-292(2015).
[17] Matsushima K, Aqil U, Mohri Y, et al. Shear strength and deformation characteristics of geosynthetic soil bags stacked horizontal and inclined. *Geosynthetics International*. **15.2**, 119-135(2008).
[18] Liu S H, FAN K W, CHEN X L W, et al. Experimental studies on interface friction characteristics of soilbags. *chinese Journal of Geotechnical Engineering*. **38.10**, 1874-1880 (2016).
[19] Mitchell J K, Soga K. *Fundamentals of soil behavior* (2005).

IOP Conf. Series: Materials Science and Engineering **275** (2017) 012035 doi:10.1088/1757-899X/275/1/012035

Comparative Research on Characteristics of the Isolation Systems with Dry Friction Damping and with Vicious Damping under Base Excitation

Junfang Hou, Min jing, Weihua Zhang, Yahui Lu and Haiwen He

Beijing Special Vehicle Research Institute, Beijing 100072, China
Email: meici010@qq.com

Abstract: As for the isolation problem of electronic equipments on vehicle, the vibration response characteristics of dry friction damping isolation system under base displacement excitation was analyzed in theory by harmonic balance method, and the displacement response was compared between the isolation systems with dry friction damping and vicious damping separately. The results show that the isolation system with small dry friction damping can't meet the demands of displacement reduction close to the natural frequency, and it can realize full-frequency vibration isolation by improving dry friction damping when the lock frequency passes beyond the resonance frequency band. The results imply that the damping mechanism of dry friction isolator can't be described only by dry friction damping, and the composite damping with dry friction and vicious damping is more appropriate.

1. Introduction

The electronic equipments on vehicle will subject to large vibration and shock environment when the vehicle running on off-road pavement. Under large vibration and shock, electronic devices may malfunction such as electric loose contact, wire distortion or dislocation, loose weld of electronic component and so on, which can reduce severely the reliability of the electronic devices. To ensure reliable work of the electronic devices on vehicle in rough environment, effective measures should be taken to isolate the vibration and absorb the shock except for strengthening of the electronic devices.

The main vibration isolators used for vibration isolation of electronic devices in armored vehicles include damping rubber type of vibration isolator and dry friction type of vibration isolator. Due to its vulnerability to high and low temperature, the damping capacity and elastic behavior of the damping rubber type of vibration isolator would get worse in extreme temperature used in armored vehicles. In addition, the rubber material is prone to aging gradually. Therefore, the damping rubber type of vibration isolator is recently replaced by dry friction type of vibration isolators, such as wire-cable vibration isolator [1], wire mesh vibration isolator [2,3] and structural dry friction vibration isolator. The wire-cable and wire mesh vibration isolators damp the vibration by the dry friction damping force between sliding wires. The structural dry friction vibration isolator [4] damps the vibration by the dry friction damping force between friction pairs constituted by metal structural parts. The theory analysis [5] and experimental research [6] was conducted by researchers.

The vibration response characteristics of dry friction damping isolation system under base displacement excitation was analyzed in this paper, and the displacement response was compared between the isolation systems with dry friction damping and vicious damping separately.

Content from this work may be used under the terms of the Creative Commons Attribution 3.0 licence. Any further distribution of this work must maintain attribution to the author(s) and the title of the work, journal citation and DOI.
Published under licence by IOP Publishing Ltd

2. Forced Vibration of Single DOF System with Dry Friction Damping Induced by Base Excitation

The typical mechanical model of wire mesh isolator and wire-cable isolator is shown in Figure1, in which the mechanical components include cubic nonlinear elastic force, vicious damping force, and dry friction force with hysteresis character. The mechanical model of the structural dry friction vibration isolator constituted by metal structural parts is shown in Figure2, in which the elastic component is linear spring and the damping component is metal friction pairs.

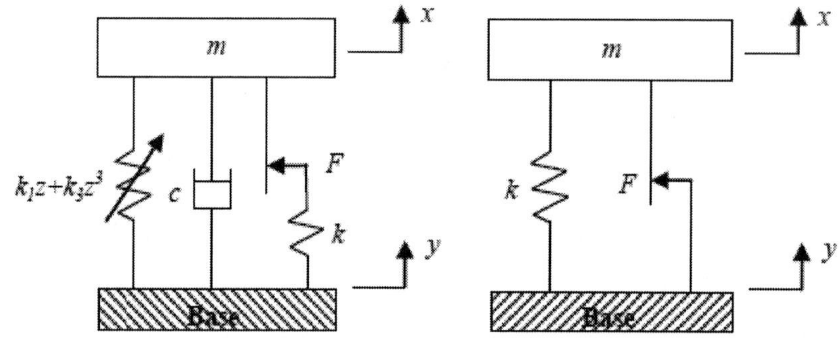

Figure1. Mechanical model of wire mesh vibration isolator; **Figure 2.** Mechanical model of structural dry friction isolator

The motion equation of the single DOF isolation system with the structural dry friction isolator under base excitation is as follows:

$$m\ddot{x}+k(x-y)+F\,\mathrm{sgn}(\dot{x}-\dot{y})=0 \tag{1}$$

And the base excitation is

$$y=y_0\sin(\omega t) \tag{2}$$

Given $x-y=z$, then

$$m\ddot{z}+kz+F\,\mathrm{sgn}(\dot{z})=m\omega^2 y_0\sin(\omega t) \tag{3}$$

The motion vibration isolation of the base is transformed into the forced vibration shown in formula (3), and the exciting force $m\omega^2 y_0\sin(\omega t)$ is the inertia force of the isolated mass induced by the displacement of the base.

The free vibration attenuates rapidly for the dry friction damping. The steady-state forced vibration of the system expressed by formula (3) is

$$z=A\sin(\omega t-\varphi) \tag{4}$$

The dry friction force $F\,\mathrm{sgn}(\dot{z})$ can be expanded into Fourier series, and the primary harmonic component is taken into consideration. Then the dry friction force is

$$F\,\mathrm{sgn}(\dot{z})\approx\frac{4F}{\pi}\cos(\omega t-\varphi) \tag{5}$$

Substitute formulas (4) and (5) into (3), then

$$(kA-m\omega^2A)\sin(\omega t-\varphi)+\frac{4F}{\pi}\cos(\omega t-\varphi)=m\omega^2 y_0\sin(\omega t) \tag{6}$$

Considering the trigonometric function transformation, then

$$\sin(\omega t) = \sin(\omega t - \varphi + \varphi) = \sin(\omega t - \varphi)\cos\varphi + \cos(\omega t - \varphi)\sin\varphi \tag{7}$$

Substitute formula (7) into (6), then

$$(k - m\omega^2)A = m\omega^2 y_0 \cos\varphi \tag{8}$$

$$\frac{4F}{\pi} = m\omega^2 y_0 \sin\varphi \tag{9}$$

According to formulas (8) and (9), the amplitude and phase of the forced vibration is

$$A = \sqrt{\frac{(m\omega^2 y_0)^2 - (4F/\pi)^2}{(k - m\omega^2)^2}} \tag{10}$$

$$\varphi = \arcsin\left(4F/m\omega^2 y_0 \pi\right) \tag{11}$$

It's seen from (9) that the steady-state vibration solution z of the system can be expressed by formula (4) on the condition that the inertia exciting force amplitude is not less than the dry friction force amplitude, that is $m\omega^2 y_0 \geq 4F/\pi$. As for the un-simplified dry friction force model $F\,\mathrm{sgn}(\dot{z})$, the isolation system is locked when $m\omega^2 y_0 \leq F$, and there is no relative displacement between the base and the isolated mass. The isolation system is unlocked when $m\omega^2 y_0 > F$. From the above analysis, the unlock frequency ω_u of the isolation system is

$$\omega_u = \sqrt{F/my_0} \tag{12}$$

According to formulas (2) and (4), the displacement of the isolated mass can be expressed as

$$\begin{aligned} x &= y_0 \sin(\omega t) + A\sin(\omega t - \varphi) \\ &= x_0 \sin(\omega t + \psi) \end{aligned} \tag{13}$$

In which

$$x_0 = \sqrt{A^2 + y_0^2 + 2Ay_0 \cos\varphi} \tag{14}$$

$$\psi = \arctan(-A\sin\varphi/(A\cos\varphi + y_0)) \tag{15}$$

And the displacement transmissibility T of the isolation system is

$$T = \frac{x_0}{y_0} = \sqrt{1 + (A/y_0)^2 + 2A\cos\varphi/y_0} \tag{16}$$

3. Comparison of the systems with dry friction damping and with viscous damping

The displacement transmissibility curves of the isolation system with dry friction damping is shown in Figure3, in which, the horizontal ordinate is frequency ratio $\gamma = \omega/\omega_0$, and $\omega_0 = \sqrt{k/m}$ is the natural frequency. The following conclusions can be summarized based on Figure3:

(1) The unlock frequency of the isolation system increases with the dry friction damping.

(2) When the unlock frequency is less than the natural frequency of the system, the displacement transmissibility increases significantly while the exciting frequency close to the natural frequency. And

$T \to \infty$ while $\gamma \to 1$.

(3) All the displacement transmissibility curves pass through the point ($\sqrt{2}$,1) with no matter how much the dry friction damping is.

(4) The non-resonant vibration isolation at all frequency band can be achieved by increasing the dry friction damping in order that the displacement transmissibility curve passes beyond the displacement enlarging frequency band when the unlock frequency larger than $\sqrt{2}$ times of natural frequency of the system. Meanwhile, the displacement isolation effectiveness at high frequency band would be reduced when continuing increasing the dry friction damping. And the lower dry friction damping is better to the displacement isolation when the unlock frequency larger than $\sqrt{2}$ times of natural frequency of the system.

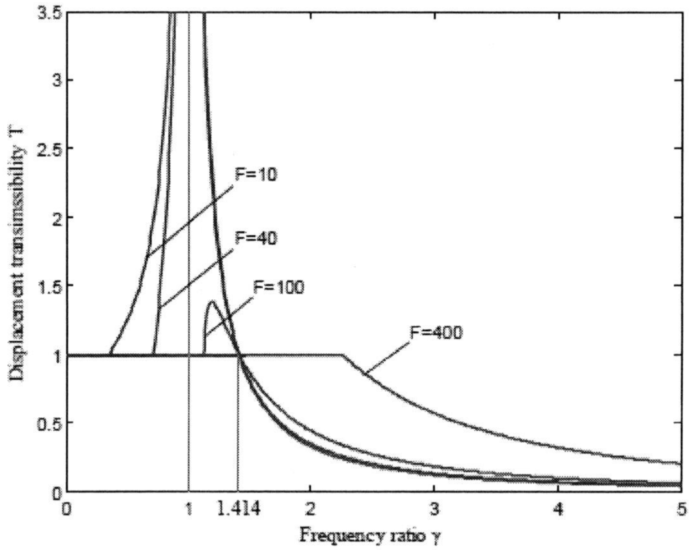

Figure 3. Displacement transmissibility curves

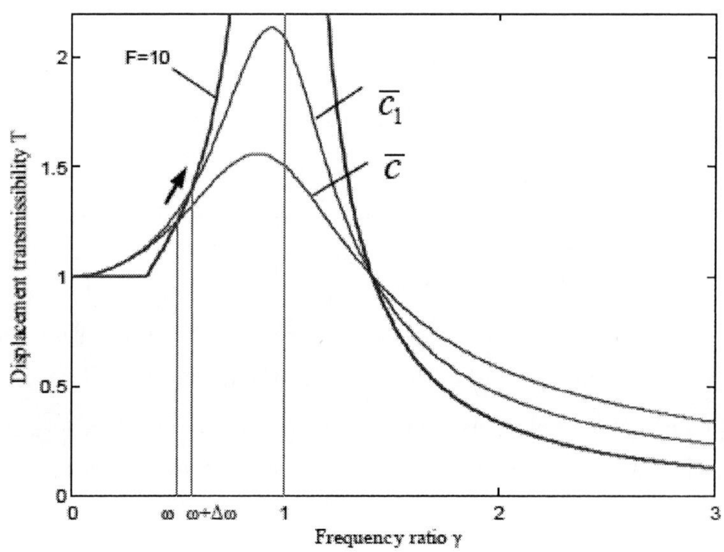

Figure 4. Transition of the displacement transmissibility curves

Compared with the vibration isolation with viscous damping, there exists locking phenomenon in the isolation system with dry friction damping, and relative displacement between the isolated mass and base exists only when the exciting frequency larger than unlock frequency. Furthermore, the displacement transmissibility increases significantly near the natural frequency of the system when the dry friction is small, which is similar to the undamped vibration isolation system.

The dry friction damping force in formula (5) is the primary harmonic component, which can be equivalent to viscous damping:

$$F\,\mathrm{sgn}(\dot{z}) \approx \frac{4F}{\pi}\cos(\omega t - \varphi) = \frac{4F}{A\omega\pi}\dot{z} \tag{17}$$

The equivalent viscous damping coefficient is

$$\overline{c} = \frac{4F}{A\omega\pi} \tag{18}$$

The exhausted energy of the equivalent viscous damping in one vibration period is

$$\Delta E_2 = \oint \frac{4F}{\pi}\cos(\omega t - \varphi)dz = 4FA \tag{19}$$

It's shown from the above analysis that the exhausted vibration energy by the equivalent viscous damping is the same as the dry friction damping.

The formula (18) shows that the equivalent viscous damping coefficient is relevant not only to the dry friction force, but also to the relative displacement amplitude and exciting frequency of the isolation system. On the condition that the exciting frequency ω is close to ω_0 and $\omega < \omega_0$, the relative displacement amplitude A increases to $A + \Delta A$ with ω increasing to $\omega + \Delta\omega$. And the equivalent viscous damping coefficient is

$$\overline{c}_1 = \frac{4F}{(A + \Delta A)(\omega + \Delta\omega)\pi} = \overline{c} - \Delta\overline{c} \tag{20}$$

Due to $\overline{c}_1 < \overline{c}$, the equivalent viscous damping coefficient decreases and the displacement transmissibility transits from the value corresponding to the viscous damping coefficient $\overline{c}$ (relevant to frequency ω) to the value corresponding to $\overline{c}_1$ (relevant to frequency $\omega + \Delta\omega$), as shown in Figure4. The displacement transmissibility trends up to infinite with ω close to ω_0.

4. Conclusion

Vibration response characteristics of the isolation system with dry friction damping under base displacement perturbation was analyzed using harmonic balance method, and the displacement response was compared between the isolation systems with dry friction damping and vicious damping separately. The results show that the isolation system with small dry friction damping can't meet the demands of displacement reduction near the natural frequency, and it can realize full-frequency vibration isolation when the lock frequency passes beyond the resonance frequency band by improving dry friction.

The damping mechanism of the dry friction type of vibration isolator used in engineering is more complex, and the dissipative mechanism cannot be described only by dry friction damping. The composite damping with dry friction and vicious damping is more appropriate for the dry friction type of isolator, and the displacement transmissibility at the frequency close to the natural frequency of the system is similar to the system with viscous damping.

5. References

[1] S.B. Ban, X.Y. Li and X.D. Jiang. Nonlinear dynamic model of the wire rope vibration isolator. : Mechanics in Engineering, 2012, 34 (1):66-69
[2] B. Zhu, Y.H. Ma, D.Y. Zhang, etc. A constitutive model of metal rubber based on hysteresis property. Acta Physica Sinica. 2012, 61 (7):474- 481
[3] Y.Y. Li, X.Q. Huang. Influence parameters of the damping performance of metal rubber material. Journal of Vibration, Measurement &Diagnosis, 2009, 29 (1):23-26
[4] H. Luo, J.S. Chen, X. l. Yan. Performance analysis and experimental verification of GWF vibration isolator. Noise and Vibration Control.2012, 32 (4): 178-182
[5] S. P. Liu, Z.S. Liu. Research on isolation character of dry friction isolator. Mechanical Manufacture. 2012, 50 (579):31-33
[6] X.Y. Yu. Technology research of self-fit-vibration isolation for electronic devices carried by vehicles. Chongqing Institute of Technology, 2008

Welding Thermal Simulation and Corrosion Study of X-70 Deep Sea Pipeline Steel

Weipeng Zhang [1], Zhuoran Li[1], Jixiang Gao[2], Zhengwu Peng[1]

[1]National Engineering Research Center of Near-Net-Shape Forming Technology for Metallic Materials, South China University of Technology, Guangzhou 510640, P. R. China

[2]College of Electromechanical Engineering, Guangdong Polytechnic Normal University, Guangzhou, 510635, P. R. China

Email: 357597978@qq.com

Abstract: Gleeble thermomechanical processing machine was used to simulate coarse grain heat affected zone (CGHAZ) of API X-70 thick wall pipeline steel used in deep sea. Microstructures and corresponding corrosion behavior of the simulated CGHAZs using different cooling rate were investigated and compared to the as-received material by scanning electron microscope and electrochemical experiments carried out in 3.5 wt. % NaCl solution. Results of this study show that the as-received samples exhibited a little bit higher corrosion resistance than the simulated CGHAZs. Among 3 sets of simulation experiments, the maximum corrosion tendency was exhibited at the $t_{8/5}$ = 20 s with the most martensite-austensite (M-A) microstructure and highest corrosion potential was shown at the $t_{8/5}$ = 60 s.

1. Introduction

Pipelines are widely used in oil and gas industry as it is the most economical and feasible way to transport fluids such as oil and gas, seawater and other materials associated with the process of exploitation, transportation, refining and storage in oil and gas industry.[1] Demand for energy all over the world is promoting non-conventional oil discoveries such as deep underwater extraction which bring many challenges so that engineering innovation, design adaptation, and selection of alternative materials and systems are needed.[2-4] The major issues in deep water production are corrosion and extreme conditions of high pressure.[5] Higher thickness to diameter ratio is the trend for pipeline steel used in deep ocean and multiple-wire submerged-arc welding which have high heat input is the most economic and effective way to connect thick wall pipeline.

Ren et al. [6] found that segregation and coarse grains lead to the high corrosion sensitivity of heat affected zone (HAZ). It is very difficult to analyze the effect of microstructure on corrosion behavior using the true weldment due to the narrow zone of HAZ in a welded joint. Usually different heat treatments procedures were used to simulate the microstructures of HAZ. With the development and application of thermal simulation technique in welding field nowadays, research on the relation between microstructures and properties for the welded joint becomes possible. [7-9]

A lot of researches has been done on the microstructure and corrosion of heat affected zone of pipeline steel [10,11], but few studied the relationship between the CGHAZ and the corrosion performance of deep sea pipeline under high input welding so far. The object of this study is to investigate the relationship between microstructure and corrosion behavior of CGHAZ of deep sea X-70 pipeline by using Gleeble thermomechanical simulator.

Content from this work may be used under the terms of the Creative Commons Attribution 3.0 licence. Any further distribution of this work must maintain attribution to the author(s) and the title of the work, journal citation and DOI.
Published under licence by IOP Publishing Ltd

2. Experimental

2.1. Materials

The test material is X-70 pipeline steel plate with a thickness of 40.5 mm. Chemical composition of the steel is shown in Table 1. The test material has high strength that yield strength and tensile strength was ~530 MPa and ~ 640 MPa, respectively. It also has an excellent Charpy v-notch impact toughness of 355 J at -30 °C.

Table1. Chemical composition of X-70 pipeline steel, in wt. %

C	Si	Mn	Ni	Cr	Mo	S	P	Co	V	Ti
0.04	0.21	1.27	0.28	0.25	0.17	0.002	0.01	0.14	0.003	0.01

2.2. Gleeble Thermomechanical Simulation

A thermal simulation test was carried out using Gleeble-3800 thermomechanical processing machine to simulate different welding heat input. The standard sample is 10 mm×10 mm×60 mm in size. Figure 1 shows the welding thermal cycle of the sample heated up to 1300 ℃ by the rate of 130 ℃/s, then, cooled down at 3 different cooling rate of $t_{8/5}$ = 20 s, 60 s and 100 s.

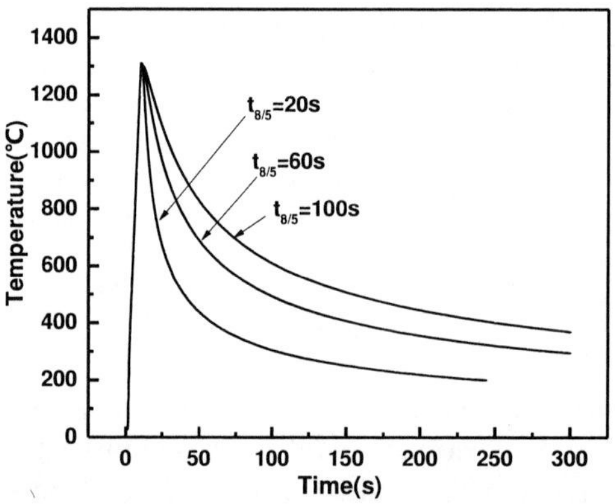

Figure 1. Schematic diagram of welding thermal cycle with different $t_{8/5}$

2.3. Electrochemical Tests

Before each single experiment, samples were ground by a wet 1500 grit SiC paper. All electrochemical experiments were carried out in 3.5 wt.% NaCl solution (100 mL) using a standard three-electrode cell (PAR), involving a saturated calomel electrode (SCE) and a platinum counter electrode. Electrochemical corrosion potential (E_{corr}) was collected by a potentiostat (IVIUM VERTEX). Potentiodynamic polarization scans (PD) were conducted with a starting potential 250 mV below the open circuit potential (OCP) under a scan rate of 0.2 mVs^{-1}. The OCP was recorded for 3600 s to reach a steady corroding condition from an open circuit setup. Values of corrosion current density (I_{corr}) were determined from Tafel plots of the polarization data. The experiments were carried out in air at room temperature. The reported values are averages from three individual measurements.

2.4. Material Characterization

Microstructure of the samples was characterized using scanning electron microscopy (SEM). All samples were prepared by standard metallographic techniques. The samples were mounted in epoxy resin at room temperature before grinding and polishing.

3. Results and Discussion

Figure 2 shows SEM microstructure of the as-received sample and CGHAZ under different thermal cycles. Microstructure of the as-received sample is mainly composed of polygonal ferrite (PF) and acicular ferrite (AF). As shown in Figure 2(b), the CGHAZ is mainly composed of bainite ferrite (BF) and few M-A island when $t_{8/5}$=20 s. Granular bainite is the main component of BF, in which a small amount of ferrite co-existing with lath bainite. The M-A island is bulk and disperses in the matrix. The granular bainite gradually enlarge, and quasi-polygonal ferrite appears with increasing $t_{8/5}$. When $t_{8/5}$= 60 s, as shown in Figure 2(c), granular bainite is the major phase and little polygonal ferrite can be clearly seen. The number of M-A island decreased significantly compared with the sample obtained by $t_{8/5}$= 20 s. The M-A island gradually disappear when $t_{8/5}$=100 s and it composed of polygonal ferrite and bainite which is larger in size.

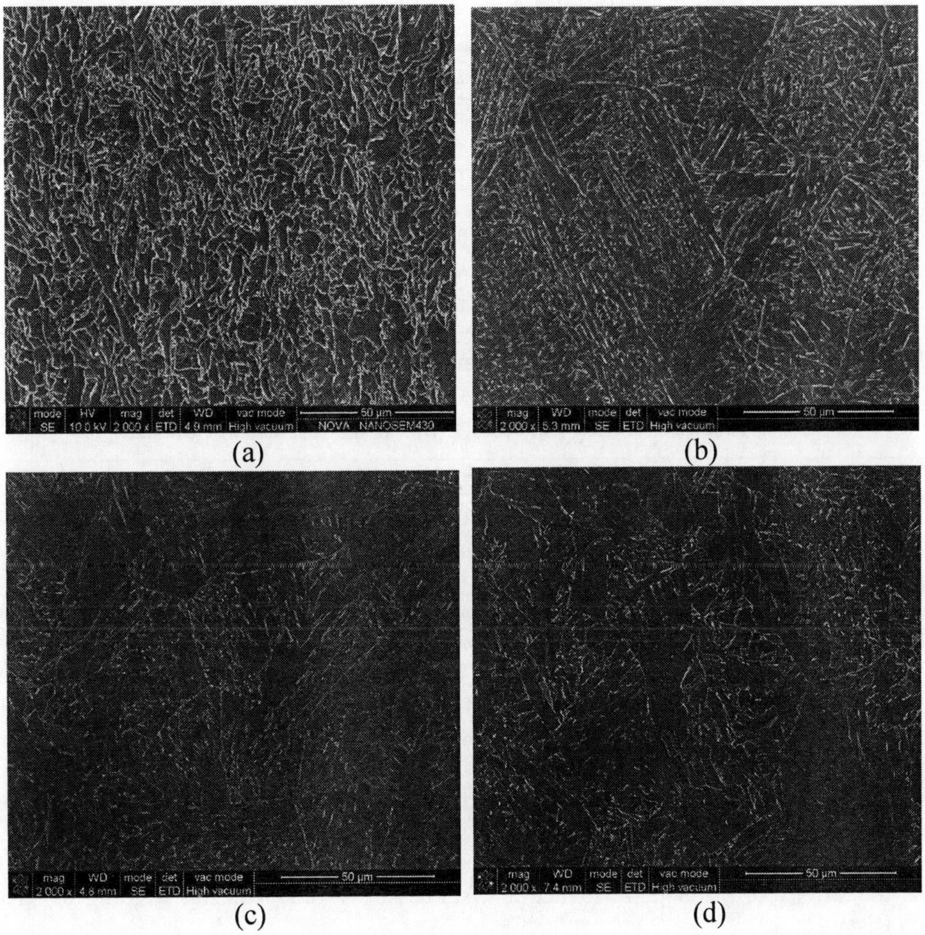

Figure 2. SEM micrographs of X-70 pipeline steel; (a) as received; and heat-treated (b) $t_{8/5}$= 20 s (c) $t_{8/5}$= 60 s(d) $t_{8/5}$= 100 s

Figure 3 depicts potentiodynamic polarization scans of the as-received sample and 3 sets of thermal simulated samples in NaCl solution. The potential E in V vs. a saturated calomel electrode (SCE) as a function of current density (i) was shown. The electrochemical corrosion experiment results are summarized in Table 2. As demonstrated in Table 2 and Figure 3, the heat affected samples showed a drop of 0.06-0.14 V compared to the E_{corr} of the as-received sample. That means the thermal simulated samples have higher tendency to corrode.

Among the 3 sets of thermal simulation experiments, the CGHAZs has the lowest corrosion potential at the $t_{8/5}$=20 s, which shows the maximum corrosion tendency and the least corrosion tendency appeared at $t_{8/5}$=60 s. The reason of the higher corrosion tendency of heat simulated samples compare to as-received sample is the changes of microstructure that uneven distribution of bainite and M-A island appeared which led to the decrease of corrosion resistance. All samples have a similar corrosion rate that their corrosion current densities are in the same order.

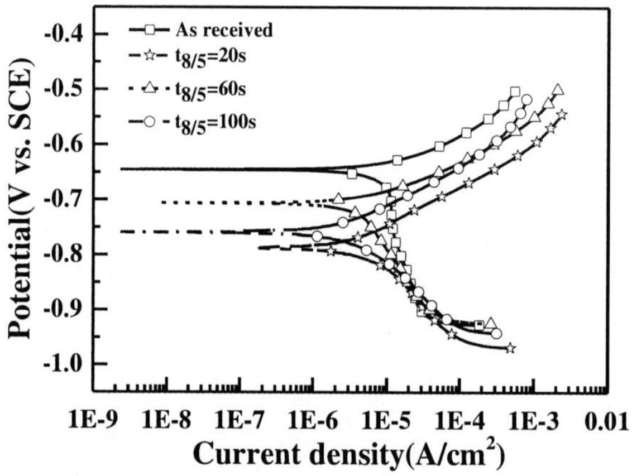

Figure 3. Potentiodynamic polarization scans of the samples in 3.5wt. % NaCl solution

Table 2. Electrochemical test results (in 3.5 wt. % NaCl) of the samples

	As received	$t_{8/5}$= 20 s	$t_{8/5}$= 60 s	$t_{8/5}$= 100 s
OCP(V)	-0.6770±0.02	-0.7183±0.02	-0.6744±0.03	-0.6915±0.04
E_{corr}(V)	-0.6433±0.03	-0.7874±0.01	-0.7051±0.02	-0.7630±0.02
i_{corr}($\times10^{-6}$A/cm^2)	1.774±0.74	1.893±1.01	3.428±1.21	1.778±0.96

4. Conclusion

(1) The microstructure of the as-received sample composed of acicular ferrite and polygonal ferrite. While, the microstructure of the heat-treated sample with the shortest $t_{8/5}$ ($t_{8/5}$= 20 s) results in the highest amount of M-A island. With the increase in $t_{8/5}$, M-A island began to disappear, microstructure transformed from bainite with fine grain to bainite and ferrite with coarse grain.

(2) As-received samples exhibited a little bit higher corrosion resistance compared to that of the simulated CGHAZs.

(3) Among 3 sets of thermal simulation experiments, microstructure of CGHAZ containing M-A island has the worst corrosion performance and should be avoided during welding to ensure good corrosion resistance.

5. Acknowledgments

The work was financially supported by the Guangzhou Science and Technology Project (no.2014Y2-00224) and Major Project of Department of Education of GuangDong Province (no.2016KZDXM046)

6. References

[1] Shahriar A, Sadiq R and Tesfamariam S 2012 *J. Loss. Prevent. Proc.* **25(3)** 505-23
[2] Ribeiro L P, Paulo C A S and Neto E A 2003 *Offshore Technology Conference* (Houston, Texas)
[3] Gao J X, Li L J, Li J F, Wang X N and Wei S J 2014 *Advanced Materials Resrearch* **997** 409-14
[4] Li X, Fan Y, Ma X, Subramanian S V and Shang C 2014 *Materials & Design* **67** 457-63

IOP Conf. Series: Materials Science and Engineering **275** (2017) 012036 doi:10.1088/1757-899X/275/1/012036

[5] Hill RT, Ramirez F A, Perez A L and Monty B A 2012 *Corros. 2012* (Salt Lake City, Utah)
[6] Ren C Q, Xian X, Wang L, Liu Y and Zheng Y P 2012 *Corros. Eng. Sci. Technol.* **47** 441-45
[7] Moeinifar S, Kokabi A H and Hosseini H R 2011 *J. Mater. Process. Technol.* **211** 368-75
[8] Sun J, Dai H and Zhang Y 2011 *Mater. Des.* **32** 1612-16
[9] Sung H K, Shin S Y, Cha W, Oh K, Lee S and Kim N J 2011 *Mater. Sci. Eng.* **528** 2250-57
[10] Ren C Q, Xian N, Wang X, Liu L and Zheng Y P 2013 *Corros. Eng. Sci. Technol.* **47(6)** 441-45
[11] Wang L W, Liu Z Y, Du C W, Wang X H and Li X G 2014 *Corros. Sci.* **85(1)** 401-10

Effect of Ultra Fast Cooling Coiling Temperature on Microstructures and Properties of X80 Pipeline Steel

Feng Zhou[1], liejun Li[1], Jixiang Gao[2], Haibo Sun[3] and Zhengwu Peng[1]

[1]South China University of Technology, China
[2]Guangdong Polytechnic Normal University, China
[3]Foshan University, China
Email:63419822@qq.com

Abstract: Effect of ultra fast cooling coiling temperature on microstructures and properties of X80 pipeline steel was studied adopting the detection means of energy dispersive spectrometer, scanning and transmission electron microscopy. Result shows that the toughness increases with the increasing of ultra fast cooling coiling temperature. The yield strength decreases firstly and then increases with the increasing of ultra fast cooling coiling temperature. The yield strength reaches the minimum value at the temperature of around 350 °C.

1. Introduction

There is a great demand on pipeline steel for oil and gas transmission, especially in P.R.China [1-3] due to its high strength and good toughness[4]. At the present, thermo-mechanical controlled processing (TMCP) is the main manufacturing technology to obtain better performance of X80 pipeline steel with refining grains [5-14].The basic mean to controlled rolling is that the big pressure rate rolling at low temperature which may beyond the equipment capability.

The core concept in ultra fast cooling which is also called NG-TMCP (new generation thermo-mechanical controlled processing) includes [15]: 1) achieving the strain accumulation in austenite during the continuous rolling at relatively high rolling temperatures; 2) ultra fast cooling after rolling to keep the work hardening of austenite; 3) stop cooling temperature near dynamic phase transition temperature; 4) cooling route control according to the requirements for microstructure and properties of the steel. The present work aims to utilize ultra fast cooling process to produce the X80 pipeline steel and investigate its microstructure characteristics and mechanical properties.

2. Experimental Procedures

The specimens were 17.5 mm-thick plates. The chemical compositions are listed in Table 1. The technical parameters of TMCP for pipeline steel are shown in Table 2. Specimens for tensile tests were taken from the rolled plates in the transverse direction. Tensile tests were conducted at room temperature at a crosshead speed of 5 mm/min on a SHT4106-1000KN servo-hydraulic machine. Drop weight tear test (DWTT) specimens were machined into the dimensions of 75 mm× 17.5mm× 305 mm (T-L). The DWTT test was performed at -15°C using a JL-40000 impact machine. Full size Charpy impact specimens were machined from the rolled plates in the transverse direction with their notch parallel to the rolling direction (T-L), and impact tests were done at tempreatures from 20°C to -100 using a ZBC-300C impact machine. Specimens for SEM observations were mechanically polished and etched by a 4 vol% nital solution or a Lepera solution. They were observed by a scanning electron microscope (Quanta 600 SEM). Specimens for transmission electron microscope (TEM, JEM – 2100F) analysis were

Content from this work may be used under the terms of the Creative Commons Attribution 3.0 licence. Any further distribution of this work must maintain attribution to the author(s) and the title of the work, journal citation and DOI.
Published under licence by IOP Publishing Ltd

mechanically ground to about 50 μm in thickness. Then, these specimens were twin-jet electropolished at a voltage of 40 V. The electrolyte consisted of 10 vol% perchloric acid and 90 vol% glacial acetic acid.

Table 1. Chemical composition of the X80 pipeline steel (wt%)

C	Si	Mn	Nb	Mo +Cr+ Ti	P	N
0.04~0.06	0.2~0.3	1.7~1.9	0.06~0.10	0.60~0.70	≤0.02	0.0073

Table 2 .The rolling process parameters for X80 pipeline steel

Code	Recrystallization zone rolling stage	Non recrystallization zone rolling stage		Ultra fast cooling stage		Laminar cooling stage	
	Start blooming rolling temperature °C	Start finishing rolling temperature °C	Finishing exit temperature °C	Exit Temperature °C	Cooling rate, °C /s	Coiling temperature °C	Cooling rate, °C /s
1#	1050	940	840	650	40	450	8
2#	1050	940	840	650	40	400	8
3#	1050	940	840	650	40	350	8
4#	1050	940	840	650	40	320	8

3. Results

3.1. Mechanical Properties

Mechanical properties are shown in Table 3. It is seen that the mechanical properties of all specimens meet the requirements of X80 pipeline steel standard. The yield ratio of Specimen 3# is the lowest. Result of drop weight tear test are shown in Table 4. It can be seen that all specimens' DWTT datum meet the requirements of X80 standard. Tear area of Specimen 4 # is relatively small. Full size Charpy impact test results are shown in Table 5. It is concluded that all specimens' mechanical properties meet the requirements of X80 standard.

Table 3 .Tensile properties of X80 pipeline steel

code	$Rt_{0.5}$, MPa	R_m,MPa	A_{50}, %	$Rt_{0.5}$,R_m
1#	652	720	38	0.91
2#	645	735	38	0.88
3#	635	745	37	0.85
4#	660	760	37	0.87

Table 4 .DWTT of X80 pipeline steel

Code	SA/%	
1#	95	95
2#	92	95
3#	90	90
4#	90	88

Table 5. CVN of X80 pipeline steel

temperature code		1#	2#	3#	4#
20°C	CVN/J	336	288	288	289
	SA /%	100	100	100	100
0°C	CVN/J	356	285	287	283
	SA /%	100	100	100	100
-20°C	CVN/J	337	280	267	259
	SA /%	100	100	100	94
-40°C	CVN/J	315	265	256	252
	SA /%	100	93	93	94
-60°C	CVN/J	303	229	245	254
	FA /%	97	80	80	92
-80°C	CVN/J	288	224	215	21
	SA /%	83	51	51	9
-100°C	CVN/J	227	155	196	13
	SA /%	65	51	51	4

From Figure1, it is observed that the impact energy of the 4# specimens under -60 °C that can satisfy the requirements of technical conditions.

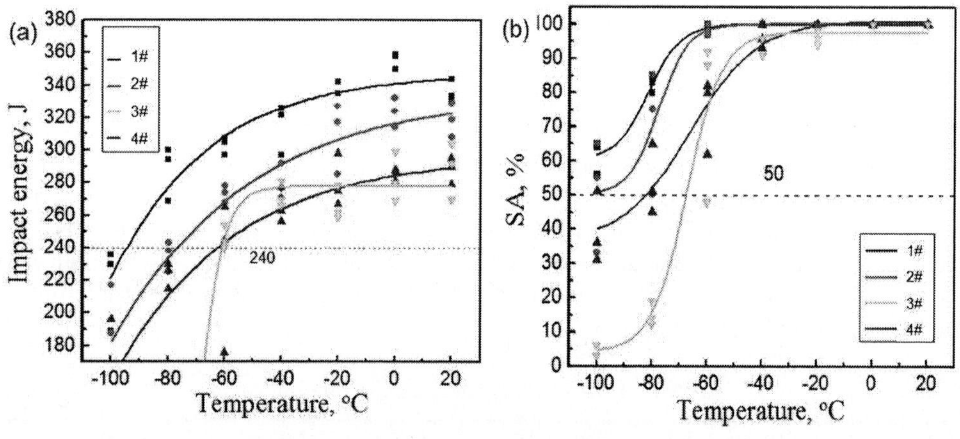

Figure 1. X80 pipeline steel impact toughness

3.2. Microstructure Characteristics

Figure2 shows SEM micrographs of the X80 pipeline steel. It can be seen microstructure of Specimen1 # is relatively bulky, There is a small amount of polygonal ferrite(PF) and M/A islands in the microstructure, Specimen 2 # mainly acicular ferrite tiny evenly, Most of M/A islands are small and evenly distributed, microstructures of Specimens 3 # and 4 # are fine, acicular ferrite matrix grain boundary in the mutual crisscross.

IOP Conf. Series: Materials Science and Engineering **275** (2017) 012037 doi:10.1088/1757-899X/275/1/012037

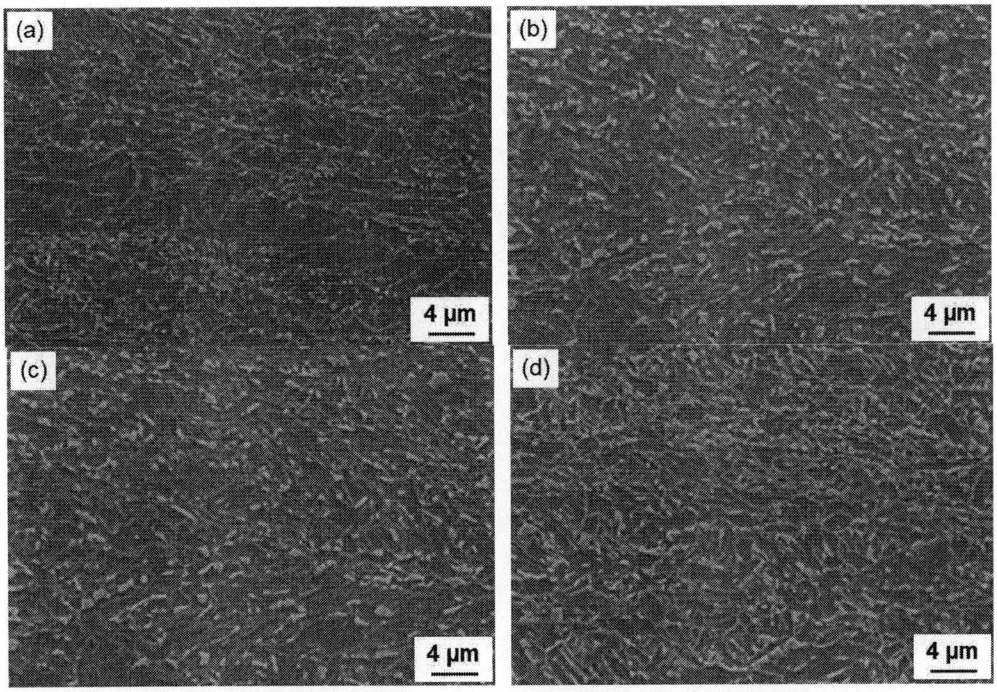

Figure 2. X80 pipeline steel impact toughness

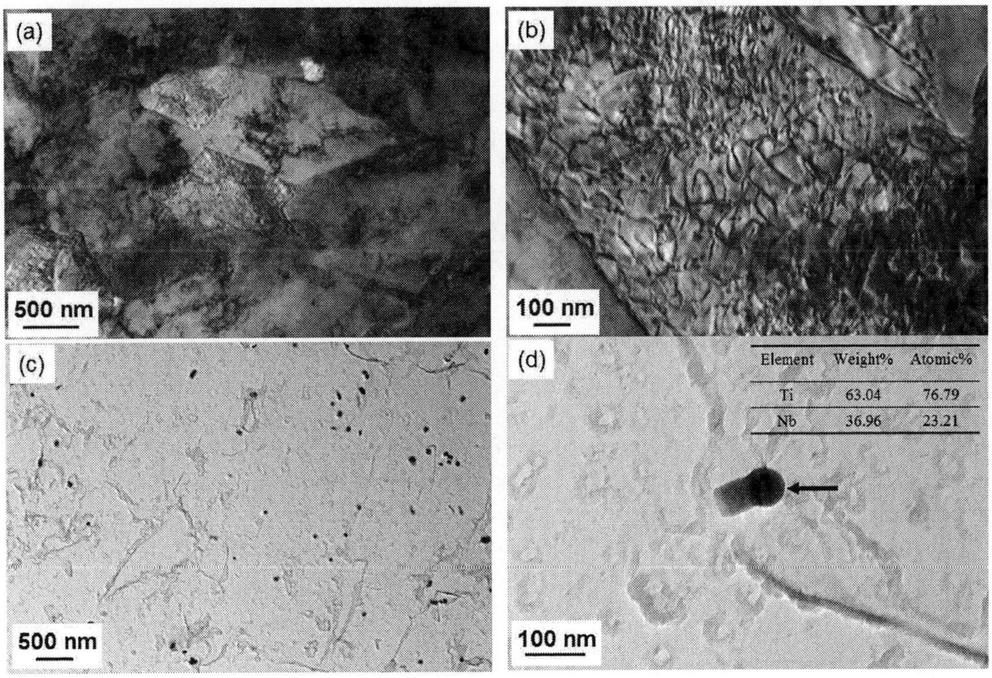

Figure 3. Specimen 2 # TEM observations

Figure 3 gives the microstructure morphology of Specimens2# and 4# for X80 pipeline steel using TEM observation.

It is seen that Specimen 2 # dislocation is developed, while precipitation is less. The size precipitation is relatively fine and uneven distributed. The types of precipitation is Ti, Nb carbonitride precipitation.

IOP Conf. Series: Materials Science and Engineering **275** (2017) 012037 doi:10.1088/1757-899X/275/1/012037

It can be known that Ti content is higher in rectangular particle, Nb content is higher in round particles. Nb/Ti ratio is 0.3.

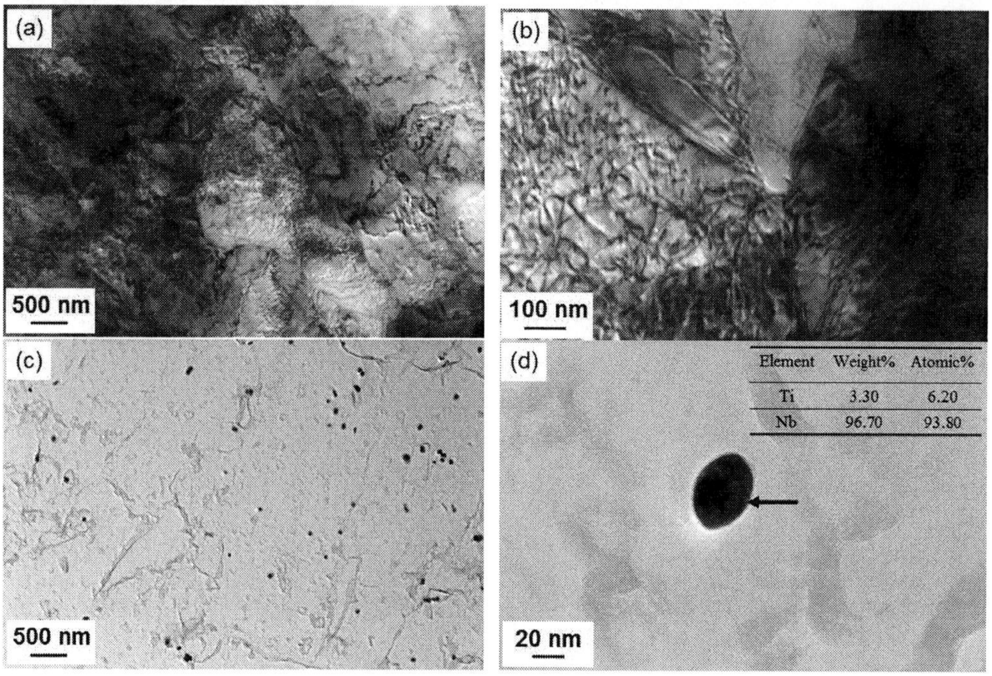

Figure 4. Specimen 4 # TEM observations

The grain boundary of Specimen 4 # has plenty of dislocation which is widely distributed in the range of observation. Figure 4(c) shows that most precipitation particle size is small (around 30nm), and Figure 4 (d) shows a single round particle whose types of precipitation are Ti, Nb carbonitride precipitation composite. Most precipitation particle size is smaller, most typical particles as shown in the picture on the right around 30 nm,A single round particles can be observed, types of precipitation mostly Ti, Nb carbonitride precipitation composite, Through the EDS analysis It can be known that Nb content is high in circular particle Ti content is low and Nb/Ti ratio of 15.1.

4. Discussion

4.1. Relationship between X80 Pipeline Steel Coiling Temperature and Mechanical Properties

The coiling temperature of the samples are shown in table 2. It is seen toughness increases with the increase of coiling temperature, and the yield strength decreases firstly and then increases for X80 pipeline steel. Moreover, at about 350 °C, the lowest yield strength presents, yield ratio firstly decreases and then increases with the rising of coiling temperature. Comprehensive analysis shows that coiling temperature at about 400 °C can deliver better comprehensive mechanical properties.

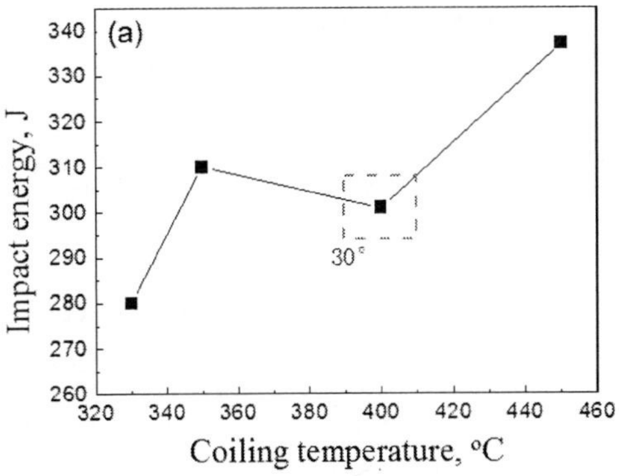

Figure 5. impact toughness to the coiling temperature

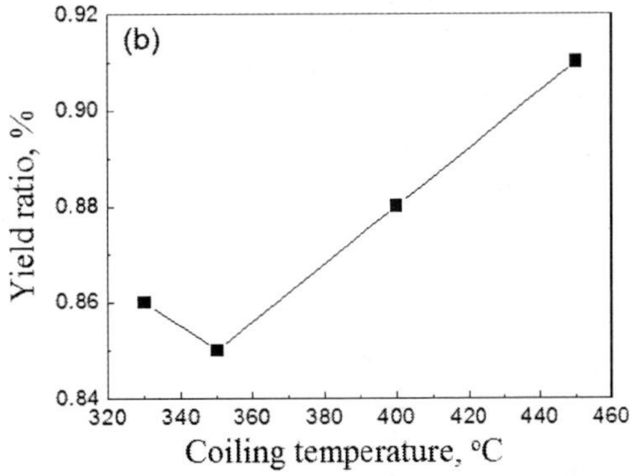

Figure 6.yield ratio to the coiling temperature

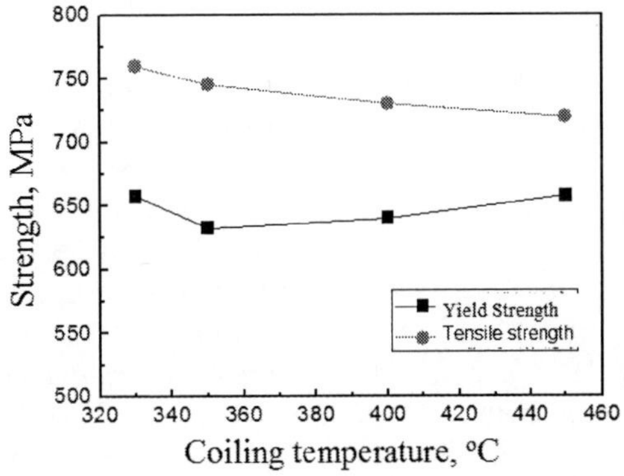

Figure 7. yield strength, tensile strength to the coiling temperature

4.2. Relationship between X80 Pipeline Steel Cooling Rate and Mechanical Properties
The cooling speed is greater than 30 degrees per second to ensure the passing through in austenite region in a tiny period of time, which results in work hardened austenite before transformation, which can lay a solid foundation for the control of phase transformation in work hardened state. The faster the cooling rate, the finer the microstructure can be obtained. Microstructure 1 # relatively is bulky.

As compared with the microstructure observation, it is seen that the M/A island near dislocation plug and launch dislocation phenomenon, dislocation in grain boundary caused the grain of work hardening and grain boundary of stress concentration. In addition, precipitation particle counterpoint dislocation pinning effect is significant. And with the increase of coiling temperature, precipitation particles get more and more, and also gradually become evenly distributed. But with the increase of coiling temperature, precipitation particle size increases gradually.

5. Conclusions

The relation between microstructure characteristics and mechanical properties of X80 pipeline steel by ultra fast cooling process was researched, and the following conclusions are drawn.

a) As the coiling temperature increases, so does the impact toughness of X80. Its yield strength decreases first and then decreases, the lowest point presents at around 350 °C, and yield ratio first decreases and then increases with the rising of coiling temperature. Comprehensive analysis shows that, at coiling temperature of about 400 °C, better comprehensive mechanical properties can be achieved.

b) Cooling speed is greater than 30°C/s to ensure the passing through in austenite region in a tiny period of time, resulting in work hardened austenite before transformation, which can lay a solid foundation for the control of phase transformation in work hardened state. The faster cooling rate, the finer microstructure.

c) Types of precipitation are mostly Ti, Nb carbonitride precipitation composite. The EDS analysis shows that Ti content is higher in rectangular particle,Nb content is higher in round particles.

d) With the increasing of coiling temperature, precipitation particles get more and more, and also gradually become evenly distributed. But with the increase of coiling temperature, precipitation particle size increases gradually.

6. Acknowledgments

The work was financially supported by Natural Science Foundation of Guangdong Province (no.2017A030313276) and Major Project of Department of Education of GuangDong Province (no.2016KZDXM046)

7. References

[1] Nayak S S, Misra R D K, Hartmann J 2008 *Mater. Sci. Eng. A* **494(1-2)** 456–463.
[2] Zhao J, Li J, Qu X Y 2011 *Sci. China Tech. Sci.* **54** 2754–2759.
[3] Corbett K T, Bowen R R, Petersen C W 2004 *International J. of Offshore and Polar Eng.* **14(1)** 75-80.
[4] Seung Y H, Sang Y S, Sunghak L 2010 *Metal. and Mater. Trans. A* **41A(2)** 329–340.
[5]Wang W, Yan W, Zhu L, 2009 *Mater. & Design* **30(9)** 436-3 443.
[6] Zhao M C, Yang K, Xiao F-R, 2003 *Mater. Sci. and Eng. A* **355(1-2)** 126-136.
[7] Xiao F-R, Liao B, Shan Y-Y 2006 *Mater. Sci. and Eng. A* **431(1-2)** 41-52.
[8] Hwang B, Lee H S, Kim Y G 2005 *Metal. and Mater. Trans. A* **36A(2)** 389-397.
[9] Xiao F, Liao B, Ren D 2005 *Mater. Characterization* **54(4-5)** 305-314.
[10] Díaz-Fuentes M and Gutiérrez I 2003 *Mater. Sci. and Eng. A* **363(1-2)** 316-324.
[11] Mingxing Z, Mokuang K 1993 *Trans. of Mater. and Heat Treatment* **14(1)** 14-19.
[12] Hwang B, Yang G K, Lee S, Kim Y M, Kim N J, & Yoo J Y 2005 *Mater. Sci. and Eng.A* **36(8)** 2107-2114.
[13] Wang C, Wu X, Liu J 2006 *Mater. Sci. and Eng. A* **438-440** 267-271.
[14] Kim Y, Kim Y G, Shin S Y 2010 *Korean J. of Metals and Mater.* **48(1)** 8-18
[15] Wang G D 2008 *Shanghai Metals* **30(2)** 1-5.

IOP Conf. Series: Materials Science and Engineering **275** (2017) 012038 doi:10.1088/1757-899X/275/1/012038

Effect of Different Connection Modes on Bolt Structural Properties of TC4 Alloy in Selective Laser Melting

Xiaodan Li[1], Shuangjun Huang[2], Liang Xu[2], Li Hui[2] and Song Zhou[2]

[1] Shenyang Aircraft Industry (Group) Co., Ltd, Lane 1, Lingbei Street, Huanggu District ,Shenyang, Liaoning, China
[2] Key Laboratory of Fundamental Science for National Defense of Aeronautical Manufacturing Process, Shenyang aerospace university, Lane 37, Daoyi Street, Shenbei District ,Shenyang, Liaoning, China.

Abstract: The bolt structural properties of selective laser melted (SLM) samples produced from TC4 powder metal has been investigated. Two different connection molds relative to single lap joint and bilateral lap joint as well as two different state of surface quality were considered. Samples and test procedures were designed in accordance with HB 5143 and HB 5287 standard. The results show that there is a strong influence of connection molds on the dynamic behavior of SLM produced TC4. The mechanical properties of bilateral lap joint are better than those of the single lap joint. Meanwhile the fatigue performance of the bilateral lap joint is much stronger than that of the single lap joint which it is a symmetrical structure of the two-shear test on both sides of the force evenly, while the single lap joint is a single shear sample of the uneven force. There are two kinds of fracture form most of which are broken in the first row of screw and a small part in the middle of the connecting plate.

1. Introduction

Selective laser melting is an additive fabrication process in which successive layers of powder are selectively melted by the interaction of a high energy density laser beam. Molten and re-solidified material forms parts, while non-melted powder remains in place to support the structure [1-4]. This layer-wise production technique offers some advantages over conventional manufacturing techniques such as high geometrical freedom, short design and manufacturing cycle time and made-to-order components. Titanium alloys have been used widely in various fields due to their good mechanical properties and corrosion resistance [5-7].

In the aerospace industry, much attention has been given to SLM of TC4. Because it shows a great potential compared to subtractive methods concerning the processing of TC4 titanium alloy which has excellent specific tensile and fatigue strengths and corrosion resistance. Forged TC4 has become increasingly popular because of its low cycle fatigue strength and crack propagation resistance, although wrought products, such as forged, extruded, as cast, or rolled materials, are widely used as structural and engine parts in aircrafts[11]. Normally, the conventionally machining of this alloy is characterized by some difficulties through a short life time of the tools and a low cutting speed. These negative aspects can be avoided using additive manufacturing methods.

With the development of aerospace technology, for reducing the mass and improving the performance of equipment effectively, the structurally large metallic components (SLMCs) manufactured by titanium alloys are increasingly applied in aircraft. However, problem of the forming metal material applied on different parts at the same component is more and more difficult [12-15].

Content from this work may be used under the terms of the Creative Commons Attribution 3.0 licence. Any further distribution of this work must maintain attribution to the author(s) and the title of the work, journal citation and DOI.
Published under licence by IOP Publishing Ltd

Therefore, the new technology of titanium alloy material properties research is particularly important. In aircraft structure, single lap joint bolt connection form has been widely adopted, and the damage form is given priority to with fittings of fatigue fracture. In this paper, TC4 titanium power was melted and formed by laser melted selection, and its structural connections were compared. The performance of bolted connecting structures under different lap forms was compared and analyzed. This study is based on the more advanced 3D printing technology, and it is different from other alloy materials. Titanium alloy is the object of study. It provides effective reference for the application of TC4 alloy unilateral lap joint in the aerospace field data support.

2. Materials and Methods

TC4 (Ti6Al4V) powder is a rich α stable element and α+β dual-phase titanium alloy powder. Table1 shows the measured element contents of TC4 powder. TC4 alloy has high mechanical strength, good fracture toughness, high hardenability and a wide forging-temperature range. TC4 alloy is usually used in aero engine compressor components. Therefore, a forging TC4 plate is used as the to-be-repaired substrate, which has circular grooves on the surface. The samples were composed by two pieces of SLM TC4 alloy plates and one base metal connecting piece, which the single/bilateral lap joint was adopted with the form of six single screw connection (M6), as shown in figure 1. There were a total of 30 samples, divided into three groups by different connection forms as follows: Group TA, Single lap joint, Ra6.3; Group TB, Single lap joint, Ra1.6; Group TC, bilateral lap joint, Ra1.6. TC4 alloy plate test section width was controlled in 25 mm, thickness in 3 mm, the connecting sheet width in 25 mm, thickness in 7 mm. the increase of the connection piece of geometry size ensured that the connection piece priority failure will not occur.

The static tensile and fatigue tests were conducted at a stress ratio of R=0.1, a sine wave of 6Hz to obtain the stress-life curves, and a constant rate loading of 1 mm/min in UTM9100. Before the test, according to the 40% of tensile failure load did the pre-tension test. During the trial, recording the real-time load and strain data ensured that there were more than 500 sampling points. Static stretching needs to ensure that the alignment is centered and the error is reduced. The load is determined by controlling the fatigue life in 180000~200000 cycle. Collect the strain date under the fixed cycle count. Unload until peak load is arrived. Record the real-time load and strain date, the fatigue cycle, failure modes and failure location. When calculating the experiment data, ensure the Degree of confidence no less than 95%, the effective number of test pieces shall not be less than four.

Table1. Chemical compositions of TC4 (wt/%)

Al	V	Fe	C	N	H	O	Ti
6.1	4.1	0.04	0.01	0.021	0.01	0.12	Bal.

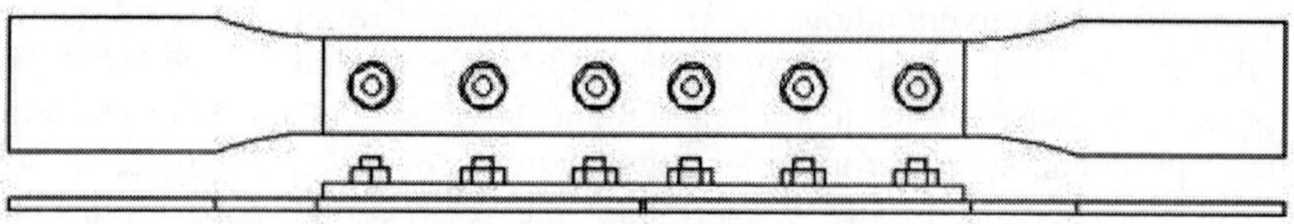

a) Single lap joint

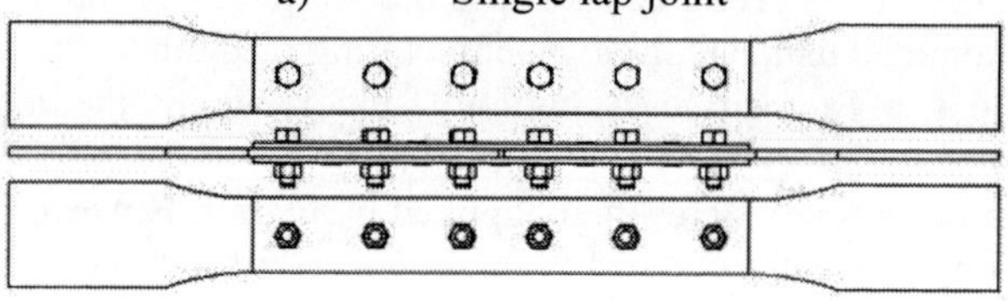

b) bilateral lap joint

Figure 1. Shape of samples

3. Results

3.1. Tensile Test

The average, standard deviation and dispersion coefficient of the test results are calculated according to the following formulas (1), (2) and (3), respectively. The calculation results are shown in Table 2. The discrete coefficients of three groups are both relatively small. Group TA and TB samples are based on the same structure and materials. The difference is that group TA samples are blown sand, and group TB samples are machined to enhance the surface quality and then blown sand. Resulting that the surface roughness is not the same which TA is Ra6.3 and TB is Ra1.6. Compared with group TA and TB samples, it can be seen that the fracture strength of TB is slightly higher than that of TA. The structure of group TB and TC samples is different which TB is unilateral lap bolt connection form and TC is bilateral lap bolt connection form. The experimental results show that the mechanical properties of the overlapped lap joint are better than those of the unilateral lap joint.

$$\overline{X}=\frac{\sum_{i=1}^{n}X_i}{n} \tag{1}$$

$$S=\sqrt{\frac{\sum_{i=1}^{n}(X_i-\overline{X})^2}{n}} \tag{2}$$

$$C_v=\frac{S}{\overline{X}}\times 100\% \tag{3}$$

Table2. The tensile test results of connecting pieces

Samples	Maximum load F/kN	Average $\overline{X}$	Standard deviation S	Discrete coefficient C_v
TA-1	38.38			
TA-2	37.51	37.95	0.6151	0.0162
TA-3	37.90			
TB-1	40.92			
TB-2	39.26	40.09	1.1738	0.0293
TB-3	40.00			
TC-1	49.95			
TC-2	51.15	50.55	0.8485	0.0168
TC-3	50.50			

Where $\overline{X}$ is the average, S is the standard deviation, C_v is the discrete coefficient, X_i is the measured value.

3.2. Fatigue Test Result

The first sample of each group was tested with 31% of static test failure load as a peak. Six samples were taken from each group to determine the peak load of the two test samples which strain collection was not necessary. But the last two samples were subjected to strain collection and the strain gauge was attached according to the test scheme to collect the trend of strain in the process of stretching near the nail hole under different cycles.

Repeated reassembly of samples may cause damage to its performance. For the titanium alloy, the crack initiation life occupies the main part of the whole fatigue life. The initial damage is caused in the vicinity of the nail hole due to the repeated assemble which greatly weaken the titanium alloy crack initiation life, So that the initial damage degree of each sample cannot be observed and quantitative analysis. Therefore, the dispersion of the experimental data is increased, and due to the limitation of the number of samples, the valid data of the group fail to obtain. But the test results still have some reference.

According to the experimental results, it can be seen that the results of TA and TB with the same structure are obviously different. In group TB, the fatigue life is lower than that of group TA under the stress peak of TA. In that case the fatigue performance of group TA is much better. At the same time, the fracture position of TA and TB can be observed, and the fracture area of TA and TB is basically the same. There are two kinds of fracture form most of which are broken in the first row of screw (fig2a) and a small part in the middle of the connecting plate (fig2b), while the fracture of TC is only presented at the first row of screws. Compared with different structures of experimental data, it can be found that the fatigue life of TC is similar to that of TA when the loading peak of TC is higher than that of TA, indicating that the fatigue performance of TC is much stronger than that of TB. TC is a symmetrical structure of the two-shear test, test plate on both sides of the force evenly while the TB is a single shear samples, the test plate on both sides of the uneven force.

Table 3. The fatigue test results of connecting pieces

Group	Number	Percentageof damage load (%)	Peak (kN)	Fatigue life (cycle)	Average (cycle)	Standard deviation	Discrete coefficient
TA	5	31	11.763	500000 227814 500000 126966 132092	248895	0.2942	0.0545
TC	6	30	15.165	211961 500000 500000 33910 246132 500000	245910	0.4538	0.0842

Note: 500000 indicates that the number of cycles is greater than 500,000

Table 4. The fatigue test results of group TB

Percentage of damage load (%)	Peak (kN)	Fatigue life(cycle)	Average(cycle)
27	10.824	299097	204364
27		109631	
28	11.225	92127	94680
28		97232	
31	12.428	72772	72772

The fatigue life of the material has a certain degree of dispersion, it is usually used to meet a certain degree of confidence in the statistical method to determine the fatigue life of a material. However, the fatigue life of the structural parts affected by factors such as the complexity of the sample geometry, the accuracy of the assembly will produce a greater dispersion resulting in more human resources, material and financial needs.

3.3. Fracture Analysis

The fatigue fracture surface is comprised of three morphologically distinct zones: the crack initiation zone, the crack propagation zone and the instant rupture zone. In the LCF, the crack initiation zone and the propagation zone occupy smaller area on the fracture surface. The fatigue fracture of the SLM TC4 titanium alloy typical connection samples can be observed in Fig. 3 in which inevitably the structure of large or small damage are occurred as a consequence of being assembled several times. Moreover the crack initiation life of the titanium alloy is greater than the crack propagation life, so the fatigue life of the samples after the injury will be greatly reduced.

The fatigue cracks mostly initiate from the samples surface or the defects inside of the samples (i.e. pores). As shown in Fig. 3, the fatigue crack nucleates at the corners of the plate samples and fatigue initiation site is obviously observed. This is because the mutual constraints between the grains at the corners and the edges are lower than the other positions. The majority of the injury appears at the edge of the plate (Fig 3a) or the edge of the bolt hole (Fig 3b). The fatigue source which is formed in the damaged surface propagates radially to the inside of the samples.

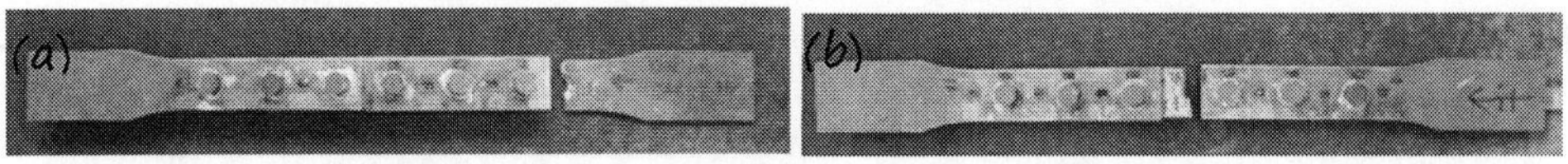

Figure2. Fracture location (a) Broken at the screw (b) Broken at the middle of the connection plate

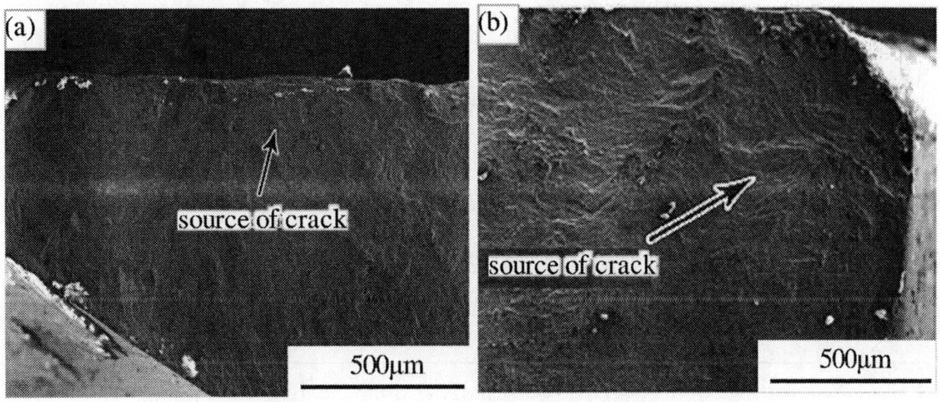

Figure3. Fracture surface morphology
(a) Crack initiates at the corner (b) Crack initiates at the bolt hole

4. Conclusions

The tensile properties are much influenced by bilateral lap form to unilateral lap form. The maximum failure load of the bilateral lap samples is much higher than that of the unilateral lap.

The consideration of connection form and surface quality has highlighted the sensitivity of mechanical properties to the microstructural and force distribution effects that arise from different lap modes. Bilateral overlap is a symmetrical double shear samples resulting in uniform force on both sides of the test plate, while the unilateral lap is a single shear samples, leading to uneven test on both sides of the test plate.

5. Acknowledgments

Authors wishing to acknowledge assistance or encouragement from colleagues, special work by technical staff or financial support from organizations should do so in an unnumbered Acknowledgments section immediately following the last numbered section of the paper.

6. References

[1] WANG Chunshui, HE Shengxin, ZHANG Erliang, et al. Analysis and Characterization of Sandblasted Surfaces Using Multiscale Analysis [J]. *Surface Technology*, 2015, 44(6):127-132.
[2] Van Hooreweder B, Moens D, Boonen R, Kruth JP, Sas P. Analysis offracture toughness and crack propagation of Ti6Al4V produced by selectivelaser melting. *Adv Eng Mater* 2012, 14: 92–7.
[3] Badrossammay M, Childs T H C.Further studies in selective laser melting of stainless and tool steel powders [J].*International Journal of Machine Tools & Manufacture*, 2007, 47(5): 779-784.
[4] Yang Yongqiang, Wang Di, Wu Weihui. Research Progress of Direct Manufacturing of Metal Parts by Selective Laser Melting [J]. *Chinese Journal of Lasers*, 2011, 38(6):54-64.
[5] Thijs L, Verhaeghe F, Craeghs T, Van Humbeeck J, Kruth JP. A study of themicrostructural evolution during selective laser melting of Ti6Al4V. *Acta Mater* 2010, 58: 3303–12.
[6] Kruth JP, Badrossamay M, Yasa E, Deckers J, Thijs L, Van Humbeeck J. Part and material properties in selective laser melting of metals. *In: Proceedings of the 16th international symposium on electromachining*. 2010.
[7] Kruth J, Froyen L, Van Vaerenbergh J, et al. Selective laser melting of iron-based powder [J]. *Journal of Materials Processing Technology*, 2004, 149(1/3):616-622.
[8] Badrossamay M, Childs T H. Further studies in selective laser melting of stainless and tool steel powders [J], *International Journal of Machine Tools & Manufacture*, 2007, 47(5):779-784.
[9] Osakada K, Shiomi M. Flexible manufacturing of metallic products by selective laser melting of powder [J]. *International Journal of Machine Tools & Manufacture*, 2006, 46(11): 1188-1193.
[10] Li Ruidi. Research on the Key Basic Issues in Selective Laser Melting of Metallic Powder [D].Hubei: *Huazhong University of Science and Technology*, 2010.
[11] V. Caina,b, L. Thijs, et al. Crack propagation and fracture toughness of Ti6Al4V alloy produced by selective laser melting [J]. *Additive Manufacturing* 2015 (5):68–76
[12] Dong Peng, Li Zhonghua, Yan Zhenyu et al. Research Status of Selective Laser Melting of Aluminum Alloys [J]. *Applied Laser*, 2015, 35(5):607-610.
[13] Zhang Xiaoli, Qi Huan, Wei Qingsong, et al. Experimental Study of Selective Laser Melted AlSi12 [J]. *Applied Laser*, 2013, 3394):391-397.
[14] Alberto Boschetto, Luana Bottini, Francesco Veniali, et al. Roughness modeling of AlSi10Mg parts fabricated by selective laser melting [J]. *Journal of Materials Processing Technology*, 2017, 241(3):154–163.
[15] ZHANG Yuan-jun. Connection Layout Analysis and Designs of High-Strength Bolts in the Large Steel Structure [J]. *Machinery Design & Manufacture*, 2012, **9**:230-231.

Surface Alloying of Au Atoms on a Ni (110) Surface: a Theoretical Study

Xin Li[1], Tianfu Gao[1], Chunhai Jiang[2] and Renzhong Huang[1*]

[1]College of Physical Science and Technology, Shenyang Normal University, Huanghe Street 253, Shenyang 110034, China
[2]School of Materials Science and Engineering, Xiamen University of Technology, Ligong Road 600, Xiamen 361024, China.
*rzhuang09@163.com

Abstract: We investigated the surface alloying of Au atoms on a Ni (110) surface using a molecular dynamics (MD) method. It is found that the concentration of deposited Au atoms and the depositing temperature determine the topological surface structures, which can be ascribed to the variation in Gibbs free energy of the system. Dimer formation on the over layer is used to explain the surface alloying of the miscible system and in good agreement with the experimental results.

1. Introduction

When one metal is deposited on another one, the deposited metal may either form islands on the substrate or alloy into the first or deeper layer of the substrate [1]. Alloying may take place at the surface whether the two metals form an alloy or not. Experiments showed [2] that interface mixing could have quite a strong influence on the electronic and magnetic properties of different systems, which promote the investigation of atomic structures of the substrate surface. The experiment of Au growth on a Cu (100) surface [3] have demonstrated that a c (2×2) Cu-Au layer is developed on the Cu (100) surface. When Au is deposited on a Ag (110) surface [4], the gold atoms can intermix with the first two Ag layers. The intermixing of Cu-Au and Au-Ag systems are characterized by the fact that these two metals are miscible. If two metals can't form stable alloys, one would expect the deposited metal to grow only on top of the substrate. However, two-dimensional surface alloys have already been found in such immiscible systems, such as Co on Cu (001) [5], Ag on Cu (001) [6] and Fe on Cr (100) [7]. It means that surface alloying might be responsible for the initial structure of metal-on-metal growth.

Nielsen *et al.*[8] investigated the nucleation and growth of Au on Ni (110) by *in situ* scanning tunneling microscopy (STM) and Rutherford backscattering spectroscopy at room temperature. The results shows that deposited Au atoms can replace surface Ni atoms forming an Au-Ni surface alloy and the surplus Ni atoms nucleate into $[1\bar{1}0]$ elongated islands. Jacobsen *et al.* [9] calculated the energetics of Au deposited on Ni surfaces using the effective medium theory and have found that the surface alloy is considerably more stable than the Au overlayer on Ni (100) or (110) surfaces. This phenomenon was modeled by Bozzolo *et al.* [10] using the Bozzolo-Ferrante-Smith (BFS) method.

In this work, in the viewpoint of kinetics and thermodynamics, we investigated the dynamical process of depositing Au atoms on a Ni (110) surface and elucidated the dependence of surface morphological structures on the Au coverage and the system temperature by using molecular dynamics (MD) simulations.

Content from this work may be used under the terms of the Creative Commons Attribution 3.0 licence. Any further distribution of this work must maintain attribution to the author(s) and the title of the work, journal citation and DOI.
Published under licence by IOP Publishing Ltd

2. Calculation Method

Atomic scale simulations are performed using Molecular Dynamics (MD) method based on tight binding theory [11]. According to N-body potential scheme, the site energy of the atom i is written as the sum of two terms, an attractive band energy E_i^b and a repulsive pair interaction E_i^r [12].

$$\mathrm{E_i = E_i^b + E_i^r} \quad (1)$$

In the framework of the second moment approximation of tight binding theory (TB-SMA), the real density of states is replaced by a schematic rectangular one having the same full width at half maximum, then

$$E_i^b = -\{\sum_j \xi_{\alpha\beta}^2 \exp[-2q_{\alpha\beta}(\frac{r_{ij}}{r_0^{\alpha\beta}} - 1)]\}^{1/2} \quad (2)$$

α and β indicate the chemical species. $\xi_{\alpha\beta}$ is an effective hopping integral and the exponent $q_{\alpha\beta}$ characterizes its distance dependence. $r_0^{\alpha\beta}$ is the equilibrium distance between the atoms i and j. The summation runs over all the neighbors j of the atom i.

The repulsive term E_i^r is described by a sum of Born-Mayer ion-ion repulsions

$$E_i^r = \sum_j A_{\alpha\beta} \exp[-p_{\alpha\beta}(\frac{r_{ij}}{r_0^{\alpha\beta}} - 1)] \quad (3)$$

Where $p_{\alpha\beta}$ is related to the bulk modulus of the metal. The estimation of the TB-SMA parameters has been made by fitting bulk (cohesive energy, the lattice parameters, the bulk modulus, the elastic constants, and mixing enthalpy) and surface properties (surface energy and surface misfit) [13,14]. The technical details and parameters of interatomic potentials can be found in the previous work [14].

3. Results and Discussion

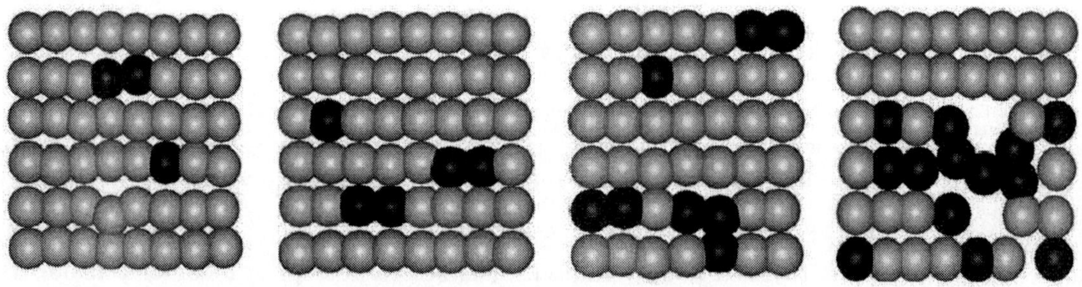

(a) (b) (c) (d)

Figure 1 Top view of the surface layer of depositing submonolayer Au atoms on a Ni (110) surface: (a) 0.09ML Au (black circles) at 300K. Dimer formation of Au atoms substituting into the Ni (110) surface layer; (b) 0.26ML Au at 300K. Appearance of dimers in the first surface layer; (c) 0.43ML Au at 300K. Appearance of trimers in the surface layer; (d) 0.26ML Au at 600K.

MD simulations are performed at a given temperature by using the Nose-Hoover method [15] to control the system temperature. The initial velocities of the atoms are assumed to be in the Boltzmann distribution at the desired temperature. We choose a cutoff distance to be two times as long as the lattice parameter of a gold crystal at 273 K, for all these simulations. In this method, the trajectory of each atom in the MD cell is followed through time and Newton's classical equations of motion with a time step of 5.0×10^{-16} s are integrated in a step-wise manner over finite time steps in terms of a fifth-order predictor-corrector method [16].

The initial MD cell consists of a 20-layer Ni (110) slab with 70 atoms in each layer. The downmost three Ni layers are fixed to keep the volume property. Varying numbers of Au atoms are deposited on the Ni (110) surface as an overlayer within the force range of its nearest surface atoms. Periodic boundary conditions are applied only in the directions parallel to the Ni (110) surface. The MD simulations are performed for different Au coverage at 300K and the results are shown in Figs. 1(a), 1(b) and 1(c).

In a low Au coverage, Au atoms would tend to form an overlayer because of its smaller surface energy than Ni. Due to the large lattice mismatch between Au (4.08 Å) and Ni (3.52 Å), the resulting overlayer is highly strained and will be a certain kind of structure in which the strain can be relieved. Similar to the experimental results [17], it can be seen from Figure 1(a) that a monomer and a dimer along the close-packed (110) direction appear in the alloying surface layer at the coverage of 0.09ML. As the Au coverage increased to 0.26ML, more Ni atoms are substituted and more dimers form as shown in Figure 1(b). At the coverage of 0.34ML, a trimer can even form by connecting a dimer together with a monomer, as shown in Figure 1(c).

Positive interface energy, in combination with the size mismatch, may lead to the formation of the alloying surface layer. Strain is partially relieved by the presence of the surface. In fact, as discussed by Tersoff [18] for systems dominated by a large size mismatch, strain effects can also be favor of the alloying at the surface even though they suppress the intermixing in the bulk. Large atoms will prefer to stay in the surface layers instead of penetrating into the bulk to decrease the system energy.

The above results demonstrate clearly that the possibility of forming dimers along the close-packed (110) direction on the Ni (110) surface at 300K is much higher than that forming a monomer. The concept of effective coordination number is used to interpret this phenomenon. Au atoms substituting into the first surface layer increase the coordination of Ni atoms and thus lower the system energy. However, the substituted Ni atoms forming Ni islands decrease its effective coordination number and thus increase the system energy. The competition of these two processes determines the morphology of Au atoms deposited on the Ni (110) surface.

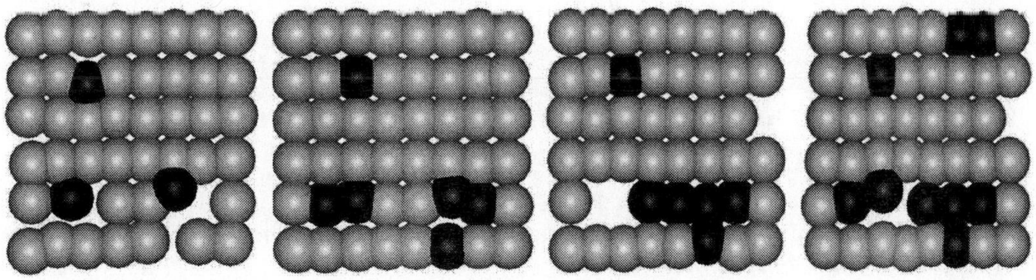

(a) (b) (c) (d)

Figure 2 Dynamical evolution process of Au deposition on a Ni (110) surface at the Au coverage of 0.57ML: (a) 50 time steps. (b) 150 time steps. (c) 350 time steps. (d) 2000 time steps.

To explain how the alloying takes place, we monitor the whole process of Au alloying into the first Ni (110) surface layer at the Au coverage of 0.57ML. As the simulation proceeds, an interesting phenomenon can be observed (shown as Figure2 (a)~(d)). Single Au atoms begin to substitute into the Ni (110) surface and form either monomer by substituting Ni atoms separately or dimer by substituting the adjacent Ni atoms. Sequentially, trimer, even Au cluster, can be seen in the alloying surface layer.

With the increase of the system temperature, the number of Au atoms entering the Ni (110) surface layer increases. Similar to the Ag-Cu system [19], the results shown in Figure 1 (b) and (d) for 0.26

IOP Conf. Series: Materials Science and Engineering **275** (2017) 012039 doi:10.1088/1757-899X/275/1/012039

ML at different system temperatures imply that the surface topological structures also depend on the system temperature. It is a competing result of energetically favorable intermixing and energetically unfavorable clustering.

To understand the driving force of surface alloying at finite temperature, one should assess Gibbs free energy (GFE) of the system. The GFE of an alloy system can be written as

$$G = H - TS \tag{4}$$

Where H is the enthalpy of the system, and S is the configuration entropy of the system.

It is assumed that there are n A atoms and (N-n) B atoms in a binary alloy consisting of N atoms, and that these atoms will diffuse spontaneously. The total number of probable arrangements is

$$\Omega = \frac{N!}{n!(N-n)!} \tag{5}$$

Then

$$S = -K \ln \Omega \tag{6}$$

When the system reaches equilibrium in a MD relaxation, we can count the number of the arrangement states, and then obtain the entropy of the equilibrium system. In fact, the atoms far away from surface layers (usually several atomic layers) in the matrix are hardly affected by the coverage of deposited solute atom and the system temperature. Here we only count the nearest-neighbor configuration number of the two topmost layers.

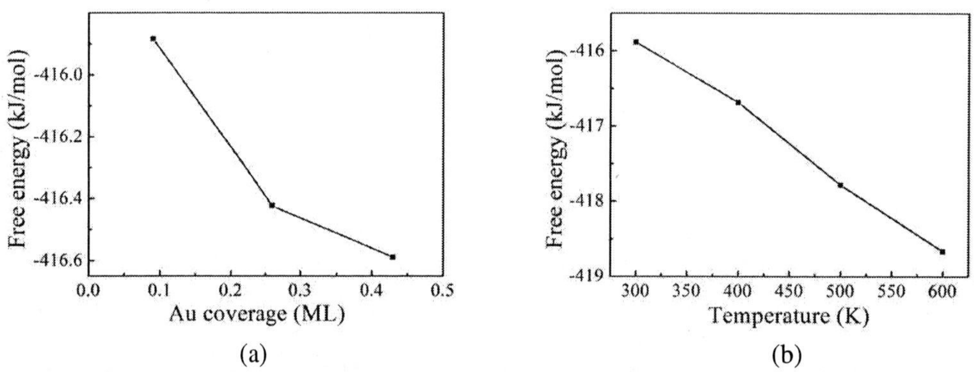

Figure 3. The Gibbs free energy of the Au-Ni system: (a) as a function of the concentration of deposited Au atoms on the Ni (110) surface at 300K; (b) as a function of its temperature at fixed Au coverage of 0.26ML.

The GFE of the Au-Ni system, as a function of the concentration of deposited Au atoms, is shown in Figure 3(a). One can find that the GFE decreases with the increase of Au coverage at lower coverage of less than 0.45ML, resulting in surface alloying. The tendency of surface alloying, however, decreases with the increase of the Au coverage. The GFE of the Au-Ni system as a function of the system temperature at 0.26ML is shown in Figure 3(b). The GFE decreases almost linearly with the increase of the system temperature. It is seen that the GFE decreases and the driving force of surface alloying can be enhanced by increasing the system temperature. Comparing Figure 3(a) with Figure 3(b), it can be concluded that the effect of temperature on surface alloying is much stronger than that of the coverage of the deposited atoms.

4. Summary

In summary, we simulated the kinetic behavior of depositing Au atoms on a Ni (110) surface using the MD method. It is found that surface alloying of Au atoms substituting into the Ni (110) surface depends on the concentration of deposited Au atoms and the system temperature. For the low Au coverage, Au atoms substitute into the Ni (110) surface and mostly appear in the form of dimer, in good agreement with the experimental results. With the increase of the deposition temperature, driving force for surface alloying is enhanced, leading to the formation of different surface topological structures, even the energetically unfavorable Au clusters. It can be attributed to the decrease in GFE of the Au-Ni system.

5. Acknowledgements

This work was supported by National Natural Science Fund of China (Grant No. 11674233).

6. References

[1] Christensen A, Ruban AV, Stoltze P, Jacobsen K W, Skriver H L, Nørskov J K and Besenbacher F 1997 *Phys. Rev. B* 56 5822-5834
[2] Mottet C, Tréglia G and Legrand B 1992 *Phys. Rev. B* 46 16018-16030
[3] Charmbliss D D and Chiang S 1992 *Surf. Sci.* 264 L187-L192
[4] Rousset S, Chiang S, Fowler D E and Chambliss D D 1992 *Phys. Rev. Lett.* 69 3200-3203
[5] Fassbender J, Allenspach R and Dürig U 1997 *Surf. Sci.* 383 L742- L748
[6] Trushin O S, Kokko K and Salo P T 1999 *Surf. Sci.* 442 420-430
[7] Choi Y J, Jeong I C, Park J Y, Kahng S J, Lee J and Kuk Y 1999 *Phys. Rev. B* 59 10918-10922
[8] Nielsen L P, Besenbacher F, Stensgaard I and Lægsgaard E 1993 *Phys. Rev. Lett.* 71 754-757
[9] Jacobsen K W, Stoltze P and Nørskov J K 1996 *Surf. Sci.* 366 394-402
[10] Bozzolo G, Ibanez-Meier R and Ferrante J 1995 *Phys. Rev. B* 51 7207-7216
[11] Rosato V, Guillope M and Legrand B 1989 *Philos. Mag. A* 59 321-336
[12] Cleri F and Rosato V 1993 *Phys. Rev. B* 48 22-33
[13] Hultgren R, Desai P D, Hawkins D T, Gleiser M, Kelly K K 1973 *Selected Values of the Thermodynamic Properties of Binary Alloys* (American Society for Metals, Metals Park, Ohio)
[14] Huang R Z, Ma X L and Wang Y M 2004 *Surf. Sci.* 549 227–236
[15] Nosé S 1984 *Mol. Phys.* 2 255-268
[16] Adler B J and Wainwright T E 1959 *J. Chem. Phys.* 31 459-466
[17] Nielsen L P, Besenbacher F, Stensgaard I and Lægsgaard E 1995 *Phys. Rev. Lett.* 74 1159-1162
[18] Tersoff J 1995 *Phys. Rev. Lett.* 74 434-437
[19] Sheng H W and Ma E 2000 *Phys. Rev. B* 61 9979-9982

IOP Conf. Series: Materials Science and Engineering **275** (2017) 012040 doi:10.1088/1757-899X/275/1/012040

Microstructure and Mechanical Properties of C/C Composite/TC17 Joints with Ag-Cu-Ti Brazing Alloy

Xiujie Cao, Ying Zhu, Wei Guo*, Peng Peng and Kaituo Ma

School of Mechanical Engineering and Automation, Beihang University, No.37 Xueyuan Road, Haidian District, Beijing, 100191, China.

*Email: gwei@buaa.edu.cn

Abstract: Carbon/Carbon composite(C/C) was vacuum brazed to titanium alloy (TC17) using Ag-Cu-Ti brazing alloy. The effects of brazing temperature on the interfacial microstructure and joint properties were investigated by energy dispersive spectrometer (EDS), a scanning electron microscope (SEM), X-ray diffraction (XRD) and Gleeble1500D testing machine. Results show that C/C composite and TC17 were successfully brazed using AgCuTi brazing alloy. Various phases including TiC, Ag(s, s), Cu(s, s), Ti_3Cu_4, TiCu, and Ti_2Cu were formed in the brazed joint. The maximum shear strength of the brazed joints with AgCuTi brazing alloy was 24±1 MPa when brazed at 860°C for 15 min.

1. Introduction

Carbon/carbon (C/C) composites have been applied in aviation and aerospace fields for their excellent physical properties , such as low density, very high thermo-elastic stability, high thermal conductivity, good thermal shock resistance, fatigue resistance, and excellent mechanical properties at high-temperature [1, 2]. C/C composite is both difficult and expensive to fabricate into large or complex shaped components due to its properties and the need of reliable integration of components into a structure requires an effective technology for their joining. Brazing between C/C composites and metal/alloy attracted more and more attention due to their wide potential engineering applications[3, 4]. There is a requirement to join them to metals/alloy in many engineering applications of C/C composite, such as TC17 titanium alloy. TC17 titanium alloy (nominal composition of Ti-5Al-4Mo-4Cr-2Sn-2Zr. for the convenience, Ti-5Al-4Mo-4Cr-2Sn-2Zr is marked as TC17 alloy in the following figures.) accounts a large proportion of the total usage of titanium alloy due to the excellent toughening and high temperature property [5].In the past decades, many processes have been developed for joining C/C composite to Titanium alloy. Due to its simplicity, cost-effectiveness and perfect adaptability of joint size and shape, brazing has been widely applied [6-9]. Two main problems of the joining of C/C composites to metals are the wetting of C/C composites by the liquid brazing alloy and the large coefficient of thermal expansion (CTE) mismatches between C/C composite ($CTE_{c/c}$=0-2 × 10-6 K-1) and titanium alloy ($CTE_{Titanium\ alloy}$ =8.4-9.9 ×10-6 K-1) which causes high residual stress, and consequently results in the degradation of joint during the cooling process[8, 9]. In order to over these problems, a soft interlayer at the interface or low-CTE reinforcements in the joining material has been applied. Song et al. [2] studied a novel Cu/TiCuZrNi composite braze used in C/C composites and Ti6Al4V brazed joint. The Cu composite interlayers could release the residual stresses and increase the mechanical properties at the joint interface. Wang et al.[10] reported a novel type of three-dimensional grapheme-reinforced Cu foam used as an interlayer for brazing seam. The extremely low CTE of graphene and the good plastic deformation capacity of Cu foam are both conductive to

Content from this work may be used under the terms of the Creative Commons Attribution 3.0 licence. Any further distribution of this work must maintain attribution to the author(s) and the title of the work, journal citation and DOI.
Published under licence by IOP Publishing Ltd

reducing the thermal residual stress during the cooling process and increasing the shear strength of the joint. Yang et al. [11] reported that the elastic deformation of Ni foam was beneficial for stress release for diffusion joining ZrB2SiC and Nb. Zhu et al. [12] used different thicknesses of Ni foam interlayer and verified that Ni foam could play the role as a buffer layer to reduce the residual stress. Moreover, many studies suggested the addition of low CTE materials(such as carbon fibers and carbon nanotubes , ceramic fibers, particles, etc.) into brazing alloy could mitigate residual stress and consequently improve the mechanical properties of a composite-metal joint [10]. Dai et al. [13] verified the B4C reinforced AgCuTi composite filler could reduce or eliminate residual stress in the interfacial region to improve the mechanical properties of the joints. Song et al. [14] used carbon nanotubes(CNTs) as reinforcements in the TiCuZrNi brazing alloy for joining C/C composites to Ti6Al4V. Zhao et al. [15] used a novel nano-Si3N4 particle as reinforcements in the AgCuTi composite filler for the brazed joint of Ti6Al4V and Si3N4 ceramic. Zhou et al. [16] verified AgCuTi composite filler strengthened by nano-Al_2O_3 improved the shear strength of brazed joint between C/C composite and TC4 alloy. The mismatch in CTE between substrates and brazing seam could be reduced due to the composite fillers by bring in fibers or tiny ceramic particles with lower CTE into traditional brazing fillers [16].

In this work, vacuum brazing of TC17 alloy and C/C composite with active Ag-Cu-Ti filler metal is performed at different temperature. The interface microstructures and evolution of the TC17/AgCuTi/C/C composite joint are analyzed and the mechanical properties of the joints are discussed.

2. Materials and Experimental Procedures

The base materials are C/C composite and TC17 alloy. C/C composites were obtained from Boyun New material Co., LTD, China, which was 2D entwined and 3D punctured. The C/C composite is a kind of carbon matrix composite reinforced by carbon fiber. The C/C composite was cut into blocks with a dimension of 5mm × 10mm × 5mm by spark cutting. Before brazing, the TC17 alloy was cut into blocks with a dimension of 5mm × 10mm × 5mm. A 50 μm thick Ag68.83-Cu26.77-Ti4.4 (wt.%) foil was used as filler metal (obtained from AECC Beijing Institute of Aeronautical Materials), which was melted in a vacuum high-frequency induction furnace using particles of titanium, copper and silver. The microstructures of C/C composite and TC17 alloy are shown in Figure1.The joining surfaces of both materials and filler foils were ground by 1000 grits silicon carbide paper and then cleaned ultrasonically for 15 min in ethanol and then dried by air blowing. The Ag-Cu-Ti foil was sandwiched between the C/C composite and TC17 alloy samples and a slight pressure was applied to maintain good contact between the joining surfaces and then the brazing assemblies were carefully placed into the vacuum furnace. The brazing assembly schematic is shown in Figure 2 (a).

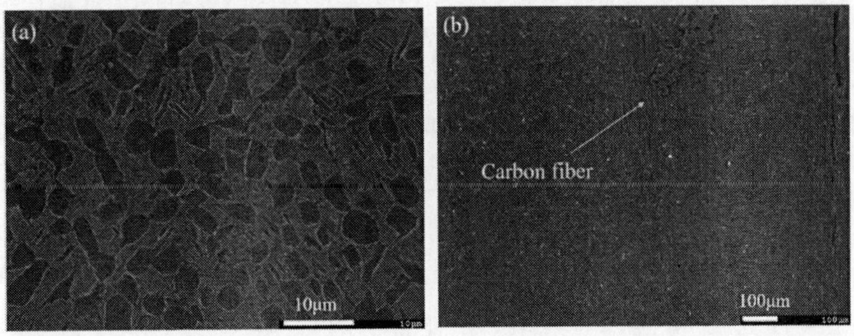

Figure 1. Microstructure of base materials: (a) TC17 alloy (b) C/C composite

The brazing was performed in a vacuum furnace under the vacuum of 1×10^{-3} Pa. The brazing temperature adopted in this study is 840°C-920°C while the temperature interval is 20℃. The brazing assemblies were first heated to 750°C at a rate of 15°C/min and held for approximately 10min, then to the brazing temperature (840°C-920°C) at a rate of 10°C/min and held at brazing temperature for 15min. Subsequently, the brazing couples were held for 15 min at brazing temperature, then cooled down to 100°C at a rate of 10°C/min. Finally, the joints were cooled down to room temperature in the vacuum furnace.

The cross-sections of the brazed joints were perpendicular to the brazed surface and prepared for metallographic analysis. The microstructures of the joints were characterized by scanning electron microscope (SEM) equipped with an energy dispersive X-ray spectrometer (EDS). Furthermore, the microstructure of TC17 side was investigated by X-ray diffraction (XRD) to identify the interfacial phases accurately. The shear test was performed at a constant speed of 0.5mm/min by the Gleeble 1500D at room temperature and the schematic diagram of shear test is shown in figure2.(b). At least five samples were used to average each set of experimental data. EDS and XRD analysis were performed to identify the interfacial phase on the fracture surfaces. The brazed specimens for XRD analysis were prepared by shear strength test. Then, the prepared specimens were polished layer by layer along the direction parallel to the brazing seam. Therefore, The XRD analysis can be executed on the surface of TC17 side.

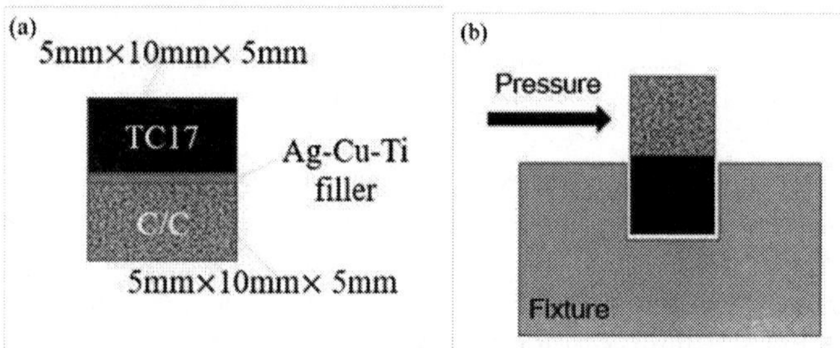

Figure 2. Schematic diagrams of (a) brazing assembly and (b) shear test experiment

3. Result and Discussion

3.1. Typical Interfacial Microstructure of TC17 /AgCuTi/ C/C Composite Joints

In this study reliable brazing of C/C composite and TC17 alloy was achieved using Ag-Cu-Ti foil. Figure3 shows the typical interfacial microstructure and the main element of C/C composite and TC17 joint brazed at 860°C for 15min. Figure 3(a) clearly presents that the interface are free of microcracks and pores and the joint exhibits sound bonding. The joint consists of one diffusion layer and three reaction layers, as shown in figure3 (a): zone I (diffusion layer adjacent to TC17 alloy), zone II (reaction zone adjacent to diffusion layer), zone III (brazing seam) and zone IV (continuous thin reaction layer adjacent to C/C composite).

Table 1. Element composition and possible phases of the regions marked in Figure3 (at.%)

	Ti	Cu	Ag	C	Possible Phase
A	93.29	6.41	0.3	-	Diffusion layer
B	67.96	29.91	2.14	-	Ti_2Cu
C	51.32	45.86	2.82	-	TiCu
D	4.22	7.42	88.36	-	Ag(s, s)
E	42.01	48.81	9.18	-	Ti_3Cu_4
F	42.73	54.67	2.6	-	Ti_3Cu_4
G	44.76	52.81	2.43	-	TiCu
H	26.93	1.35	5.26	66.46	TiC

Figure3 (b) and (c) show the magnified microstructures of TC17/AgCuTi interface and AgCuTi/C/C composite interface, respectively. It can be seen that there exist one thin layer(layer H) near C/C composite side and three reaction layers(Layer B, C and F) near TC17 side. In the center of brazing zone, there are three different phases: gray phase(marked by F), white matrix phase (marked by D) and dark block phase(marked by G).The EDS result of A-H zones were listed in Table 1.

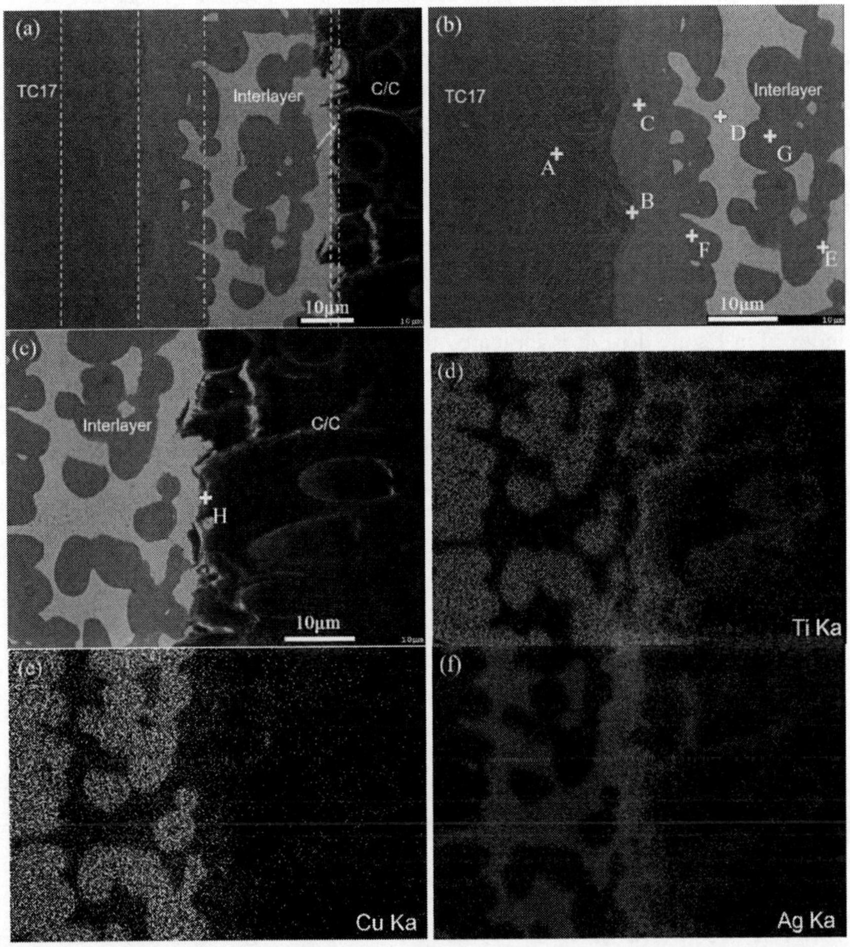

Figure 3. Back-scattered electron image of C/C composite/AgCuTi/TC17 joint (a) the brazed joint at 860℃ for 15min, (b) AgCuTi/TC17 interface, and (c) C/C composite/AgCuTi interface(d) interface between C/C composite and interlayer and EDS element distribution maps of Ti (e) EDS element distribution maps of Cu (f) EDS element distribution maps of Ag

According to the EDS result and Ti-Cu binary alloy diagram[17], the three reaction layers in zone II are probably Ti_2Cu (layer B), TiCu(layer C) and Ti_3Cu_4 intermetallic phase (layer F). The content of element Ti and Cu indicates that diffusion and reactions occur between base material and brazing filler. Along layer A, layer B,layer C,and layer F, the content of element Ti decreases while the element of Cu increases gradually. This indicates that diffusion and reactions between base material and brazing filler. In the center of brazing zone, the white phase(marked by D) is Ag-based solid solution (abbreviated as Ag(s,s)) white the dark phase (marked by G) and gray phase (marked by E) are probably TiCu and Ti_3Cu_4. Moreover, the layer H is very thin and mainly composed of Ti(~26.93 at.%) and C (~66.46 at.%). The content of C element in layer H is higher than other layer. One reason is that

C element from the C/C composite diffused inte layer H during brazing process. According the thermodynamic data manual [18], the Gibbs free energy of TiC formation are represented as follows:

$$Ti+C \rightarrow TiC \qquad (1)$$

$$\Delta G^0(KJ/mol)=-183.1+0.01T$$

The Gibbls free energy of forming TiC calculated by Formula (1) is about -172KJ/mol at 860°C .The Gibbs free energy of forming TiC is negative. Therefore, element Ti and C can react with each other at 860°C. The formation of TiC promotes the wettability of brazing filler on C/C composite, which is in accordance with literatures reported by other researchers[19].According to Ag-Cu binary alloy phase diagram[17], during the brazing process, the AgCu eutectic melt firstly, and then the element Ti can dissolve in it. The dissolved Ti diffused toward to C/C composite, and finally reacted with the element C to form TiC reaction layer.

To further identify the phase constitutions, XRD analysis for C/C compositer/TC17 brazed joint is performed.Figure4(a) shows the XRD pattern of the surface of joint brazed at 860°C for 15 min after shear tests. The peaks of TiC phase could be indexed in the profile. Figure4(b) shows the XRD pattern of the surface of joint brazed near to diffusion layer after polishing process. The peaks of Ti_2Cu phase can only be indexed in Figure 4(b). It is indicated that Ti_2Cu is only formed near TC17 alloy side. This is accordance with the EDS result. Cu(s,s) is indexed in Figure4. According to EDS result, Cu(s,s) cannot be identified accurately from layer A to F. In accordance with literatures reported by other researchers on brazing carbon materials [4, 14, 19], Cu(s,s) probably exists in layer G and/or layer E. As analyzed above, the typical interfacial microstructure of brazed joint is TC17/diffusion layer/Ti_2Cu+TiCu+Ti_3Cu_4/Ag(s,s)+Cu(s,s)+TiCu/TiC/C/C composite.

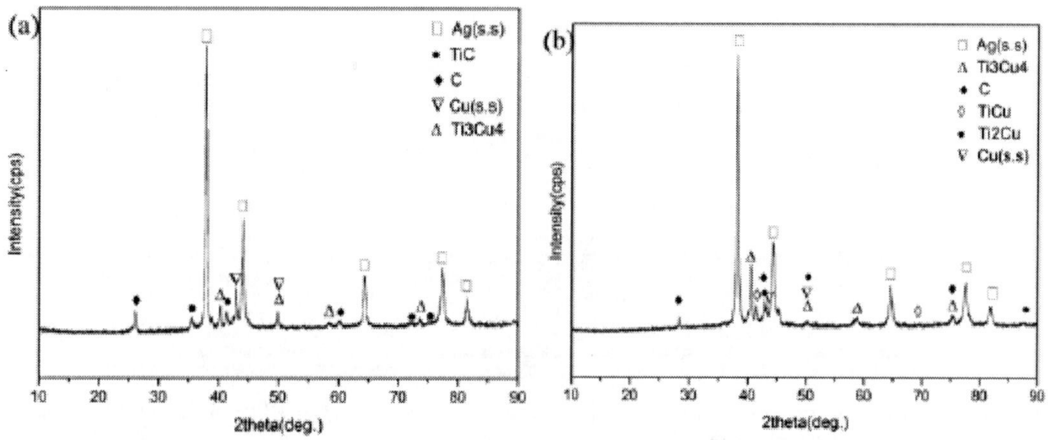

Figure 4. XRD pattern from the surface of joint brazed at 860°C for 15 min after shear tests (a) layers at TC17 side (b) layers near to diffusion layer after polishing process

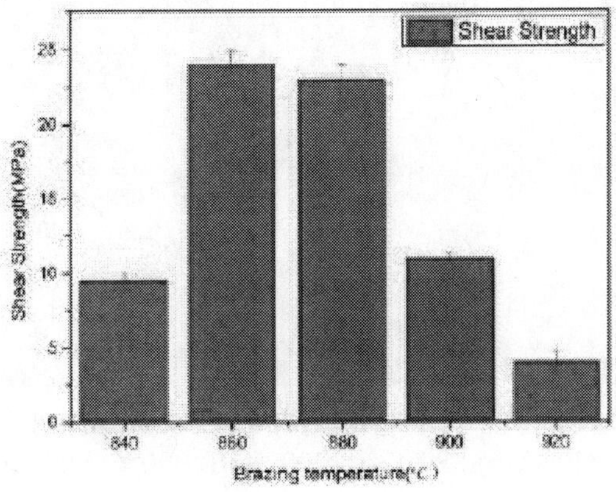

Figure 5. Effect of brazing temperature on the shear strength of TC17 AgCuTi/ C/C composite joints

3.2. Effect of Brazing Temperature on the Interfacial Microstructure of TC17 /AgCuTi/ C/C Composite Joints

Figure5 illustrates the interfacial microstructures of TC17 /AgCuTi/ C/C composite joints brazed at 840°C, 880°C, 900°C and 920°C, for 15 min. It can be seen the interfacial microstructure has significantly changed with the increase of brazing temperature from 840 to 920°C. Because of the effect of capillarity at high temperature, the molten filler is expelled out of the joint gap. Therefore, the width of brazing seam decreases with the increasing temperature. On the contrary, the increasing temperature is beneficial to the intensive diffusion and dissolution between brazing filler and TC17 alloy. So the thickness of diffusion increases with the increasing temperature. On the other hand, Ti_2Cu and TiCu layers close to TC17 side thicken gradually with the increasing temperature. However, Ti_3Cu_4 layer firstly thickens when the brazing temperature is blow 900°C and decreases significantly when the temperature exceeds 900°C as shown in Figure5.

According to the literature[16], it can be explained that when the temperature reaches 920°C, the excessive growth of TiCu phase in the brazing seam which depletes element Ti and Cu. Therefore, Ti_3Cu_4 layer almost disappears. High magnified SEM image of TiC layer was obtained (the inserted images in Figure5) to investigate the thickness of TiC layer changes when the brazing temperature increases. With the increasing temperature, the TiC layer thickens gradually. The TiC layer is thin at lower brazing temperature (840°C). Further increasing the brazing temperature to 900°C and 920°C, as shown in Figure 5(c) and (d), the formation of massive TiC layer is detected which deteriorates the property of brazed joint.

With the increase of temperature from 840°C to 920°C, Ag(s,s) phase changes obviously. When the brazing temperature reaches to 920°C, Ag(s,s) phase changes into an integrated layer. According to the literature [20], the Ag-Ti system exhibits a weak compound formation tendency compared to the Cu-Ti system. Because of the effect of high temperature, the content of TiCu increases obviously.

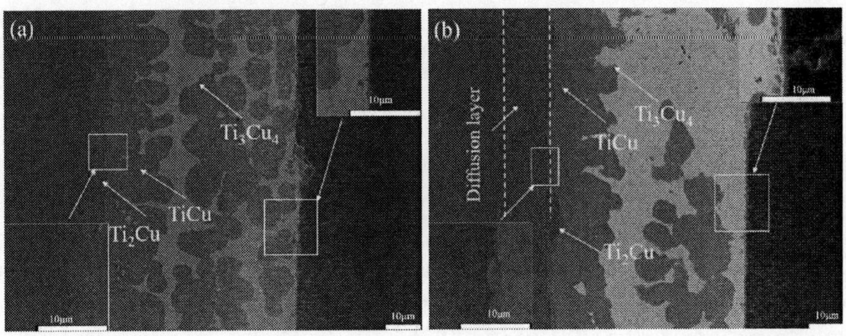

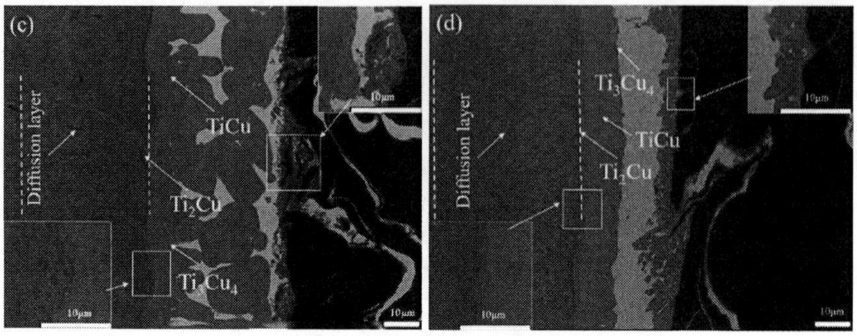

Figure 6. Interfacial microstructure of TC17 /AgCuTi/ C/Ccomposite joints brazed at different temperatures for 15 min: (a) 840°C (b) 880°C (c) 900°C (d) 920°C

3.3. Mechanical Property Evaluation of the TC17 /AgCuTi/ C/C Composite Joints

The effect of the brazing temperature on the shear strength of the joints is shown in Figure6. When the brazing temperature is relative low (840°C), the reaction at brazing joint is insufficient and the diffusion layer is very thin. Thus, the shear strength of the joint is lower. As the brazing temperature increases, the reaction at brazing joint increases greatly and the shear temperature increases as well. When the brazing temperature is 860°C, the shear strength reaches a maximum of about 24MPa. Further increasing the brazing temperature, the shear strength of joints is stabilized (about 23MPa). When the brazing temperature reaches 900°C and 920°C, the shear strength of joints decreases significantly.

4. Conclusion

Brazing of C/C composite to TC17 alloy was performed at 840, 860,880,900 and 920°C for 15 min by using AgCuTi brazing filler alloy. The interfacial microstructure and effects of brazing temperature and mechanical property of brazed joints are investigated in this study. The main findings are summarized as follows.

1) The TC17 alloy and C/C composites are successfully joined with Ag-Cu-Ti active filler metal by vacuum brazing.
2) The typical interfacial microstructure of the brazed joint is TC17/diffusion layer/Ti_2Cu+TiCu+Ti_3Cu_4/Ag(s,s)+Cu(s,s)+TiCu/TiC/C/C composite.
3) With the increase of brazed temperature, the thickness of diffusion layer, Ti_2Cu layer, TiCu layer and TiC layer increase and the width of brazing seam decreases gradually respectively.
4) The maximum shear strength of the joint is 24 ± 1MPa brazed at 860°C for 15 min.

5. Acknowledgment

We are grateful the financial support to this study from National Natural Science Foundation of China (Grant No. 11372019) and National International Collaborative Science and Technology Research Project China (Grant No. 2013DFR50590).

6. References

[1] G. Kou, L.-j. Guo, Z.-q. Li, J. Peng, J. Tian, C.-x. Huo, Microstructure and flexural properties of C/C-Cu composites strengthened with in-situ grown carbon nanotubes, Journal of Alloys and Compounds 694 (2017) 1054-1060.
[2] X. Song, H. Li, V. Casalegno, M. Salvo, M. Ferraris, X. Zeng, Microstructure and mechanical properties of C/C composite/Ti6Al4V joints with a Cu/TiCuZrNi composite brazing alloy, Ceramics International 42(5) (2016) 6347-6354.

[3] T. Wang, T. Ivas, C. Leinenbach, J. Zhang, Microstructural characterization of Si3N4/42CrMo joint brazed with Ag–Cu–Ti + TiNp composite filler, Journal of Alloys and Compounds 651 (2015) 623-630.
[4] K.X. Zhang, W.K. Zhao, F.Q. Zhang, L.L. He, New wetting mechanism induced by the effect of Ag on the interaction between resin carbon and AgCuTi brazing alloy, Materials Science and Engineering: A 696 (2017) 216-219.
[5] Y. Yang, H. Zhang, H. Qiao, Microstructure characteristics and formation mechanism of TC17 titanium alloy induced by laser shock processing, Journal of Alloys and Compounds 722 (2017) 509-516.
[6] D. Fan, J. Huang, B. Cui, J. Yang, S. Chen, X. Zhao, Interfacial Behavior and Its Effect on Mechanical Properties of Cf/SiC Composite/TiAl6V4 Joint Brazed with TiZrCuNi, Journal of Materials Engineering and Performance 26(3) (2017) 1114-1121.
[7] J.M. Shi, J.C. Feng, H. Liu, X.Y. Tian, L.X. Zhang, Vacuum brazing of the Gr/2024Al composite and TC4 alloy using AgCuTi filler alloy with Ni-Al interlayer as auxiliary heat source, Journal of Alloys and Compounds 694 (2017) 672-681.
8] J.M. Shi, J.C. Feng, X.Y. Tian, H. Liu, L.X. Zhang, Interfacial microstructure and mechanical property of ZrC-SiC ceramic and Ti6Al4V joint brazed with AgCuTi alloy, Journal of the European Ceramic Society 37(8) (2017) 2769-2778.
[9] S. Zhang, Y. Yuan, Y. Su, X. Song, Interfacial microstructure and mechanical properties of brazing carbon/carbon composites to stainless steel using diamond particles reinforced Ag-Cu-Ti brazing alloy, Journal of Alloys and Compounds 719 (2017) 108-115.
[10] Z. Wang, G. Wang, M. Li, J. Lin, Q. Ma, A. Zhang, Z. Zhong, J. Qi, J. Feng, Three-dimensional graphene-reinforced Cu foam interlayer for brazing C/C composites and Nb, Carbon 118 (2017) 723-730.
[11] W.Q. Yang, P. He, T.S. Lin, C.B. Song, R.S. Li, D.C. Jia, Diffusion bonding of ZrB2-SiC and Nb using dynamic compressed Ni foam interlayer, Mat Sci Eng a-Struct 573 (2013) 1-6.
[12] Y. Zhu, D. Qi, W. Guo, H. Kang, P. Qu, The braze joint between Al2O3 to 1Cr18Ni9Ti using a nickel foam, Welding in the World 59(4) (2015) 491-496.
[13] X. Dai, J. Cao, Z. Chen, X. Song, J. Feng, Brazing SiC ceramic using novel B4C reinforced Ag–Cu Ti composite filler, Ceramics International 42(5) (2016) 6319-6328.
[14] X.-r. Song, H.-j. Li, X. Zeng, Brazing of C/C composites to Ti6Al4V using multiwall carbon nanotubes reinforced TiCuZrNi brazing alloy, Journal of Alloys and Compounds 664 (2016) 175-180.
[15] Y.X. Zhao, X.G. Song, C.W. Tan, S.P. Hu, J. Cao, J.C. Feng, Microstructural evolution of Si 3 N 4 /Ti6Al4V joints brazed with nano-Si 3 N 4 reinforced AgCuTi composite filler, Vacuum 142 (2017) 58-65.
[16] Y.H. Zhou, D. Liu, H.W. Niu, X.G. Song, X.D. Yang, J.C. Feng, Vacuum brazing of C/C composite to TC4 alloy using nano-Al2O3 strengthened AgCuTi composite filler, Materials & Design 93 (2016) 347-356.
[17] T.B. Massalski, J.L. Murray, L.H. Bennett, H. Baker, Binary alloy phase diagrams, American Society for Metals1990.
[18] P.W. Atkins, Physical Chemistry, Universty Press*2006.
[19] Q. Miao, W. Ding, Y. Zhu, Z. Chen, Y. Fu, Brazing of CBN grains with Ag-Cu-Ti/TiX composite filler - The effect of TiX particles on microstructure and strength of bonding layer, Materials & Design 98 (2016) 243-253.
[20] H. Wang, J. Cao, J. Feng, Brazing mechanism and infiltration strengthening of CC composites to TiAl alloys joint, Scripta Materialia 63(8) (2010) 859-862.

IOP Conf. Series: Materials Science and Engineering **275** (2017) 012041 doi:10.1088/1757-899X/275/1/012041

High Strain Rate Response of 7055 Aluminum Alloy Subject to Square-spot Laser Shock Peening

Rujian Sun, Ying Zhu, Liuhe Li, Wei Guo and Peng Peng

School of Mechanical Engineering and Automation, Beihang University, Beijing, 100191, China

E-mail address: sunrujian@buaa.edu.cn

Abstract: The influences of laser pulse energy and impact time on high strain rate response of 7055 aluminum alloy subject to square-spot laser shock peening (SLSP) were investigate. Microstructural evolution was characterized by OM, SEM and TEM. Microhardness distribution and in-depth residual stress in 15 J with one and two impacts and 25 J with one and two impacts were analyzed. Results show that the original rolling structures were significantly refined due to laser shock induced recrystallization. High density of microdefects was generated, such as dislocation tangles, dislocation wall and stacking faults. Subgrains and nanograins were induced in the surface layer, resulting in grain refinement in the near surface layer after SLSP. Compressive residual stresses with maximum value of more than -200 MPa and affected depths of more than 1 mm can be generated after SLSP. Impact time has more effectiveness than laser pulse energy in increasing the magnitude of residual stress and achieving thicker hardening layer.

1. Introduction

Laser shock peening (LSP), also called laser shock processing, is an innovative surface treatment technique, which has been widely employed to enhance fatigue, corrosion and wear resistance of metallic components served in aerospace engineering [1-3]. Comparing to shot peening (SP), the compressive residual stress layer introduced by LSP is 2-5 times thicker than that generated by SP [4]. LSP applies a high power density (in GW cm^{-2} range), ultra-short duration (tens of nanoseconds) pulse laser to the surface of metallic materials, forming plasma that reaches a temperature higher than 10000°C and generating a high pressure shock wave (in GPa range). The interaction between shock wave and target material will generate high density of microdefects, induce grain refinement and modify near surface residual stress field [5].

According to the shape of laser spot, LSP can be divided into circular-spot laser shock peening (CLSP) and square-spot laser shock peening (SLSP) [6]. In a typical SLSP process, laser energy distributes homogeneously in the square spot after adjusting spot shape using beam homogenizer. SLSP can treat a large area effectively with the overlapping ratio less than 10% [7]. Cao et al. reported that surface morphology produced by square laser spots was smoother than that produced by circular spots [8]. In addition, square laser spot can effectively avoid the phenomenon of residual stress hole. Hu et al. studied the non-homogeneity of residual stress field and found that the residual stress drop in a square laser spot was less severe than that in a circular laser spot [9]. However, little open literature pays attention to the microstructural evolution, mechanical property and experimental residual stress field of materials under SLSP.

Content from this work may be used under the terms of the Creative Commons Attribution 3.0 licence. Any further distribution of this work must maintain attribution to the author(s) and the title of the work, journal citation and DOI.
Published under licence by IOP Publishing Ltd

In this study, SLSP was carried out to investigate high strain rate response of 7055 aluminum alloy, focusing on microstructural evolution, microhardness distribution and residual stress field. Different laser pulse energy and impact times were examined to offer a better understanding of material response under SLSP. It is believed that this study can provide a new insight for widening LSP technique in both scientific and industrial application.

2. Experiments

2.1. Experimental Procedure

In this study, 7055 aluminum alloy was selected in this work with a full chemical composition listed in Table.1. It is initially developed by Alcoa in 1980s, and is mainly applied as envelope of horizontal tail and other parts. The samples were cut into a rectangular plate with dimensions of 50 mm×30 mm×5 mm (length × width × thickness). Before SLSP, all samples were first ground with SiC paper from 150 # to 2000 # and then rinsed with deionized water by ultrasonic vibration. Shortly after the preparation, SLSP was performed in water confinement using a Q-switched Nd:YAG laser operating in a wavelength of 1064 nm, pulse width of 15 ns and a repetitive frequency of 1 Hz. The focused laser presents a 4 mm square spot after beam homogenization. All specimens were protected from the thermal effects of SLSP by a 100 μm aluminum foil (made by 3M). Two kinds of pulse energies of 15 and 25 J and impact time of one and two were selected for SLSP experiments. The laser path was a zigzag covering an area of 40 mm × 20 mm. The overlapping rate (ov %) represents percentage of the overlapped length in center distance of two adjacent laser spot. Thereby, an overlapping rate of 15% was adopted. The laser path and overlapping illustration are shown in Figure 1.

Table 1. Chemical compositions of 7055 aluminum alloy (wt %)

Cu	Mg	Zn	Zr	Fe	Mn	Si	Al
2.0 - 2.6	1.8 - 2.3	7.6 - 8.4	0.08 - 0.25	≤0.15	≤0.05	≤0.10	Bal.

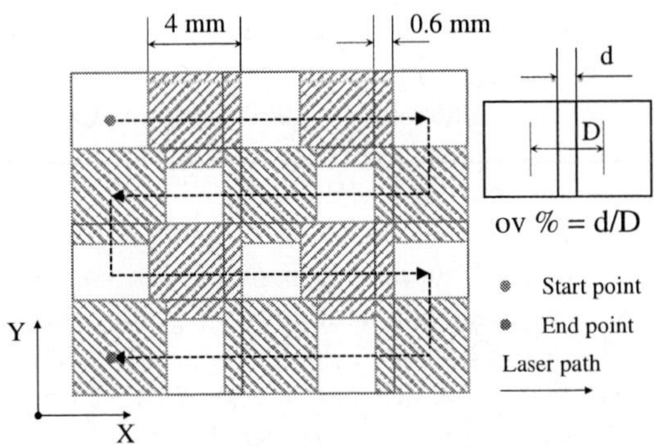

Figure 1. Laser path illustrating an overlapping rate of 15%.

2.2. Material Characterization

Optical Microscope (ASA1, Zeiss) and Scanning Electron Microscopy (SEM) were used to observe microstructure features of samples before and after SLSP. Prior to the observation, cross section of each sample was ground to 2500 # and then mechanically polished before etched by Keller's reagent (95 mL H_2O, 2.5 mL HNO_3, 1.5 mL HCl and 1 mL HF). In addition, Polarized Light Microscopy (PLM) was employed to reveal the grain structure after anodic oxidation with Barker's reagent (97.5 mL H_2O, 2.5 mL HBF_4). Transmission electron microscope (TEM, JEM-2100, JEOL), operated at 200 kV, was employed to investigate microstructural evolution of samples after SLSP. The sample was cut into a 0.5 mm thickness sheet and afterwards mechanically ground and polished down to about 100 μm thickness. The 100 μm sheet was punched out to 3 mm diameter discs, followed by ion-milling by Ar + bombardment with proper incident angles.

Micro-vickers hardness tester (FUTURE-TECH, FM-800) was adopted to study the variety of microhardness from SLSP treated surface down to the base metal with a step distance of 100 μm, test weight of 200 g, and dwell time of 10 s. To eliminate the deviation, each measure was repeated for three times and the average values were presented in this study. Prism residual stress measurement system based on incremental hole drilling method was applied to measure residual in the depth direction of specimen before and after SLSP. It is a destructive residual stress measurement. Residual stresses were measured from the surface to a depth of 1 mm.

3. Results and Discussion

3.1. Microstructure

Figure 2 presents PLM and SEM images of the sample before SLSP, which show clear rolling structures. Longish grains in both PLM and SEM images are confirmed.

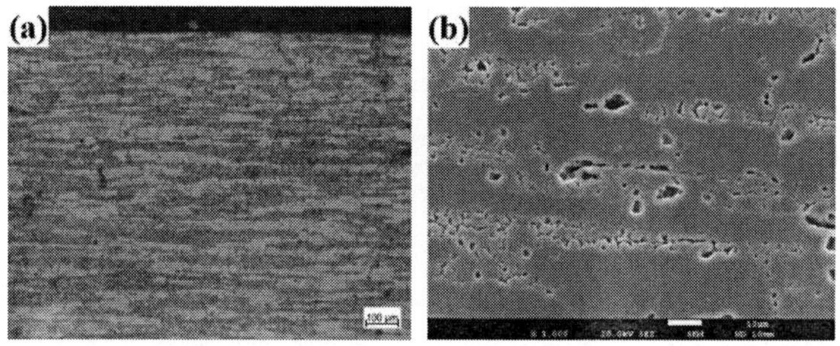

Figure 2. Microstructure before SLSP. (a) OM image and (b) SEM image.

Figure 3 and Figure 4 illustrate PLM and SEM images of the samples after SLSP, respectively. Figure 3(a) and (b) expound microstructural evolution under 15 J with one and two impacts. The continuous longish grains broke into intermittent smaller grains and further into randomly oriented fragmentary grains after one and two impacts. Similar results were reported by Trdan et al. who found randomly distributed smaller grains on the LSPed surface layer and oriented longish grains in the center of sample [10] Figure 3(c) and (d) show microstructural evolution under 25 J with one and two impacts. The density and the length of such longish grains decreased when increasing laser energy from 15 to 25 J. In addition, recrystallization of longish grains can also be observed. Randomly oriented net-like grains can be discovered after 25 J SLSP with two impacts. Lu et al. point out that a severe plastic deformation layer was generated near the top surface during LSP, and the grains in the severe plastic deformation layer were significantly refined [11].

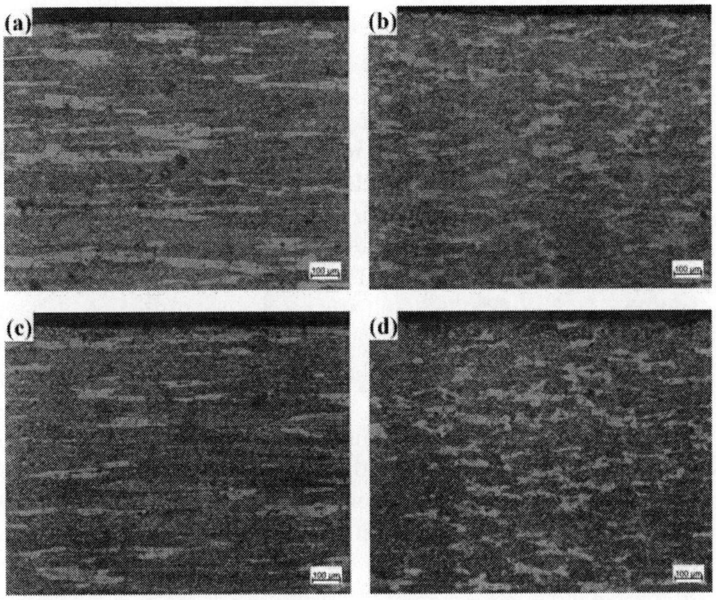

Figure 3. Microstructural evolution after SLSP under PLM. (a) 15 J with one impact; (b) 15 J with two impacts; (c) 25 J with one impact and (d) 25 J with two impacts.

Figure 4 shows SEM images, which confirm the presence of recrystallized grains in samples after SLSP. Longish grains still can be observed in Figure 4(a). However, the longish grain broke into intermittent smaller grains with the grain sizes varying from 2 μm to 20 μm, as shown in Figure 4(b), when the sample was subject to 15 J SLSP. When the samples are treated by 25 J with one and two impacts, small equiaxed grains can be obtained. The average grain in Figure 4(d) is about 5-8 μm.

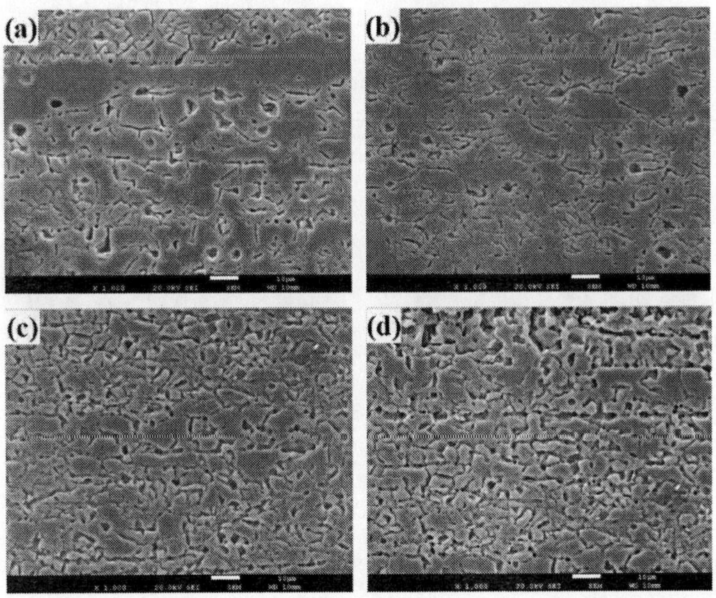

Figure 4. Microstructural evolution after SLSP under SEM. (a) 15 J with one impact; (b) 15 J with two impacts; (c) 25 J with one impact and (d) 25 J with two impacts.

Hence, it can be concluded from both PLM and SEM images that SLSP is an effective method to generate refined grains in the treated region, which can be favorable to the improvement on the fatigue life of 7055 aluminum alloy.

Figure 5 shows typical TEM images of samples treated by SLSP with laser energy of 15 J. From Figure 5(a), it can be clearly observed that a large number of microdefects randomly distributed in the grain, such as dislocation tangles and dislocation walls. However, no obvious observation of dislocation lines, which can be attribute that the surface layer was severely deformed during the SLSP process that dislocation lines slipped and multiplicated into dislocation tangles and wall [12]. Figure 5(b) present the image taken from the zone [A] in Figure 5(a). Stacking faults can be discovered in the edge of dislocation tangle and subgrain boundary between the dislocation tangle and original grain can also be detected. Figure 5(c) is the high-resolution transmission electron microscopy (HRTEM) image taken from ellipse [B] in Figure 5(b). Typical stack faulting structure spacing 0.606 nm can be observed in region [I] and periodic dislocation of atom arrangement in region [II] spaced 0.202 nm. Figure 5(d) presents details in ellipse [C] in Figure 5(b), in which a number of nanograins can be observed with its corresponding selected area electron diffraction (SEAD) being concentric circles. Its HRTEM is shown in Figure 5(e), in which the sizes of nanograins are about 20 to 30 nm. In addition, subgrains and subgrain boundary can be found in Figure 5(f), which are the very first step for the formation of refined grains.

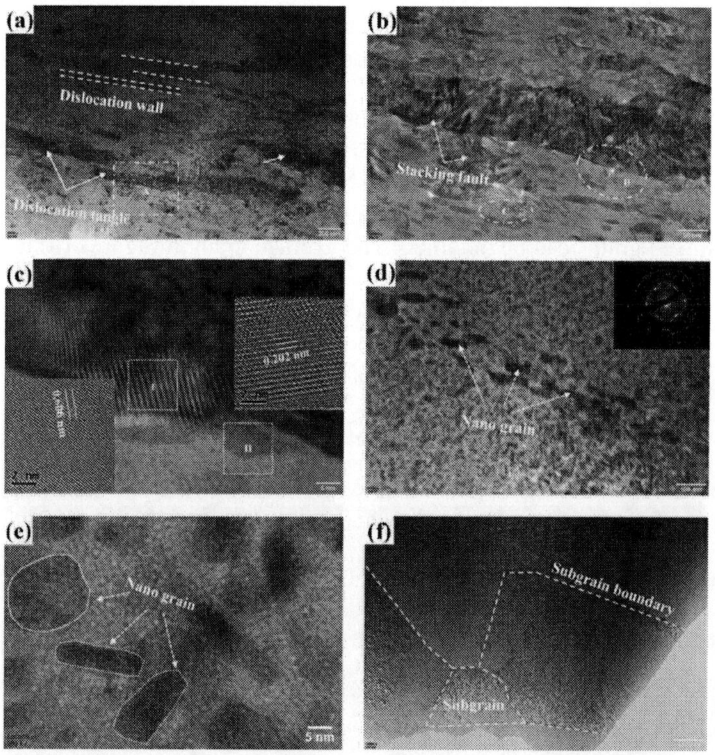

Figure 5. Typical TEM images after SLSP. (a) dislocation structures; (b) detail of zone [A] illustrating stack faults; (c) HRTEM taken from ellipse [B] in Figure 5(b); (d) detail of ellipse [C] in Figure 5(b) illustrating namograins; (e) HRTEM of nanograins and (f) surgrains and subgrain boundary.

3.2. Microhardness

Figure 6 presents a comparison of in-depth microhardness of samples before and after SLSP. It can be clearly found that SLSP can introduce a hardening layer within the top surface. It is also worth noting that the microhardness increment in SLSP is lower than that in conventional CLSP. This is due to low magnitude of peak pressure resulted from different laser energy distribution in the spot. The laser energy distributes as Gaussian function in a circular spot, while it distributes homogeneously in a square spot. The net result is that the peak pressure in the center of a square spot is lower than that in a circular spot. As for surface microhardness, increasing laser pulse energy to 15 J and 20 J, surface microhardness increased by approximately 10 HV and 20 HV respectively comparing with original surface microhardness, while increasing impact time from one to two resulted in slight change of surface microhardness both in 15 J and 25 J SLSP. According to Hall-patch theory [13, 14], HV is proportional to $\rho^{1/2}$ (ρ, dislocation density). Therefore, the enhancement in microhardness is due to the generation of high-density dislocation.

As for the depth of hardening affected layer, increasing laser pulse energy to 15 J and 20 J, an affected depth of 780 μm and 1100 μm were obtained respectively. This is because SLSP-induced shock wave transmitted deeper in 25 J and interacted more intensively with the target material. When increasing impact time of peening in both 15 J and 25 J, the affected depths in both condition exceeded 1.3 mm. The affected depth generated by a square spot is deeper than that of a circular spot because the shock in a square spot expands like a planar front, which attenuates at a rate of 1/r, while in a circular spot the shock wave behaves like a sphere, which results in attenuation at a rate of $1/r^2$ [5, 10].

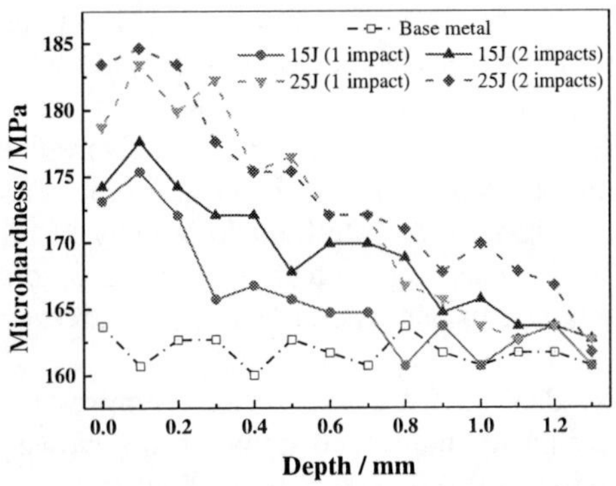

Figure 6. In-depth microhardness distribution after 15 J and 25 J SLSP with one and two impacts.

3.3. Residual Stress

Figure 7 exhibits the comparison of in-depth residual stresses after 15 J and 25 J SLSP with one and two impacts. It is clearly shown that compressive residual stresses with maximum value of several hundred MPa and affected depths of several hundred microns to more than 1 mm have been introduced in the surface and subsurface. The peak compressive residual stress increased from -120 MPa to -166 MPa, when increasing laser energy from 15 J to 25 J. Moreover, when increasing impact time from one to two in 15 J and 25 J, the compressive residual stress increased by 61.6% and 42.0% (from -120 MPa to -194 MPa and from -166 MPa to -234 MPa), respectively. The lower increment in 25 J can be attributed to the degree of residual stress saturation, which means the residual stress generated in 25 J is more saturated than that in 15 J with one impact. The more the generated stress is, the lower increment of the residual stress will be. As for the residual stress affected depth, it increased from ~700 μm to ~1 mm, when increasing laser pulse energy from 15 J to 25 J. The similar increments can also be achieved when increasing impact time from one to two in both 15 J and 25 J SLSP. Noting that the peak compressive residual stress in 15 J with two impacts is higher than that in 25 J with one impact, which proves that increasing impact time is a more effective method in acquiring high value residual stress. It is also worth noting that the peak compressive residual stresses are not located in the surface, but a depth between 100 μm to 200 μm. This is due to the release wave generated by boundary effect focalized to its center, causing reverse plastic strain and the stress drop [9].

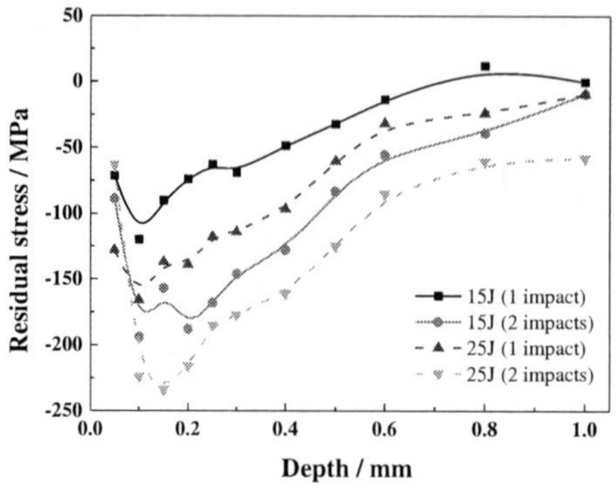

Figure 7. In-depth residual stress distribution after 15 J and 25 J SLSP with one and two impacts

4. Conclusions

Square-spot laser shock peening on 7055 aluminum alloy with laser energies of 15 J one and two impacts and 25 J one and two impacts has been investigated. The main findings are listed below.

- Microstructural PLM and SEM observations revealed that microstructures were significantly refined due to laser shock induced recrystallization. The longish rolling grains broke into intermittent smaller grains and further into randomly oriented fragmentary grains.
- Microdefects such as dislocation tangles, dislocation wall and stacking fault were generated in the surface layer after SLSP. Subgrains and nanograins were also obtained in the surface layer. These microdefects played a significant role in surface layer grain refinement.
- A higher value of surface microhardness can be achieved by increasing laser pulse energy, while increasing impact time only cause slight change in surface microhardness. However, thicker hardening layers can be obtained by increasing laser pulse energy or impact time. The hardening layer exceeded 1.3 mm in 25 J SLSP with two impacts.
- A higher value of peak residual stresses and thicker compressive residual stress layers can be achieved by increasing laser pulse energy and impact time. However, increasing impact time can be a more effective method in acquiring high peak residual stress.

Acknowledgement

The authors would like to acknowledge the support of this research work from the National International Collaborative Science and Technology Research Project China (Grant No. 2013DFR50590), and National Natural Science Foundation of China (Grant No. 11372019).

References

[1] Montross C S, Wei T, Ye L, Clark G and Mai Y W 2002 Laser shock processing and its effects on microstructure and properties of metal alloys: a review *International Journal of Fatigue* **24** 1021-36

[2] Nalla R K, Altenberger I, Noster U, Liu G Y, Scholtes B and Ritchie R O 2003 On the influence of mechanical surface treatments—deep rolling and laser shock peening—on the fatigue behavior of Ti–6Al–4V at ambient and elevated temperatures *Materials Science & Engineering A* **355** 216-30

[3] Yang J M, Her Y C, Han N and Clauer A 2001 Laser shock peening on fatigue behavior of 2024-T3 Al alloy with fastener holes and stopholes *Materials Science & Engineering A* **298** 296-9

[4] Ding K and Ye L 2006 *Laser shock peening: performance and process simulation*: Woodhead Publishing)

[5] Sun R, Li L, Zhu Y, Zhang L, Guo W, Peng P, Li B, Guo C, Liu L and Che Z 2017 Dynamic response and residual stress fields of Ti6Al4V alloy under shock wave induced by laser shock peening *Modelling & Simulation in Materials Science & Engineering* **25** 065016

[6] Ding K 2003 Three-dimensional Dynamic Finite Element Analysis of Multiple Laser Shock Peening Processes *Surface Engineering* **19** 351-8
[7] Kim T, Hwang S, Hong K H and Yu T J 2016 Analysis of the Square Beam Energy Efficiency of a Homogenizer Near the Target for Laser Shock Peening *Journal of the Optical Society of Korea* **20** 407-12
[8] Cao Z W, Gong S L and Gao Y 2013 Characterization of TC17 Titanium Alloy Treated by Square-Spot Laser Shock Peening *Advanced Materials Research* **654** 2378-83
[9] Hu Y, Gong C, Yao Z and Hu J 2009 Investigation on the non-homogeneity of residual stress field induced by laser shock peening *Surface & Coatings Technology* **203** 3503-8
[10] Trdan U, Skarba M and Grum J 2014 Laser shock peening effect on the dislocation transitions and grain refinement of Al–Mg–Si alloy *Materials Characterization* **97** 57-68
[11] Lu J Z, Luo K Y, Zhang Y K, Cui C Y, Sun G F, Zhou J Z, Zhang L, You J, Chen K M and Zhong J W 2010 Grain refinement of LY2 aluminum alloy induced by ultra-high plastic strain during multiple laser shock processing impacts *Acta Materialia* **58** 3984-94
[12] Lainé S J, Knowles K M, Doorbar P J, Cutts R D and Rugg D 2017 Microstructural characterisation of metallic shot peened and laser shock peened Ti–6Al–4V *Acta Materialia* **123** 350-61
[13] Chu J P, Rigsbee J M, Banaś G and Elsayed-Ali H E 1999 Laser-shock processing effects on surface microstructure and mechanical properties of low carbon steel *Materials Science & Engineering A* **260** 260-8
[14] Lu J Z, Wu L J, Sun G F, Luo K Y, Zhang Y K, Cai J, Cui C Y and Luo X M 2017 Microstructural response and grain refinement mechanism of commercially pure titanium subjected to multiple laser shock peening impacts *Acta Materialia* **127** 252-66

IOP Conf. Series: Materials Science and Engineering **275** (2017) 012042 doi:10.1088/1757-899X/275/1/012042

Effect of Bamboo Viscose on the Wicking and Moisture Management Properties of Gauze

Abdul R Akbar[1, 2], Siwei Su[1], Bilal Amjad[2], Yingjie Cai[1, 3] and Lina Lin[1, 2, 3 *]

[1]Hubei Provincial Engineering Laboratory for Clean Production and High Value Utilization of Bio-based Textile Materials, Wuhan Textile University, Wuhan 430073, China
[2]College of Textile Science and Engineering, Wuhan Textile University, Wuhan, 430200, China
[3]Engineering Research Centre for Clean Production of Textile Dyeing and Printing, Ministry of Education, Wuhan Textile University, Wuhan, 430073, China
Email: linalin@wtu.edu.cn

Abstract: Bamboo viscose or regenerated cellulose fibers were used to check their absorbency properties effect on the wicking and moisture management in gauzes. Bamboo viscose and cotton fibers were spun into five different yarn samples with different fiber proportion by ring spinning. Fifteen different gauze samples were made of these yarn samples. The gauze samples were subjected to wicking test to check the wicking ability. Water vapor transmission test was applied to check the vapor transmission rate. These tests were applied to measure the effectiveness of bamboo viscose, cotton and blended gauze samples in wound healing. Pure bamboo gauzes and gauzes with high content of bamboo fiber, i.e. 75B:25C and 50B:50C, shows better wicking and vapor transmission properties. It makes gauzes with high bamboo viscose suitable for wound care applications because of moisture absorbency.

1. Introduction

The presence of the naturally occurring antimicrobial bio-agent, named as bamboo "Kun" makes bamboo fiber anti-microbial, order resistive and antifungal. "Kun" decreases the unpleasant odour causing bacteria on fabric and human skin [1]. Regenerated bamboo fibers from bamboo pulp have same processing method as ordinary viscose fiber so called as bamboo viscose fibers [2].

The medical application of bamboo has long history. Due to presence of phenolic compounds in chemical continents bamboo have shown antibacterial activities against both Gram-positive and Gram-negative bacteria. Some studies reported antimicrobial activity of charcoal and bamboo vinegar as well [2, 3]. Industry stakeholders presented different bamboo fabrics claiming their anti-microbial and eco-friendly behaviour. Some claimed the presence of hydroxyl group (-OH) in bamboo fibers but they all are unable to provide scientific evidences. Repeated technological analysis has shown that fibre has a fineness and whiteness grade close to normal bleached viscose also possess strong durability, stability and tenacity [3, 4]. Bamboo fabric is characterized by its good hygroscopicity, excellent permeability, soft feel, easiness to straighten and dye and splendid colour effect of pigmentation [3, 4]. Moisture management is an essential feature of gauze fabric regarding its wound management capability. A quantitative antibacterial capability test for bamboo fabrics at China Industrial Testing Centre (CTITC) showed that 100% bamboo fabric exhibits 99.8% antibacterial kill rate. Studies by the Japan Textile

Content from this work may be used under the terms of the Creative Commons Attribution 3.0 licence. Any further distribution of this work must maintain attribution to the author(s) and the title of the work, journal citation and DOI.
Published under licence by IOP Publishing Ltd

Inspection Association (JTIA) revealed long-term antibacterial efficacy of greater than 70% bamboo fabric.

Gauze, a thin, lightweight fabric having open woven structure is usually made of natural fibers such as cotton, linen and silk. It can be also made of synthetic fibers. Gauze fabric may be manufactured through one of two methods: the simple leno weave and the plain weave but manufacturing of gauze by leno weaving method is mostly used. Through this process, the threads of the fabric cross over one another, interlacing with one or more filler threads to give the fabric a sheer appearance while making it stronger. This type of weave makes the fabric lightweight and stronger while still allowing light and air to pass through it freely [3]. Plain woven fabric is created on a loom by interlacing the warp threads tightly with the weft threads. The woven fibers of this weave produce a flat, tight surface suitable for printing and other finishes. The plain weave is the simplest to produce, making it the least expensive [5].

Wounds heal better in a moist environment. Nerve endings are protected and skin layers repair at a faster rate producing less scarring than in dry wounds. As part of the normal healing process wounds release fluid (or exudate). Too much exudate (maceration) or too little exudate (desiccation) can interfere with wound healing; therefore, it is important to manage moisture levels, particularly in chronic wounds. Selection of a dressing normally gauze is based on specific wound characteristics; the presence or absence of necrotic tissue, bacterial load, exudate volume and need for a bacterial barrier [6]. Therefore, the optimal dressing should be able to control exudate without desiccating the wound surface, act as a bacterial barrier, allow a traumatic removal with no dressing left in the wound, and provide moisture vapor permeability sufficient to prevent over hydration of the wound and surrounding skin. There are a multitude of dressings on the market and clinicians are frequently confused as to which type of dressing is the best for a particular wound. Effective dressing selection requires both accurate wound assessment and current knowledge of available dressings and these gauzes may contribute in that aspect.

These unique antimicrobial properties, natural effects of sterilization and bacteriostasis of bamboo fiber make potential of using bamboo fiber for making sanitary materials include bandage, mask, surgical clothes, nurse's wears, on application in sanitary material absorbent pads, food packing and so on. In the medical scope, it can be processed into the products of bamboo fiber gauze, operating coat, and nurse dress, etc. The finished products do not need to be added with any artificial synthesized antimicrobial agent, so it will not cause the skin allergy phenomena. Thus this experiment is conducted to check the potential of bamboo fiber gauzes in moisture management and wicking abilities and to compare these properties of bamboo, cotton and cotton & bamboo blended gauzes.

2. Experimental

2.1. Materials and Machine

Bamboo viscose (regenerated bamboo) and cotton fibers needed to make five different samples of yarn. Different machines were used in spinning process of yarn samples included blowroom, card, draw frame, simplex, ring machine, and winding machine. Sample loom was used to weave fifteen different plain 1/1 gauze fabric samples. Jigger machine was used to scouring and bleaching gauze samples. For tensile strength test of gauzes "universal tensile strength tester" was used. Wicking test performed with the help of red dye solution. Lea strength tester, wrapping real, twist tester were used for testing yarn properties.

2.2. Methods

Ring spinning technique was used to prepare 100% cotton, 100% bamboo, 50:50 cotton: bamboo, 75:25 cotton: bamboo and 25:75 cotton: bamboo yarn samples of 30 Ne count (Table 1). These yarn samples were weaved into different fifteen gauze fabric samples with the help of sample loom.

On the jigger scouring and bleaching was done and then wicking test is performed with the help of red dye solution. Water vapor transmission test was applied to check the ability of gauze fabric samples of vapor transmission. The purpose of wicking and water vapor transmission test was to check and compare the results of gauzes to manage the healing of wound with the help of wicking and vapor transmission abilities. Fifteen different gauze samples can be seen in Table 2.

IOP Conf. Series: Materials Science and Engineering **275** (2017) 012042 doi:10.1088/1757-899X/275/1/012042

2.3. Lea Strength Test

A lea strength tester measures the strength of one lea (120 yds.) of yarn. Strength is a measure of the steady force necessary to break a material and is measured in pound. The lea strength tester works in constant rate of extension. The maximum strength which was achieved was of the 100% bamboo yarn.

Table 1. Developed yarn samples of Ne 30 count.

S. No.	Samples	Lap Weight (Ounce/yard)	Roving Count	TPI	CLSP
1	100% Cotton	14	1.2	28.3	2009.60
2	75% Cotton : 25% Bamboo	9	1.1	25.3	2207.17
3	25% Cotton : 75% Bamboo	15	1.13	23.4	2309.13
4	50% Cotton : 50% Bamboo	17	1.19	22.3	1795.85
5	100% Bamboo	22	1.17	21.22	2376.12

Table 2. Fifteen gauze samples.

S. No.	WARP	WEFT
1	50C:50B	100C
2	50C:50B	100B
3	50C:50B	50C:50B
4	25C:75B	50C:50B
5	25C:75B	100C
6	25C:75B	100B
7	100C	100C
8	100C	50C:50B
9	100C	100B
10	100B	100B
11	100B	50C:50B
12	100B	100C
13	75C:25B	50C:50B
14	75C:25B	100B
15	75C:25B	100C

2.4. Wicking Test

Wicking test performed to evaluate the wicking ability of vertically or horizontally aligned test specimens, exposed to distilled or deionized water over a period of time, where wicking is influenced by gravity [6]. The rate (distance per unit of time) liquid travels along and/or through a vertical fabric specimen was visually observed, manually timed and recorded at specified intervals. Test specimens of 5 cm x 1 cm were cut and dipped 1cm into the red dye solution and after one minute the level penetration of dye solution on the surface of sample was noted.

2.5. Water Vapor Transmission Test

It is the measurement of the amount of moisture vapor in grams that pass (breathe or permeate) through an area of fabric in 24 hours at specified temperature and humidity [6]. 200 g water was filled into the glasses which then were sealed with the test specimen. Then weight of water was measured before placing in controlled atmosphere. Periodically weighed and noted down the values to calculate the gram of water vapor transmitted through the samples in 24 hrs.

IOP Conf. Series: Materials Science and Engineering **275** (2017) 012042 doi:10.1088/1757-899X/275/1/012042

3. Results and Discussion

3.1. Twist per Inch (TPI) of Yarn Samples

Twist is an important factor to impart strength in yarn. The TPI values of yarn samples are shown in figure 1. Due to lower strength of cotton fibers the cotton yarn with 30 Ne need more twist to be imparted than same count bamboo yarn. Bamboo fibers also have greater length compared to cotton other than greater strength. The increase in ratio of the bamboo fiber decreases the twist insertion because of the addition of bamboo fibers with longer lengths.

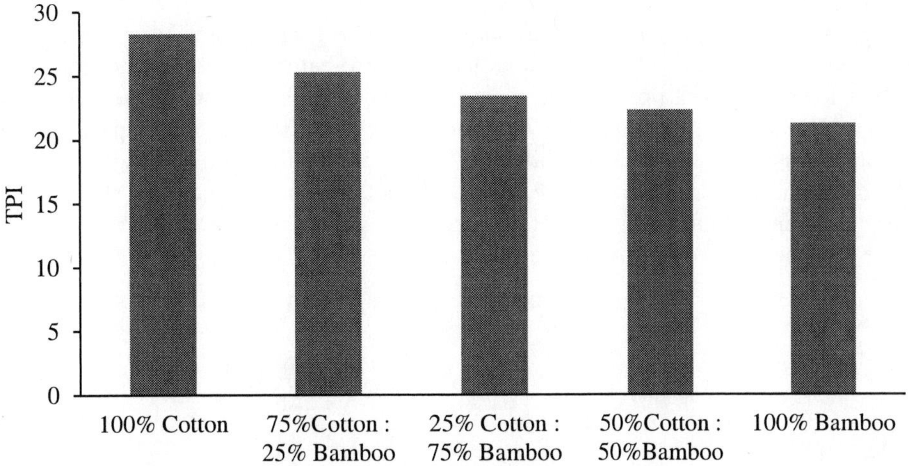

Figure 1. Twist per inch of developed 30 Ne yarn samples.

Lea strength of all five samples was measured on lea strength tester. The figure 2 shows that the 100% bamboo yarn has the highest strength because of the higher strength of fibers is better than cotton. Higher the contents of bamboo fibers in the yarn higher the strength.

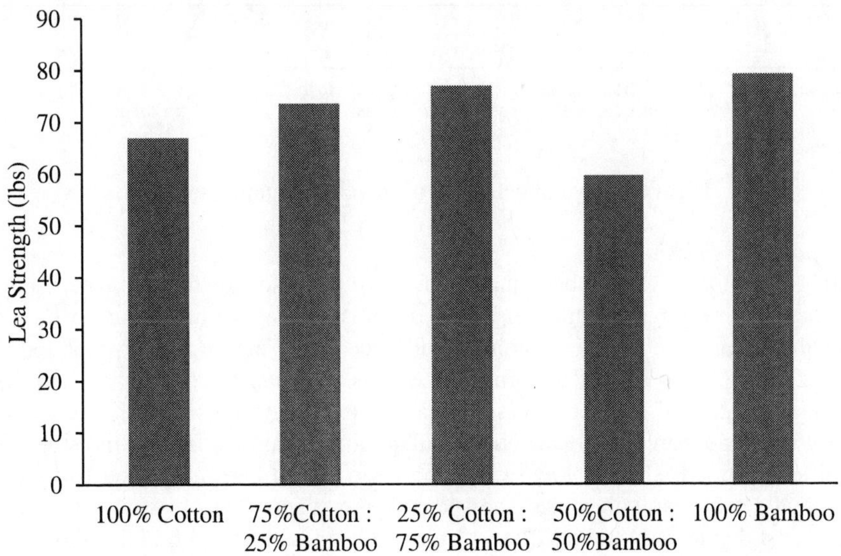

Figure 2. Lea strength of developed 30 Ne yarn samples.

3.2. Wicking

Absorption of exudate by wound dressings from wound is very important for wound healing because if proper wound management is not done, it can cause oppression, adhesion and swelling. All the fifteen were tested and results are shown in figure 3. Wicking ability of samples with 100% bamboo fiber is showing highest value due to the ability of bamboo fibre's to wick the exudate from wound while gauze made from 100% cotton fibers shows the lowest wicking value. Blended samples with high ratio of bamboo fiber i.e. samples with 100% bamboo in warp etc. showed high wicking value. Results from wicking test showed that by using 100% bamboo yarn or yarn with high ratio of bamboo fiber: we can make gauzes which would have high ability to wick exudate from wound which is very important to heal the wound.

The better wicking properties by fabric samples with higher contents of bamboo fiber can be attributed to the lower twist levels in the yarns having more bamboo fibers proportion. The more the bamboo fibers lower the twist imparted, as discussed above, leads to more capillaries inside the yarn between fibers. Higher twist increases the yarn compactness and reduces inner spaces between fibers. Lower the capillaries lower the wicking ability and large size capillaries allow more fluid to be passed through them. The fabric sample with 100% bamboo fibers in both weft and warp shown highest values of 2.83 cm of dye penetration, this is due to larger capillaries in bamboo yarns and also due to the natural ability of the bamboo fibers of absorbing moisture.

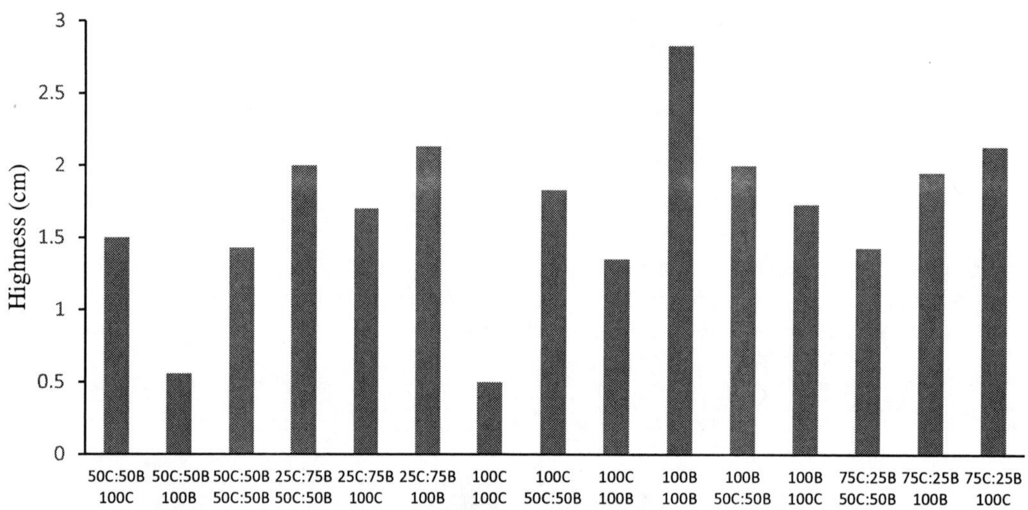

Figure 3. Dye penetration of developed gauze samples.

3.3. Water Vapor Transmission

Moisture vapor transmission should be a quality of gauze because moisture management of a wound is an important factor in healing a wound. High rate of moisture vapor transmission would minimize the bacterial growth so that there will be less chance of infection. Weight (g) of evaporated water through developed gauze samples is shown in figure 4. Results of water vapor transmission test show that samples with higher ration of bamboo fibers in it have better vapor transmission rate. This is due to low twist in yarns with higher bamboo fibers and natural quality of bamboo fiber to moisture absorption and better transmission property. Due to low twist in bamboo yarn, fibers are packed with less complexity into the yarn that's why bamboo gauze showed a higher vapor transmission rate. Also high vapor transmission rate is due to the bamboo fiber structure. Other samples with high ratio of bamboo fibers also show a good water vapor transmission rate. Results of this test clearly give an idea of good moisture vapor transmission rate of bamboo gauzes which would be greatly helpful to understand the ability of bamboo gauzes of wound healing.

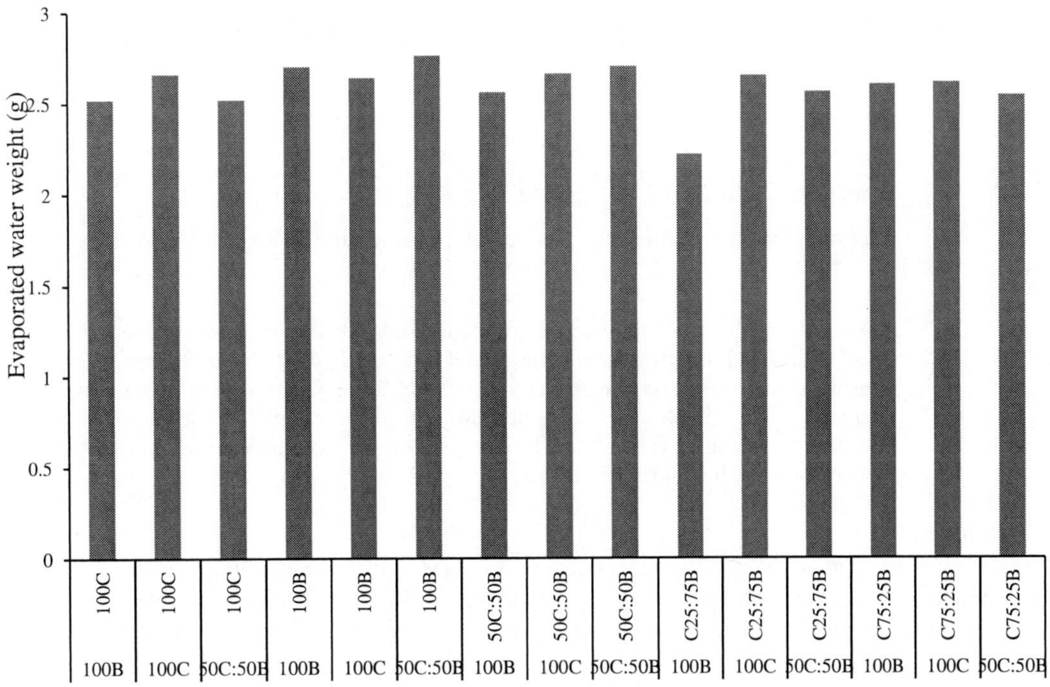

Figure 4. Weight of evaporated water through developed gauze samples.

4. Conclusion

Gauzes made of 100% bamboo found to be very impressive in wicking properties. Blended samples having high ratio of bamboo fiber (75B: 25C) shows good wicking properties. Also 100% bamboo fiber gauzes show highest vapor transmission rate which shows the potential of bamboo gauzes in wound care. Samples with high ratio of bamboo fiber shows good vapor transmission rate. 100% bamboo gauzes and gauzes having high ratio of bamboo fiber shows good wicking and vapor transmission rate; these gauzes can be used for effective wound care.

5. Acknowledgement

This work was financially supported by the China National Textile & Apparel Council (2013 "Textile Vision" Applied Basic Research, 2013-153); Hubei Province Science and Technology Support Program (Grant No. 2013BAA043) and the Collaborative Innovation Plan of Hubei Province for Key Technology of Eco-Ramie Industry (2014-8).

6. References

[1] Erdumlu N and Ozipek B 2008 *Fibres Text. East. Eur.* **16** 43
[2] Xia L and Chun D 2012 Proc. of the 55th Int. Convention of Society of Wood Science and Technology 1
[3] Shanmugasundaram O L and Gowda R V 2011 *Fiber. Polym.* **12** 15
[4] Qamar M, Amin N, Mahmood N, Ahmad I and Abass M, 2015 *J. Text. Sci. Eng.* **5** 41
[5] Tsuzuki A, Kanwar T and Wang R K 2012 *J. Text. I.* **103** 844
[6] Gericke J. van 2010 *JFECS* **38** 63

The Optical Dielectric Model of Cu Thin Film and Its Verification

Guozhong Lai*, **Xiong Liang and Jing Lv**

School of Physics and Electromechanical Engineering, Longyan University, Longyan 364012, Fujian, P. R. China.zhglai55@163.com

Abstract: Cu film was deposited on transparent substrate by magnetron sputtering a Cu target. The Drude-Lorentz optical dielectric model was used to fit the measured reflectance and transmittance data in wavelength range of 300-2200nm. The optical constants, as well as the thickness of the Cu film were obtained. Moreover, the calculated optical constants and the thickness were confirmed by the TFCalc software. It was suggested that the Drude-Lorentz model is suitable for Cu film.

1. Introduction

Copper has been studied over a century because of its extensive applications from metallurgy to semiconductor. Among all the applications, the most attractive use of copper is interconnect. [1] In computer and microelectronics industry, copper plays an important role in the newest development of chip design. Aluminum is being replaced by copper which is used to be the conductor of choice in transistor systems. [2] These applications have increased the requirements for characterizing the physical properties of copper thin film, especially the optical properties related to dielectric constant. It is important during the design of an optical multilayer interference system that the optical constants of all the materials used in it be known. [3] However, with the development of micromation devices, the thickness of interconnect Cu film is reduced to submicron and nanoscale. With the decreases of the thickness, the specific surface area of film increases dramatically, and so the performances of Cu film will be different from copper bulk. Hence, it is necessary and important to determine optical constants of Cu thin film accurately.

In fact, some investigators have devoted to the optical and electrical properties of Cu film. M. Ph. Stoll [4] gave the principle of a polarimetric method for the study of the optical properties of metals, and the method is illustrated with measurements on copper. T. Smith [5] has measured the optical constants of copper and nickel polycrystals at wavelengths 546.1 and 632.8 nm. Ahuja et al. [6] used a non-interacting s-d band model to evaluate $\varepsilon(\omega)$ in copper and calculated $\varepsilon(\omega)$ gave reasonable agreement with measured optical conductivity, where $\varepsilon(\omega)$ is the dielectric function which is bound up with the optical constants. Ya. B. Soskovets et al. [7] have investigated the effect of the structure of smooth copper surfaces on the accuracy of determination of the optical constants of copper. Gong Jun-Bo et al. [8] determined the thickness and optical constants accurately by using newly developed ellipsometry combined with transmittance iteration method in a wavelength range of 300 nm -1000nm, and analyzed the thickness dependence of optical constants of thin oxidized Cu films. M. M. Alam et al.2 have presented a technique to deposit high strength and highly conductive copper thin films on glass substrates at room temperature.

In this work, Cu film was deposited on transparent substrate by magnetron sputtering a Cu target. Using the measured reflectance and transmittance data, associated with the Drude- Lorentz optical

Content from this work may be used under the terms of the Creative Commons Attribution 3.0 licence. Any further distribution of this work must maintain attribution to the author(s) and the title of the work, journal citation and DOI.
Published under licence by IOP Publishing Ltd

IOP Conf. Series: Materials Science and Engineering **275** (2017) 012043 doi:10.1088/1757-899X/275/1/012043

dielectric model, the optical constants, as well as the thickness of the Cu film were calculated. Moreover the calculated constants were confirmed by the TFCalc software. It was proved that the Drude-Lorentz model is suitable for Cu film.

2. Experimental Details

Cu film was deposited on quartz substrate by sputtering a Cu target (99.99%, purity) with ultra-high vacuum magnetron sputtering system (JGP560B II). The substrate, 0.5mm thick, 2.5cm × 2.5cm, was cleaned in acetone and ethanol. The target-to-substrate distance was about 6.0cm. Prior to the deposition, the Cu target was pre-sputtered for 5 min to remove surface contaminations. The base vacuum level was 4.0×10^{-4}Pa. The flow rate of Ar was 25 sccm and the pressure during deposition was kept at ~0.9 Pa. The substrate temperature was 200°C. The deposition time was 10 min, and the sputtering power was 90 W.

The crystal structure of the sample, namely the Cu thin film together with the substrate, was investigated using an X-ray diffractometer (XRD) (MiniFlex II) with Cu K radiation. The normal incidence transmittance (T) and reflectance (R) of the sample were measured in wavelength range of 300 – 2200 nm by a spectrophotometer (Perkin – Elmer Lambda-950) equipped with an integrating sphere.

3. Results and Discussions

3.1. Crystal Structure of the Sample

XRD pattern of the Cu film was shown in Figure 1. Comparing the results with data from JCPDS international diffraction database, it was found that the peaks at 43.7oand 50.9o correspond to Cu (1 1 1) and Cu (2 0 0), respectively, with (1 1 1) as the preferred orientation. It was the same as our previous results, [9] where the Cu films were deposited on quartz substrates by thermal evaporation.

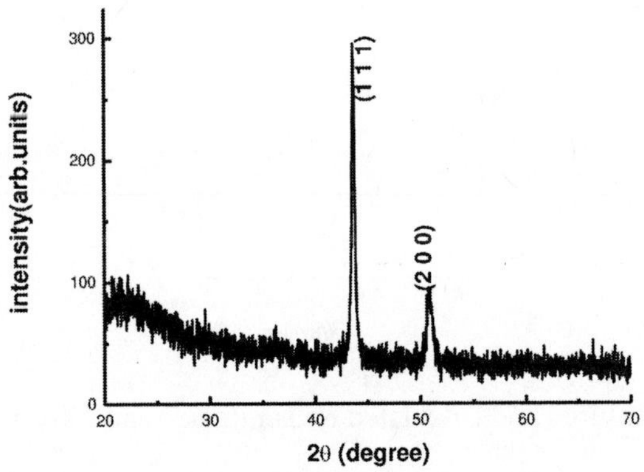

Figure 1. XRD pattern of the Cu film.

Also, it was consistent with the results reported by other authors, 1,2 and must be attributed to the energy of the Cu particles hitting on the surface of the substrates.[10]

3.2. Optical Constants and Thickness of the Cu Thin Film

Commonly, the optical constants are retrieved by simulating the transmittance and (or) reflectance based on dielectric models and by using a nonlinear least squares fitting procedure. [11] The as-deposited Cu film is a metal film. Accordingly, the Drude - Lorentz model was used to determine the optical constants of the Cu film from the reflectance and transmittance data, which combines the Drude model with the Lorentzian oscillator model and has been applied in metal film by Baron et al.

[12] The dielectric function of the Drude - Lorentz model can be expressed as [13]

$$\varepsilon(\omega)=\varepsilon_1+i\varepsilon_2=(n+ik)^2=\varepsilon_\infty+\frac{(\varepsilon_s-\varepsilon_\infty)\omega_0^2}{\omega_0^2-\omega^2+i\Gamma\omega\omega_0}-\frac{\omega_D^2}{\omega^2-i\omega_\tau\omega} \quad (1)$$

Where ε_∞ and ε_s represent the high- and low-frequency dielectric constants, respectively. Accordingly, ω_0 is oscillator resonance frequency, Γ is the damping factor of the oscillator, ω_D is the plasma resonance frequency, ω_τ is the collision frequency, and ω is the angular frequency of incidence wave.

Based on the optical dielectric function in Eq. (1), a computer program was written in FORTRAN for calculating the refractive index (*n*), extinction coefficient (*k*), and its thickness (*d*) of the Cu film by simulating the reflectance and transmittance data. A merit function *f* was defined as

$$f(d,n(\lambda),k(\lambda))=\sum_\lambda[T_{meas}(\lambda)-T_{calc}(\lambda)]^2+\sum_\lambda[R_{meas}(\lambda)-R_{calc}(\lambda)]^2 \quad (2)$$

Where λ was wavelength, $T_{meas}(\lambda)$ and $R_{meas}(\lambda)$ referred to the measured transmittance and reflectance, respectively. $T_{calc}(\lambda)$ and $R_{calc}(\lambda)$ referred to the calculated transmittance and reflectance, respectively. In the calculation process, *d* and the parameters of dielectric function were adjusted so that the calculated spectrum was close to the measured spectrum to full extent and the merit function had the minimum.

Figure 2 showed the measured and calculated transmittance and reflectance of the sample. It could be seen that the calculated transmittance and reflectance were in good agreement with the measured ones.

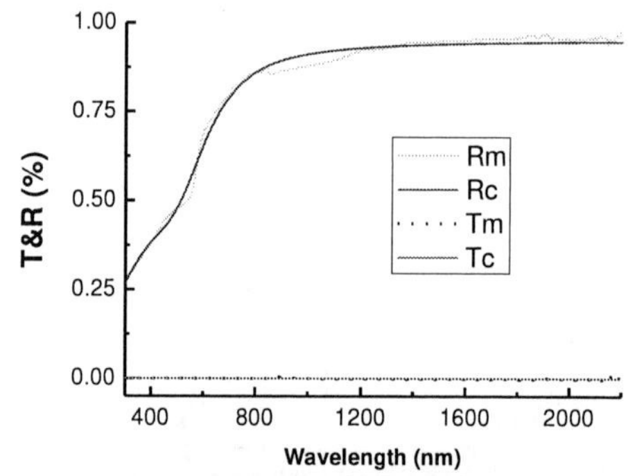

Figure 2. The measured and calculated transmittance and reflectance of the sample.

Based on the measured data in Figure 2, optical constants n and k and the thickness d of the Cu film can be obtained by using eqs. (1) and (2), and a computer program. Figures 3 (a) and (b) displayed the calculated refractive index and extinction coefficient of the Cu film, respectively. The calculated results indicated that d of the Cu thin film was 123 nm. The optical constants of Cu thin film obtained from handbook [14] were also presented in Figures 3. It was found that the relationships of both the refractive index and the extinction coefficient with the wavelength were similar to the ones from handbook, respectively. Both n and k increase with an increase in wavelength in the infrared region, due to Cu being a good conductor. [15] On the other hand, it was evident that our refractive indexes were larger than those from handbook. This must be attributed to sputtering technique. The film used in the handbook was obtained by evaporating. So the density of our film was large than that of handbook, resulting in the increase of refractive index.

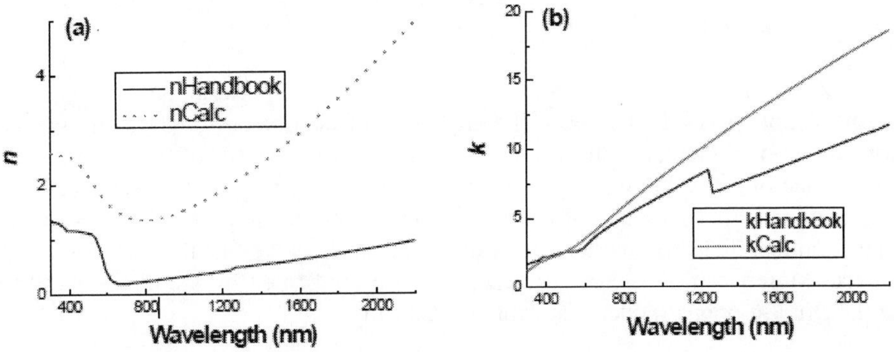

Figure 3. The calculated (a) refractive index and (b) extinction coefficient of the Cu film and the ones obtained from handbook.

3.3. Verifying the Optical Dielectric Model

TFCalc is one of the most commonly used tools for the design and analysis of optical thin film. Put the optical constants both of quartz and Cu thin film calculated in Sec. 3.2, as well as its calculated thickness, into the TFCalc software, the reflectance (R_{Tfc}) and transmittance (T_{Tfc}) of sample were obtained once again. Due to the transmittance went near to zero, only the measured reflectance and the one calculated from TFCalc were shown in Figure 4.

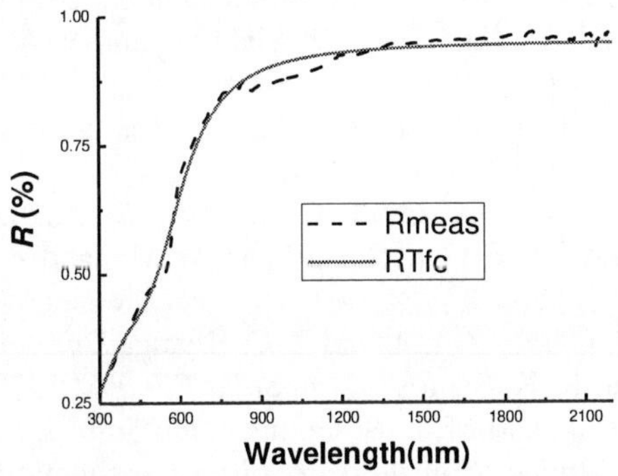

Figure. 4. The measured reflectance and the one coming from TFCalc.

It could be seen from Figure 4, the reflectance resulting from TFCalc was in good agreement with the measured one in all over the wavelength range of 300 – 2200 nm. So, it could be received definitely that the Drude - Lorentz model was suitable for the Cu thin film. Moreover, it was comprehensible. The Drude - Lorentz model basically made of the contributions of the high frequency lattice dielectric constant, ε_∞, which represents the contribution to $\varepsilon(\omega)$ outside of the simulated range, the Lorentz oscillator, i.e. the second term of the complex dielectric function, which describes the behavior of the dispersion law in the near UV spectral region and characterizes the dielectric properties of Cu thin film, determines the variation of the optical band-gap energy, and the Drude oscillator, i.e. the third term of the complex dielectric function describes the free-carriers in the near-IR spectra and, as a consequence, the conductive properties of the Cu films. In fact, the Drude - Lorentz model, namely eq. (1), is one of the oldest but the most used classical dispersion law [16] and has also been applied to investigate the optical properties of indium tin oxide (ITO) thin films. [17]

4. Conclusions

Cu film was deposited on transparent substrate by magnetron sputtering a Cu target. The crystal structure of the sample was investigated using an X-ray diffractometer (XRD) (MiniFlex II) with Cu K radiation. The normal incidence transmittance (T) and reflectance (R) of the sample were measured in wavelength range of 300-2200 nm by a spectrophotometer (Perkin – Elmer Lambda-950) equipped with an integrating sphere. Using the measured reflectance and transmittance data, associated with the Drude- Lorentz optical dielectric model, the optical constants, as well as the thickness of the Cu film were calculated. Moreover the calculated constants were confirmed by the TFCalc software. It was proved that the Drude-Lorentz model is suitable for Cu film.

5. Acknowledgements

This work was financially supported by the Natural Science Foundation of Fujian Province of China (2014J01016).

6. References

[1] H. Y. Chen, K. H. Yang, and F. Y. Ouyang, *ICEP-IAAC Proceedings*, 596-600 (2015).
[2] M. M. Alam, Md. Nasrul Haque Mia, R. Hasan1, M. Shahinuzzaman1, M. K. Islam1, and Khan. M. Nasir Uddin, *Mater. Sci. Appl.* 6 (2015) 753.
[3] J. A. Dobrowolski, F. C. Ho, and A. Waldorf, *Appl. Opt.* 22 (1983) 3191.
[4] M. Ph. Stoll, *J. Appl. Phys.* 40 (1969) 4533.
[5] T. Smith, *J. Opt. Soc. Am.* 67 (1977) 48.
[6] A. K. Ahuja, S. Auluck and S. K. Joshi, *Solid State Commun.* 38(1981) 275.
[7] Ya. B. Soskovets, A. Ya. Khairullina, T. A. Zhevlakova, and V. A. Tolmachev, *Opt. Spectrosc.* 97 (2004) 951.
[8] J. B. Gong, W. L. Dong, R. C. Dai, Z. P. Wang, Z. M. Zhang, and Z. J., *Chin. Phys. B* 23 (2014) 087802.
[9] G. Z. Lai, H. Q. Lan, S. Z. Lin, Y. Qu, and F. C. Lai, *Surf. Rev. Lett.* 20 (2013) 1350011.
[10] Q. X. Zhao, F. Bian, Y. Zhou, Y. F. Gao, S. B. Wang, L. Ma, and Z. Yan, B. T. Liu, *Mater. Lett.* 62 (2008) 4140.
[11] F. C. Lai, L. M. Lin, R. Q. Gai, Y. Z. Lin, and Z. G. Huang, *Thin Solid Films*, 515 (2007) 7387.
[12] W. B. Lance, N. Jason, and K. K. Santosh, *Thin Solid Films* 515 (2007) 3363.
[13] A. Khodorov, M. Piechowiak, and M. J. M. Gomes, *Thin Solid Films* 515 (2007) 7829.
[14] E. D. Palik, *Handbook of Optical Constants of Solids.* (Academic Press, San Diego, USA, 1998), pp.283-286.
[15] J. Chen, X. S. Ge, *Int. J. Thermophys.* 21(2000) 269.
[16] M. Kadi, A. Smaali, and R. Outemzabet, *Surf. Coat. Technol.* 211 (2012) 45.
[17] S. D'Elia, N. Scaramuzza, F. Ciuchi, C. Versace, G. Strangi, and R. Bartolino, *Appl. Surf. Sci.* 255 (2009) 7203.

IOP Conf. Series: Materials Science and Engineering **275** (2017) 012044 doi:10.1088/1757-899X/275/1/012044

Application and Prospect of Superconducting High Gradient Magnetic Separation in Disposal of Micro-fine Tailings

Changqiao Yang, Suqin Li[1], Zijie Guo and Jiawei Kong

[1]University of Science and Technology Beijing, No.30 Xueyuan Road, Haidian District, Beijing, China.
School of Metallurgical and Ecological Engineering, University of Science and Technology Beijing, Beijing, China.E-mail: lisuqin@metall.ustb.edu.cn

Abstract: Magnetic separation technology is playing an increasingly important role in the field of environmental protection such as waste gas, waste water and solid waste treatment. As a new type of solid waste treatment technology, superconducting high gradient magnetic separation (HGMS) is mainly applied in the separation of micro-fine weakly magnetic particles because of the advantages of high separation efficiency, energy saving, simple equipment and easy automation. In this paper, the basic principle of superconducting HGMS was firstly introduced, then the research status of scholars at home and aboard on the disposal of micro-fine tailings were summarized. Finally, the direction of development for HGMS was put forward.

1. Introduction

Mineral resources are the important material foundation for the survival and development of human society. About 85% of the raw materials come from mineral resources in the process of rapid development of modern industry, and the demand for mineral resources is increasing rapidly[1]. Due to the complicated mineral composition and the backward mining technology, the resource loss is very serious in the exploitation of mineral resources, and a large amount of tailings are came out. In China, the average grade of iron ore is only 30%, 70% of the mineral resources will become tailings; mineral separation of copper, zinc and other mineral will produce more than 90% tailings; the tailings of gold, tungsten and other precious metals can be as high as 99%[2]. According to statistics, by 2015, China's iron tailings has reached more than 8 billion tons, and showed an increasing trend year by year [3]. The stockpiled tailings not only take up a lot of cultivated land, pollute water body, destroy the ecological environment, but also induce serious security risks. In order to maintain the safety of the tailings dam and reduce the environmental damage caused by tailings, miners enterprise and government every year put a lot of money and manpower, but still difficult to avoid disaster accidents. For example, on September 8, 2008, a particularly significant dam break accident had been happened in New Tower Mining Co., Ltd which located in Xiangfen, Shanxi Province. Accident discharged capacity of 268000 m^3, mud area of 0.3 km^2, affecting the lower reaches of the mining area of about 500 m of office buildings, markets and some houses, the death toll was 277, the injured was 33 and 4 persons were missing, the direct economic losses of 96.192 million yuan. On November 23, 2015, Longxing Antimony Mining Co., Ltd which located in Xihe County, Longnan City, Gansu Province, occurred tailings leakage, and resulted in an abrupt environmental events across Gansu, Shaanxi, Sichuan three provinces. It serious impacted on people's production and living water. After assessment, the event caused a total economic loss of 6120.79 million Yuan. According to statistics, since 2000, the national

Content from this work may be used under the terms of the Creative Commons Attribution 3.0 licence. Any further distribution of this work must maintain attribution to the author(s) and the title of the work, journal citation and DOI.
Published under licence by IOP Publishing Ltd

total of more than 70 cases of tailings accidents had occurred. It brought a great harm to the survival of the people. According to the United States Clark University of Public Health Assessment Group research results show that the hazards of tailings accident was ranked 18th in the 93 kinds of accidents hazards[4]. Therefore, how to effective comprehensive utilization the tailings has become a hotspot in the field of scientific research workers under the severe situation of the shortage of mineral resources and the double pressure of the safety and damage of ecological environment caused by tailings.

From the point of view of resource utilization and law of conservation of mass, there is no "waste" in the world, and all substances can be participated in the next material cycle [5,6]. At present, it is generally believed that the tailings are a kind of resource that is put in the wrong place, and the tailings can be reclaimed by scientific separation and recycle technology. Comprehensive utilization of tailings can not only alleviate the increasingly scarce resources, but also can reduce the ecological and environmental pollution problems, to bring huge economic benefits, also is the mining enterprises take the direction of sustainable development[7].

Micro-fine particles easily lead to mineral particles on the surface of hydration film, due to minimal particle size, and have an important effect on its dispersion and aggregation in aqueous solution[8,9]. At the same time, it is extremely easy to slime. The traditional processing such as gravity separation, magnetic separation and flotation to deal with these micro-fine minerals are difficult to achieve satisfactory results. When used to treat such ore, it could not be effectively recovered, and caused a large number of useful mineral loss [10]. The superconducting high gradient magnetic separation (HGMS) technology has become one of the most promising new tailings recycling technology with unique separation principle and many advantages. With the continuous development of technology theory and equipment, it has been expanded and successfully applied in industrial solid waste and mine tailings. It has unique advantages in dealing with micro-fine weakly magnetic particles and non-magnetic particles in solid waste. In this paper, the development of superconducting HGMS was firstly introduced and the research status was summarized, then the direction of development for HGMS was analyzed.

2. Development of Superconducting HGMS

Magnetic separation is physical separation method based on the difference in magnetic properties of various ore minerals or materials [11]. The schematic diagram of the separator is shown in Figure 1[12]. Application of magnetic separation technology began in 1792, the British scholars used magnetic separation technology for iron ore selection. They got a good mineral processing indicator, and subsequently declared the relevant patents. In 1845, the United States invented the industrial magnetic separator, by the 1920s, various types of magnetic separators successive came out, magnetic separation technology in the field of mineral processing continue to get development and improvement. But at this time the magnetic separation technology can only be separated from large and strong magnetic particles, while the separation of paramagnetic and weakly magnetic particles was powerless, magnetic separation technology was greatly limited[13].

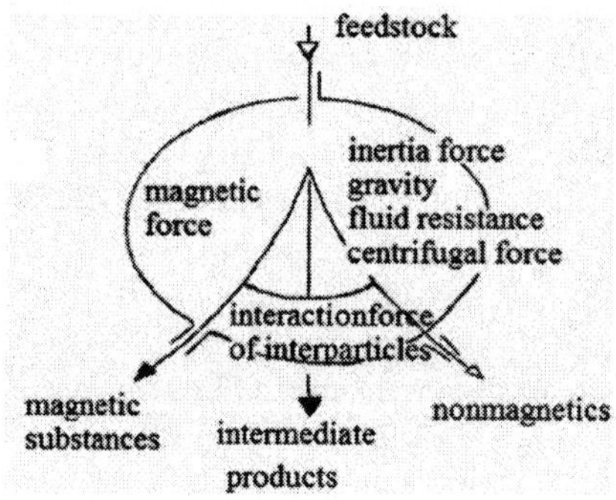

Figure 1. Schematic diagram of magnetic separation

IOP Conf. Series: Materials Science and Engineering **275** (2017) 012044 doi:10.1088/1757-899X/275/1/012044

In the design of the magnetic separator, the magnetic field strength and the magnetic field gradient are the two most important variables that affect the separation effect. Magnetic field gradient refers to the magnetic field strength in the direction of the magnetic field rate of change. Figure 2 illustrates two different magnetic field morphologies. Case A represents a uniform magnetic flux pattern without a magnetic field gradient, and the field strengths at any point are substantially the same. Hence, the magnetic particles entering the magnetic field will be attracted to the magnetic flux line and then remain fixed. Case B represents a high gradient magnetic field pattern in which the flux is converged. When such a magnetic flux passes through a smaller area, the magnetic field strength will increase significantly. The magnetic particles entering the magnetic field will not only be attracted to the magnetic flux line, but will also move to the region with the largest magnetic flux density, usually at the top of the bottom pole piece. In short, the magnetic field absorbs the particles while the magnetic field gradient moves the particles [14].

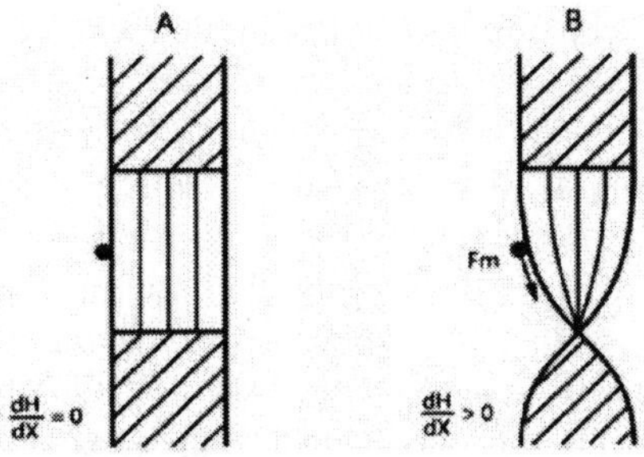

Figure 2. The illustration of two kinds of magnetic field gradient

In the early 1970s, the American scholars raised up the HGMS technology research, they used the interaction of different magnetic particles and magnetic matrix (usually ferromagnetic metal wire) which are in magnetic field. Ferromagnetic metal wire in the magnetic field produces a high gradient magnetic field with uneven distribution of magnetic field intensity. Magnetic particles are subjected to magnetic forces and they are easily separated from non-magnetic particles. As the HGMS device can be large-scale, rapid separation of magnetic particles, so the application scope of HGMS technology has gone beyond the traditional magnetic mineral processing, now into the waste gas, waste water and other environmental protection areas. However, because of magnetic field intensity constraints, the slurry flow velocity is low, processing capacity is small, non-magnetic and weakly magnetic particles are difficult to separate [15].

Usually the main magnetic material (ferromagnetic material) of magnetic separator has magnetic saturation limit (2 T), so the magnetic flux density of conventional magnetic separator has no more than 2 T. To break this limitation, need to change the magnetic material from the ferromagnetic to superconductor. When the material in the superconducting state, the resistance is zero, so in a fine superconducting wire, can pass a lot of current, no heat loss, which can be get more than 2 T super magnetic field[16]. After a long period of exploration, superconducting technology came into being, the use of superconducting coil instead of the conventional copper coil, can be up over to 10 T high gradient magnetic field [17], so that the scope of application greatly expanded, not only can handle magnetic particles, weakly magnetic or even non-magnetic particles can also be adsorbed and separated.

The development of superconducting HGMS technology makes it possible for magnetic separation to be used for separation and recycle of valuable materials from micro-fine tailings. As the superconducting operation of zero resistance, low power consumption, less area demanded and high degree of automation. At the same time, it is the most potential new technology under the energy saving emission reduction situation, and has the advantages of environmentally friendly. Hence, superconducting HGMS technology will have a good prospect of industrial application.

3. Application Status of Superconducting HGMS in Micro-Fine Tailings

At present, cryogenic superconductivity high gradient magnetic separator is widely used. According to the presence of magnetic matrix, they can be divided into superconducting high gradient magnetic separator (with magnetic matrix) and superconducting open gradient magnetic separator (without magnetic matrix). While the former accounted for the majority. According to the transmission medium they can be divided into dry and wet superconducting high gradient magnetic separator [18].

3.1. The Application Instance of Dry Superconducting High Gradient Magnetic Separator

The United States Auburn University and Oak Ridge National Laboratory (ORNL) conducted dry HGMS of coal desulfurization experiment and achieved success in 1976. Then they started to make dry high gradient magnetic separator research work. The main processes are: 1,low speed airflow--gravity feed dry HGMS process[19]; 2, airflow feed dry HGMS process[20]; 3, circulating air fluidized bed feed dry HGMS process[21].

In the Iron and steel industry, the number of iron particles in the dust, which produced by the top bloom oxygen converter and electric arc furnace steelmaking, is particularly large. Take Guangzhou Iron & Steel Company as an example, converter steelmaking dust is strong magnetic substance, the specific magnetic susceptibility of dust is greater than 38×10^{-6} m^3/kg. But the particle size of the dust is very small, -74 μm particles size accounted for 96.85%. The separation efficiency is not satisfactory using traditional magnetic separation technology. As the high magnetic field intensity and high magnetic field gradient advantages, the HGMS technology can make the magnetic particles in the dust be subjected to great magnetic force and easily overcome the viscous force of the airflow, thus effectively capturing the micro-fine magnetic particles in the dust. Wu Xiuwen et al.[22] took the dust which from Guangzhou Iron & Steel Company as the research object. They carried out dust removal experiments to investigate the use of dry HGMS technology. The experimental results determine the optimum process parameters i.e., magnetic flux density of 0.52 T, magnetic matrix filling ratio of 2%, gas flow of 12 m^3/h. The total dust removal efficiency can reach 94.7%.

The grain size of a potassium feldspar ore in Lianzhou City Guangdong Province is -45 μm, the content of Fe_2O_3 is 0.41%, and mechanical iron is 0.1%. It mainly composed of ferrous manganese ore which is weakly magnetic mineral. Feng Dingwu et al.[23] employed a dry superconducting high gradient magnetic separator to remove metal impurity from feldspar. The potassium feldspar powder is dried and transported by wind to a separator. When the powder passes through the dispenser nozzle, the gas expands so that the particles are sufficiently dispersed. Dispersed magnetic particles are captured by magnetic matrix by vibrating superconducting high gradient magnetic separator. Under the condition of magnetic flux density of 1.0 T, particle flow velocity of 2.0 m/s, magnetic matrix filling ratio of 3.8% and magnetic matrix vibration frequency of 30 Hz. The content of Fe_2O_3 in magnetic concentrate decreased from 0.41% to 0.27%, the recovery rate of nonmagnetic material was 80.8%.

3.2. The Application Instance of Wet Superconducting High Gradient Magnetic Separator

Currently in the field of industrial production the superconducting high gradient magnetic separator is mainly wet type. The schematic diagram of common wet type separator is shown in Figure 3[24]. The advantage of wet HGMS is that it is easy to disperse for pulp and high separation efficiency. The shortcomings are: 1, dry materials need to add water into pulp; 2, magnetic concentrate need dehydration and drying, micro-fine particles are easy to loss during dehydration process; 3, some dry process does not allow the introduction of wet operation [25,26].

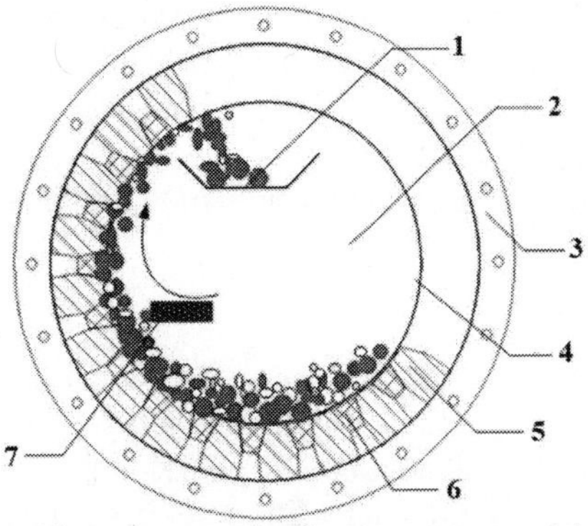

1-concentrate trough, 2-magnetic drum, 3-iron yoke, 4-separating cylinder,
5-the main magnetic system, 6-deputy magnetic, 7-feed device

Figure 3 Schematic diagram of wet superconducting high gradient magnetic separator

Zhejiang Huangyan Lead Zinc Plant produces 100 tons of tailings per day, the tailings (without any treatment) are stacked directly in tailings impoundment, which seriously threaten the survival of downstream residents. Main components of tailings are silicate and containing aluminum minerals, such as quartz, sericite, illite, etc. Those can replace feldspar for ceramic energy-saving raw material and make the porcelain temperature reduced by about 100 °C. However, ceramic energy-saving material requests the content of Fe_2O_3+MnO less than 1.5%, while the content of Fe_2O_3+MnO in Huangyan tailings is 14.03%, which is far beyond the standard. The presence of Fe and Mn is mainly pyrite, hematite, rhodochrosite, manganese calcite, which can be easy to remove by superconducting HGMS. Peng Shiying[27] carried out a laboratory study on the tailings, the results confirm the optimal parameters i.e., magnetic flux density of 1.3 T, slurry flow velocity of 2.32 cm/s, slurry concentration of 10%, dispersant (sodium hexametaphosphate) of 2.5 kg/t, magnetic matrix filling ratio of 5%. The content of Fe_2O_3 in magnetic concentrate has decreased from 6.66% to 1.05% and MnO from 6.82% to 0.41%. Obviously, superconducting HGMS is a very effective way to remove Fe and Mn impurities from the tailings. The content of Fe_2O_3+MnO in magnetic concentrate is less than 1.5%, which meets the requirement of ceramic energy-saving raw materials.

Wen Haitao et al. [28] study on the utilization of low grade iron tailings by superconducting HGMS Technology. The samples are mainly weakly magnetic hematite and goethite with particle size of -45 μm. They carried out single factor experiment to investigate the effect of magnetic flux density, slurry concentration and slurry flow velocity on separation efficiency. Under the optimal parameters: magnetic flux density of 1.2 T, slurry concentration of 25 g/L and slurry flow velocity of 500 mL/min, the iron content of magnetic concentrate increased from 32% to 65%, which can meet the requirement of blast furnace iron making. The experiment result is satisfactory and has been provided an effective way to utilization of iron tailings.

Chen Luzheng et al.[29] analyzed the effect of parameters on separation efficiency in a laboratory scale with the micro-fine particle hematite (-74 μm). The results of investigation indicate that changes in the magnetic flux density and the rotation speed of matrix had the most significant influences on the performance. An increase in the rotation speed improved concentrate grade but reduced iron recovery, while the reverse was observed for changing the magnetic flux density. A too low feed flow rate resulted in an extremely high recovery but a lower concentrate grade. They obtained a magnetic concentrate, in which the iron content increased from 35.02% to 60.98%, and the concentrate recovery rate reached 84.28% under the optimum parameters of magnetic flux density of 0.7 T, ore feed rate of 50 g and slurry flow velocity of 9 L/min. It is proved that superconducting HGMS technology is feasible for recycle and enrichment of weakly magnetic iron tailings.

Dwari R.K. et al. [30] carried out the investigation on extraction iron from low grade siliceous iron ore by superconducting HGMS. The main components of the samples are magnetite, hematite, goethite, etc., and the gangue minerals are silicate and carbonate minerals, the particle size is -100 μm. Under the conditions of magnetic flux density of 0.2 T, slurry concentration of 20% and slurry flow velocity of 2 L/min, they obtained a good mineral processing indicator with concentrate iron grade of 66% and yield of 88%.

4.Future Prospects

With the continuous technological breakthrough and rapid development of industrial practice, superconducting HGMS technology shows more and more advantages in kaolin purification, mineral dressing, coal desulfurization, sewage treatment and other fields. In the field of mineral dressing, superconducting HGMS can capture effectively the micro-fine weakly magnetic particles and even separate the micro-fine particles which remaining in the tailings after conventional magnetic separation [31]. In the field of coal desulfurization, it can effectively reduce the soot of environmental pollution. In the field of sewage treatment, the superconducting HGMS technology has achieved remarkable results in the purification of industrial wastewater (such as papermaking, chemical manufacturing, textile, thermal power plant) and domestic sewage. It can be recycled sewage, reduce environmental pollution while also save water resources. Superconducting HGMS technology has a tendency to use in conjunction with other technologies, such as biological water treatment technology, membrane separation technology. It can recycle the valuable secondary resources and achieve the environmental benefits and social benefits of the harmonious unity. Superconducting HGMS technology after decades of research and development, has begun to go out of the laboratory and into the industrial production stage, it will gradually replace the existing standard industrial magnetic separator. Separator is expected to develop in the next few years as following aspects. 1, Low energy consumption, 2, High field strength and high gradient, 3, Equipment light weight, small size.

Although the superconducting HGMS technology has shown great advantages and broad application prospects, there are the limitations and technical difficulties such as the remanence of magnetic matrix. Hence, superconducting HGMS technology will be subject to certain restrictions in practice. In the future, the following three aspects will be the focus of research. 1, In order to improve the magnetic field gradient, must choose high magnetic saturation of the magnetic matrix. On the other hand, remanence of magnetic matrix, it is difficult to rinse the magnetic particles adsorbed by magnetic matrix and affect the efficiency of the next cycle. Research workers need to be considered how to solve this contradiction. 2, Due to the magnetic matrix selection had some technical difficulties, and it would increase operating costs, research workers should increase the intensity of magnetic matrix materials research and achieve lower costs. 3, Strengthen the exploration of scientific research such as the surface modification of magnetic seed, the interaction between magnetic seed and fine weakly magnetic particles, magnetic seed recovery process and the design and optimization of magnetic filter, etc.

5. References

[1] Z.L. Zhao, Q. Gao, L.F. Wang, F.Y. Han and W. Liang, 2016, *Metal World*, 4, 77-80.
[2] L. Liu, Q.Q. Hao and Z.G. Hao, 2013, *Geology and Exploration*, 49, 437-443.
[3] Y.G. Huang, 2013, *Resources & Industries*, 15, 40-44.
[4] W.Q. Tian, 2003, *Labour Protection*, 9, 20-23.
[5] Z.D. Li and Q.L. Zhou, 2003, *Mitigation of Mine Resource Crisis--Proceedings of the Symposium on China 's Resource Crisis*, 50-56.
[6] R.C. Gupta, P. Mehra and B.S. Thomas, 2017, *Journal of Materials in Civil Engineering*, 29, 04016274-7.
[7] F.B. Ono, E.S. Penido, R. Tappero and D. Sparks, 2016, *Environmental Geochemistry and Health*, 38, 1083-1096.
[8] X.X. Yu and Y.H. Wang, 2014, *Optics Express*, 22, 177-182.

[9] H. Cheng, Y.H. Wang, H.W. Dai, J.B. Han and X.C. Li, 2015, *The Journal of Physical Chemistry C*, 119, 3288-3292.
[10] Y. Yu, F.S. Niu, F.J. Li and G. Wu, 2009, *Metal Mine*, 391, 62-65.
[11] R.K. Dwari, D.S. Rao and P.S.R. Reddy, 2013, *International Journal of Mining Science and Technology*, 23, 1-5.
[12] S.Q. Li, M.F. Wang, Z.A. Zhu, Q. Wang, X. Zhang, H.Q. Song and D.Q. Cang, 2012, *Separation and Purification Technology*, 84, 56-62.
[13] L.Y. Li, Y. Chen, L.Y. Liu, B.J. Jia, Q.Z. Feng, Areyiguli and X.X. Wu, 2014, *Academic Proceedings of the Chinese Society of Environmental Science in 2014*, 4600-4604.
[14] Q. Zhang, J.G. Zhang, 2014, *Technology Innovation and Application*, 14, 94-95.
[15] S.N. Podoynitsyn, O.N. Sorokina, and A.L. Kovarski, 2016, *Journal of Magnetism and Magnetic Materials*, 397, 51–56.
[16] C.L. Jiang, 1994, *Magnetic Separation Theory and Technology*, 122-123.
[17] J.A. Ritter, A.D. Ebner, K.D. Daniel and K.L. Stewart, 2004, *Journal of Magnetism and Magnetic Materials*, 280, 184-201.
[18] Y. Nakai, K. Senkawa, F. Mishima, Y. Akiyama and S. Nishijima, 2011, *IEEE Transactions on Applied Superconductivity*, 21, 2063-2067.
[19] E.C. Hise, 1979, *Separation of dry crushed coals by HGMS*, Report No. ORNL-4471.
[20] J. Iannicelli,1976, *IEEE Transaction on Magnetics*, 12, 436-443.
[21] T. Oda, Y. Kunisue and S. Masuda, 1983, *IEEE Transaction on Magnetics*, 19, 2121-2123.
[22] X.W. Wu, H.X. Lan and Y.P. Yan, 2010, *Journal of Dongguan University of Technology*, 17, 83-87.
[23] D.W. Feng, Z.Y. Sun and J. Chen, 1995, *Non-Metal Mines*, 18, 22-24.
[24] S.L. Zeng, W.L. Zeng, L.Y. Ren, D.Q. An and H.Y. Li, 2015, *Minerals Engineering*, 71, 21-26.
[25] T.Y. Ying, S. Yiacoumi and C. Tsouris, 2000, *Chemical Engineering Science*, 55, 1101-1113.
[26] A. Ditsch, S. Lindenmann, P.E. Laibinis, D. Wang and T.A. Hatton, 2005, *Industrial & Engineering Chemistry Research*, 44, 6824-6836.
[27] S.Y. Peng, 1995, *The Chinese Journal of Nonferrous Metals*, 5, 23-27.
[28] H.T. Wen, S.Q. Li, C.Q. Zhang and W. Wei, 2014, *Advance materials research*, 968, 168-172.
[29] L.Z. Chen, J.W. Zeng, C.Q. Guan, H.F. Zhang and R.Y. Yang, 2015, *Minerals Engineering*, 78, 122-127.
[30] R.K. Dwari, D.S. Rao and P.S.R. Reddy, 2013, *International Journal of Mining Science and Technology*, 23, 1-5.
[31] L.N. He, 2014, *Cryogenics & Superconductivity*, 41, 55-58.

Fabrication and Characteristics of Al/PTFE Multilayers and Application in Micro-initiator

Yuxin Zhang, Hongchuan Jiang, Xiaohui Zhao, Wanli Zhang and Yanrong Li
State Key Laboratory of Electronic Thin Films and Integrated Devices, University of Electronic Science and Technology of China, Chengdu 610054, China
Email: pzh_zyx@126.com

Abstract: In this paper, a micro-initiator was designed and fabricated by integrating Al/PTFE multilayers with a Cu film bridge. The regularity layer structure and interface composition of Al/PTFE multilayers was analysed by transmission electron microscope and X-ray photoelectron spectroscopy, respectively. The heat release reaction in Al/PTFE multilayers can be triggered with reaction temperature of 430 °C, and the overall heat of reaction is 3192 J/g. Al/PTFE multilayers with bilayer thickness of 200 nm was alternately deposited on a Cu film bridge to improve the electric explosion performances. Compared to Cu film bridge, the Al/PTFE /Cu integrated film bridge exhibits improved performances with longer explosion duration time, more violent explosion phenomenon and larger quantities of ejected product particles.

1. Introduction

In recently years, the investigations for nano-energetic materials have triggered extensive interest for their various applications including welding and soldering [1], micro-initiator [2], initiation of secondary reactions [3], and airbags [4], et al. Nano-energetic materials composited of fuel and oxidizer in nanoscale exhibit better performances in many aspects such as rapid energy release, high energy density, and more elements to choice compared to conventional energetic materials. There are many ways to fabricate nano-energetic materials including power mixing, Magnetron sputtering, sol-gel, high-energy ball milling and so on [5]. Among these methods, alternately depositing two or more different multilayers exhibits superiority over others for maximum interfacial contact area, minimum distance among different layers and easy integration with microelectronic and mechanical systems. Integrating nano-multilayers with a metal film bridge initiator provides a fascinating method to improve the electric explosion performances of micro-initiator with compact size, and the performances are tunable upon changing the bilayer thickness and the number of layers. A lot of efforts have been devoted on integrating nano-multilayers with a micro-initiator such as Al/Ni [6, 7], Al/CuO [2, 8], Al/NiO [9], and B/Ti [10, 11]. Compared with these reactive nano-multilayers, Al/PTFE multilayers system could release gas during the reaction, which is beneficial for increasing the performances of micro-initiator. Moreover, Al/PTFE have an ultrahigh energy density of 21 KJ/cm^3 [12].

In this paper, a micro-initiator (Al/PTFE /Cu integrated film bridge) was designed and fabricated by integrating Al/PTFE multilayers with a Cu film bridge. The structure and interface composition of the Al/PTFE multilayers was characterized by TEM and XPS analysis, respectively. The electric explosion performance of Cu film and Al/PTFE /Cu integrated film bridge were characterized through

Content from this work may be used under the terms of the Creative Commons Attribution 3.0 licence. Any further distribution of this work must maintain attribution to the author(s) and the title of the work, journal citation and DOI.
Published under licence by IOP Publishing Ltd

an electric ignition measurement system, and high speed camera was utilized to record the reaction processes.

2. Experimental Section

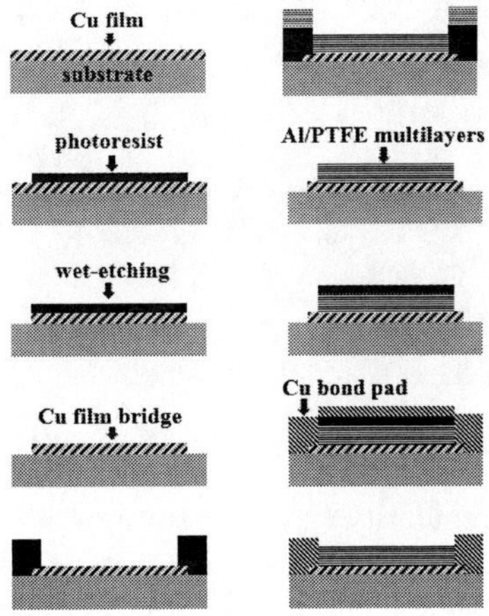

Figure 1. Schematic representation of fabrication process for Al/PTFE /Cu integrated film bridge.

Figure 1 shows the manufacture process of Al/PTFE/Cu integrated film bridge. Before deposition, an alumina substrate was cleaned in an ultrasonic bath with acetone, alcohol, and deionized water for 10 min sequentially. Then, the substrate was dried by nitrogen gas and heat treated at 150 °C for 20 min. A Cu target (99.995 wt. %) with a diameter of 100 mm was applied for sputtering. 2 μm thick Cu film was deposited on the alumina substrate, and the sputtering temperature, pressure and power of 100 °C, 0.55 Pa and 100 W, respectively. After that, the Cu film spin-coated with photoresist (PRI-4000A) was patterned with photolithography through a designed mask. Then, the exposed Cu film was directly wet-etched in $Fecl_3$ solution.

Afterwards, the reversal photoresist (AZ5200) was spin-coated onto the Cu film bridge, which was exposed twice to generate a reentrant profile. Then, Al/PTFE multilayers with bilayer thickness of 200 nm (Al, 50 nm; PTFE, 150 nm) were alternately deposited on the Cu film bridge from Al and PTFE targets by DC magnetron sputtering and DF magnetron sputtering. The sputtering power and pressure for Al film and PTFE film were set at 0.5 Pa, 100 W and 1.5 Pa, 300 W, respectively. The gross thickness of Al/PTFE multilayers was 2 μm. After removing the developed photoresist, two Cu bound pads were deposited on two sides of Al/PTFE multilayers and lead wires were soldered for the connection to the voltage source.

The periodic layer structure and interface composition of the Al/PTFE multilayers were verified by TEM and XPS, respectively. The heat release properties of Al/PTFE multilayers were tested by DSC with a heating rate of 10 °C/min from 50 to 800 °C in flowing nitrogen, The electric explosion performances were characterized through an electric ignition measurement system, and high speed camera was utilized to record the reaction processes.

IOP Conf. Series: Materials Science and Engineering **275** (2017) 012045 doi:10.1088/1757-899X/275/1/012045

3. Results and Discussion

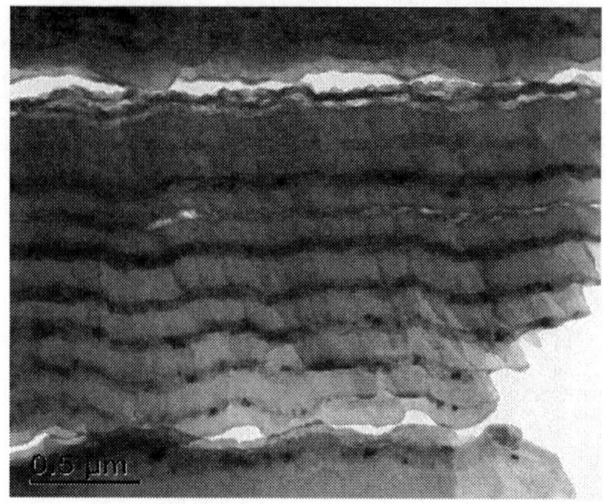

Figure 2. Cross-sectional TEM image of Al/PTFE multilayers.

Figure 2 reveals the periodic and regularity layer structure of as-deposited Al/PTFE multilayers. The Al layer and PTFE layer can be clearly defined, and the thickness is controlled. We also can see some surface roughness between Al layer and PTFE layer, which are beneficial for the intermetallic reaction between Al layer and PTFE layer to release energy.

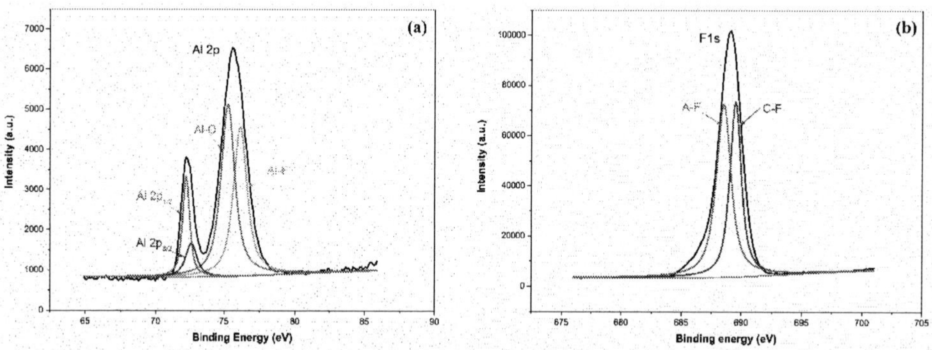

Figure 3. XPS spectra of Al 2p core level (a) and F 1s core level (b).

The interface composition of the Al/PTFE multilayers were investigated by XPS analysis, as shown in Figure 3. The Al 2p core level of the sample with 1 nm PTFE film deposition on Al layer. Two peaks of the Al 2p core level appeared at BE of 72.2 and 75.6 eV. The peak at 72.2 eV of the Al 2p spectra of Al 2p1/2 and Al 2p3/2 is Al film. The peak at 75.6 eV of the Al 2p core level was weight than the Al 2p core level of metallic Al, which could be ascribed to the electronic feature induced by the deposition of PTFE film on Al layer. Meanwhile, compared to PTFE film, the F 1s core level of the sample with 1 nm PTFE film deposition on Al layer shifted as well. These results indicates that the Al/PTFE multilayers is composed of Al layer, A-F bond (inert layer) and PTFE layer assembling in a metastable system. The Al-F bond existed between Al Layer and PTFE layer is important component to keep high energy density of Al/PTFE multilayers.

Figure 4 reveals the heat release characteristics of Al/PTFE multilayers with bilayer thickness of 200 nm. There are only one major exothermic peaks at 535°C, and the onset reaction temperature is 430 °C. The total reaction heat is calculated at 3192 J/g through integrating the positive exothermic heat flow. The high heat energy release is conductive to enhance the electric explosion performances of Cu film bridge.

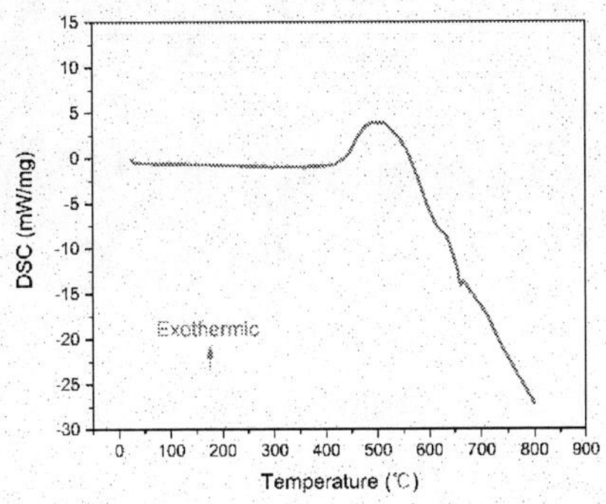

Figure 4. DSC trace for Al/PTFE multilayers using a heating rate of 10 °C/min in flowing nitrogen.

For the electric explosion analysis of Cu film bridge and Al/PTFE /Cu integrated film bridge, A discharge capacitor (1800 V, 0.22 μF) was adopted to apply voltage. Figure 5 reveals the experimental results of Voltage-current plots. After applying voltage to film bridge, the resistance and temperature of film bridge rise instantly due to the rapidly increasing current density. When the film bridge begin to vaporize, the voltage stops increasing and reaches the maximal value. Thus, the maximal voltage is regarded as the electron explosion voltage, and the time is regarded as ignition delay. The explosion voltage is measured to be 1575 V for Cu film bridge, and 1655V for Al/PTFE /Cu integrated film bridge. The ignition delay is 78 ns for Cu film bridge, and 71 ns for Al/PTFE /Cu integrated film bridge. The explosion voltage and ignition delay for two types of film bridges are similar, this might be explained that the room temperature resistivity of Cu film is far less than that of PTFE layers in static, most of the current is expected to pass through the Cu film bridge.

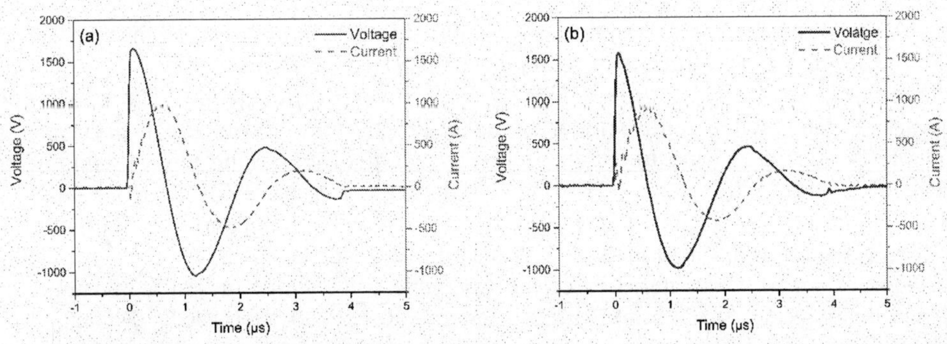

Figure 5. Voltage and current plots of electron explosion processes: (a) Cu film bridge, (b) Al/PTFE /Cu integrated film bridge

As shown in Figure 6, a high speed camera was utilized to record the reaction dynamic processes of film bridges during the voltage-current tests. The interval between adjoining pictures is 50 μs. The specific flame structures in various electron explosion stages were obviously observed. We can observe a fiercer electric explosion phenomenon accompanied with larger quantities of ejected product particles occurred on the Al/PTFE/Cu film bridge. The explosion duration time is about 300 μs for Cu film bridge and 350 μs for Al/PTFE/Cu film bridge. It can be definitely concluded that the self-propagation exothermic reaction of Al/PTFE multilayers is triggered during the electric explosion process, and the reaction can improve the explosion performances of Cu film bridge substantially.

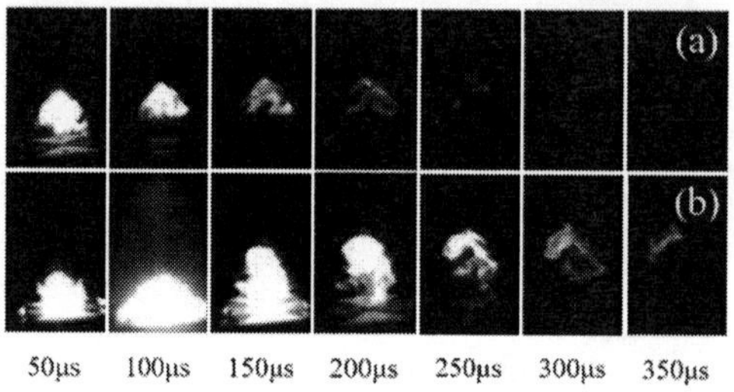

Figure 6. Explosion processes of Cu film bridge (a) and Al/PTFE/ Cu integrated film bridge (b).

4. Conclusion

In this work, an integrated film initiator is realized by integrating Al/PTFE multilayers with a Cu film bridge on alumina substrate. From the comparison of the electric explosion characteristics of the Cu film bridge and Al/PTFE /Cu integrated film bridge, it can be concluded that the combination of Al/PTFE multilayers with Cu film bridge can improve the electric explosion performances substantially. In addition, the electric explosion performances of Al/PTFE /Cu integrated film bridge could be tuned by simply altering the films thickness, and the Al/PTFE /Cu integrating film bridge can be fabricated by standard microfabrication technique that allow batch fabrication and high level of integration.

5. References

[1]. Shen, J. and Y.C. Chan, Research advances in nano-composite solders. Microelectronics Reliability, 2009. 49(3): p. 223-234.
[2]. Taton, G., et al., Micro-chip initiator realized by integrating Al/CuO multilayer nanothermite on polymeric membrane. Journal of Micromechanics & Microengineering, 2013. 23(10): p. 3210-3216.
[3]. Morris, C.J., et al., Rapid initiation of reactions in Al/Ni multilayers with nanoscale layering. Journal of Physics and Chemistry of Solids, 2010. 71(2): p. 84-89.
[4]. Baginski, T.A., S.L. Taliaferro and W.D. Fahey, Novel Electroexplosive Device Incorporating a Reactive Laminated Metallic Bridge. Journal of Propulsion & Power, 2001. 17(1): p. 184-189.
[5]. Zhou, X., et al., Nanostructured Energetic Composites: Synthesis, Ignition/Combustion Modeling, and Applications. ACS Applied Materials & Interfaces, 2014. 6(5): p. 3058-3074.
[6]. Qiu, X., et al., A micro initiator realized by reactive Ni/Al nanolaminates. Journal of Materials Science: Materials in Electronics, 2012. 23(12): p. 2140-2144.
[7]. Zhang, Y., et al., Characteristics of the Energetic Micro-initiator Through Integrating Al/Ni Nano-multilayers with Cu Film Bridge. Nanoscale Research Letters, 2017. 12(1): p. 38.
[8]. Zhou, X., et al., Influence of Al/CuO reactive multilayer films additives on exploding foil initiator. Journal of Applied Physics, 2011. 110(9): p. 094505.
[9]. Yan, Y.C., et al., Characteristics of the Energetic Igniters Through Integrating Al/NiO Nanolaminates on Cr Film Bridge. Nanoscale Research Letters, 2015. 10(1): p. 1-7.
[10]. Yan, Y., et al., Characteristics of the Energetic Igniters Through Integrating B/Ti Nano-Multilayers on TaN Film Bridge. Nanoscale Research Letters, 2015. 10(1): p. 244.
[11]. Zhang, Y., et al., B/Ti nano-multilayers as effective heat energy source for enhanced micro-initiator. Vol. 117. 2017. 617-621.
[12]. Wang, J., et al., Design and fabrication of energetic superlattice like-PTFE/Al with superior performance and application in functional micro-initiator. Nano Energy, 2015. 12(Supplement C): p. 597-605.

IOP Conf. Series: Materials Science and Engineering **275** (2017) 012046 doi:10.1088/1757-899X/275/1/012046

Structure and Photoluminescence Properties of β-Ga2O3 Nanofibres Synthesized via Electrospinning Method

Chao Sun, Jinxiang Deng*, Le Kong, Liang Chen, Zhen Shen, Yisen Cao, Hao Zhang and Xiaoran Wang

College of Applied Science of Beijing University of Technology, Beijing 100124, China

Email: jdeng@bjut.edu.cn

Abstract: This paper reported the β-Ga2O3 nanofibres which fabricated by electrospinning, and then calcining in oxygen at 750, 850, 950 and 1050°C. The structure and properties of β-Ga2O3 nanofibers have been studied though kinds of methods such as XRD, Photoluminescence (PL) spectrum, Raman spectrum, Scanning electron microscope (SEM) and FT-IR. The diameters of these nanofibres are from 60 to 130nm and the lengths of these nanofibres are about couple millimetres. The spectrum of PL which excitation at 365nm gave us the information that the emission peak of these β-Ga2O3 nanofibres is about 470nm, it may be coursed by the various defects including the vacancies of gallium and oxygen and the gallium-oxygen vacancy pairs as well, and observed that with the increasing of the annealing temperature, the emission peaks have a small bule swifting, and the crystallinity become better at the same time.

1. Introduction

β-Ga2O3 is a very important semiconducting material [1], and it has kinds of crystalline structures called α, β, γ, δ and ε phases. It has a band gap of 4.9 eV [2]. Nanosize β-Ga2O3 owns special properties of conduction and semiconductor which indicate applications in the devices of optoelectronic such as flat panel displays, photoelectric converter, and Optical limiter for UV oxide because of its surface area/volume ratio [3]. It also can be used to fabricate solar-blind devices due to its solar-blind properties under deep ultraviolet environment [4-8]. And the bigger surface area/volume ratio is the better for the application of gas sensors. Such as S.H.Park, et al. [9] observed CO gas sensing properties through synthesis surface nitriding Ga2O3 nanowires. And Shinji Nakagomi, et al.[10] found that the sensors fabricated by a β-Ga2O3 thin film could detect 100ppm H2 in 20% O2/N2 at 400°C. Therefore, it is necessary to get nanostructure β-Ga2O3. Now days, β-Ga2O3 of nanostructure could be synthesized by several kinds of methods like simple precipitation technique [11], electrochemical reaction [12], atmospheric-pressure CVD [13], metal organic chemical vapor deposition (MOCVD) [1], and thermal evaporation method[14]. But, these methods need complicate equipments, catalysis, or introduce impurity atom, and the prepared β-Ga2O3 including the vestigital of the catalysis [15]. Therefore, it is necessary to fabricate β-Ga2O3 nanofibres which have a high purity. However, there are no reports on the effect of annealing temperature on the structure, photoluminescence properties and purity of β-Ga2O3 nanofibres by electrospinning. In this paper, a kind of simple and reliable method of fabricating β-Ga2O3 nanofibres is recommended—electrospinning. Electrospinning is a simple and less costly method to synthesis nanofibers with high draw ratio. Kinds of nanofibres of materials could be generated with it including

Content from this work may be used under the terms of the Creative Commons Attribution 3.0 licence. Any further distribution of this work must maintain attribution to the author(s) and the title of the work, journal citation and DOI.
Published under licence by IOP Publishing Ltd

polymer, polymer or inorganic composites, and inorganic nanofibres. Through sintering of the as-prepared compound fibers carefully, β-Ga2O3 nanofibres have been fabricated successfully.

2. Experiment

Ga2O3 nanowires were electrospun from precursor which was prepared by the solution of Ga(NO3)3·6H2O, and polyvinylpyrrolidone(PVP, Sigma Aldrich, Mw≈1300000), PVP is an aggregation of composite materials because it has good ethanol and water solubility and the compatibility with nitrate. The precursor solution containing 1.0g Ga(NO3)3, 6.5 ml C2H5OH, 5.5 ml H2O, and 4.0 g PVP respectively. And then the solution was stirred for 12h directly. After that, the viscous solutions was obtained. Then the viscous solution was put into a injector with a hollow needle, The process of electrospinning was shown in Figure. 1. Positive pole of a high-voltage power supply and the point of the syringe needle were concatenated while the negative pole was connected to the plate of the collector. Voltages connected to the tip and the collector plate was 15 kV and -15KV respectively, and the distance between the needle tip and the collector was 25 cm. After all these parameters were set up, the solution would be electrospun next. The deposition continued for 1h so that the dense mats would be well obtained. After spinning, the fibers which prepared would be dried at 80°C for 3h. And then the fibres would be calcined at different temperatures to get different samples.

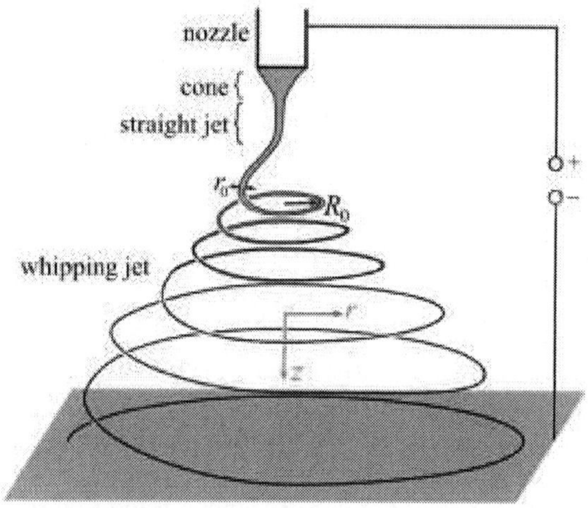

Figure1. Process of electrospinning

The crystalline structure of the fibres were examined by X-ray diffraction (XRD) using General motors XD-3 with Cu target. Raman spectrum were detected on a Horiba jobin yvon T64000 Confocal Raman spectrometer using 532nm excitation laser. The FT-IR spectrum has been detected during the range of 4000-400 cm-1 with a Bruker Vertex 80V FTIR spectrometer instrument. The morphology of the nanofibers were characterized by the SEM(JSM-6510). The photoluminescence (PL) spectra were measured at room temperature with a Edinburgh FLSP920 spectrophotometer. The excitation light was the monochromatic light from a Xenon short Arc lamp with a wavelength of λ=365nm.

3. Result and Discussion

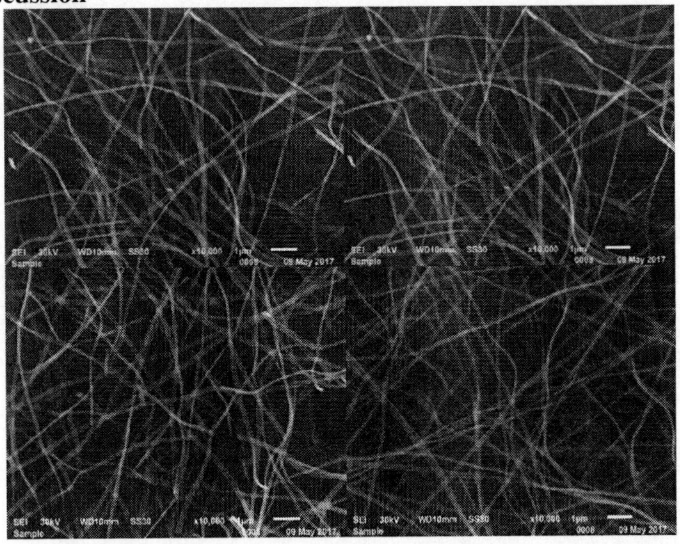

Figure. 2. SEM images of the β-Ga2O3 calcined nanofibres at : (a)750°C, (b) 850°C, (c) 950°C, (d) 1050°C

Figure.2 shows the typical SEM images of the β-Ga2O3 nanofibers which calcined at different temperatures and the annealed fibres are distribution randomly. And the surface of the fibres is very slick. There are several factures in the 850°C and 950°C samples maight be caused by the evaporation of the solvent, There are no great particles attached on the surface of the nanofibres. And, the width of the fibres is in the range of 60nm-200nm, and the average diameter of the β-Ga2O3 nanofibres is about 120nm. And the length are several micrometers. It is observed that the calcined temperature do not have an obvious effect on both the surface topography and the crystalline phase of the fibres.

Figure. 3 shows the XRD pattern of β-Ga2O3 nanofibres which annealed at 750, 850, 950, 1050 °C. The lattice constants of the crystalline phase are a0 = 1.2227 nm, b0 = 0.3039 nm, c0 = 0.5808nm. All the peaks are matched well to β-Ga2O3 and the peaks are (201), (-202), (020) respectively [16]. The XRD pattern signs that the annealed nanofibres was β-Ga2O3 surely. And with the increasing of the annealing temperature, all the peaks become higher and sharper indicate a higher degree of crystallization. No other impurity phases are detected, so the samples we obtained were pure without other dephasigns.

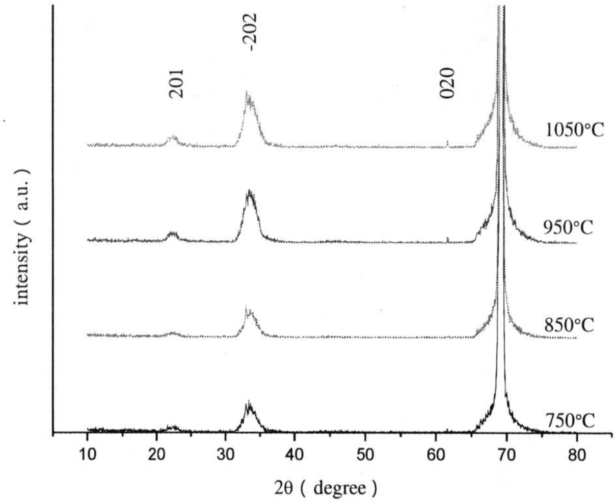

Figure 3. XRD pattern of β-Ga2O3 nanofibres annealed at 750, 850, 950, 1050 °C

Figure.4 shows the PL spectrum of the β-Ga2O3 nanofibres annealed at 750, 850, 950, 1050 °C respectively under excitation of 365nm. The dominant emission located at approximately 470 nm in the blue region. And the emission (470 nm) of blue band may be produced by the recombination of an electron of a donor originate from oxygen vacancies (VO) and a hole from an acceptor originate from gallium vacancies (VGa) or the gallium-oxygen electron hole pair (VO, VGa) [17][18]. It is reported that the peaks of the blue emission will enhance with the increase of the gallium-oxygen vacancy pairs [16][19]. Figure.4 shows that the blue emission peaks decrease with the increase of the annealing temperature. It may result from the reduce of the gallium-oxygen vacancy pairs caused of the reacting between O2 and as-prepared samples more fully. As for the blue shift of the emission spectrum, it probablely results of the quantum size effect [20] caused by the small tapering off of the nanowires which cannot be observed easily.

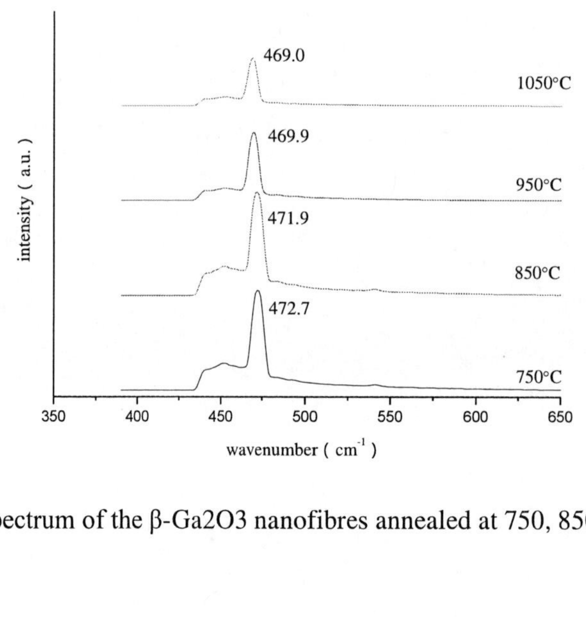

Figure 4. PL spectrum of the β-Ga2O3 nanofibres annealed at 750, 850, 950, 1050 °C

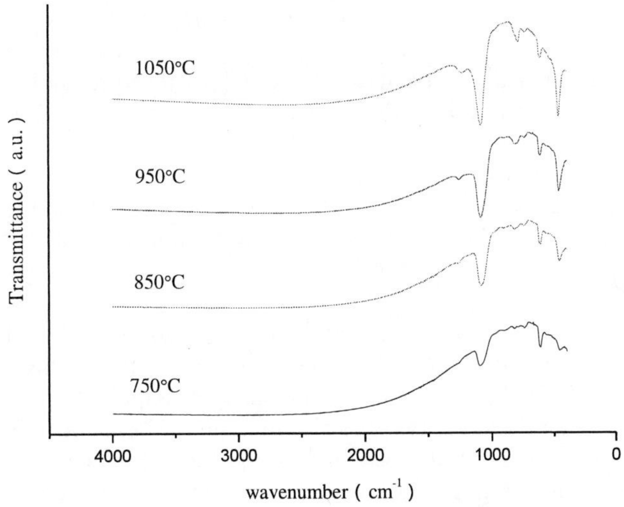

Figure 5. FT-IR spectrum of the β-Ga2O3 nanofibres calcined at 750, 850, 950, 1050 °C

The peaks located at 613 and 459 cm^{-1} in Figure.5 assigned to Ga–O–Ga and Ga–O vibration of β-Ga2O3 respectively, instead of 650 and 480 cm^{-1} of the peaks in the α-Ga2O3 [21]. and the peak at 1089 cm^{-1} belongs to asymmetric stretching vibration absorption of Si-O-Si [22]. it's obviously observed that with the increase of the annealing temperature, the infrared absorption is becoming stronger which indicated that the content of Ga–O–Ga and Ga–O increased. It signify that with the increase of the annealing temperature, oxygen vacancies were filled with oxygen elements to form Ga–O–Ga and Ga–O bonds.

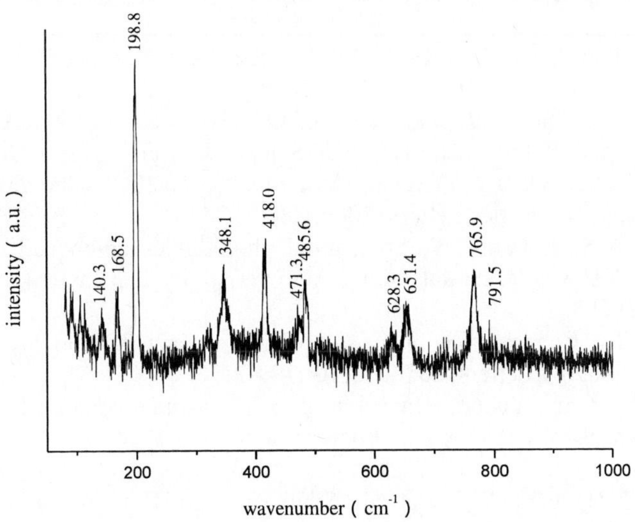

Figure 6. Raman spectrum of β-Ga2O3 nanofibres

Figure. 6 shows typical micro-Raman spectra of β-Ga2O3 nanofibres. The Raman spectra show 11 clear Raman peaks. And the positions of the peaks locate at 140.3, 168.5, 198.8, 348.1, 418.0, 471.3, 485.6, 628.3, 651.4, 765.9, 791.5 (cm^{-1}). These peaks are very peaked and narrow which sign that the crystallinity is very well and these peaks belong to the C2h group symmetry, implying a unit cell with two formula units, Ga2O6 octahedra and GaO4 tetrahedra [16]. The peaks of the Raman spectrum can be divided into three kinds such as low (below 200 cm^{-1}), mid (between 300 $cm^{-1}$and 500 cm^{-1}), and high (between 500 cm-1and 800 cm^{-1}) frequency modes. The low modes are related to the low-frequency libration and translation of tetrahedron–octahedron chains [23]. The mid modes are relate to the mid-frequency deformation of Ga2O6 octahedra. The last high modes attributed to the high frequency and result of stretching and bending of GaO4 tetrahedra [24].

4. Conclusion

In summary, we have studied the structure and photoluminescence properties of β-Ga2O3 which synthesized via electrospinning. SEM show the Morphology of the fibres, XRD and Raman confirmed the monoclinic phase of β-Ga2O3 nanowires, indicate that the samples we fabricated are pure. The PL spectrum of the β-Ga2O3 nanofibres exhibits the blue emission peaks about 470nm under excitation of 365nm at room temperature and the mechanisms of the blue shift and photoluminescence of the β-Ga2O3 nanofibres were discussed. FT-IR spectrum indicated that oxygen vacancies decreased with the annealing temperature increasing.

5. Acknowledgment

Our project was supported by the Funding for the Development Project of Beijing Municipal Education Commission of Science and Technology, China (Grant No. KZ201410005008) and the Natural Science Foundation of Beijing City, China (Grant No. 4102014).

6. Reference

[1] Kumar S, Tessarek C, Christiansen S, et al. A comparative study of β-Ga 2 O 3, nanowires grown on different substrates using CVD technique [J]. Journal of Alloys & Compounds, 2014, 587:812-818.

[2] PaiChun Chang, Zhiyong Fan, WeiYu Tseng,et al. β-Ga2O3 nanowires: Synthesis, characterization, and p-channel field-effect transistor [J]. Applied Physics Letters, 2005, 87(22):431.

[3] Gai, Ligang, Jiang, Haihui, tian, Yan, et al. Low-temperature synthesis of β-Ga2O3 nanorods on SBA-15 microparticles by solvothermal method [J]. Nanotechnology, 2006, 17(23):5858.

[4] Weng W Y, Hsueh T J, Chang S J, et al. A, β-Ga2O3 /GaN Hetero-Structured Solar-Blind and Visible-Blind Dual-Band Photodetector [J]. IEEE Sensors Journal, 2011, 11(6):1491-1492.

[5] P. Feng, J. Y. Zhang, Q. H. Li,et al. Individual β-Ga2O3 nanowires as solar-blind photodetectors[J]. Applied Physics Letters, 2006, 88(15):7433.

[6] Suzuki R, Nakagomi S, Kokubun Y. Solar-blind photodiodes composed of a Au Schottky contact and a β-Ga2O3 single crystal with a high resistivity cap layer [J]. Applied Physics Letters, 2011, 98(13):293.

[7] Weng W Y, Hsueh T J, Chang S J, et al. A Solar-Blind, β-Ga2O3 Nanowire Photodetector [J]. IEEE Photonics Technology Letters, 2010, 22(10):709-711.]

[8] Yuehua An, Xulong Chu, Yuanqi Huang,et al. Au plasmon enhanced high performance β-Ga2O3, solar-blind photo-detector [J]. Progress in Natural Science: Materials International, 2016, 26(1):65-68.

[9] Park S H, Kim S H, Park S Y, et al. Synthesis and CO gas sensing properties of surface-nitridated Ga2O3 nanowires [J]. Rsc Advances, 2014, 4(108):63402-63407.

[10] Nakagomi S, Sai T, Kokubun Y. Hydrogen gas sensor with self temperature compensation based on β-Ga2O3 thin film [J]. Sensors & Actuators B Chemical, 2013, 187:413-419.

[11] Shan J J, Li C H, Wu J M, et al. Shape-controlled synthesis of monodispersed β-gallium oxide

crystals by a simple precipitation technique[J]. Ceramics International, 2017, 43(8):6430-6436.

[12] Feng L, Li Y, Su X, et al. Growth and characterization of spindle-like Ga 2 O 3, nanocrystals by electrochemical reaction in hydrofluoric solution[J]. Applied Surface Science, 2016, 389:205-210.

[13] Tomoaki Terasako, Yuki Kawasaki, Masakazu Yagi. Growth and morphology control of β -Ga2O3, nanostructures by atmospheric-pressure CVD [J]. Thin Solid Films, 2016, 620:23-29.

[14] Q.N. Abdullah, F.K. Yam, K.H. Mohmood, et al. Free growth of one-dimensional β-Ga2O3, nanostructures including nanowires, nanobelts and nanosheets using a thermal evaporation method[J]. Ceramics International, 2016, 42(12):13343-13349.

[15] Fu L, Liu Y, Hu P, et al. Ga2O3 Nanoribbons: Synthesis, Characterization, and Electronic Properties [J]. Chemistry of Materials, 2003, 15(22):4287-4291.

[16] Kumar S, Sarau G, Tessarek C, et al. Study of iron-catalysed growth of β-Ga2O3 nanowires and their detailed characterization using TEM, Raman and cathodoluminescence

[17] Blasse G, Bril A. Some observations on the luminescence of β-Ga2O3 [J]. Journal of Physics & Chemistry of Solids, 1970, 31(4):707-711.

[18] Harwig T, Kellendonk F, Slappendel S. The ultraviolet luminescence of β-Ga2O3 [J]. Journal of Physics & Chemistry of Solids, 1978, 39(6):675-680.

[19] Feng Q, Li F, Dai B, et al. The properties of gallium oxide thin film grown by pulsed laser deposition [J]. Applied Surface Science, 2015, 359(3):847-852.

[20] Cheng Y, Liang H, Shen R, et al. Band gap broadening and photoluminescence properties investigation in Ga2O3, polycrystal [J]. Journal of Materials Science Materials in Electronics, 2013, 24(8):2750-2754.

[21] Cho S, Lee J, Park I Y, et al. New simple synthesis route of GaN powders from gallium oxyhydroxide [J]. Materials Science & Engineering B, 2002, 95(3):275-278.

[22] Elnaggar M E, Hassabo A G, Mohamed A L, et al. Surface modification of SiO2 coated ZnO nanoparticles for multifunctional cotton fabrics. [J]. Journal of Colloid & Interface Science, 2017, 498:413-422.

[23] ZHAO Jian-Guo, et al. Structural and Photoluminescence Properties of β-Ga2O3 Nanofibres Fabricated by Electrospinning Method [J]. Chinses Physics Letters, 2008, 25(10):3787-3789.

[24] Kumar S, Dhara S, Agarwal R, et al. Study of photoconduction properties of CVD grown β-Ga2O3, nanowires [J]. Journal of Alloys & Compounds, 2016, 683:143-148.

Enhanced Light Absorptions of Pyramid Square Grating Structure for Crystalline Silicon Solar Cells

Lingxiao Wu, Aqing Chen, Zhenguo Ji and Qinan Mao

College of Materials & Environmental Engineering, Hangzhou Dianzi University, Hangzhou 310018, P R China. Email: aqchen@hdu.edu.cn

Abstract: A special structure of pyramid square grating (PSG) is designed to reduce the light reflection of crystalline silicon solar cells. Reflectance and optical absorptions which are calculated by Finite Difference Time Domain (FDTD) methods suggested that the PSG structures had highly efficient anti-reflection effects from 300 to 400 nm. Moreover, the photocurrent density of PSG calculated in the wavelength range of 300-1000 nm has been improved by 45.09 % compared to that of planar structure, under the air mass 1.5 (AM1.5) solar illuminations.

1. Introduction

The crystalline silicon solar cells are still predominant in the commercial solar cells although many other cells including organic solar cells [1–3], thin film solar cells [4] and nanowire-based cells[5] have been developed. But the high cost of crystalline silicon solar cells is the main barrier for the further development. In order to cut the cost, numerous works have been done to improve the performance of silicon-based solar cells. Various technologies have been developed to improve the performance of solar cells [6,7]. Significantly enhancing the optical absorption is an effective approach of cutting the cost of single crystalline solar cells. The methods of improving optical absorption generally include light trapping and anti-reflection effects. Anti-reflection effects can be achieved by textured surface [8] and hybrid nanostructure [9].New methods for light trapping in solar cells are to use the metallic nanoparticles to enhance the absorption of the cells due to the coupling of surface plasmon and the light [10]. Moreover, Bragg mirrors of porous silicon were used as the back reflector for light trapping [11]. However, these methods of improving the absorption are not only expensive but also complex for mass production except for texturing the surface of cells. Moreover, improving the absorption of solar cells in broader wavelength range is important for the enhancement their performance.

In this present work, a novel structure of pyramid square grating (PSG) was proposed. The optical properties of PSG structures were calculated by Finite Difference Time Domain (FDTD) methods. The PSG structures have a low reflectance and strong absorptions in the wavelength range of 300 to 1000 nm.

2. Calculation Modes

Figure. 1 shows the simulated PSG structure model where Λ and d are the period and the width of silicon pillars, respectively. The period Λ varies from 1.0 μm to 2.0 μm. The silicon pillar width d varies from 0.5 to 1 μm. In order to obtain the enhanced anti-reflection effects of PSG, we calculated the optical properties of PSG structures with different periods by the finite difference on time domain (FDTD) methods, using a freely available software package [12].The calculated PSG structure is schematically shown in fig. 1. Both the silicon thickness and the depth of grating are 10 μm.

Content from this work may be used under the terms of the Creative Commons Attribution 3.0 licence. Any further distribution of this work must maintain attribution to the author(s) and the title of the work, journal citation and DOI.
Published under licence by IOP Publishing Ltd

IOP Conf. Series: Materials Science and Engineering **275** (2017) 012047 doi:10.1088/1757-899X/275/1/012047

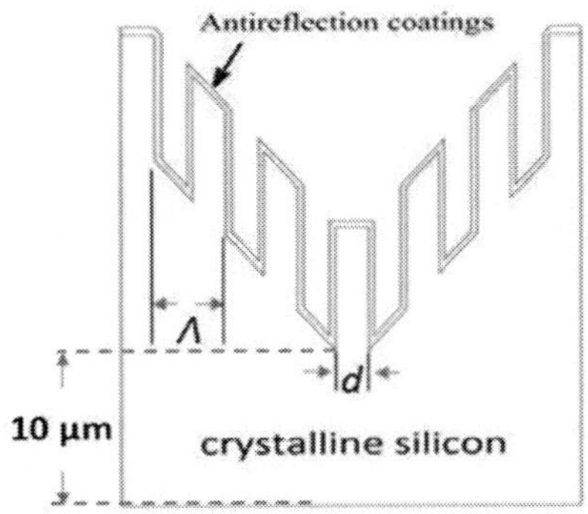

Figure 1. Schematic diagram of PSG structure (a), top-view SEM image of the pyramid groove array (b), top-view (c) and side-view (d) SEM images of a pyramid groove.

3. Results and Analyses

Figure. 2 (a) shows the reflectance of the PSG structure as a function of wavelength when $\Lambda = 1000$ nm and $d = 500$ nm (solid line).For the square grating, the reflectance is larger than 40% in the wavelength range of 300 to 600 nm, but decreases sharply from 600 to 1000 nm. The low reflectance in the wavelength range of 600 to 1000 nm is due to the fact that the silicon pillars with the width of 500 nm act as the effective scattering centers coupling with the light [13]. Compared to the reflectance of square grating, the reflectance of the PSG is lower than that of square grating structures in the wavelength range of 300 to 1000 nm. Especially, in the wavelength range of 600 to 1000 nm, the reflectance of the PSG is smaller than 0.05. Furthermore, compared to the reflectance of commercial solar cells, as shown in fig. 2 (a), it is interesting to note that the PSG structures have a lower reflectance in the wavelength range of 300 to 400 nm, clearly suggesting the PSG has enhanced anti-reflection effects. The reasons for the low reflectance of PSGs can be clearly illustrated further by the electric field intensity. As shown in fig. 2 (b), the electric field intensity distributions in the PSG and square grating at the wavelength of 903 nm and 420 nm. From fig. 2 (b), it is also observed that the light is significantly focused and intensely scatted by PSG structures, which ultimately leads to the enhancement of absorptions.

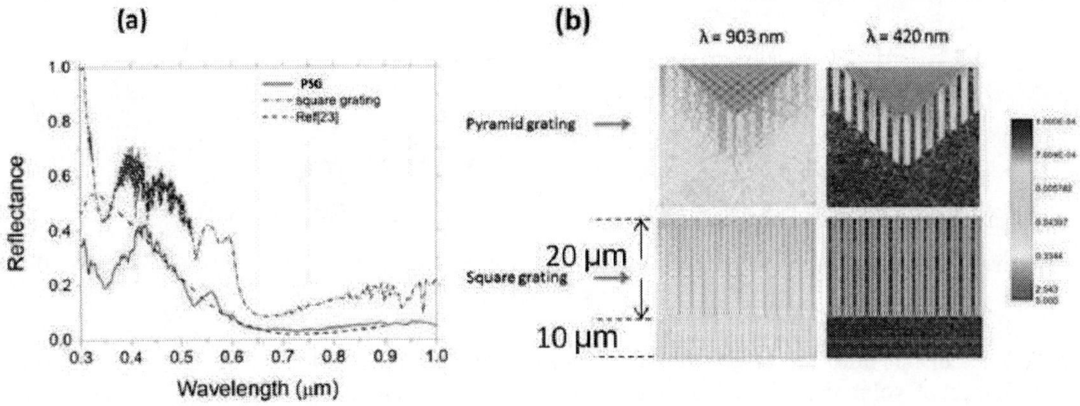

Figure 2. (a) the reflectance of PSG (red solid line) and square grating (blue dot-dash line) from 300 nm to 1000 nm and a traditional crystalline silicon solar cell with SiNx ARC [15] (black dash line) and (b) Calculated electric field intensity distributions at 903 nm and 420 nm for PSG and square grating.

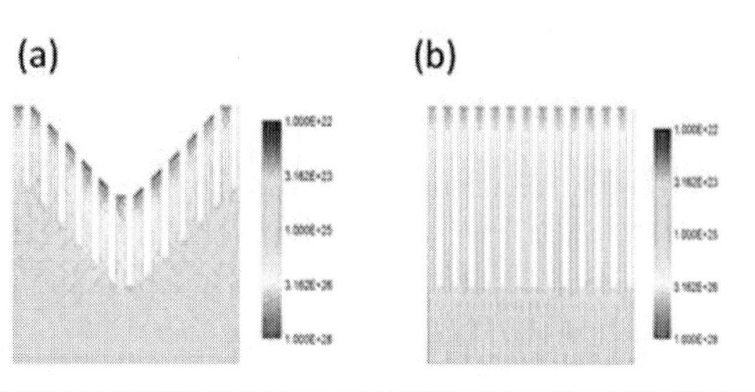

Figure 3. (a) and (b) show the absorptance of PSG structures with different period and silicon pillar width, respectively. The absorptance in the wavelength range of 300 to 1000 nm is enhanced slightly by increasing the period of PSG. Because the increase of the PSG period results in the higher-order diffraction modes which is obtained by the following grating diffraction equation

$$n_1 sin\theta_1 = n_2 sin\theta_2 = \frac{m\lambda}{a} \quad (1)$$

where n_1 and n_2 are the refractive index of air and silicon, respectively, m is the diffracted order, θ_1 and θ_2 are the angles of propagation in air and silicon, λ is the wavelength and a is the grating period. The light, hence, couples to the diffraction orders propagating outside the escape cone of Si [19], which results in the strong light absorption. In addition, decreasing the width of silicon pillars enhances the absorption of PSG, as shown in fig. 5 (b). The absorptance of PSG with the silicon pillar width of 500 nm is higher than that of square grating with the silicon pillar width of 1000 nm. Since the light strongly interact with the silicon pillars in the subwavelength scale.

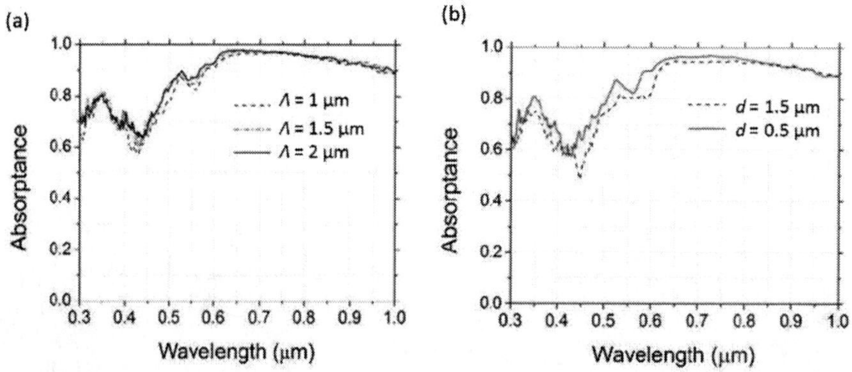

Figure 4. the absorptance of PSG structures with different period Λ (a) and different silicon pillar width d (b)

To gain insight into the effects of the enhanced anti-reflection effects and light trapping on the performance of solar cells based on the PSG, the photocurrent density in this structure has been calculated. Assuming that all the photo-generated electron-hole pair contributes to the photocurrent. So, the photocurrent density J_{ph} can be given by

$$J_{ph} = e \int A(\lambda) AM1.5(\lambda) d\lambda \quad (2)$$

Where e is the charge on an electron, $A(\lambda)$ is the absorption spectrum obtained by FDTD simulation and the AM1.5 (λ) is the standard AM1.5G spectrum. Table I lists the photocurrent density of different light-trapping structure. By fabricating square grating, the photocurrent density is raised by 4.21 mA/cm^2. As expected, the photocurrent density of the PSG increases as the period increases from 1000 to 2000 nm. The photocurrent density (34.46 mA/cm^2) of the PSG with the period of 2000 nm increases by 45.09 % and by 23.24% compared to the planar structure and the square grating structure, respectively.

Table I. the short circuit current density Jph generated by different structures

Light trapping structure	Parameters		Jph (mA cm-2)
	a (nm)	b (nm)	
Planar	N.A	N.A	23.75
Square grating	1000	500	27.96
PSG	1000	500	33.86
	2000	500	34.46

4. Conclusion

In summary, the PSG structures have significantly enhanced anti-reflection effects in the wavelength range of 300 to 400 nm. FDTD calculation results indicate that photocurrent density generated in the PSG solar cells is as high as 34.46 mA/cm2 which is 90% of the ideal photocurrent density generated by AM1.5 solar illumination in the wavelength range of 300 to 1000 nm.

5. Acknowledgements

This study was funded by the Natural Science Foundation of Zhejiang Province of China (Grant No. LQ17B030004) and National Natural Science Foundation of China (Grant No. 61372025)

6. References

[1] A.J. Maclachlan, T. Rath, U.B. Cappel, S. a Dowland, H. Amenitsch, A. Knall, C. Buchmaier, G. Trimmel, J. Nelson, S. a Haque 2015 Adv. Funct. Mater. 409–420.
[2] B. Qi, J. Wang 2012, J. Mater. Chem. 24315–24325.
[3] G. Li, V. Shrotriya, J. Huang, Y. Yao, T. Moriarty, K. Emery, Y. Yang 2005 Nat. Mater. 4 864–868.
[4] K.L. Chopra, P.D. Paulson, V. Dutta, 2004 Prog. Photovoltaics Res. Appl. 12 69–92.
[5] E.C. Garnett, M.L. Brongersma, Y. Cui, M.D. McGehee 2011 Annu. Rev. Mater. Res. 41 269–295.
[6] A. ur Rehman, S.H. Lee, E.G. Shin, S.H. Lee 2015 Mater. Lett. 161 181–184.
[7] Y. Wang, X. Zhang, B. Han, L. Bai, H. Zhao, F. Yang, J. Liang, Q. Huang, X. Chen, Y. Zhao 2015 Appl. Surf. Sci. 355 14–18.
[8] M.L. Addonizio, L. Fusco, A. Antonaia, F. Cominale, I. Usatii 2015 Appl. Surf. Sci. 357 651–658.
[9] Y. Liu, W. Zi, S. (Frank) Liu, B. Yan 2015 Sol. Energy Mater. Sol. Cells. 140 180–186.
[10] H.A. Atwater, A. 2010 Nat. Mater. 9 205–213.
[11] H.Y. Seba, T. Hadjersi, N. Zebbar, 2015 Appl. Surf. Sci. 350 57–61
[12] A.F. Oskooi, D. Roundy, M. Ibanescu, P. Bermel, J.D. Joannopoulos, S.G. Johnson, 2010 Comput. Phys. Commun. 181 687–702
[13] E. Garnett, P. Yang, 2010 Nano Lett. 10 1082–1087
[14] C.H. Sun, P. Jiang, B. Jiang, 2008 Appl. Phys. Lett. 92 6–9.
[15] T.K. Chong, J. Wilson, S. Mokkapati, K.R. Catchpole, 2012 J. Opt. 14 24012.

PLZT Ceramic Driving Rotary Micro-mirror Based on Photoelectric-electrostatic Mechanism

Yujuan Tang[1], Zhong Yang[1], Yusong Chen[2] and Xinjie Wang[2]

[1]School of Intelligent Science and Control Engineering, Jinling Institute of Technology, Nanjing 211169, China
[2]School of Mechanical Engineering, Nanjing University of Science and Technology, Nanjing 210094, China
Email: yjtang@jit.edu.cn

Abstract: Based on the anomalous photovoltaic effect of PLZT, a rotary micro-mirror driven by hybrid photoelectric-electrostatic actuation of PLZT ceramic is proposed. Firstly, the mathematical modelling of coupled multi-physics fields of PLZT ceramic is established during illumination and light off phases. Then, the relationship between the rotation angle and the photovoltage of PLZT ceramics is established. In addition, the feasibility of rotary micro-mirror with hybrid photoelectric-electrostatic driving is verified via closed-loop control for photo-induced voltage of PLZT ceramic. The experimental results show that the photo-induced voltage of PLZT ceramics has good dynamic control precision using on-off closed-loop control method.

1. Introduction

Driving and control system of micro-mirror based on the micro-opto-electro-mechanical system (MOEMS) technology has the features of small size, light weight, low power consumption and compatible with large scale integrated circuit manufacturing technology. It is widely applied in optical communication, biomedical imaging, optical precision measurement and other military and civil fields. Combined with different micro-mirror driving device, MOEMS micro-mirror can achieve the intensity and phase modulation of the incident light as well as the gating of light path, which contribute micro-mirror to have many functions such as optical switching, scanning and detecting. At present, there are some popular driving modes such as piezoelectric, electromagnetic, and electrostatic driving, etc. [1]. The above-mentioned driving methods have their own advantages and disadvantages so that they can be used in different operating environment and technical fields according to their characteristics. However, all of them need to be driven by an external power. It is not only susceptible to electromagnetic interference, but also has more limitations working in the clean operating space, vacuum environment and other independent working environment. Compared with the above-mentioned driving mode, optical drive technology has the advantages of driving cleaning, such as no electromagnetic interference, non-contact remote optical control and wireless energy transmission, which is an ideal driving mode for MOEMS micro mirror in an independent operating environment [2]. PLZT ceramic can produce anomalous photovoltage effect and photo-induced deformation in the near-ultraviolet band, and can produce the voltage of several kV/cm and displacement of photo-induced deformation in the polarization direction. Therefore, the photovoltage and photo-induced deformation effect of PLZT ceramics can be applied in the driving mode for MOEMS micro-mirror, which can offset the drawbacks of traditional micro-mirror driving mode and have a wide application prospect.

Content from this work may be used under the terms of the Creative Commons Attribution 3.0 licence. Any further distribution of this work must maintain attribution to the author(s) and the title of the work, journal citation and DOI.
Published under licence by IOP Publishing Ltd

The mathematical model of PLZT ceramic with coupled multi-physics fields is a basis for the theoretical analysis of MOEMS rotary micro-mirror with PLZT ceramics. The mathematical model of PLZT ceramic has been studied by some scholars in recent years. In 1983, RC charging circuit model was proposed to describe the photoelectric field intensity in polarized direction by Brody et al [3], but did not analyze the change of the electric field after the light turned off. In 1993, Fukuda et al [4] experimentally confirmed that the photostrictive effect of PLZT ceramic is the result of multi-physical fields coupling, however, the coupling relationship of different physical fields has not been modeled. In 2005, Shih et al [5] proposed a photostrictive constitutive model of PLZT ceramic; however, they ignored the impacts of thermal expansion on the electric field. Considering the effect of thermal expansion on the photoelectric fields, Wang and Huang et al [6, 7] proposed a mathematical model of photovoltage and photo-induced deformation based on coupled multi-physical fields in 2014. Meanwhile, they experimentally verified the established mathematical model and deduced the light-electric-thermo-mechanical coupling constitutive equation of PLZT ceramic deformation during light off stage. In addition, Zheng et al used the finite element method to construct the constitutive model in 0-1 and 0-3 polarized PLZT ceramic [8]. Jiang et al established an electrical model of hybrid photostrictive/piezoelectric actuation mechanism [9], which provides a theoretical basis for the closed-loop control of PLZT ceramics in micro-driving fields.

Aiming at the shortcomings of traditional MOEMS micro-mirror driving method, a rotary micro-mirror driven by hybrid photoelectric-electrostatic driving of PLZT ceramic is proposed. Firstly, working principle of the rotary micro-mirror mechanism based on hybrid photoelectric-electrostatic driving is present. Then the relationship between the rotary angle and the photo-generated voltage of PLZT ceramic is established. Finally, the feasibility of rotary micro-mirror with hybrid photoelectric-electrostatic driving is verified via experiment of closed-loop control for the photo-generated voltage of PLZT ceramic.

2. Working Principle of the Rotary Micro-mirror with Photoelectric-electrostatic Driving

The rotary mechanism of micro-mirror can make a lot of advantages for micro-mirror such as a large adjustable angle range, the fast response speed and the small volume. A novel rotary micro-mirror with hybrid photoelectric-electrostatic driving based on the anomalous photovoltage effect of PLZT ceramic is shown in figure 1. The micro-mirror is supported by the support column and the flexible support beam. One driving electrode is plated on the non-reflective surface (i.e. lower surface) of the micro-mirror and the other driving electrode is plated on one the upper surface of the base. The driving electrodes are respectively connected to the electrodes of the PLZT ceramic. When the PLZT ceramic is irradiated by the high-energy ultraviolet light, the photovoltage of PLZT ceramic is applied on the driving electrodes of the micro-mirror. Based on the electrostatic driving force between the driving electrodes, the micro-mirror is rotationally driven.

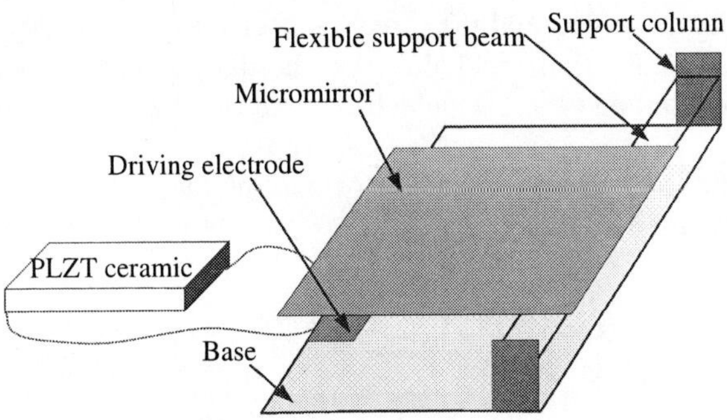

Figure 1. Rotary micro-mirror with hybrid photoelectric-electrostatic driving

As a mature rotary micro-mirror technology which has been widely used in MOEMS micro-mirror, the electrostatic twisting mechanism has lots of advantages, such as simple structure, fast response, no assembly error, small size and easy control [10]. If two rotation mechanisms in two directions are required, the rotary micro-mirror mechanism with hybrid photoelectric-electrostatic driving based on PLZT anomalous photovoltage effect can be respectively arranged in the x and y directions on the moving platform of the micro-mirror translation mechanism. It is also possible to arrange numerous rotating micro-mirror mechanisms proposed in this paper on the moving platform of the micro-mirror to form the array of optical control micro-mirror.

3. Multi–physics Fields Coupling Model of Photo voltage of PLZT Ceramics

Photovoltage can be generated when the PLZT ceramic is irradiated by ultraviolet light. Afterwards, the photo-induced deformation is produced based on the inverse piezoelectric effect. However, the photovoltage of the PLZT ceramic is not proportional to the photo-induced deformation which is far behind the photovoltage. The reason is that the thermal expansion caused by the increase of PLZT ceramic temperature under the UV light cause the hysteresis. Therefore, the photostrictive effect of PLZT ceramic is a comprehensive result of coupled multi-physics field such as anomalous photovoltaic effect, photothermal effect, pyroelectric effect, thermal expansion effect and piezoelectric effect [11]. Figure 2 shows the coupling relationships of multi-physics fields of PLZT ceramic photostrictive effect.

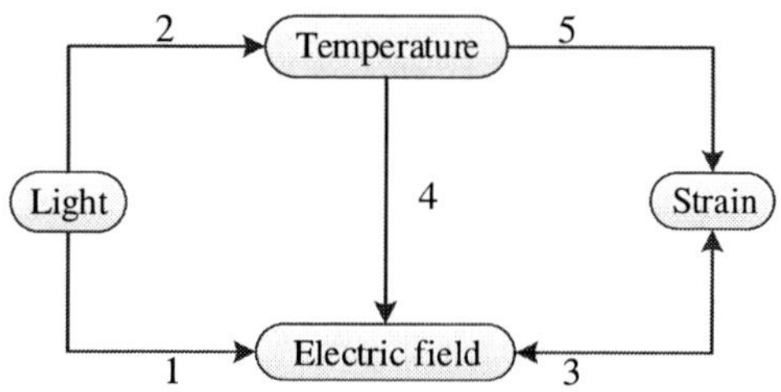

(1)Anomalous photovoltaic effect, (2) Photothermal effect, (3) Piezoelectric effect, (4) Pyroelectric effect, (5) Thermal expansion effect

Figure 2. The coupling relationships of multi-physics fields of PLZT ceramic

As shown in figure 2, a part of the light energy is converted to thermal energy by the photothermal effect (light energy to thermal energy conversion), which causes the rising temperature. The majority of electric energy is generated in the poling direction based on the anomalous photovoltaic effect (light energy to electric energy conversion) and the pyroelectric effect (thermal energy to electric energy conversion). Another portion of electric energy is produced in opposite direction with the poling direction based on thermal expansion and the direct piezoelectric effect (mechanical energy to electric energy conversion). The total strain includes the thermal expansion strain (thermal energy to mechanical energy conversion) and the piezoelectric strain (total electric energy to mechanical energy conversion).

According to the coupling mechanism analysis of photovoltage effect, the photovoltage expression of PLZT ceramic can be written as [12]:

$$\begin{aligned} V(t) &= V_p(t) + V_\theta(t) - \beta V_e(t) \\ &= V_s(1-e^{-\frac{t}{\tau_1}}) + (\frac{AP}{C_p} - \beta\frac{\lambda D_e}{d_{3i}Y_a})\Delta T_s(1-e^{-\frac{t}{\tau_\theta}}) \end{aligned} \tag{1}$$

where d_{3i} indicates the piezoelectric constant of the PLZT ceramic; V_p is photovoltage and $V_p(t)=V_S(1-e^{-\frac{t}{\tau_1}})$, where V_s is saturated photovoltage, τ_1 is the illumination time constant; V_θ is pyroelectric and $V_\theta(t)=\frac{AP\Delta T_S}{C_p}(1-e^{-\frac{t}{\tau_\theta}})$, where A is the electrode area of PLZT ceramic, P is the pyroelectric coefficient of PLZT ceramic [13], ΔT_S is the maximal temperature variation, τ_θ is the thermal time constant; λ is the thermal stress coefficient of PLZT ceramic; Y_a is the elastic modulus of PLZT ceramic; V_e is Thermal deformation voltage and $V_e(t)=\frac{\lambda D_e}{d_{3i}Y_a}\Delta T_S(1-e^{-\frac{t}{\tau_\theta}})$, where De is the distance between the two electrodes of PLZT ceramic and β is the conversion coefficient of thermal deformation and electric field.

After UV light is switched off, the residual voltage and deformation of PLZT ceramics gradually decrease to the initial state of the PLZT before illuminated. Investigating the trend of PLZT ceramics in the light-off phase can help further study of photo-deformation characteristics of PLZT ceramics and promote the engineering application of PLZT ceramics. The residual voltage after turning off the UV light can be obtained as:

$$\begin{aligned} V_d(t) &= V(t_0)+V_{\theta-d}(t)-\beta_2 V_{e-d}(t) \\ &= V(t_0)-(\frac{AP}{C_p}-\frac{\beta_2\lambda D_e}{d_{3i}Y_a})\Delta T_{s-d}+(\frac{AP}{C_p}-\frac{\beta_2\lambda D_e}{d_{3i}Y_a})\Delta T_{s-d}e^{-\frac{t}{\tau_d}} \end{aligned} \quad (2)$$

where ΔT_{s-d} is the maximal temperature decrement during light-off phase and $\Delta T_{s-d}=T(t_0)-T_0$; $T(t_0)$ is the temperature of PLZT ceramic when the ultraviolet light is switched off; τ_d is the dark time constant. β_2 is the conversion coefficient of thermal deformation and electric field during light switched off.

4. Rotation Angle of the Micro-mirror with Hybrid Photoelectric-electrostatic Driving

As shown in figure 3, it is a cross-sectional schematic view of the rotary micro-mirror mechanism with hybrid photoelectric-electrostatic driving, where h_m is the vertical distance between the flexible support beam and the driving electrode on the base; l_m, w_m and t_m are the length, width and thickness of the flexible support beam respectively; a_m is the width of the micro-mirror. Thus, with respect to the rotary axis, the position of the driving electrode on the base (i.e., s_1 and s_2) should satisfies the relation of $s_1=\alpha_1 a_m$, $s_2=\alpha_2 a_m$.

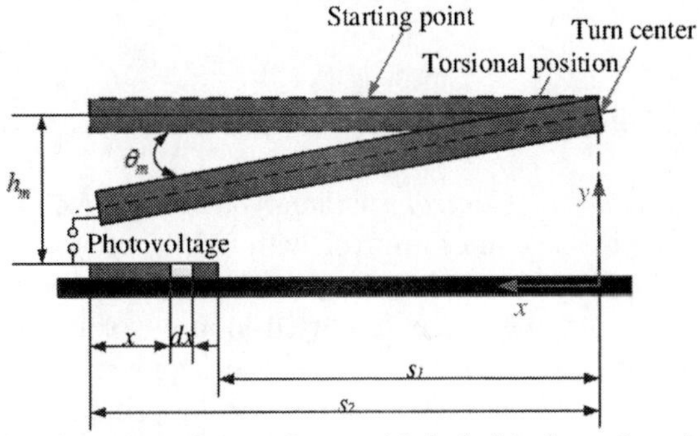

Figure 3. Cross profile of rotary micro-mirror with hybrid photoelectric-electrostatic driving

When the PLZT ceramic is irradiated by the UV light, a photo-induced voltage of several kV/cm is generated and applied to the driving electrodes of the micro-mirror via the lead wire. Micro-mirror will be rotationally driven due to electrostatic force, therefore, the electrostatic moment M_e is obtained as [14]:

$$
\begin{aligned}
M_e &= \int_{\alpha_1 a_m}^{\alpha_2 a_m} F dx = \int_{\alpha_1 a_m}^{\alpha_2 a_m} x \frac{\varepsilon L_m V^2}{2(h_m - x\tan\theta_m)^2} dx \\
&= \frac{\varepsilon L_m V^2}{2\theta_m^{\ 2}} [\frac{1}{1-(\alpha_2 a_m / h_m)\theta_m} - \frac{1}{1-(\alpha_1 a_m / h_m)\theta_m} + \ln(\frac{1-(\alpha_2 a_m / h_m)\theta_m}{1-(\alpha_1 a_m / h_m)\theta_m})]
\end{aligned} \quad (3)
$$

where ε is the conductivity of air.

In order to simplify the electrostatic moment expression of the rotary micro-mirror, suppose $\Theta = \theta_m / \theta_{m-\max}$, where $\theta_{m-\max}$ is the maximum twist angle of the rotary micro-mirror, $\theta_{m-\max} = h_m / a_m$ is obtained from the geometric relationship, the electrostatic moment M_e can be simplified as [14]:

$$
M_e = \frac{\varepsilon L_m V^2}{2\theta_{m-\max}^2 \Theta^2} [\frac{1}{1-\alpha_2\Theta} - \frac{1}{1-\alpha_1\Theta} + \ln(\frac{1-\alpha_2\Theta}{1-\alpha_1\Theta})] \quad (4)
$$

According to the theory of torsional deformation in the mechanics of materials [15], when the micro-mirror is rotated under the action of electrostatic force, the recovery torque produced by the flexible support beam is:

$$
M_r = S_0\theta_m \quad (5)
$$

where S_0 is the torsional stiffness of the flexible support beam and $S_0 = 2GI_p / l_m$, where G is the shear elastic modulus of support beam.

The torsional cross-section modulus of the flexible support beam of the micro-mirror is [16]:

$$
I_p = t_m w_m^3 [\frac{1}{3} - 0.21\frac{w_m}{t_m}(1 - \frac{w_m^4}{t_m^4})] (w_m \le t_m) \quad (6)
$$

According to the principle of moment balance in material mechanics, it can be obtained that $M_e = M_r$. So the relationship between the PLZT ceramic photovoltage and the rotation angle of the micro-mirror can be written as:

$$
V = k_0 \sqrt{\frac{1}{2}\{\frac{\Theta^3}{[1/(1-\alpha_2\Theta)] - [1/(1-\alpha_1\Theta)] + \ln[(1-\alpha_2\Theta)/(1-\alpha_1\Theta)]}\}} \quad (7)
$$

where $k_0 = \sqrt{[\frac{S_0\theta_{m-\max}^2}{\varepsilon L_m}]}$.

5. Close-loop Control for Photovoltage of PLZT Ceramic

5.1. Experimental Setup of Close-loop Control for Photovoltage of PLZT Ceramic

In this section, the feasibility of rotary micro-mirror with hybrid photoelectric-electrostatic driving is verified via closed-loop control for photo-induced voltage of PLZT ceramic. By using the light shutter to control the UV light on and off, the dynamic closed-loop control of photovoltaic voltage can be achieved. Experimental setup of closed-loop control for photovoltage of PLZT ceramic is illustrated in figure 4. Firstly, start the software of the servo control system written by LabVIEW in the computer. Afterwards, enter the target value of voltage and initialize the light shutter to "ON". When the UV light is switched on, PLZT ceramic produces photovoltage, which is measured by a high impedance voltmeter and the data is acquired and processed by servo control program. Afterwards, the voltage data are transferred to the optical shutter control thread. Then via serial port, the control command is inputted to the light shutter. Finally, closed-loop servo control for photovoltage of PLZT ceramic is achieved. Since the closed-loop control scheme is carried out based on the photovoltage of PLZT ceramic is real-time measured and feedback, so test program for the high impedance voltmeter is written and developed using LabVIEW.

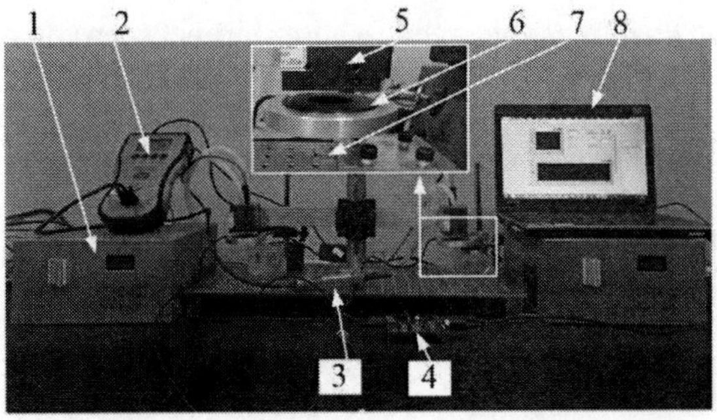

(1) UV light controller, (2) High impedance voltmeter, (3) Probe of high impedance voltmeter, (4) Controller of optical shutter, (5) UV probe, (6) Optical shutter, (7) PLZT ceramic, (8) Computer

Figure 4. Experimental setup of close-loop control for photovoltage of PLZT ceramic

5.2. Experimental Results of Closed-loop Control of PLZT Ceramic Photovoltage

The closed-loop photovoltage control experiment of single PLZT ceramic is carried out, where the sampling period is set to 300 ms and the target voltage is set to 1000V. Figure 5 shows the experimental curve of the closed-loop photovoltage control of single PLZT ceramic with light intensity of 50mW/cm^2 and 100mW/cm^2. As illustrated in figure 5(a), the photovoltage of PLZT ceramic takes about 18.3s to increase to the target value when the light intensity is 50mW/cm^2. Afterwards, the photovoltage fluctuates around the target voltage, and the average fluctuation height average wave height f_1 is 23V. As shown in figure 5(b), the PLZT ceramic photovoltage takes about 8.1s to reach the target voltage when the light intensity is 100mW/cm^2 UV light. The average fluctuation height f_1 is 54V.

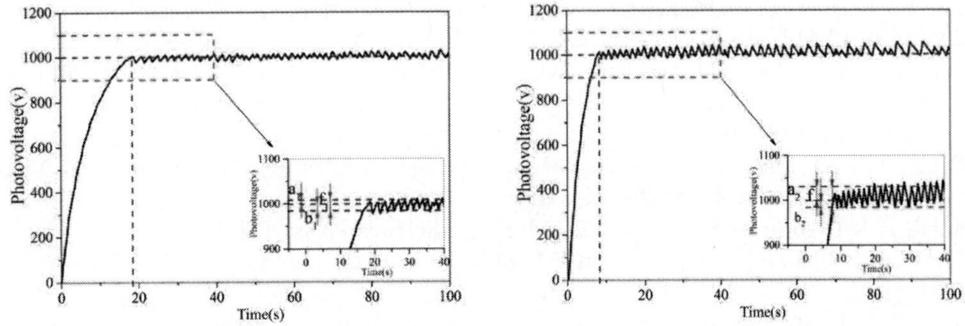

(a) Light intensity is 50mW/cm^2 (b) Light intensity is 100mW/cm^2

Figure 5. The closed-loop photovoltage control curve of single PLZT ceramic

The experimental results verify the feasibility of hybrid photoelectric-electrostatic driving micro-mirror. Because of the limitation of the sampling frequency of the high impedance voltmeter, voltage data is acquired with certain time delay. Improving the sampling frequency of the high impedance voltmeter and the control algorithm will be carried out to achieve a better control effect in our future research.

6. Conclusions

The rotary micro-mirror with hybrid photoelectric-electrostatic driving is proposed based on anomalous photovoltaic effect in PLZT ceramics exposed to high-energy ultraviolet light. The photovoltage equations in illumination phase and light-off phase are obtained. Based on the multi-physics coupling model of PLZT ceramics, the relationship between the rotation angle of the rotary micro-mirror and the photovoltage of the PLZT ceramic is given and analysed. In addition, the closed-loop experimental study on the photovoltage of PLZT ceramics under different light intensities is carried out.

The experimental results show that the time of the photovoltage of PLZT ceramic reaching the target value of 1000V under the illumination of 50mW/cm^2 and 100mW/cm^2 is 18.3s and 8.1s respectively. That means the time of PLZT ceramic photovoltage reaches the target value decreases with the increase of light intensity. In other words, the response speed of the closed-loop control system can be improved with stronger UV light. In addition, the fluctuation amplitude of the photovoltage around the target voltage increases gradually when the light intensity increases. The larger fluctuation amplitude of the photovoltage will impact the control accuracy for the micro-mirror. Therefore, our next work is to eliminate fluctuation using variable light intensity control or improving the sampling frequency of the high impedance voltmeter.

Acknowledgment

This study was supported by the National Natural Science Foundation of China (Grant no. 51505204).

References

[1] Lu A J, Zhang Z P, Huang Z J 2016 *Optics and Precision Engineering* **24** 400-406.
[2] Zhu Y, Liu W, Jia K, et al. 2011 *Sensors and Actuators A: Physical* **167** 495-501.
[3] Brody P S. 1983 *Ferroelectrics* **50** 27-32.
[4] Fukuda T, Hattori S, Arai F, Matsuura H, Hiramatsu T, Ikeda Y and Maekawa A 1993 *Proceeding of IEEE Robotics and Automation Conference* 618-23.
[5] Shih H R, Tzou H S and Saypuri M 2005 *Journal of Sound and Vibration* **284** 361-378.
[6] Huang J H, Wang X J and Wang J 2015 *Smart Materials and Structures* **25** 025002.
[7] Huang J H, Wang X J and Wang J2015 *Optics and Precision Engineering* **23** 760-768.
[8] Zheng S J, Tong LY and Luo Q T 2015 *International Journal of Applied Electromagnetics and Mechanics* **49** 513-530.
[9] Jiang J, Li X N, Yue H H and Deng Z Q 2017 *Journal of Mechanical Engineering* **53** 65-71.
[10] Huang J M, Liu A Q, Deng Z L, Zhang Q.X. Ahna J and Asundid 2004 *Sensors and Actuators A: Physical* **115** 159-167.
[11] Wang X J, Lu F, Liu Y F and Huang J H 2016 *Optics and Precision Engineering* **24** 2505-14.
[12] Fridkin V M 1979 *Photoferroelectrics* (New York: Springer).
[13] Zhong W L 1996 *Ferroelectric physics* (Beijing: Science Press) 487-488.
[14] Zhang X M, Chau F S, Quan C, Lam Y L and Liu A Q 2001 *Sensors and Actuators A: Physical* **90** 73-81.
[15] Liu H W 2010 *Mechanics of materials* (5th Edition) (Beijing: Higher Education Press)
[16] Yeow J T W, Yang V X D, Chahwan A, Gordon ML,Qi B, Vitkin I A, Wilson B C and Goldenberga A A 2005 *Sensors and Actuators A: Physical* **117** 331-340.